MODERN FILTER DESIGN

Active RC
and Switched Capacitor

**PRENTICE-HALL SERIES IN ELECTRICAL
AND COMPUTER ENGINEERING**

Leon O. Chua, Editor

MODERN FILTER DESIGN

Active RC and Switched Capacitor

M.S. GHAUSI

John F. Dodge Professor of Engineering
Oakland University

K.R. LAKER

Member of Technical Staff
Supervisor, Networks and Signalling Group
Bell Telephone Laboratories

PRENTICE-HALL, INC., Englewood Cliffs, NJ   07632

Library of Congress Cataloging in Publication Data

GHAUSI, MOHAMMED SHUAIB.
   Modern filter design.

   (Prentice-Hall series in electrical and computer
engineering)
      Includes bibliographical references and index.
      1. Electric filters, Active.   2. Electronic
circuit design.   I. Laker, K. R.   II. Title.
III. Series.
TK7872.F5G43   621.3815'324   81-1104
ISBN 0-13-594663-8   AACR2

Editorial/production supervision
   and interior design by Karen Skrable
Manufacturing buyer: Joyce Levatino
Cover design by Carol Zawislak

Printed in the United States of America

10 9 8 7 6 5 4 3 2 1

Prentice-Hall International, Inc., *London*
Prentice-Hall of Australia Pty. Limited, *Sydney*
Prentice-Hall of Canada, Ltd., *Toronto*
Prentice-Hall of India Private Limited, *New Delhi*
Prentice-Hall of Japan, Inc., *Tokyo*
Prentice-Hall of Southeast Asia Pte. Ltd., *Singapore*
Whitehall Books Limited, *Wellington, New Zealand*

To Our Families: *Marilyn, Nadjya, and Simine;*
and Mary Ellen, John,
Christopher, and Brian

Contents

chapter 5

High-Order Filter Realization

277

chapter 6

Active Switched Capacitor Sampled-Data Networks

376

Preface

The book is intended as a text for senior undergraduates and/or first-year graduate students to be covered in a one-semester course. The book is also intended for self-teaching and reference for practicing engineers.

Chapter 1 covers basic properties and classifications of systems together with filter transmission and approximations. Frequency- and time-domain, continuous-time and discrete-time signal transmissions are considered. The latter is important in the analysis and design of switched capacitor networks discussed later in the book. Chapter 2 deals entirely with op amps. In this book the op amp is considered (for obvious reasons) as the main active element, and other elements, such as gyrators and frequency-dependent negative resistors, are op amp-derived circuits. The discussion of the op amp includes both bipoler and MOS integrators. Chapter 3 gives various definitions of sensitivity as used in the literature. Statistical sensitivity measures, which are becoming more popular and are found to be useful in practical design, as they give good correlation with Monte Carlo analysis, are treated in detail. Chapter 4 discusses continuous-time second-order active sections which are predominantly used as blocks in active filter design. The biquadratic sections considered include single- and multiple-op amp, active-RC, active-R, and active-C realizations. Chapter 5 provides a complete treatment of high-order filter design. Multiple-loop feedback design methods are considered in some detail, as these configurations provide the best sensitivity performance in a properly designed high-order filter. The merits and disadvantages of various design methods are considered so that the designer is made aware of the compromises that must be made in a practical design. Chapter 6 is concerned with the analysis and design of switched capacitor filters. This type of filters is rapidly becoming very popular, as they are compatible with

MOS large-scale integration. Switched capacitor recursive filters are recent developments that fully utilize the advantage provided by MOS LSI, and hence hold future promise in filter design.

Three appendices are included, namely, selected topics in passive-network properties; tables of classical filter functions; and, op amp terminologies and selected data sheets. These appendices are intentionally kept brief to avoid duplication of material found in several books.

Each chapter is complemented by references and problems. The references used are books and published papers. Only available and pertinent references are given. No attempt has been made to cite the original papers or to be complete. The problems are used to provide the reader with exercise material in order to gain a better understanding of the text and to extend the design methods to cover practical situations.

Although active filters have reached maturity, the need for better performance, lower sensitivity, and reduced cost has motivated several recent advances in the state of the art. These advances include computer-aided design (CAD) methods for coupled or multiple-loop feedback topologies used to realize low-sensitivity high-order filters, active-R and active-C networks for monolithic integrated-circuit, high-frequency filters, and switched capacitor networks to realize precision, monolithic integrated-circuit, audio-frequency filters. Thus, one of the purposes of this book is to provide discussions in a tutorial manner, for students as well as practicing engineers, of these and other recent advances which heretofore have appeared only in technical articles scattered in many journals. To the best of our knowledge, some of this material, especially that dealing with switched capacitor networks, appears for the first time in a textbook in a detailed manner. Although we discuss recent advances, we do, of course, include the bread-and-butter active-filter realization methods that have stood the test of time. The book is not, then, just another state-of-the-art text, but one that provides a comprehensive tutorial presentation of active-filter technology from a practical standpoint as well as from the state-of-the-art point of view.

The authors wish to acknowledge the contributions of many authors whose results appear in this text but whose names may or may not appear. We would like especially to thank Professors Adel Sedra and Rolf Schauman, who reviewed the complete manuscript and offered many helpful suggestions and thoughtful advice. Dr. Pradeep Padukone and Dr. Ali Gonuleren also helped in proofreading the manuscript. Some of the original research results of the authors appearing in this book were supported by National Science Foundation grants, for which we wish to express our gratitude.

Some of the content of this book, in particular Chapter 6, received inspiration from my (K.R.L.) work in the Signal Processing and Integrated Circuit Design Department at Bell Laboratories. Significant parts of Chapter 6, which covers the area of switched capacitor networks, are based on collaborative work with Dr. Paul Fleischer. We are indebted to K.R.L.'s former department head, Mr. Carl Simone, and present department head, Dr. Dan Stanzione, for providing their encouragement and for providing the stimulating evironment that influenced the presentation of many of the concepts covered in this book.

A personal note of gratitude goes to Mr. Joe Friend for his encouragement, his careful reading of the various versions of the manuscript, and his helpful suggestions throughout the writing of this book.

Last but not least, we express our gratitude and love to our families for their understanding and patience throughout the development of this manuscript.

Bloomfield Hills, Michigan   M.S. GHAUSI
Staten Island, New York   K.R. LAKER

chapter one

....

Filter Transmission
and Related Topics

1.1 INTRODUCTION

An *electrical wave filter* can be defined as an interconnected network of electrical
components, such as resistors, capacitors, inductors, and transistors, which
operates on or processes applied electrical signals. The applied electrical signal
is referred to as the *input signal* or *excitation*. The product of the processing
performed by the network on the excitation is referred to as the *output signal*
or *response*. The excitation and response will differ according to the processing
or filtering performed by the network. As we shall see shortly, there are several
means by which we can represent and specify this filtering operation. In this
book we are concerned with the analysis, specification, design, and realization
of electrical filters for the frequency range $f \leq 500$ kHz, with emphasis on
audio-frequency ($f \leq 20$ kHz) applications. It is in this frequency range where
the greatest benefits are derived from the use of active components in electrical
filter networks.

Filtering in the general sense is a selection process. The output of a filter
is then a selected subset of the input. In electrical filtering the object is to
perform frequency-selective transmission. We know from elementary signal
theory that any periodic wave of period $2\pi/\omega_R$ can be represented by its Fourier
series expansion

$$s(t) = a_0 + \sum_{k=1}^{\infty} a_k \cos k\omega_R t + \sum_{k=1}^{\infty} b_k \sin k\omega_R t \qquad (1\text{-}1)$$

where the coefficients a_k and b_k define the harmonic content of the wave $s(t)$.
The frequency-independent term a_0 is referred to as the *dc component*. When a

harmonically rich $s(t)$ is applied to a filter, the filtering process will serve to alter the magnitude of the coefficients a_k and b_k; some coefficients may be greatly attenuated, thus defining a stop band. The selective enhancing, attenuating, and removal of the harmonic components of $s(t)$ is what electrical filtering is all about. In a linear filter, the filtering operation cannot yield an output that has more harmonic components than the input. Thus, given an input and a desired output, it is the task of the filter designer to determine:

1. The filtering operation required.
2. The network to implement the filtering operation.
3. The network component values that realize the specific filtering operation.

The choice of network implementation is almost always determined by economic considerations and technology. Because of advances in technology, what was considered the best economical filter implementation 10 years ago, perhaps only one year ago, may not be the most economical implementation today. In Fig. 1-1 is shown the evolution of filter technology in the Bell System, for

Fig. 1-1 Bell System progress in filter technology, depicting the evolution of voice-frequency ($f <$ 4 kHz) filters from 1920 to 1980. Shown on the crescent is the hardware required to realize a second-order section with (a) passive LC, (b) discrete component active-RC, (c) and (d) hybrid thin-film active-RC technologies. Shown in the oval are examples of monolithic filters: (e) an analog switched-capacitor building block capable of realizing up to 11 second-order sections and (f) a digital signal processor capable of realizing up to 37 second-order sections with an 8-kHz sampling rate.

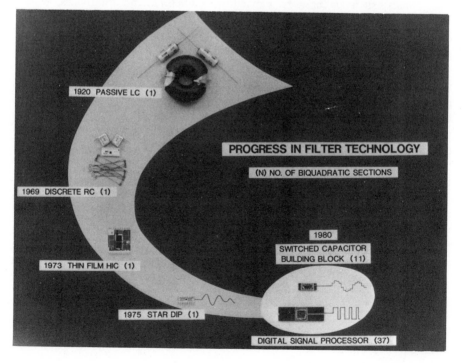

2

voice-frequency ($f < 4\,\text{kHz}$) applications over a nearly 60-year span. What is shown in each stage of Fig. 1-1 is the hardware to implement a second-order filtering function. Although the evolution depicted is specific to the Bell System, it is to a good approximation representative of the evolution industry wide. From 1920 to the latter 1960s the majority of voice-frequency filters were realized as discrete *RLC* networks. However, it was recognized in the 1950s that size and eventual cost reductions could potentially be achieved by replacing the large costly inductors with active networks. That is, a network comprised of resistors, capacitors, and transistors could be made to resonate like a tuned *RLC* network. These active networks, referred to as *active-RC networks*, remained essentially research curiosities until the mid-1960s, when good-quality active components such as operational amplifiers became inexpensive and readily available. Although size has not been significantly reduced, filtering and amplification were achieved simultaneously. In the early 1970s the economic potential envisioned for active-*RC* filters began to be realized with batch-processed thin-film hybrid integrated circuits (HICs). The HIC shown in Fig. 1-1, composed of two thin-film capacitors, nine thin-film resistors, and one silicon integrated-circuit (SIC) operational amplifier, is 1.05×1.00 in. This circuit represented about a factor-of-2 cost reduction over the equivalent passive-*RLC* realization. By 1975, thin-film technology had advanced such that the 1.05×1.00 in. HIC could be reduced in size to fit into a small 16-pin dual-in-line package (DIP). Today, using switched-capacitor and digital filter techniques, very high order filters can be realized as microminiature silicon integrated-circuit chips. As examples we show in Fig. 1-1 two packaged integrated circuit chips; namely, the digital signal processor (DSP) and the switched capacitor building block (SCBB). The DSP implements up to 37 second order sections of digital filtering where as the SCBB implements up to 11 second order sections of analog filtering. A detailed description of the SCBB is reserved for Chapter 6. The reader interested in pursuing the area of digital filters is referred to references [B9, B10, and B11].

As the evolution depicted in Fig. 1-1 has demonstrated, the use of active networks has enabled engineers to utilize the advances in integrated-circuit technology to implement low-cost, microminiature voice-frequency filters. Although active filters can be used throughout the voice-frequency range ($f < 4\,\text{kHz}$), voice-frequency applications account for tens of millions of the active filters that are produced yearly throughout the world. Aside from their obvious size and weight advantages over equivalent passive-*RLC* implementations shown in Fig. 1-1, active filters provide the following additional advantages:

1. Increased circuit reliability because all processing steps can be automated.

2. In large quantities the cost of integrated-circuit active filters are much lower than equivalent passive filters.

3. Improvement in performance because high-quality components can be readily manufactured.

4. A reduction in parasitics because of smaller size.

5. Active filters and digital circuitry can be integrated onto the same silicon chip. This advantage has been fully realized with active switched-capacitor filters.

In addition to these advantages, which stem from their physical implementation, there are other advantages which are circuit-theoretic in nature, namely:

1. The design and tuning processes are simpler than those for passive filters.
2. Active filters can realize a wider class of filtering functions than passive filters. Some properties of passive networks are reviewed in Appendix A.
3. Active filters can provide gain; in contrast, passive filters often exhibit a significant loss.

With these advantages there are some drawbacks to active network implementations. They are:

1. Active components have a finite bandwidth, which limits most active filters to audio-frequency applications. In Chapter 4 we see that by properly using this intrinsic band limiting, this limitation can be significantly reduced. In contrast, passive filters do not have such an upper-frequency limitation and they can be used up to approximately 500 MHz.
2. Passive filters are less affected by component drifts in manufacture or drifts due to environmental changes. This phenomenon, referred to as *sensitivity*, will be seen to be an important criterion for comparing similar filter realizations. Much of the disadvantage experienced by active-network implementations stems from the relatively large number of components, required to realize an active filter, as compared to equivalent passive realizations. In Chapter 5 we show techniques for minimizing this disadvantage.
3. Active filters require power supplies, whereas passive filters do not. In this era of energy conservation it is particularly important that active devices be used in a most efficient manner. For this reason the reader will be encouraged to design active filters with the minimum number of active devices consistent with precision performance.

It can be said that in voice and data communication systems, the economic and performance advantages of active filters far outweigh these disadvantages. Testimony to this fact is the general acceptance and usage of active filters throughout the telecommunications industry.

It is assumed that the reader has a working knowledge of network theory and has had some exposure to linear system theory. In this chapter we review some of the fundamental concepts and tools needed to characterize a continuous linear filter. We also take the opportunity to introduce the reader to the fundamental concepts of sampled-data systems and the z-transform as these concepts are used in the analysis, specification, and design of sampled-data switched capacitor filters.

1.2 CLASSIFICATION OF SYSTEMS

A schematic "black-box" representation of a system is shown in Fig. 1-2. On the left side of the box we symbolically show the input to the system or the excitation. As the arrow indicates, the input represents information (e.g., an electrical signal) injected into the system. The output or response, shown symbolically on the right side of the box, represents the systems reaction to its excitation.

Fig. 1-2   Schematic representation of a system.

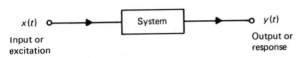

This output may be observable to the outside (e.g., voltmeter, oscilloscope, etc.), or it may serve as the input to yet another system. Generally, the input $x(t)$ and output $y(t)$ are functions of time and are called *signals*. The black box may, for example, represent a linear or a nonlinear network.

In order to describe a system, let us define the various classes of systems.

1.2.1 Linear and Nonlinear Systems

A system is *linear* if it satisfies the principle of superposition. Mathematically, a system is linear if and only if

$$f(\alpha x_1 + \beta x_2) = \alpha f(x_1) + \beta f(x_2) \tag{1-2}$$

where α and β are arbitrary constants and x_1 and x_2 are input signals. For example, if an input $x_1(t)$ yields an output $y_1(t)$, which we may write symbolically as $x_1(t) \rightarrow y_1(t)$, and if $x_2(t) \rightarrow y_2(t)$, then from Eq. (1-2) we write

$$\alpha x_1(t) + \beta x_2(t) \longrightarrow \alpha y_1(t) + \beta y_2(t) \tag{1-3}$$

To be more germane to the material in this text, it is stated without proof that any system governed by linear differential or difference equations is linear. Only linear systems are considered in this text.

1.2.2 Continuous-Time, Discrete-Time, and Sampled-Data Systems

A system is said to be a *continuous-time* or a *continuous analog system* if input x and output y are capable of changing at any instant of time. We make the fact of continuous change evident by writing x and y as functions of the continuous-time variable t; that is,

$$x = x(t) \quad \text{and} \quad y = y(t) \tag{1-4}$$

In discrete-time and sampled-data systems, the input and output signals change at only discrete instants of time. The values of the signals between instants are of no interest in discrete-time systems. In sampled-data systems, which are actually analog systems, the signals are usually held constant between (sampling) instants. More will be said later about the difference between discrete-time and sampled-data systems. However, we can represent sampled-data input and output signals as functions of discrete variables kT:

$$x = x(kT) \quad \text{and} \quad y = y(kT) \tag{1-5}$$

where k is an integer and T is the time duration between samples. The signal formats for each of these system types is shown in Fig. 1-3. Note that the amplitude of the output may be larger, smaller, or equal to the input, depending on the type and design of the filter.

An important mathematical distinction between continuous-time and sampled-data (also discrete-time) systems is the fact that continuous-time systems are characterized by differential equations, whereas discrete-time systems are characterized by difference equations. We will explore this difference in greater detail later.

1.2.3 Time-Invariant and Time-Varying Systems

A system is said to be *time-invariant* if the shape of the response to an input applied at any instant of time depends only on the shape of the input and not on the time of application. In mathematical terms, a system is linear and time-invariant if

$$x(t + \tau) \longrightarrow y(t + \tau) \tag{1-6}$$

for all $x(t)$ and all $\tau > 0$. Similarly, for a discrete-time system, if we have

$$x(kT) \longrightarrow y(kT) \tag{1-7}$$

then the system is time-invariant if and only if

$$x[(k - n)T] \longrightarrow y[(k - n)T] \tag{1-8}$$

for any $x(kT)$ and n.

For a causal system, the response cannot precede the excitation; that is, if the excitation is applied at some time t_0 or mT, then the response is zero for $t < t_0$ or $k < m$.

The equilibrium conditions describing a lumped-element linear time-invariant system are characterized by linear differential or difference equations with constant coefficients—in contrast with distributed systems, which are

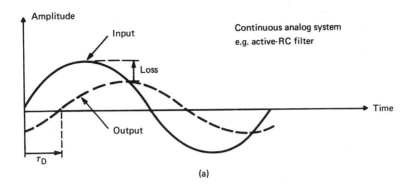

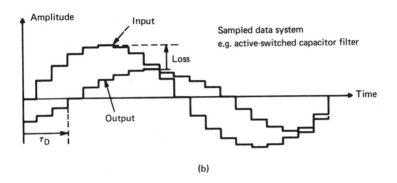

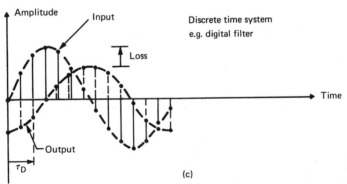

Fig. 1-3 Signal formats for continuous analog, sampled-data analog, and discrete-time systems.

described by partial differential equations. By a *lumped-element system* we mean a system comprised of lumped passive elements and active devices. Lumped sampled-data systems will also contain samplers or switches. Most of the networks considered in this text are time-invariant. However in Chapter 6 we will see that switched-capacitor sampled-data networks are, in general, time-varying systems. Fortunately, after some mathematical manipulation, time-varying switched-capacitor networks can be treated like time-invariant networks, albeit networks of increased complexity.

1.3 TECHNIQUES FOR THE ANALYSIS OF CONTINUOUS, LINEAR, TIME-INVARIANT SYSTEMS

In Section 1.2 we stated that continuous, linear, time-invariant systems, composed of lumped passive elements and active devices, are characterized by linear differential equations with constant coefficients. Thus, a single-input, single-output lumped linear time-invariant system can be characterized in general by the following input–output relationship:

$$b_n \frac{d^n y(t)}{dt^n} + b_{n-1} \frac{d^{n-1} y(t)}{dt^{n-1}} + \ldots + b_0 y(t) = a_m \frac{d^m x(t)}{dt^m} + a_{m-1} \frac{d^{m-1} x(t)}{dt^{m-1}}$$
$$+ \ldots + a_0 x(t) \tag{1-9}$$

where $x(t)$ and $y(t)$ designate the input and output variables, respectively. In electronic networks, $x(t)$ and $y(t)$ will be current $i(t)$ and/or voltage $v(t)$ variables. For a given input and initial conditions $y(0)$, $dy(0)/dt, \ldots$, $d^{n-1} y(0)/dt^{n-1}$, the solution to Eq. (1-9), namely the output $y(t)$, is completely determined.

Let us now state some definitions of terms commonly used when referring to lumped linear systems. The *zero-input response* of the system is the response obtained when the input is identically zero. Such a response is not necessarily zero, because there may be initial charges on the capacitors an/or initial fluxes in the inductors. The *zero-state response* is the response obtained for an arbitrary input when all initial conditions are identically zero. It follows then, for a linear system, that the complete response is equal to the sum or superposition of the zero input and zero-state responses.

In continuous linear time-invariant systems, Laplace transform techniques can be used to transform differential equations with constant coefficients into linear algebraic equations. This transformation greatly facilitates the analysis of the system and provides us with immediate insight into the behavior of the system. Noting that the Laplace transform of the nth derivative of some time function y is given by

$$\mathscr{L}\left[\frac{d^n y(t)}{dt^n}\right] = s^n Y(s) - s^{n-1} y(0) - s^{n-2} \frac{dy(0)}{dt} - \ldots - \frac{d^{n-1} y(0)}{dt^{n-1}} \tag{1-10}$$

we can transform Eq. (1-9) term by term to yield

$$(b_n s^n + b_{n-1} s^{n-1} + \ldots + b_0) Y(s) + \mathrm{IC}_y(s)$$
$$= (a_m s^m + a_{m-1} s^{m-1} + \ldots + a_0) X(s) + \mathrm{IC}_x(s) \tag{1-11}$$

where in $\mathrm{IC}_y(s)$ and $\mathrm{IC}_x(s)$ we have lumped together all terms involving the initial values for y and x and their first $n - 1$ derivatives. In Eq. (1-11) $Y(s)$ is the *Laplace transform* of the zero-state response. A useful concept in the analysis and synthesis of linear networks is the *network function*, which is the ratio of two Laplace-transformed terminal or port variables. When the two variables ratioed are the input $X(s)$ and the output $Y(s)$, the network function is referred to as the *transfer function* $H(s)$. Thus, the transfer function can be defined in terms of the Laplace-transformed excitation $X(s)$ and the zero-state response $Y(s)$:

$$H(s) = \frac{\mathcal{L}[\text{zero-state response } y(t)]}{\mathcal{L}[\text{excitation } x(t)]} = \frac{Y(s)}{X(s)}$$

$$= \frac{a_m s^m + a_{m-1} s^{m-1} + \ldots + a_0}{b_n s^n + b_{n-1} s^{n-1} + \ldots + b_0} \qquad (1\text{-}12)$$

where $m \leq n$ for any realizable practical network. Note that, when $X(s) = 1$ or $x(t) = \delta(t)$, $H(s) = Y(s)$ is the impulse response.

When (a) $Y(s)$ and $X(s)$ are voltage transforms [i.e., $V_{out}(s)$ and $V_{in}(s)$ respectively], $H(s)$ is referred to as a *voltage transfer function*.

(b) $Y(s)$ and $X(s)$ are current transforms [i.e., $I_{out}(s)$ and $I_{in}(s)$ respectively], $H(s)$ is referred to as a *current transfer function*.

(c) $Y(s)$ and $X(s)$ are voltage and current transforms [i.e., $V_{out}(s)$ and $I_{in}(s)$], $H(s)$ is referred to as a *transfer impedance function*.

(d) $Y(s)$ and $X(s)$ are current and voltage transforms [i.e., $I_{out}(s)$ and $V_{in}(s)$], $H(s)$ is referred to as a *transfer admittance function*.

There is yet another type of network function, called a *driving-point function*. Driving-point functions relate the voltage at a terminal or port to the current at the same terminal or port. For the input port, the *driving-point impedance function* is defined as

$$Z_{in}(s) = \frac{V_{in}(s)}{I_{in}(s)} \qquad (1\text{-}13a)$$

and the *driving-point admittance function* is defined as the reciprocal of the driving-point impedance function:

$$Y_{in}(s) = \frac{I_{in}(s)}{V_{in}(s)} \qquad (1\text{-}13b)$$

The output driving point functions are defined in a similar way.

If the network function, $H(s)$, is factored, the following alternative representation is obtained:

$$H(s) = \frac{a_m(s - z_1)(s - z_2)\ldots(s - z_m)}{b_n(s - p_1)(s - p_2)\ldots(s - p_n)} = \frac{N(s)}{D(s)} \qquad (1\text{-}14)$$

In Eq. (1-14) the roots of the numerator polynomial $z_1, z_2, \ldots, z_m$ are referred to as the *zeros* of $H(s)$ because $H(s) = 0$ when $s = z_i$. The roots of the denominator polynomial $p_1, p_2, \ldots, p_n$ are referred to as the *poles* of $H(s)$ because

$H(s) = \infty$ when $s = p_i$. The poles and zeros can be plotted in the complex s-plane, where $s = \sigma + j\omega$ as shown in Fig. 1-4. Note that, since the coefficients of $H(s)$ are all real, imaginary and complex poles and zeros occur in conjugate pairs. For stability all poles must lie in the left half-plane. Recall from elementary transform theory that networks having only left half-plane poles yield responses,

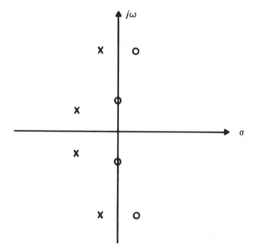

Fig. 1-4 Poles and zeros plotted in the complex s-plane.

for bounded inputs, which decay with time. When the poles lie on the $j\omega$-axis and are simple, the network oscillates, and when the poles lie in the right half-plane, the responses grow exponentially with time. These properties are illustrated graphically for various pole positions in Fig. 1-5. When zeros lie on or to the left of the $j\omega$ axis (i.e., no right-half-plane zeros), $H(s)$ is referred to as a *minimum phase function*.

Active filters are typically specified by the voltage transfer function $H(s) = V_{out}(s)/V_{in}(s)$, sometimes more specifically called the *voltage gain function*. Under steady-state conditions (i.e., $s = j\omega$), the voltage gain function can be written as

$$H(j\omega) = |H(j\omega)|\, e^{j\phi(\omega)} \qquad (1\text{-}15)$$

where $|H(j\omega)|$ is the magnitude or gain function and $\phi(\omega)$ is the phase function. We may also express Eq. (1-15) as

$$H(j\omega) = e^{-[\alpha(\omega) + j\beta(\omega)]} \qquad (1\text{-}16a)$$

where α is the loss and β is the phase, that is,[1]

$$\alpha(\omega) = -\ln|H(j\omega)| \qquad \text{nepers} \qquad (1\text{-}16b)$$

$$\beta = -\phi(\omega) \qquad \text{radians} \qquad (1\text{-}16c)$$

[1] 1 neper = $(20 \log_{10} e)$ dB = 8.686 dB and 1 radian = 57.296 deg.

Filter Transmission and Related Topics

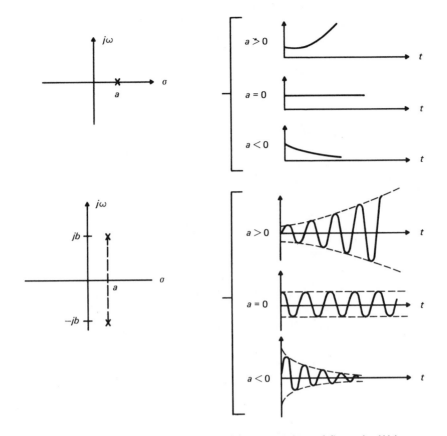

Fig. 1-5 s-Plane pole and time-domain representations of first-order $H(s) = 1/(s + a)$ and second-order $H(s) = 1/[(s + a)^2 + b^2]$ for $a > 0$, $a = 0$, and $a < 0$.

Also, the *group delay function*, which represents the time delay experienced by each component of the input spectrum, is defined as

$$\tau(\omega) = -\frac{d\phi(\omega)}{d\omega} \tag{1-16d}$$

Usually, the magnitude response for an active filter is specified as gain, in decibels:

$$G(\omega) = 20 \log_{10} |H(j\omega)| \qquad \text{dB} \tag{1-17a}$$

and phase, in degrees:

$$\beta(\omega) = -\phi(\omega) \times 57.296 \qquad \text{degrees} \tag{1-17b}$$

It is not uncommon to see active filters specified by a loss function $X(s)/Y(s)$ (in terms of voltages $V_{in}(s)/V_{out}(s)$) or in steady state $X(j\omega)/Y(j\omega)$. The use of loss functions is a carry-over from passive filter design. In this book we will work exclusively with gain functions.

One particularly important class of network functions is the second-order transfer function

$$H(s) = \frac{a_2 s^2 + a_1 s + a_0}{s^2 + b_1 s + b_0} = \frac{a_2(s + z_1)(s + z_2)}{(s + p_1)(s + p_2)} \tag{1-18}$$

This function, referred to as the *biquadratic function*, serves as the building block for a wide variety of active filters. In fact, Chapter 4 is devoted entirely to active-filter realizations of this function.

Alternatively, for complex poles and zeros, where $z_2 = z_1^*$ and $p_2 = p_1^*$ (the superscript * denotes a complex conjugate), we can express Eq. (1-18) as

$$H(s) = \frac{K[s^2 + [2\,\text{Re}(z_1)]s + \text{Im}(z_1)^2 + \text{Re}(z_1)^2]}{s^2 + [2\,\text{Re}(p_1)]s + \text{Im}(p_1)^2 + \text{Re}(p_1)^2} \tag{1-19a}$$

$$= \frac{K[s^2 + (\omega_z/Q_z)s + \omega_z^2]}{s^2 + (\omega_p/Q_p)s + \omega_p^2} \tag{1-19b}$$

where Re ($\cdot$) and Im ($\cdot$) denote the real part and the imaginary part, respectively. Using Eq. (1-19), we can establish the following information regarding the character of the gain function $|H(j\omega)|$. First, the dc gain

$$20\log_{10}|H(j0)| = 20\log_{10}\left(K\frac{\omega_z^2}{\omega_p^2}\right) \tag{1-20}$$

and the asymptotic gain as $\omega \longrightarrow \infty$ is given by

$$20\log_{10}|H(j\infty)| = 20\log_{10}(K) \tag{1-21}$$

The gain function reaches its maximum value approximately at the pole frequency ω_p, where

$$\omega_p = \sqrt{\text{Im}(p_1)^2 + \text{Re}(p_1)^2} \tag{1-22}$$

which is the radial distance from the origin to the pole location. Note that for high Q and when $\omega_p \gg \omega_z$ or $\omega_p \ll \omega_z$, the position of the gain maximum is virtually unaffected by the position of the zeros. The zero frequency, ω_z, determines the point at which the gain function is minimum. The zero frequency is related to the zero locations by the expression

$$\omega_z = \sqrt{\text{Im}(z_1)^2 + \text{Re}(z_1)^2} \tag{1-23}$$

which is the radial distance from the origin to the zero location. The sharpness of the maximum or bump at ω_p is determined by the *pole quality factor* Q_p, where

$$Q_p = \frac{\omega_p}{2\,\text{Re}(p_1)} = \frac{\sqrt{\text{Im}(p_1)^2 + \text{Re}(p_1)^2}}{2\,\text{Re}(p_1)} \tag{1-24}$$

Also, the depth of minimum at $s = j\omega_z$ is determined by the *zero quality factor* Q_z, where

$$Q_z = \frac{\omega_z}{2\,\mathrm{Re}(z_1)} = \frac{\sqrt{\mathrm{Im}(z_1)^2 + \mathrm{Re}(z_1)^2}}{2\,\mathrm{Re}(z_1)} \qquad (1\text{-}25)$$

In many cases $Q_z = \infty$; that is, $\mathrm{Re}(z_1) = 0$ and $\omega_z = \mathrm{Im}(z_1)$ defines a null in gain (i.e., attenuation is infinite).

1.4 TYPES OF FILTERS

As discussed in Section 1.1, a filter is an electrical network that provides frequency-weighted transmission. Filters are typically categorized according to the functions they perform, e.g., low-pass, high-pass, band-pass, band-reject, all-pass, or delay equalizer. In this section we define each of these filter types and the manner in which they are specified.

1.4.1 Low-Pass

The function of the *low-pass* (LP) *filter* is to pass low frequencies from dc to some specified cutoff frequency and to attenuate high frequencies. This filter is specified by its cutoff frequency, ω_c, stop band (SB) frequency ω_s, dc gain, passband (PB) ripple, and stop band attenuation. The filter passband is defined as the frequency range $0 \le \omega \le \omega_c$, the stop band as the frequency range $\omega \ge \omega_s$, and the transition band (TB) as the frequency range $\omega_c < \omega < \omega_s$. These specifications are shown graphically in Fig. 1-6a. Note that the specified filter must lie within the unshaded region in Fig. 1-6a.

A second-order function that realizes a low-pass gain characteristic is given by the following transfer function:

$$H(s) = \frac{K\omega_p^2}{s^2 + (\omega_p/Q_p)s + \omega_p^2} \qquad (1\text{-}26)$$

A sketch of the corresponding gain function is given in Fig. 1-6b, and the pole–zero plot for Eq. (1-26) is given in Fig. 1-6c. Note that $H(s)$ has a pair of zeros at $s = \infty$ and the dc gain $|H(j0)|$ is K. Comparing Eq. (1-26) for $\omega \gg \omega_p$ and Fig. 1-6b, we observe that $|H(j\omega)|$ decreases as $1/\omega^2$ or -40 dB/decade. We can extrapolate this observation for an Nth-order all-pole low-pass function, where $|H(j\omega)|$ rolls off for high frequencies at a rate of $-N \times 20$ dB/decade.

1.4.2 High-Pass

The function of the *high-pass* (HP) *filter* is to pass high frequencies above some specified cutoff frequency ω_c and to attenuate low frequencies from dc to some specified stop-band frequency ω_s. Here the high-pass filter is specified in much the same manner as the low-pass filter. The high-pass filter specification is shown graphically in Fig. 1-7a. Again the specified filter response is to lie

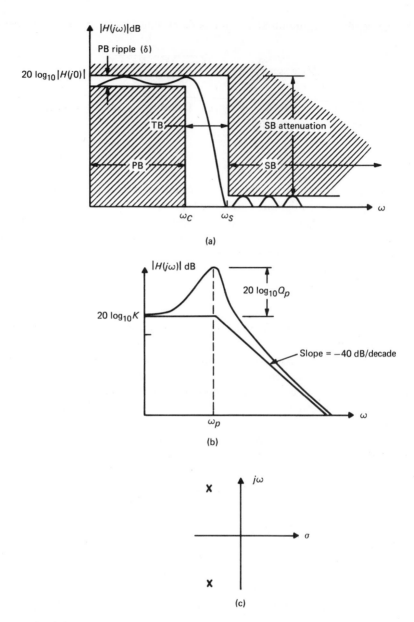

Fig. 1-6 (a) Low-pass filter specification. (b) Second-order low-pass filter gain response. (c) s-Plane pole–zero plot for the second-order low-pass filter.

within the unshaded area. In principle, the passband of the high-pass filter extends to $\omega = \infty$. In practice, however, the passband is limited in active filters by the finite bandwidth of the active devices and parasitic capacitances which nature provides at no extra cost. As a result, the gain of the high-pass filter will eventually roll off at high frequencies, as illustrated in Fig. 1-7a.

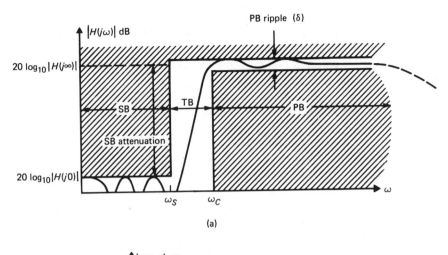

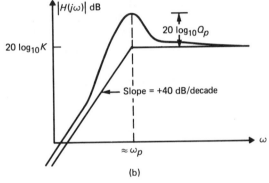

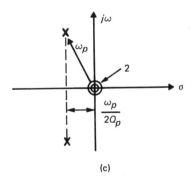

Fig. 1-7 (a) High-pass filter specification. (b) Second-order high-pass filter gain response. (c) s-Plane pole–zero plot for the second-order high-pass filter.

A second-order function which realizes a high-pass gain characteristic is given by the following transfer function:

$$H(s) = \frac{Ks^2}{s^2 + (\omega_p/Q_p)s + \omega_p^2} \tag{1-27}$$

A sketch of the corresponding gain function is given in Fig. 1-7b, and the pole–zero plot for Eq. (1-27) is given in Fig. 1-7c. Here $|H(j\omega)|$ increases as $1/\omega^2$ for low frequencies, which corresponds to a low-frequency slope of 40 dB/decade. Also since a pair of zeros are located at dc, the attenuation at dc is infinite. Note that in Eq. (1-27), $|H(j\infty)| = K$.

1.4.3 Band-Pass

The function of the *band-pass* (BP) *filter* is to pass a finite band of frequencies while attenuating both lower and higher frequencies. This filter has both a lower stop band, SB_L, and an upper stop band, SB_H. In general, the band-pass filter will not be symmetrical, and attenuation in the lower and upper SBs will be different. Similarly, the upper and lower transition bands TB_L and TB_H need not be the same (i.e., in general, $\omega_{SH}/\omega_{CH} \neq \omega_{CL}/\omega_{SL}$). Of course, such a band-pass filter specification can be met with a symmetric band-pass filter; however, a symmetric filter will provide an overdesign in one stop band. This typically implies that the nonsymmetric specification can be met with a lower-order nonsymmetric filter. However, as we will see in Chapter 5, it is often much easier to design a geometrically symmetric band-pass filter. The band-pass filter specification is shown in detail in Fig. 1-8a.

A second-order function that realizes a band-pass gain characteristic is given by the following transfer function:

$$H(s) = \frac{K(\omega_p/Q_p)s}{s^2 + (\omega_p/Q_p)s + \omega_p^2} \tag{1-28}$$

A sketch of the corresponding gain function is given in Fig. 1-8b, and the pole–zero plot is given in Fig. 1-8c. For $Q_p \gg 1$, $|H(j\omega)|$ in Eq. (1-28) is approximately symmetric about ω_p. The roll-off for both low and high frequencies is 20 dB/decade. The zero at the origin provides infinite attenuation at dc.

1.4.4 Band-Reject

The function of the *band-reject* (BR) *filter* is to attenuate a finite band of frequencies while passing both lower and higher frequencies. As a result, this filter has both lower- and upper-frequency passbands, PB_L and PB_H. The specification for this filter, shown graphically in Fig. 1-9a, is similar to the band-pass case. Like the high-pass filter, in reality the upper passband is limited due to the band-limited active devices and parasitics. The resulting high-frequency roll-off is shown by a dashed curve in Fig. 1-9a.

A second-order function that realizes a band-reject gain characteristic is given by the following transfer function:

$$H(s) = \frac{K(s^2 + \omega_z^2)}{s^2 + (\omega_p/Q_p)s + \omega_p^2} = \frac{K(s^2 + \omega_z^2)}{s^2 + (\omega_z/Q_p)s + \omega_p^2} \tag{1-29}$$

The corresponding gain response and pole–zero plots are given in Fig. 1-9b

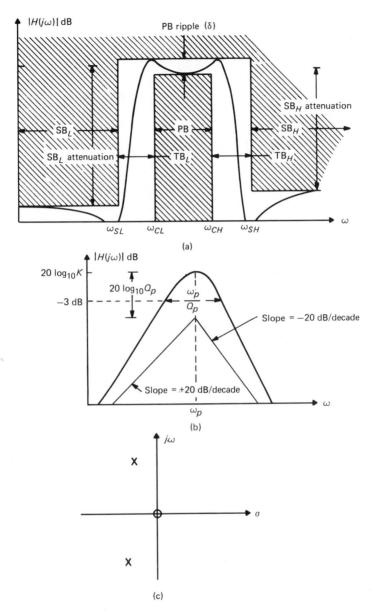

Fig. 1-8 (a) Band-pass filter specification. (b) Second-order band-pass filter gain response. (c) s-Plane pole–zero plot for the second-order band-pass filter.

and c, respectively. Note that to achieve a symmetric band-reject gain response, $\omega_p = \omega_z$. In this response the attenuation is infinite at ω_z and Q_p controls the sharpness of the notch. That is, the higher the Q_p, the sharper the transition to the notch. Also, $|H(j0)| = |H(j\infty)| = K$. As shown in Figs. 1-10 and 1-11, low-pass notch (LPN) and high-pass notch (HPN) gain responses can be

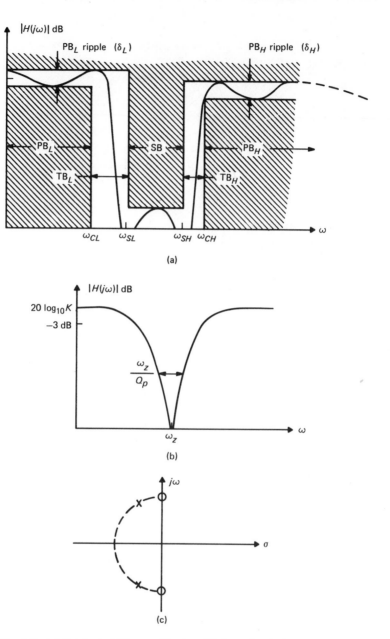

Fig. 1-9 (a) Band-reject filter specification. (b) Second-order band-reject filter gain response. (c) s-Plane pole–zero plot for the second-order band-reject filter.

achieved by offsetting ω_p from ω_z. For a low-pass notch, $\omega_p < \omega_z$, and for a high-pass notch, $\omega_p > \omega_z$. In both cases the attenuation is infinite at ω_z. Observe that a combination of a LPN and a HPN yield a fourth-order band-pass filter like that sketched in Fig. 1-8a. Also, in practice, high-order BR filters usually use LPNs and HPNs.

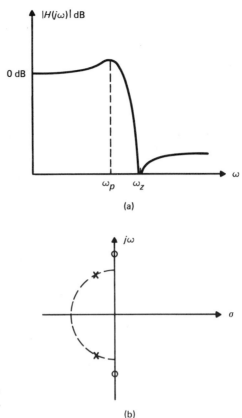

Fig. 1-10 (a) Second-order low-pass notch filter gain response. (b) s-Plane pole–zero plot.

1.4.5 All-Pass and Delay Equalization

Thus far we have discussed gain-shaping filters in which the gain or magnitude response was specified. We have not yet concerned ourselves with the phase or delay response of a filter. In voice and audio applications, ignoring the phase and delay behavior is usually justifiable because the human ear is insensitive to changes in phase or delay with frequency. However, in video and digital transmission applications, the phase changes introduced by a filter can cause intolerable distortions in the shape of the time-domain video or digital signal. Recall that phase $\phi(\omega)$ and delay $\tau(\omega)$ are related according to

$$\tau(\omega) = -\frac{d\phi(\omega)}{d\omega} \tag{1-30}$$

The compensation applied to correct for delay distortions introduced by gain-shaping filters and other parts of the transmission system is referred to as *delay equalization*. The delay-shaping filter that performs this compensation is called a *delay equalizer*.

To achieve distortionless video transmission, the phase must be linear or,

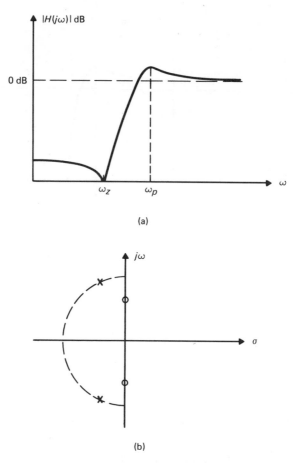

Fig. 1-11 (a) Second-order high-pass notch filter gain response. (b) s-Plane pole–zero plot.

according to Eq. (1-30), the delay must be flat for all frequencies. For example, a pulse subjected to a flat delay characteristic will be translated in time by τ_0 seconds but otherwise transmitted without waveform distortion. Mathematically, this property is expressed as follows:

$$y(t) = Kx(t - \tau_0) \tag{1-31}$$

Where K is a constant. According to Eq. (1-31), the output is an exact replica of the input but attenuated (or amplified) depending on the value of K and delayed by τ_0 seconds. Taking the Laplace transform of Eq. (1-31) yields

$$Y(s) = Ke^{-s\tau_0}X(s) = H(s)X(s) \tag{1-32a}$$

where the gain response is

$$|H(j\omega)| = K \tag{1-32b}$$

the phase response is

$$\phi(\omega) = -\omega\tau_0 \qquad (1\text{-}32c)$$

and the delay response is

$$\tau(\omega) = \tau_0 \qquad (1\text{-}32d)$$

In general, the delay characteristics of gain-shaping filters are not flat and may, therefore, need to be corrected. For example, consider the phase and delay characteristics for a second-order low-pass function of the form given in Eq. (1-26). The phase function is given by

$$\phi(\omega) = -\tan^{-1}\frac{\operatorname{Im} D(j\omega)}{\operatorname{Re} D(j\omega)}$$

$$= -\tan^{-1}\frac{\omega\omega_p}{Q_p(\omega_p^2 - \omega^2)} \qquad (1\text{-}33)$$

where $D(j\omega)$ is the denominator of $H(j\omega)$ in Eq. (1-26). Then according to Eq. (1-30), the delay $\tau(\omega)$ is obtained by differentiating Eq. (1-33) with respect to ω: i.e.,

$$\tau(\omega) = \frac{\omega_p}{Q_p}\frac{\omega^2 + \omega_p^2}{(\omega_p^2 - \omega^2)^2 + \omega^2\omega_p^2/Q_p^2}$$

$$= \frac{1}{\omega_p Q_p}\frac{(\omega/\omega_p)^2 + 1}{[1 - (\omega/\omega_p)^2]^2 + (\omega/\omega_p)^2/Q_p^2} \qquad (1\text{-}34)$$

The delay characteristic in Eq. (1-34) is plotted in Fig. 1-12 for $Q_p = 0.5$, $1/\sqrt{3}$, 1, and 10. Note that for $Q_p > 1/\sqrt{3}$, the delay characteristic is peaked, with the peak becoming more pronounced as the Q_p increases. Thus, high-Q gain-shaping filters will yield severe delay distortion [i.e., significant departure from the ideal flat delay given in Eq. (1-32c)]. Note that for $Q_p = 1/\sqrt{3}$, the delay is "maximally flat." The maximally flat delay character will be pursued further in the next section.

The purpose of the delay equalizer is then to introduce sufficient delay shaping to the gain-shaping filter to make the total delay as flat as possible. In addition, the delay equalizer must not alter the gain response provided by the gain-shaping filter. Thus, the gain of the delay equalizer must be flat over the band of interest. Such a filter is referred to as an *all-pass filter*.

A second-order all-pass filter can be described by the transfer function

$$H(s) = \frac{s^2 - (\omega_p/Q_p)s + \omega_p^2}{s^2 + (\omega_p/Q_p)s + \omega_p^2} \qquad (1\text{-}35)$$

The complex poles and zeros for Eq. (1-35) are symmetrical about the $j\omega$ axis, as shown in Fig. 1-13. Note that since the transmission zeros are in the right half-plane, the all-pass function is a nonminimum phase function. The gain and phase for the all-pass function are sketched in Fig. 1-14. Observe that the gain is ideally flat over the infinite frequency band $0 \leq \omega \leq \infty$. However, as

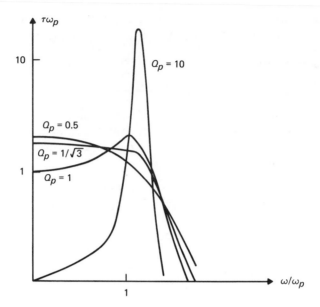

Fig. 1-12 Delay characteristics for a second-order low-pass filter with Q_p = 0.5, $1/\sqrt{3}$, 1, and 10.

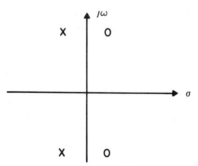

Fig. 1-13 s-Plane pole–zero plot for a second-order all-pass network.

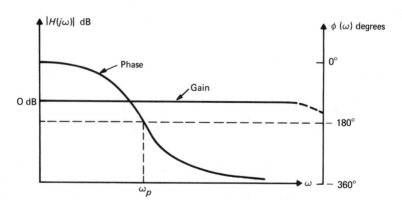

Fig. 1-14 Gain and phase responses for a second-order all-pass network.

noted earlier, because of the frequency characteristics of the active elements and parasitics, an all-pass function will be band-limited in practice. Also, the phase at $\omega = \omega_p$ is $-180°$. The delay for a second-order all-pass function is given by

$$
\begin{aligned}
\tau(\omega) &= \frac{2(\omega_p/Q_p)(\omega_p^2 + \omega^2)}{(\omega_p^2 - \omega^2)^2 + \omega_p^2\omega^2/Q_p^2} \\
&= \frac{(2/\omega_p Q_p)[1 + (\omega/\omega_p)^2]}{[1 - (\omega/\omega_p)^2]^2 + (\omega/\omega_p)^2/Q_p^2}
\end{aligned}
\tag{1-36}
$$

Comparing Eqs. (1-34) and (1-36), we observe that

$$
\tau_{\mathrm{AP}}(\omega) = 2\tau_{\mathrm{LP}}(\omega)
\tag{1-37}
$$

Thus, the delay characteristics in Fig. 1-12, when multiplied by 2, represent the delay responses for the all-pass function with corresponding Q_p.

1.5 APPROXIMATION METHODS FOR FILTER DESIGN

In this section we consider techniques for approximating ideal filter transmission characteristics. Here we confine our discussion to the low-pass case; however, in Section 1.6 we show that the approximations developed in this section apply to other filter types via simple frequency transformations. Approximation, in general, is a broad topic and in this chapter we consider only the commonly used classical approximations. For a detailed derivation of the approximations discussed below, the reader is referred to references [B2, B3, and B7]. Furthermore, we restrict ourselves to an all-pole approximation, at least for the time being.

1.5.1 Butterworth Low-Pass Filters

Consider a realizable magnitude function that approximates the ideal low-pass transmission by

$$
|H(j\omega)|^2 = \frac{H^2(0)}{1 + (\omega/\omega_c)^{2n}} \qquad n = 1, 2, 3, \ldots
\tag{1-38}
$$

Since $|H(j\omega)|$ is a magnitude function, it is approximated with an even function, as given by Eq. (1-38). The expression in Eq. (1-38) is arrived at by requiring that the magnitude function be *maximally flat* at $\omega = 0$ (see [B2]). This implies that lower-order derivatives of the function are set equal to zero (see Prob. 1.6). Naturally, the approximation improves as n increases. Note in Eq. (1-38) that $\omega_{3\mathrm{dB}} = \omega_c$ and the largest error within the band occurs at $\omega = \omega_c$, which is 3 dB. We normalize ω_c to unity, without loss of generality, to determine the pole locations corresponding to Eq. (1-38):

$$
|H(s)|^2 = H(s)H^*(s) = \frac{H^2(0)}{1 + (-1)^n s^{2n}}
\tag{1-39}
$$

where * denotes a complex conjugate. The location of the poles of $H(s)$ are found from

$$s^{2n} = (-1)^{n+1} \tag{1-40a}$$

and by retaining only the left-half-plane poles in Eq. (1-40a), we obtain

$$s = \exp\left[j\left(\frac{2k - 1 + n}{2n}\right)\pi \right] \qquad k = 1, 2, \ldots, n \tag{1-40b}$$

It is seen that all the poles are located on a unit circle as shown in Fig. 1-15.

Fig. 1-15   Roots of Butterworth polynomials for orders 2, 3, and 4.

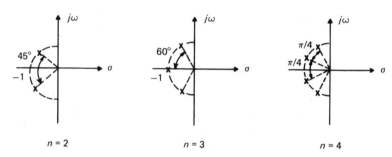

n = 2   n = 3   n = 4

The transfer function $H(s)$, for an nth-order all-pole low-pass response, is written

$$H(s) = \frac{H(0)}{s^n + b_{n-1}s^{n-1} + b_{n-2}s^{n-2} + \ldots + b_1 s + 1} \tag{1-41}$$

The Butterworth polynomials, which define the denominator of $H(s)$ in Eq. (1-41), for even and odd order n are given by

$$n \text{ even:} \quad \prod_{k=1}^{n/2} [s^2 + (2\cos\theta_k)s + 1] \qquad \theta_k = \frac{(2k - 1)\pi}{2n} \tag{1-42a}$$

$$n \text{ odd:} \quad (s + 1)\prod_{l=1}^{(n-1)/2} [s^2 + (2\cos\theta_l)s + 1] \qquad \theta_l = \frac{l\pi}{n} \tag{1-42b}$$

The numerical coefficients of the first four polynomials in the denominator of Eq. (1-41) are listed in Table 1-1. (See Appendix B for higher-order polynomials

TABLE 1-1   Butterworth ploynomials

n	Polynomial
1	$s + 1$
2	$s^2 + \sqrt{2}s + 1$
3	$s^3 + 2s^2 + 2s + 1 = (s + 1)(s^2 + s + 1)$
4	$s^4 + 2.613s^3 + 3.414s^2 + 2.613s + 1 = (s^2 + 0.765s + 1)(s^2 + 1.848s + 1)$

up to $n = 10$.) The order of the Butterworth filter is determined by the specification of attenuation beyond the passband. The magnitude responses for the Butterworth filters for $n \le 10$ are shown in Fig. 1-16.

Fig. 1-16 Gain responses of Butterworth filters for $n \le 10$.

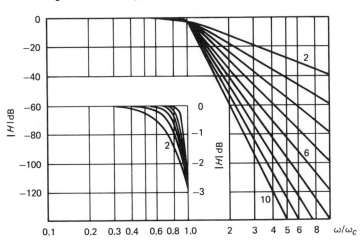

EXAMPLE 1-1

Determine the order of the Butterworth filter so that at $\omega/\omega_c = 2.0$, the magnitude is down by at least 40 dB. From Eq. (1-38),

$$20 \log \frac{|H(0)|}{|H(j2\omega_c)|} \ge 40 \text{ dB}$$

i.e.,

$$20 \log(1 + 2^{2n})^{1/2} \ge 40 \text{ dB}$$

Solving for n yields $n > 6$. Hence, a seventh-order filter is required (i.e., $n = 7$).

1.5.2 Chebyshev Low-Pass Filters

An approximation that distributes the error evenly throughout the passband in an oscillating manner is called the *Chebyshev* or *equal-ripple approximation*. In this approximation the largest error within the band is minimized and the magnitude decreases monotonically within the stop band [B2, B3]. The Chebyshev low-pass magnitude is given by

$$|H(j\omega)|^2 = \frac{H_0^2}{1 + \epsilon^2 C_n^2(\omega/\omega_c)} \tag{1-43}$$

The passband ripple (δ) (see Figs. 1-6a to 1-9a), expressed in decibels, is related to ϵ according to

$$\epsilon^2 = 10^{0.1\delta} - 1 \tag{1-44}$$

The parameter ϵ is a real constant taking on values less than 1, and the Chebyshev polynomials $C_n (\omega/\omega_c)$ are defined by

$$C_n\left(\frac{\omega}{\omega_c}\right) = \begin{cases} \cos\left(n \cos^{-1}\dfrac{\omega}{\omega_c}\right) & 0 \leq \dfrac{\omega}{\omega_c} \leq 1 \tag{1-45a} \\[2mm] \cosh\left(n \cosh^{-1}\dfrac{\omega}{\omega_c}\right) & \dfrac{\omega}{\omega_c} > 1 \tag{1-45b} \end{cases}$$

The first few Chebyshev polynomials and the recursion formula are given in Table 1-2.

TABLE 1-2   Chebyshev polynomials

$$C_0\left(\frac{\omega}{\omega_c}\right) = 1$$

$$C_1\left(\frac{\omega}{\omega_c}\right) = \frac{\omega}{\omega_c}$$

$$C_2\left(\frac{\omega}{\omega_c}\right) = 2\left(\frac{\omega}{\omega_c}\right)^2 - 1$$

$$C_3\left(\frac{\omega}{\omega_c}\right) = 4\left(\frac{\omega}{\omega_c}\right)^3 - 3\left(\frac{\omega}{\omega_c}\right)$$

$$C_n\left(\frac{\omega}{\omega_c}\right) = 2\left(\frac{\omega}{\omega_c}\right)C_{n-1}\left(\frac{\omega}{\omega_c}\right) - C_{n-2}\left(\frac{\omega}{\omega_c}\right)$$

It should be noted from Table 1-2 that only for n odd in Eq. (1-43), $H_0 = H(0)$. To find the pole location of Eq. (1-43) we have, with $s = j(\omega/\omega_c)$,

$$|H(s)|^2 = H(s)H^*(s) = \frac{H_0^2}{1 + \epsilon^2 C_n^2(s/j)} \tag{1-46}$$

It can be shown [B2, B3] that the pole locations of a Chebyshev low-pass filter lie on an ellipse.

The denominator polynomials for the all-pole transfer function,

$$H(s) = \frac{H_0}{s^n + a_{n-1}s^{n-1} + a_{n-2}s^{n-2} + \ldots + a_1 s + a_0} \tag{1-47}$$

are listed in Table 1-3 for $\frac{1}{2}$-dB, 1-dB, and 2-dB ripple magnitudes and n up to 4. (See Appendix B for higher-order polynomials up to $n = 10$.) Extensive tables for various ripple values are listed in the literature (e.g., see [B3]). For convenience, these denominator polynomials are also given in factored form.

There are two parameters in the design of Chebyshev filters: the ripple (δ) and the amount of attenuation (A) beyond the passband, which determines

TABLE 1-3

n	Polynomial

1/2-dB ripple ($\epsilon = 0.3493$)

1 $s + 2.863$
2 $s^2 + 1.425s + 1.516$
3 $s^3 + 1.253s^2 + 1.535s + 0.716 = (s + 0.626)(s^2 + 0.626s + 1.142)$
4 $s^4 + 1.197s^3 + 1.717s^2 + 1.025s + 0.379 = (s^2 + 0.351s + 1.064)(s^2 + 0.845s + 0.356)$

1-dB ripple ($\epsilon = 0.5088$)

1 $s + 1.965$
2 $s^2 + 1.098s + 1.103$
3 $s^3 + 0.988s^2 + 1.238s + 0.491 = (s + 0.494)(s^2 + 0.490s + 0.994)$
4 $s^4 + 0.953s^3 + 1.454s^2 + 0.743s + 0.276 = (s^2 + 0.279s + 0.987)(s^2 + 0.674s + 0.279)$

2-dB ripple ($\epsilon = 0.7648$)

1 $s + 1.308$
2 $s^2 + 0.804s + 0.637$
3 $s^3 + 0.738s^2 + 1.022s + 0.327 = (s + 0.402)(s^2 + 0.369s + 0.886)$
4 $s^4 + 0.716s^3 + 1.256s^2 + 0.517s + 0.206 = (s^2 + 0.210s + 0.928)(s^2 + 0.506s + 0.221)$

the order (n). In general, for specified δ and n, the complex poles $\sigma_k + j\omega_k$ for $H(s)$ [Eq. (1-47)] are given [B4] by

$$\sigma_k = -\frac{1}{2}\sin\left(\frac{\pi}{2}\frac{1+2k}{n}\right)\left[\left(\frac{1}{\epsilon}+\sqrt{\frac{1}{\epsilon^2}+1}\right)^{1/n} - \left(\frac{1}{\epsilon}+\sqrt{\frac{1}{\epsilon^2}+1}\right)^{-1/n}\right]$$

(1-48a)

$$\omega_k = \frac{1}{2}\cos\left(\frac{\pi}{2}\frac{1+2k}{n}\right)\left[\left(\frac{1}{\epsilon}+\sqrt{\frac{1}{\epsilon^2}+1}\right)^{1/n} + \left(\frac{1}{\epsilon}+\sqrt{\frac{1}{\epsilon^2}+1}\right)^{-1/n}\right]$$

(1-48b)

where $k = 0, 1, 2, \ldots, n - 1$ for left-half-plane poles. We note that the 3-dB bandwidth of a Chebyshev low-pass filter is given by

$$\frac{\omega_{3dB}}{\omega_c} = \cosh\left(\frac{1}{n}\cosh^{-1}\frac{1}{\epsilon}\right)$$

(1-49)

where ω_c is the ripple (ϵ) bandwidth.

The magnitude responses for Chebyshev filters for $n \leq 10$ and 1-dB ripple are shown in Fig. 1-17. Similar response curves for 2-dB and 0.5-dB ripples can be found in [B5].

EXAMPLE 1-2

Determine the order of the Chebyshev filter required to achieve a passband ripple $\delta = 0.5$ dB and a loss of 60 dB at $\omega/\omega_c = 2.0$. The loss, A, at $\omega > \omega_c$ is obtained from Eqs. (1-43) and (1-45b):

$$A \leq 10\log\left[\frac{|H(0)|^2}{|H(j\omega)|^2}\right] = 10\log\left[1 + \epsilon^2\cosh 2\left(n\cosh^{-1}\frac{\omega}{\omega_c}\right)\right]$$

(1-50)

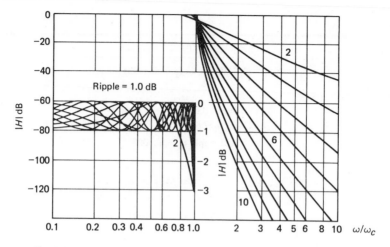

Fig. 1-17 Gain responses for 1-dB ripple Chebyshev filters for $n \leq 10$.

Substituting for ϵ^2, according to Eq. (1-44), and solving for n yields

$$n \geq \frac{\cosh^{-1}\{[(10^{0.1A} - 1)/(10^{0.1\delta} - 1)]^{1/2}\}}{\cosh^{-1}(\omega/\omega_c)} \tag{1-51}$$

For $A = 60$ dB, $\delta = 0.5$ dB, and $\omega/\omega_c = 2$, we obtain, from Eq. (1-51),

$$n \geq 5.57$$

Since a real filter requires that n be an integer, we choose $n = 6$.

●

1.5.3 Bessel (or Thompson) Low-Pass Filters

In Section 1.4 we observed that for distortionless transmission,

$$H(s) = Ke^{-s\tau}$$

which provides linear phase or flat delay. In other words, the output of such a network is an exact replica of the input (perhaps amplified or attenuated by K), delayed by τ seconds.

An all-pole approximation to $e^{-s\tau}$ may be written for nth order as

$$H(s) = \frac{b_0}{s^n + b_{n-1}s^{n-1} + \ldots + b_1 s + b_0} \tag{1-52}$$

where the coefficients b_i are obtained by setting the second- and higher-order derivatives of the phase function to zero. This procedure follows analogously to that used to derive the coefficients for the all-pole Butterworth low-pass

Filter Transmission and Related Topics

filters [B2, B3]. A simpler approach is indicated in Prob. 1.7. The polynomials so obtained are called *Bessel polynomials* and the filters are referred to as either Bessel or *Thompson* or *maximally flat delay* (MFD) *filters*. The first few Bessel polynomials are given in Table 1-4 with τ normalized to unity (i.e., $b_1 = b_0$).

TABLE 1-4 Bessel polynomials

n	Polynomial
1	$s + 1$
2	$s^2 + 3s + 3$
3	$s^3 + 6s^2 + 15s + 15 = (s + 2.322)(s^2 + 3.678s + 6.460)$
4	$s^4 + 10s^3 + 45s^2 + 105s + 105 = (s^2 + 5.792s + 9.140)(s^2 + 4.208s + 11.488)$

(The reader is referred to Appendix B for higher orders up to $n = 10$.) Note in Table 1-4 that the polynomials are normalized, so that $\tau(0) = 1$ sec. The 3-dB band-width of the Thompson filters is given by the following approximate formula:

$$\tau\omega_{3dB} \approx \sqrt{(2n - 1)\ln 2} \qquad n \geq 3 \qquad (1\text{-}53)$$

The magnitude responses of Thompson filters for $n \leq 10$ are shown in Fig. 1-18. The delay characteristics of the Thompson filters for $n \leq 10$ are

Fig. 1-18 Gain responses for Thomson (Bessel) filters for $n \leq 10$.

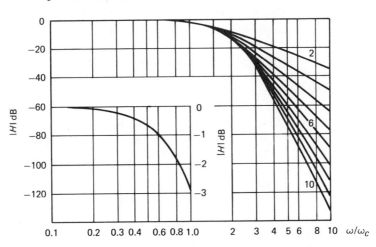

shown in Fig. 1-19. It should be noted that the normalizations in these two figures are such that ω_{3dB} is unity and not τ. This is done so that one could compare the response with those of Butterworth filters. Note also that the delay is flat as expected; hence, the name maximally flat delay filters. The delay characteristics of Butterworth and Chebyshev filters for $n > 2$ exhibit a peaking near $\omega = \omega_c$. For example, the delay value of a tenth-order Butterworth filter

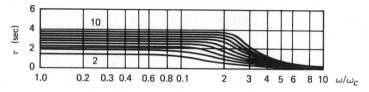

Fig. 1-19   Delay responses for Thomson (Bessel) filters for $n \leq 10$.

at $\omega = \omega_c$ is almost twice the value at dc frequency. Note, however, that the Thompson filters are inefficient in terms of gain selectivity. For example, the attenuation for a fourth-order Butterworth low-pass filter is about 50 dB for $\omega/\omega_c = 4$, whereas a seventh-order Thompson filter is required to achieve this attenuation. For this reason it is usually preferable to realize video and digital transmission filters with a combination of a gain-shaping filter and a delay equalizer. The gain-shaping filter, perhaps a Butterworth or δ-ripple Chebyshev filter, is designed to meet only the gain specifications as per Fig. 1-6. Then a sufficient number of all-pass second-order sections are added to compensate for the difference in the delay characteristic of the gain-shaping filter and the delay specification. Delay equalization and all-pass filters were discussed in Section 1.4.

1.5.4 Elliptic (or Cauer) Low-Pass Filters

In previous subsections we considered approximations by all-pole functions. Naturally, this restriction is not necessary. The all-pole functions were considered because of their ease of realization. We now briefly consider a general approximation having poles and zeros [B3]. One widely used classical pole-zero approximation is the *elliptic* approximation, also referred to as the *Cauer* approximation. In an elliptic filter the transmission function has zeros on the $j\omega$-axis which are distributed across the stop band. Because of the $j\omega$-axis zeros, the filters will provide steep skirts or narrow transition bands, and thus for a given selectivity (which is usually given by the ratio $\omega_{60\text{dB}}/\omega_{6\text{dB}}$ or $\omega_{30\text{dB}}/\omega_{3\text{dB}}$), lower-order filters can be used than those derived from Chebyshev or Butterworth approximations. An elliptic approximation for a typical specification is illustrated in Fig. 1-20. The magnitude function in this case may be written

$$| H(j\omega) |^2 = \frac{1}{1 + \epsilon^2 R_{n,\omega_s}^2(\omega/\omega_p)} \tag{1-54}$$

where $R_{n,\omega_s}(\Omega)$, with $\Omega = \omega/\omega_p$, is the ratio of two polynomials of the form

$$R_{n,\omega_s}(\Omega) = \begin{cases} \displaystyle\prod_{i=1}^{n/2} \frac{[\Omega^2 - (\omega_i/\omega_p)^2]}{[\Omega^2 - (\omega_s/\omega_i)^2]} & \text{for } n \text{ even} \tag{1-55a} \\[2em] \displaystyle\Omega \prod_{j=1}^{(n-1)/2} \frac{[\Omega^2 - (\omega_j/\omega_p)^2]}{[\Omega^2 - (\omega_s/\omega_j)^2]} & \text{for } n \text{ odd} \tag{1-55b} \end{cases}$$

Note that the poles and zeros of $R_{n,\omega_s}(\Omega)$ exhibit geometric symmetry about $\sqrt{\omega_s \omega_p}$. The quantity ϵ^2 in Eq. (1-54) determines the passband ripple as it did

Filter Transmission and Related Topics

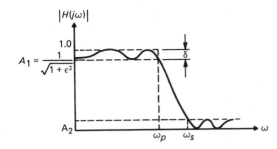

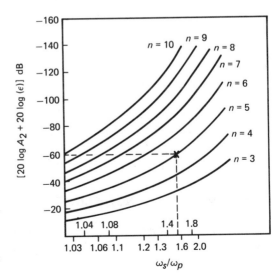

Fig. 1-20 Typical gain response of a Cauer filter.

previously for the Chebyshev filters. An elliptic filter is specified by the inband ripple $\delta = 20 \log (1 - A_1)$, the minimum stop-band attenuation $-20 \log (A_2)$, the selectivity ω_s/ω_p, and the order n (see Fig. 1-20). The total number of peaks and valleys in the passband is equal to the order n. Given these specifications, the determination of poles and zeros for $H(s)$ generally requires computer aids. Such an exercise is beyond the scope of this book. Tables of poles and zeros for a wide variety of cases can be found in [B6]. A sample of normalized elliptic low-pass filter functions is given in Appendix B. In these tables the frequencies are normalized to the passband edge frequencies (i.e., $\Omega = \omega/\omega_p$). The form of the transfer function for an elliptic filter is observed to be

$$H(s) = \frac{K \prod_{i=1}^{m} (s^2 + a_i)}{s^n + b_{n-1}s^{n-1} + b_{n-2}s^{n-2} + \ldots + b_1 s + b_0} \qquad (1\text{-}56)$$

where in Appendix B, K is chosen such that the peak gain is unity.

To determine, for a given A_2, ω_s, and ϵ (where $\epsilon = \sqrt{10^{0.1\delta} - 1}$), the required order n, the curves in Fig. 1-21 can be particularly helpful [B7]. For example, if $\delta = 0.5$ dB, $20 \log A_2 \leq -50$ dB, and $\omega_s/\omega_p = 1.5$, the required elliptic filter order, obtained from Fig. 1-21, is $n = 5$.

Fig. 1-21 Curves for determining required elliptic filter order (n) from out-of-band attenuation (A_2) and in-band ripple (ϵ) specifications. (From Sedra and Brackett [B7], with permission.)

One important example of the use of elliptic filters are the channel bank filters used in pulse code modulation (PCM) telephone systems. A typical transfer function for this application is the fifth-order elliptic function [P5].

$$H(s) = \frac{K(s^2 + \omega_1^2)(s^2 + \omega_2^2)}{(s + p_0)(s + p_1)(s + p_1^*)(s + p_2)(s + p_2^*)} \qquad (1\text{-}57)$$

where
$\omega_1 = 2.92 \times 10^4$ rad/sec
$\omega_2 = 4.32 \times 10^4$ rad/sec
$p_0 = 1.68 \times 10^4$ rad/sec
$p_1, p_1^* = (-0.97 \pm j1.75) \times 10^4$ rad/sec
$p_2, p_2^* = (-0.236 \pm j2.24) \times 10^4$ rad/sec

The gain response for this filter is sketched in Fig. 1-22. More will be said about this filter in Section 1.10 Finally, it is noted that a second-order elliptic transfer function is of the form of the low-pass notch function given in Fig. 1-10 and Eq. (1-28) for $\omega_p < \omega_z$.

Fig. 1-22 Fifth-order elliptic filter response for the transfer function in Eq. (1-57).

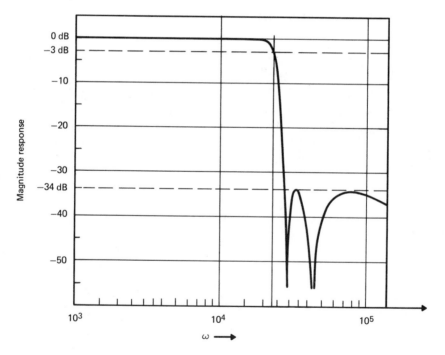

1.6 FREQUENCY TRANSFORMATIONS

So far we have considered only transfer functions for meeting low-pass specifications. We shall now show how we can derive other filter designs, such as high-pass, band-pass, and band-reject, using an initial low-pass design and

Filter Transmission and Related Topics

simple frequency transformations. Let us, for simplicity, normalize frequency such that the cutoff frequency ω_c of the low-pass function is unity. The low-pass frequency domain will be called p and s is the frequency domain of interest.

1.6.1 Low-Pass to High-Pass Transformation

From given low-pass characteristics, we can obtain high-pass characteristics by the following low-pass to high-pass transformation:

$$p = \frac{1}{s} \qquad (1\text{-}58)$$

For example, a low-pass third-order Butterworth filter that is given in the p domain by

$$H(p) = \frac{K}{p^3 + 2p^2 + 2p + 1} \qquad (1\text{-}59)$$

can be transformed to a high-pass filter by the use of Eq. (1-58):

$$H(s) = \frac{K}{(1/s)^3 + 2(1/s)^2 + 2(1/s) + 1} = \frac{Ks^3}{s^3 + 2s^2 + 2s + 1} \qquad (1\text{-}60)$$

Note that Eq. (1-60) is a high-pass response with the same passband flatness as that of the Butterworth low-pass filter.

A simple example of such a transformation is RC:CR transformation as shown in Fig. 1-23a and b. In other words a low-pass circuit that consists of R_i

Fig. 1-23 (a) Second-order low-pass active-RC filter. (b) RC:CR admittance transformed second-order high-pass active-RC filter.

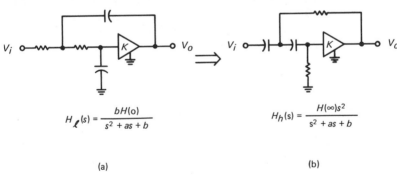

$$H_\ell(s) = \frac{bH(o)}{s^2 + as + b}$$

(a)

$$H_h(s) = \frac{H(\infty)s^2}{s^2 + as + b}$$

(b)

and C_i and voltage amplifiers can be transformed from a low-pass circuit to a high-pass circuit by replacing each resistor R_i in the original network by a capacitor of value $1/R_i$ farads and each capacitor C_j by a resistor of value $1/C_j$ ohms. The original and the transformed networks are obviously related. The relationships in the voltage or current ratios are of the form shown in Eq. (1-58) and can be verified by the reader (Prob. 1.11).

1.6.2 Low-Pass to Band-Pass Transformation

A low-pass characteristic can be transformed to a band-pass characteristic by the low-pass to band-pass transformation given by [B2]:

$$p = \frac{s^2 + \omega_0^2}{(\omega_{3dB})s} = \frac{Q(s_n^2 + 1)}{s_n} \tag{1-61}$$

where $Q = \omega_0/\omega_{3dB}$ is the *quality factor*, $s_n = s/\omega_0$, and ω_0 is the *center frequency*, defined by

$$\omega_0 = \sqrt{\omega_u\omega_l} \simeq \frac{\omega_u + \omega_l}{2} \quad \text{for } \frac{\omega_0}{\omega_{3dB}} \gg 1 \tag{1-62a}$$

$$\omega_{3dB} = \omega_u - \omega_l \tag{1-62b}$$

The quantities are shown in Fig. 1-24d. As an example, consider a normalized first-order filter

$$H(p_n) = \frac{K}{p_n + 1} \tag{1-63a}$$

Substitution of Eq. (1-61) into Eq. (1-63a) yields

$$H(s_n) = \frac{(K/Q)s_n}{s_n^2 + (1/Q)s_n + 1} \tag{1-63b}$$

$H(s_n)$ is observed to be identical in form to the second-order band-pass function in Eq. (1-27). The denormalized pole–zero configurations corresponding to Eq. (1-63a) and Eq. (1-63b) are shown in Fig. 1-24a and b. It should be noted that low-pass to band-pass transformation always results in symmetric BP filters.

In band-pass filters the *quality factor* Q is an important parameter. In many applications Q is high (i.e., $Q \gg 1$). Such filters are referred to as *narrow-band filters* (i.e., $\omega_{3dB} \ll \omega_0$). For a narrow-band filter the response is symmetric about the center frequency, ω_0. In *LC* filters, the low-pass to band-pass transformation is simply achieved in a circuit by using the admittance transformation in Fig. 1-25. In Chapters 2 and 4 we show that the use of inductors is not necessary. Since in integrated circuits practical values of inductors are not available, we are therefore concerned mainly with active filters using *RC*, or *R*, or *C*, or switched-*C* networks and amplifiers. The simulation of inductive immittances and the realization of resonant circuits with active networks are treated in Chapters 2 through 6.

1.6.3 Low-Pass to Band-Reject Transformation

The low-pass to band-reject transformation is given by

$$p = \frac{(\omega_{3dB})s}{s^2 + \omega_0^2} = \frac{s_n}{Q(s_n^2 + 1)} \tag{1-64}$$

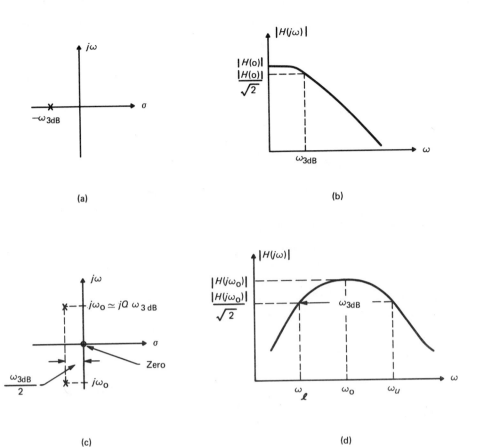

(a)

(b)

(c)

(d)

Fig. 1-24 First-order low-pass (a) s-plane plot, (b) gain response and the low-pass to band-pass transformed second-order band-pass, (c) s-plane plot, (d) gain response.

Fig. 1-25 Low-pass to band-pass admittance transformations.

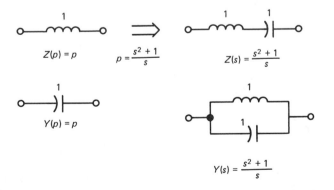

$$Z(p) = p$$

$$p = \frac{s^2 + 1}{s}$$

$$Z(s) = \frac{s^2 + 1}{s}$$

$$Y(p) = p$$

$$Y(s) = \frac{s^2 + 1}{s}$$

where $s_n = s/\omega_0$. For example, the band-reject transfer function corresponding to the low-pass transfer function in Eq. (1-63a) is given by

$$H(s_n) = \frac{K(s_n^2 + 1)}{s_n^2 + s_n(1/Q) + 1} \qquad (1\text{-}65)$$

Note that $s_n = j1$ and $|H(j1)| = 0$; such filters are referred to as *notch filters* with normalized null frequency at $\omega_0 = 1$.

1.7 TIME-DOMAIN CONSIDERATIONS

Recall from Section 1.4.5 that the response of a system is undistorted when the delay is flat at all frequencies. For a distortionless transmission, the output is an exact replica of the input as given by Eq. (1-31):

$$y(t) = Kx(t - \tau) \qquad (1\text{-}66)$$

However, in any practical situation, distortions do occur and one can only approximate a distortionless transmission. In the following subsections we consider certain terminologies that are used to define deviations from an ideal transmission.

1.7.1 Step Response

The *step response*, the response $y(t)$ for a step input $x(t) = u(t)$, for a distortionless transmission system is given by

$$y(t) = Ku(t - \tau) \qquad (1\text{-}67)$$

However, the realization of Eq. (1-67) requires an ideal transfer function of $H(s) = Ke^{-s\tau}$, which with lumped elements can only be approximated. If time-domain distortion is an important factor, one approximation that can be used is the Bessel filter.

Let us now consider the response of a low-pass filter for a step input. A typical response is shown in Fig. 1-26. The quantity γ, which is the difference between the peak value and the final value, is called the *overshoot* and is given in

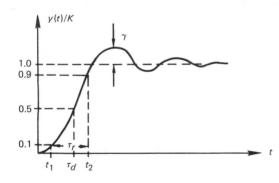

Fig. 1-26 Typical step response for a stable linear system.

Filter Transmission and Related Topics

percent. τ_d is the *delay time*, which is the time required for the step response to reach 50% of its final value. τ_r, the *rise time*, is the time required for the step response to rise from 10% to 90% of its final value. For example, consider the simple one-pole case

$$H(s) = \frac{1}{s + p_1}$$

The step response is given by

$$y(t) = (1 - e^{-p_1 t})u(t) \tag{1-68}$$

In this case there is no overshoot, $\gamma = 0$, and we find $\tau_d = 0.69/p_1$ and $\tau_r = 2.2/p_1$. In general, for a filter with negligible overshoot, $\gamma \leq 5$ percent, the following empirical result holds:

$$\tau_r \omega_{3dB} \simeq 2.2 \quad \text{or} \quad \tau_r f_{3dB} \simeq 0.35 \tag{1-69}$$

Step responses of various low-pass Butterworth, Chebyshev (1/2 dB ripple) and Thomson filters are shown in Fig. 1-27a–c, respectively.

1.7.2 Impulse Response

When the input is an impulse [i.e., $x(t) = \delta(t)$], the ideal impulse response is given by

$$y(t) = h(t) = \mathcal{L}^{-1}(Ke^{-s\tau}) = K\delta(t - \tau) \tag{1-70}$$

Since in Eq. (1-70), $y(t)$ is the derivative of the response in Eq. (1-67), one can obtain the impulse response from the step response, and vice versa. Impulse responses, corresponding to those in Fig. 1-27a–c, are given in Fig. 1-28a–c.

Precise calculations of τ_r and τ_d for a given filter response are usually time consuming. A more convenient method for such calculations can be used if the definitions of rise time and delay time are cast in different form. Elmore's definitions for rise and delay times result in considerable simplification in characterizing time-domain responses.

1.7.3 Elmore's Definitions of Rise Time and Delay Time [B2, B8]

Elmore's definitions of delay time and rise time are very convenient for computation and are sometimes used in literature. These are defined as follows:

$$\tau_D = \int_0^\infty th(t)\,dt \tag{1-71}$$

$$\tau_R = \left[2\pi \int_0^\infty (t - \tau_D)^2 h(t)\,dt\right]^{1/2} \tag{1-72a}$$

$$= \sqrt{2\pi}\left[\int_0^\infty t^2 h(t)\,dt - \tau_D^2\right]^{1/2} \tag{1-72b}$$

where $h(t)$ is the impulse response normalized such that $\int_0^\infty h(t)\,dt = 1$. Capital-letter subscripts are used to differentiate these definitions from those given in Section 1.7.1.

Fig. 1-27 Step responses for (a) Butterworth filters, (b) Chebyshev filters (1/2 dB ripple), and (c) Thomson (Bessel) filters for $n \le 10$. [From Henderson and Kautz [P2])

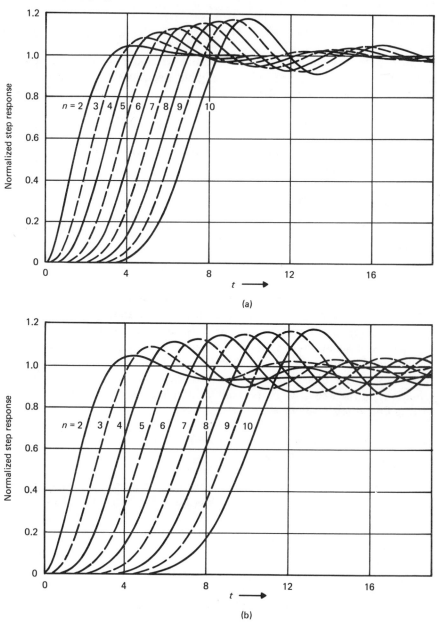

(a)

(b)

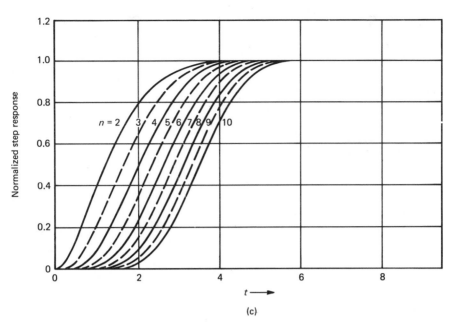

Fig. 1.27 cont.

The interpretation of Eqs. (1-71) and (1-72) is shown in Fig. 1-29a and b. The ease of application of the foregoing definitions is illustrated below. Consider the overall normalized transfer function

$$H(s) = \frac{1 + a_1 s + a_2 s^2 + \ldots + a_m s^m}{1 + b_1 s + b_2 s^2 + \ldots + b_n s^n} \qquad n \geq m \qquad (1\text{-}73)$$

The transfer function is related to the normalized impulse response by

$$H(s) = \int_0^\infty h(t) e^{-st} \, dt \qquad (1\text{-}74)$$

Equation (1-74) can be expanded into a power series as:

$$\begin{aligned} H(s) &= \int_0^\infty h(t) \left(1 - st + \frac{s^2 t^2}{2!} + \ldots \right) dt \\ &= 1 - s\tau_D + \frac{s^2}{2!} \int_0^\infty t^2 h(t) \, dt + \ldots \qquad (1\text{-}75) \\ &= 1 - s\tau_D + \frac{s^2}{2!} \left(\frac{\tau_R^2}{2\pi} + \tau_D^2 \right) + \ldots \end{aligned}$$

From Eq. (1-73) one gets, by direct division,

$$H(s) = 1 - (b_1 - a_1)s + (b_1^2 - a_1 b_1 + a_2 - b_2)s^2 + \ldots \qquad (1\text{-}76)$$

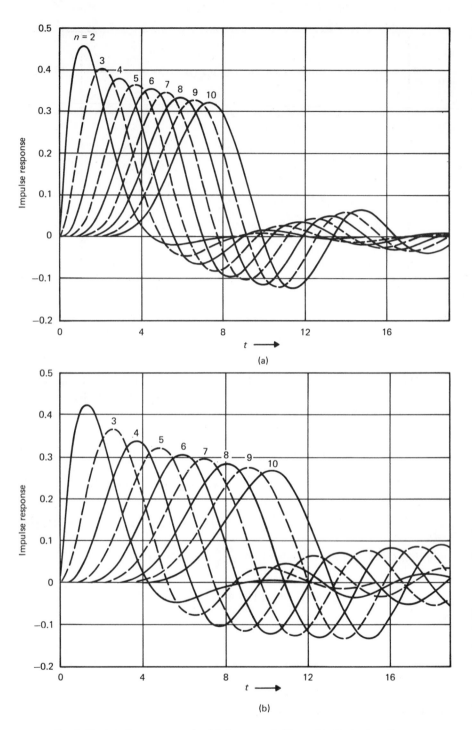

Fig. 1-28 Impulse responses for (a) Butterworth filters, (b) Chebyshev filters (1/2 dB ripple), and (c) Thomson (Bessel) filters for $n \leq 10$. (From Henderson and Kautz [P2])

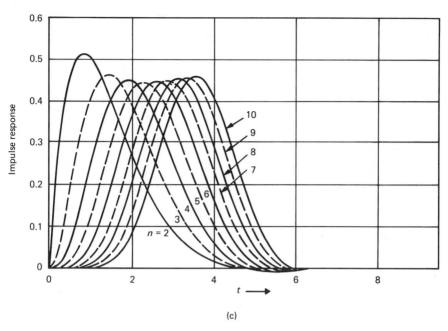

(c)

Fig. 1.28   cont.

A comparison of Eqs. (1-76) and (1-75) yields

$$\tau_D = b_1 - a_1 \tag{1-77}$$

$$\tau_R = \{2\pi[b_1^2 - a_1^2 + 2(a_2 - b_2)]\}^{1/2} \tag{1-78}$$

Note that the application of Eqs. (1-77) and (1-78) is valid only when the values of a_i and b_i are such that the step response is monotonic (i.e., nonovershooting) [B2].

●

EXAMPLE 1-3

Consider the response of a four-pole Thomson filter. The transfer function from Table 1-4, rewritten in the form of Eq. (1-73), is

$$H_n(s) = \frac{1}{1 + s + \frac{3}{7}s^2 + \frac{2}{21}s^3 + \frac{1}{105}s^4}$$

From Eqs. (1-77) and (1-78), we have

$$\tau_D = 1 \qquad \tau_R = 0.92$$

whereas the classical definitions—the 50% delay time and the 10% to 90% rise times—are

$$\tau_d = 0.98 \qquad \text{and} \qquad \tau_r = 1.05$$

●

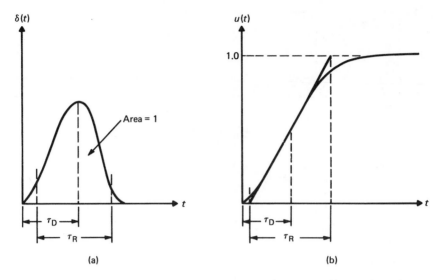

Fig. 1-29 Impulse and step responses illustrating Elmore's definitions of rise and delay times.

The ease of computation of these quantities using Elmore's definitions are obvious. For higher-order systems this advantage becomes even more attractive. It can be shown (Prob. 1.14) that for k monotonic cascaded stages, the overall delay and rise times are given by

$$(\tau_D)_o = \sum_{i=1}^{k} (\tau_D)_i \tag{1-79}$$

$$(\tau_R)_o = \left[\sum_{i}^{k} (\tau_R)_i^2 \right]^{1/2} \tag{1-80}$$

1.8 NORMALIZATION

In this book frequency and impedance will frequently be normalized for convenience and without loss of generality. Under an appropriate change of scale, the tedium of computations with large numbers and powers of 10 can be reduced to simple numerical operations. Normalization is particularly useful in performing pencil-and-paper calculations. However, if a computer or a hand calculator is used in the analysis and design, normalization may not provide any advantage. A change in the frequency scale is referred to as *frequency normalization* and a change in the magnitude scale is referred to as *impedance normalization*. If we scale the radian frequency by Ω_0 and the resistors by R_0, the normalized frequency and passive component values are given by

$$s_n = \frac{s}{\Omega_0} \qquad \text{and} \qquad R_n = \frac{R}{R_0}$$

$$C_n = R_0 \Omega_0 C \qquad \text{and} \qquad L_n = \frac{L \Omega_0}{R_0}$$

Filter Transmission and Related Topics

EXAMPLE 1-4

To illustrate normalization, consider for example a second-order Butterworth filter to be designed for a cutoff frequency $\omega_{3dB} = 2\pi(10^4)$ rad/sec, using the circuit shown in Fig. 1-30, where a noninverting amplifier of gain K is used. The

Fig. 1-30 Example active-RC circuit,

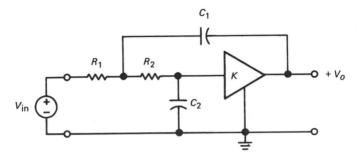

voltage gain at dc is not specified. The voltage transfer function of the circuit can be readily determined to be

$$\frac{V_o}{V_i}(s) = \frac{K/R_1 R_2 C_1 C_2}{s^2 + s\left(\dfrac{1}{R_1 C_1} + \dfrac{1}{R_2 C_1} + \dfrac{1-K}{R_2 C_2}\right) + \dfrac{1}{R_1 R_2 C_1 C_2}} \tag{1-81}$$

For $R_1 = R_2$ and $C_1 = C_2 = C$, we have

$$\frac{V_0}{V_i}(s) = \frac{K/R^2 C^2}{s^2 + s\left(\dfrac{3-K}{RC}\right) + \dfrac{1}{R^2 C^2}} \tag{1-82}$$

We let $\Omega_0 = 2\pi(10^4)$ rad/sec and $R_0 = 1$ kΩ. For a second-order Butterworth filter, the normalized denominator is

$$s^2 + \sqrt{2}\,s + 1 \tag{1-83}$$

Hence, equating the coefficients of Eqs. (1-82) and (1-83) yields

$$\frac{1}{R^2 C^2} = 1 \quad \text{and} \quad \frac{3-K}{RC} = \sqrt{2} \tag{1-84}$$

In this example we have more degrees of freedom than the specifications require. Hence, in addition to equal-valued R's and C's, we arbitrarily choose $R_n = 1$. From Eq. (1-84), $C_n = 1$ and $K = 1.586$. The actual element values are then $R = R_n R_0 = 1$ kΩ, $C = C_n/R_0\Omega_0 = 0.0159$ μF.

1.9 TECHNIQUES FOR THE ANALYSIS OF LINEAR TIME-INVARIANT SAMPLED-DATA SYSTEMS

A sampled-data system operates on signal samples $x(k\tau)$ which are obtained by sampling a continuous signal at τ intervals of time. Thus, the $k\tau$, for $k = 0$, $1, 2, \ldots$, correspond to the discrete instants of time the continuous signal is sampled. In a digital sampled-data system, the system function is completely characterized by the input and output samples. Of course, in a digital system the samples are quantized and expressed as digital words. In an analog sampled-data system, an analog output signal must be constructed from the output samples. This *construction* or *reconstruction process*, as it is commonly called, is inherent in such analog sampled-data systems as switched capacitor filters and charge-coupled-device (CCD) filters. As we will demonstrate shortly, the system function for analog sampled-data systems is affected by the reconstruction process.

It should be noted that a complete treatment of sampled-data systems would consume an entire book and is beyond the scope of this text. What we seek here is to provide sufficient elementary background material to support the treatment of sampled-data switched-capacitor filters in Chapter 6. Those readers interested in a more in-depth treatment of the subject are referred to [B9], [B10], and [B11].

1.9.1 The Sampling Process

An ideal impulse sampler would extract exact values of the signal $x(t)$ at the sampling instants $t = k\tau$ to provide the sampled sequence $[x(k\tau)]$, where τ is the sampling period and k is an integer (i.e., $k = 0, 1, 2, 3, \ldots$). In practice, ideal sampling is not possible but can be approximated closely. One convenient way to view sampling is to consider the sampled signal $x^{\#}(t)$ to be the product of the continuous signal $x(t)$ and a periodic sampling function $s(t)$, as shown in Fig. 1-31a. We shall, for the time being, use superscript $^{\#}$ to denote a sampled function. Later this notation will be dropped and a signal represented as $x(k\tau)$ will be understood to be a sampled signal.[2] We note that this sampling process represents a form of amplitude modulation, where $s(t)$ is the carrier signal and $x(t)$ the modulating signal.

Let us first consider the case of ideal impulse sampling, as shown in Fig. 1-31b. Here the sampling function is a periodic train of impulses. We shall refer to this sampling function as $s_\delta(t)$, where

$$s_\delta(t) = \sum_{k=-\infty}^{\infty} \delta(t - k\tau) \tag{1-85}$$

Performing the multiplication shown in Fig. 1-31a, we write the following

---

[2]The notation can be further simplified by deleting the explicit reference to the sampling interval τ [i.e., $x(k) = x(k\tau)$].

Filter Transmission and Related Topics

expression for the sampled $x(t)$:

$$x^{\#}(t) = x(t)s(t) \qquad (1\text{-}86a)$$

i.e., $$x^{\#}(t) = x(t) \sum_{k=-\infty}^{\infty} \delta(t - k\tau) = \sum_{k=-\infty}^{\infty} x(t)\delta(t - k\tau) \qquad (1\text{-}86b)$$

Fig. 1-31   Sampling viewed as a modulation process.

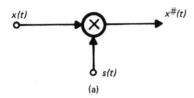

(a)

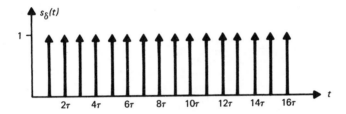

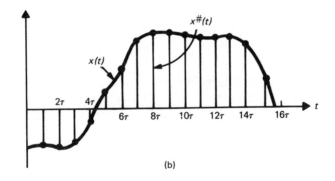

(b)

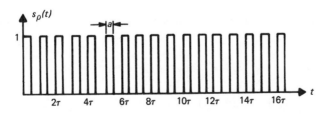

(c)

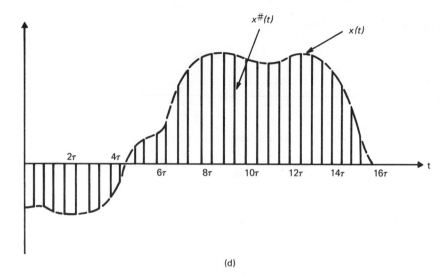

(d)

Fig. 1-31   cont.

Since the delta function is defined as

$$\int_{-\infty}^{\infty} \delta(t - k\tau)\, dt = 1$$

$$\delta(t - k\tau) = 0 \qquad \text{for } t \neq k\tau$$

we can rewrite $x^{\#}(t)$ as

$$x^{\#}(t) = \sum_{k=-\infty}^{\infty} x(k\tau)\delta(t - k\tau) \tag{1-87}$$

The term $x^{\#}(t)$ is an idealized analog signal comprised of a train of impulses of amplitude corresponding to the samples of $x(t)$: namely, $x(k\tau)$. In contrast, finite pulse sampling is illustrated in Fig. 1-31c. More will be said about finite pulse sampling later.

Since the sampling function is periodic, it can be represented by the Fourier series

$$s(t) = \sum_{k=-\infty}^{\infty} C_k e^{jk\omega_s t} \tag{1-88}$$

when $\omega_s = 2\pi/\tau$ is the sampling frequency in radians per second and C_k are the Fourier coefficients:

$$C_k = \frac{1}{\tau} \int_{-\tau/2}^{\tau/2} s(t)e^{-jk\omega_s t}\, dt \tag{1-89}$$

For impulse sampling, where $s(t) = s_{\delta}(t)$ according to Eq. (1-85), evaluating C_k yields

$$C_k = \frac{1}{\tau} \int_{-\tau/2}^{\tau/2} s_{\delta}(t)e^{-jk\omega_s t}\, dt = \frac{1}{\tau} \tag{1-90}$$

Filter Transmission and Related Topics

Substituting Eq. (1-88) into Eq. (1-86a) yields

$$x^{\#}(t) = x(t) \sum_{k=-\infty}^{\infty} C_k e^{jk\omega_s t} = \sum_{k=-\infty}^{\infty} C_k x(t) e^{jk\omega_s t} \qquad (1\text{-}91)$$

Let us now examine the frequency-domain implications of Eq. (1-91). We know that the Fourier transform of a bounded continuous signal $x(t)$ is expressed as

$$F[x(t)] = X(j\omega) \qquad (1\text{-}92)$$

However, the Fourier transform of the sampled signal is

$$X^{\#}(j\omega) = F[x^{\#}(t)] = F\left[\sum_{k=-\infty}^{\infty} C_k x(t) e^{jk\omega_s t} \right]$$

$$= \sum_{k=-\infty}^{\infty} C_k F[x(t) e^{jk\omega_s t}] \qquad (1\text{-}93)$$

$$= \sum_{k=-\infty}^{\infty} C_k X(j\omega - jk\omega_s)$$

From Eq. (1-93) we see that the sampling operation has introduced new spectral components which are translations of the base-band spectrum $X(j\omega)$ to integer multiples or harmonics of the sampling frequency ω_s. Each translation of $X(j\omega)$ is scaled according to the Fourier coefficients of the periodic sampling function $s(t)$. Substituting Eq. (1-90) into Eq. (1-93) we observe, for ideal impulse sampling, that the translations of $X(\omega)$ are of equal amplitude.

To view the practical importance of Eq. (1-93), let us sketch the spectrum for a continuous band-limited signal $x(t)$, where

$$X(j\omega) = 0 \qquad \text{for } |\omega| > \omega_c$$

This spectrum is sketched in Fig. 1-32a. Let us first consider the spectrum that results from sampling $x(t)$ at a sampling frequency $\omega_s < 2\omega_c$. This spectrum, obtained from Eq. (1-93), is sketched in Fig. 1-32b. Here we observe that the translations of $X(j\omega)$ overlap. This overlapping, referred to as *aliasing*, introduces an ambiguity into $X^{\#}(j\omega)$ and prevents the eventual recovery of the base-band spectrum $X(j\omega)$. When $x(t)$ is sampled at $\omega_s > 2\omega_c$, the translations do not overlap, as shown in Fig. 1-32c. In this case the base-band spectrum $X(j\omega)$ can be fully recovered by passing $X^{\#}(j\omega)$ through a low-pass filter of the form shown in Fig. 1-33. We note that Figs. 1-32c and 1-33 serve as a graphical statement of the important sampling theorem attributed to Shannon [P3]. This theorem states:

A function $x(t)$ that has a Fourier spectrum $X(j\omega)$ such that $X(j\omega) = 0$ for $|\omega_c| \leq \omega_s/2$ is uniquely described by a knowledge of its values at uniformly spaced time instants, τ instants apart ($\tau = 2\pi/\omega_s$).

In the communications literature, $2\omega_c$ is called the *Nyquist rate*.

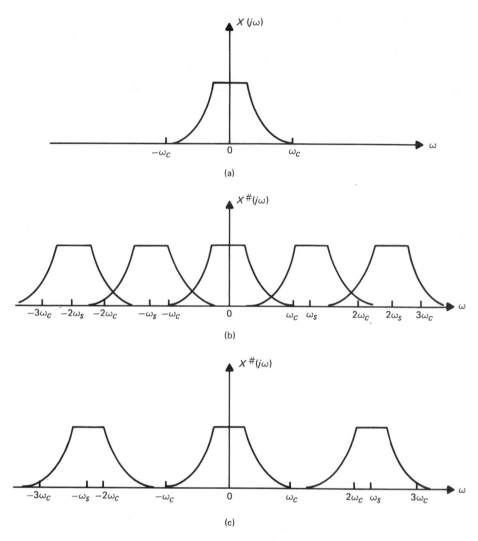

Fig. 1-32 Spectra for (a) continuous base-band signal $x(t)$, (b) $x(t)$ sampled at $\omega_s < 2\omega_c$, and (c) $x(t)$ sampled at $\omega_s > 2\omega_c$.

The filter characterized in Fig. 1-33 is often referred to as a *reconstruction filter*. Clearly, the greater the separation between ω_s and ω_c, the wider the transition band (TB) for the reconstruction filter. As demonstrated in Section 1.4, the widening of the required TB translates into a lower-order filter realization.

The more realistic finite-pulse sampling model leads to a more intuitively acceptable result than the situation depicted in Fig. 1-32. In this case the translations of $X(j\omega)$ are actually attenuated by the Fourier coefficients C_k defined

Filter Transmission and Related Topics

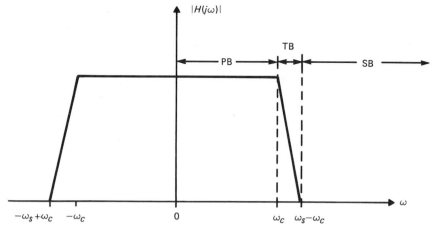

Fig. 1-33 Reconstruction filter.

according to Eq. (1-89). Evaluating C_k for the finite pulse $s(t)$ of width a, as shown in Fig. 1-31c, yields

$$C_k = \frac{1}{\tau} \int_{-a/2}^{a/2} e^{-jk\omega_s t}\, dt = \frac{a}{\tau} \frac{\sin (k\omega_s a/2)}{k\omega_s a/2} \tag{1-94}$$

Equation (1-94) then places the sin $(\alpha)/\alpha$ envelope, where $\alpha = k\omega_s a/2$, onto the equal amplitude $X(j\omega)$ translations in Fig. 1-32c. The modulation in Fig. 1-31d is often referred to as *pulse amplitude modulation* (PAM). Note that when the PAM $s(t)$, illustrated in Fig. 1-31c, is at unity, $x^{\#}(t)$ tracks $x(t)$.

1.9.2 The z-Transform and Its Use in the Analysis of Sampled Systems

The use of the z-transform in the analysis of discrete-time and sampled systems is motivated for many of the same reasons that make the Laplace transform so useful in the analysis of continuous-time systems. That is, by using the z-transform, the difference equations that characterize these discrete-time systems are transformed into linear algebraic equations, which are much simpler to solve. Furthermore, by studying the z-domain poles and zeros, the designer obtains, by inspection, insight into the behavior of the system. Also, the gain and phase responses for a disctete-time or sampled system are readily obtained from the z-domain transfer function. Thus, sampled systems, like continuous systems, can be specified in terms of their frequency-domain requirements, as discussed in Section 1.3.

To see how the z-transform naturally arises in the mathematical description of discrete-time systems, let us again consider the impulse sampled signal $x^{\#}(t)$ given in Eq. (1-87):

$$x^{\#}(t) = \sum_{k=-\infty}^{\infty} x(k\tau)\delta(t - k\tau) \tag{1-95}$$

Taking the Laplace transform of $x^{\#}(t)$ yields

$$X^{\#}(s) = \mathcal{L}[x^{\#}(t)] = \sum_{k=-\infty}^{\infty} x(k\tau)e^{-ks\tau} \qquad (1\text{-}96)$$

As we will demonstrate shortly, it is most convenient mathematically to make the following substitution in Eq. (1-96):

$$z = e^{s\tau} \qquad (1\text{-}97)$$

which yields

$$X(z) = \sum_{k=-\infty}^{\infty} x(k\tau)z^{-k} \qquad (1\text{-}98)$$

$X(z)$ is referred to as the *z-transform* of $x^{\#}(t)$, where z is the z-transform complex variable. Note that the z and Laplace transforms are related according to Eq. (1-97). For sinusoidal steady state, $s = j\omega$ and $z = e^{j\omega\tau}$. As with the Laplace transform, we can define both a one-sided and a two-sided z-transform. These definitions are, respectively,

$$Z_{\text{I}}[x(k)] = X_{\text{I}}(z) = \sum_{k=0}^{\infty} x(k\tau)z^{-k} \qquad (1\text{-}99\text{a})$$

$$Z_{\text{II}}[x(k)] = X_{\text{II}}(z) = \sum_{k=-\infty}^{\infty} x(k\tau)z^{-k} \qquad (1\text{-}99\text{b})$$

Thus, $X(z)$ in Eq. (1-98) represents the two-sided z-transform of the sequence $x(k\tau)$. Throughout the remainder of this book we will assume all signals to begin at $t = 0$ and the one sided z-transform will be used. The one sided z-transform will then be referred to as $Z[x(k\tau)]$, with the subscript I omitted.

●

EXAMPLE 1-5

Determine the z-transform for the unit step $x(t) = u(t)$, where

$$u(t) = \begin{cases} 0 & \text{for } t < 0 \\ 1 & \text{for } t \geq 0 \end{cases} \qquad (1\text{-}100)$$

Thus, $x(k\tau) = 1$ for all $k \geq 0$, with τ the sampling period. Using Eq. (1-99a) the z-transform is determined to be

$$X(z) = \sum_{k=0}^{\infty} z^{-k} \qquad (1\text{-}101)$$

Equation (1-101) is recognized to be a geometric series, which converges for $|z| > 1$ to

$$X(z) = \frac{1}{1 - z^{-1}} \qquad (1\text{-}102)$$

EXAMPLE 1-6

Determine the z-transform for $x(t) = e^{-at}u(t)$ sampled at τ intervals of time:

$$x(k\tau) = e^{-ak\tau} \quad \text{for } k \geq 0 \qquad (1\text{-}103)$$

Substituting Eq. (1-103) into Eq. (1-99a) yields the following z-transform:

$$X(z) = \sum_{k=0}^{\infty} e^{-ak\tau}z^{-k} \qquad (1\text{-}104)$$

$$= \frac{1}{1 - e^{-a\tau}z^{-1}} \quad \text{for } |z| > e^{-a\tau} \qquad (1\text{-}105)$$

●

A single input/single output, linear, time-invariant, discrete-time, or sampled system is represented by a difference equation of the form

$$y(k\tau) + \sum_{n=1}^{N} b_n y[(k - n)\tau] = \sum_{n=0}^{M} a_n x[(k - n)\tau] \qquad (1\text{-}106)$$

where sequence $y[(k - n)\tau]$ is the output and sequence $x[(k - n)\tau]$ is the input. In Eq. (1-106), M and N are finite nonnegative integers. There are two general classes of discrete-time systems, each distinguished by particular properties of Eq. (1-106). If $b_n = 0$ for all n, the system is nonrecursive. This nonrecursive system is also referred to as a $M + 1$ tap transversal filter or a finite-duration impulse response (FIR) filter. If $N \geq 1$, $b_n \neq 0$, the system is referred to as an Nth-order recursive system or an infinite impulse response (IIR) filter.

Taking the z-transform of both sides of Eq. (1-106) yields (with the assistance of the time-shift property of the z-transform: namely, $Z[x(k - n)] = z^{-n}X(z)$ where $k = 0, 1, 2, \ldots$)

$$Y(z)\left(1 + \sum_{n=1}^{N} b_n z^{-n}\right) = X(z) \sum_{n=0}^{M} a_n z^{-n} \qquad (1\text{-}107)$$

so that

$$H(z) = \frac{Y(z)}{X(z)} = \frac{\sum_{n=0}^{M} a_n z^{-n}}{1 + \sum_{n=1}^{N} b_n z^{-n}} \qquad (1\text{-}108)$$

$H(z)$ is the pulse transfer function and serves a role in sampled systems analogous to that of $H(s)$ in continuous, time-invariant systems. The inverse z-transform of $H(z)$, namely, $h(k)$, is referred to as the *unit response* and is analogous to the impulse response in continuous systems. The numerator and denonominator of $H(z)$ can be factored:

$$H(z) = \frac{b_0(1 - \beta_1 z^{-1})(1 - \beta_2 z^{-1})\ldots(1 - \beta_m z^{-1})}{(1 - \alpha_1 z^{-1})(1 - \alpha_2 z^{-1})\ldots(1 - \alpha_n z^{-1})} \qquad (1\text{-}109)$$

where $z = \alpha_i$ and $z = \beta_i$ define, respectively, the poles and zeros of the pulse transfer function $H(z)$.

In order to gain insight from the poles and zeros of $H(z)$, let us examine in greater detail the mapping $z = e^{s\tau}$. As demonstrated previously, the Fourier transform $X^{\#}(j\omega)$ of an impulse sampled function $x^{\#}(t)$ is comprised of repeated translations of the base-band response $X(j\omega)$. The repetition rate for $X(j\omega)$ is the sampling rate ω_s. Thus, knowledge of the base-band response $X(j\omega)$ over the frequency range $-\omega_s/2 \leq \omega \leq \omega_s/2$ is sufficient to determine $X^{\#}(j\omega)$ for all ω (recall $\omega_s = 2\pi/\tau$). The mapping $z = e^{s\tau}$ maps this frequency strip uniquely into the z-plane. To see the nature of the mapping, let $s = \sigma + j\omega$, yielding

$$z = e^{\sigma\tau}e^{j\omega\tau} \tag{1-110a}$$

Since $|z| = e^{\sigma\tau}$, clearly

$$|z| \begin{cases} < 1 & \text{for } \sigma < 0 \\ = 1 & \text{for } \sigma = 0 \\ > 1 & \text{for } \sigma > 0 \end{cases} \tag{1-110b}$$

Thus, points on the $j\omega$-axis, in the s-domain, get mapped onto the unit circle defined by $z = e^{j\omega\tau}$, points in the left-half s-plane get mapped inside the unit circle, and points in the right-half s-plane get mapped outside the unit circle. Lines parallel to the $j\omega$-axis get mapped into circles of the form $|z| = e^{\sigma\tau}$, and lines parallel to the σ-axis get mapped into rays of the form $\arg(z) = \omega\tau$, radiating from $z = 0$. The origin of the s-plane corresponds to $z = 1$ and the σ-axis corresponds to the positive $\text{Re}(z)$-axis. As ω varies between $-\omega_s/2$ and $\omega_s/2$, $\arg(z) = \omega\tau$ varies between $-\pi$ and π radians. Thus, successive horizontal strips of the s-plane bounded by $-\omega_s/2 + k\omega_s \leq \omega \leq \omega_s/2 + k\omega_s$ for $k = 0, 1, \ldots$, get mapped on top of each other in the z-plane. Many of these concepts are illustrated in Fig. 1-34.

By plotting the z-domain poles and zeros for a system function $H(z)$, a great deal of information can be derived about the unit response by inspection. Consider, for example, the first-order transfer function

$$H(z) = \frac{1}{1 - az^{-1}} \tag{1-111}$$

There are six different categories of time sequences $h(k)$ that can arise, depending on the location of the pole $z = a$ in Eq. (1-111):

1. If $a > 1$, the sequence diverges (unstable system).
2. If $a = 1$, we have a sequence of 1's (unstable system).
3. If $0 \leq a < 1$, we have a sequence of positive monotonically decreasing numbers (stable system).
4. If $-1 < a \leq 0$, we have a sequence of numbers alternating in sign and decreasing in magnitude (stable system).

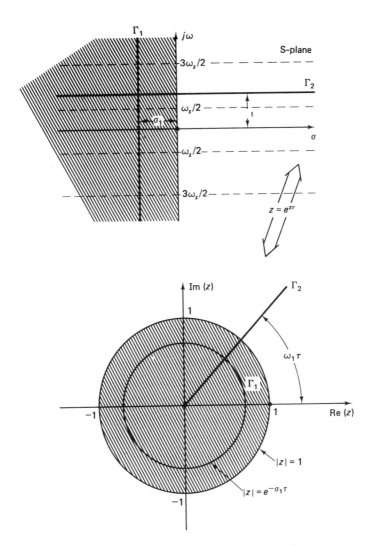

Fig. 1-34   Mapping between the s- and z-domains.

5. If $a = -1$, we obtain a sequence of 1's alternating in sign (unstable system) at a rate $\omega_s/2$.

6. If $a < -1$, we have sequence of numbers alternating in sign and increasing in magnitude (unstable system).

These six cases are sketched in Fig. 1-35. In general, stable poles will be inside the unit circle. Also, $j\omega$-axis zeros, which as we observed in Section 1.3 provide transmission nulls in the gain responses, lie on the unit circle. To obtain the gain and phase responses from $H(z)$, we substitute $z = e^{j\omega\tau}$ and evaluate the resulting function at specific values of ω. The gain and phase

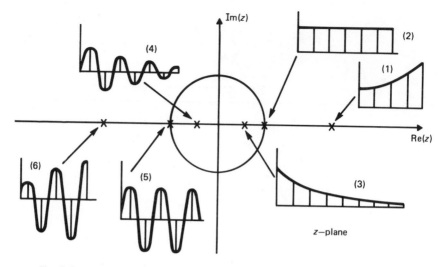

Fig. 1-35 $h(k)$ corresponding to the first-order discrete-time system characterized by $H(z) = z^{-1}/(1 - az^{-1})$ for (1) $a > 1$, (2) $a = 1$, (3) $0 < a < 1$, (4) $-1 < a < 0$, (5) $a = -1$, and (6) $a < -1$.

responses, respectively, are given by

$$G(\omega) = 20 \log \left[|H(z)| \Big|_{z=e^{j\omega\tau}} \right] \quad \text{dB} \tag{1-112a}$$

$$\phi(\omega) = \tan^{-1} \frac{\text{Im } H(z)}{\text{Re } H(z)} \Big|_{z=e^{j\omega\tau}} \quad \text{rad} \tag{1-112b}$$

For example, if $H(z)$ has the form

$$H(z) = \frac{K \prod_{i=1}^{M} (1 - a_i z^{-1})}{\prod_{i=1}^{N} (1 - b_i z^{-1})} \tag{1-113a}$$

$$= \frac{K z^{N-M} \prod_{i=1}^{M} (z - a_i)}{\prod_{i=1}^{N} (z - b_i)} \tag{1-113b}$$

then

$$G(\omega) = 20 \log \left(\frac{|K| \prod_{i=1}^{M} |e^{j\omega\tau} - a_i|}{\prod_{i=1}^{N} |e^{j\omega\tau} - b_i|} \right) \tag{1-114}$$

and

$$\phi(\omega) = (N - M)\omega\tau + \sum_{i=1}^{M} \arg(e^{j\omega\tau} - a_i) - \sum_{i=1}^{N} \arg(e^{j\omega\tau} - b_i) \tag{1-115a}$$

$$= (N - M)\omega\tau + \sum_{i=1}^{M} \tan^{-1} \left(\frac{\sin \omega\tau}{\cos \omega\tau - a_i} \right) - \sum_{i=1}^{N} \tan^{-1} \left(\frac{\sin \omega\tau}{\cos \omega\tau - b_i} \right) \tag{1-115b}$$

Geometrically, the factors of $H(\omega)$ (i.e., $e^{j\omega\tau}$, $e^{j\omega\tau} - a_i$, and $e^{j\omega\tau} - b_i$) can be represented as vectors drawn from $z = 0$, a_i, and b_i to the point $\omega\tau$ on the unit circle. Hence, the magnitudes and angles in Eqs. (1-114) and (1-115) can be determined graphically in much the same way as magnitudes and angles are determined for s-plane poles and zeros of continuous systems. With a bit of practice, magnitude and phase responses for discrete and sampled systems can be sketched directly from a z-domain pole–zero plot.

●

EXAMPLE 1-7

As an illustrative example let us sketch the gain response, $G(\omega)$, for the pole–zero plot given in Fig. 1-36. The following expression for the gain response can be written directly from the pole–zero plot:

$$G(\omega) = 20 \log \left(\frac{|e^{j\omega\tau} - e^{j\omega_1\tau}||e^{j\omega\tau} - e^{-j\omega_1\tau}|}{|e^{j\omega\tau} - re^{j\omega_1\tau}||e^{j\omega\tau} - re^{-j\omega_1\tau}|} \right) \tag{1-116a}$$

$$= 20 \log \left(\frac{d_2 \cdot d_2^*}{d_1 \cdot d_1^*} \right) \tag{1-116b}$$

Fig. 1-36 Pole–zero plot for Example 1-7.

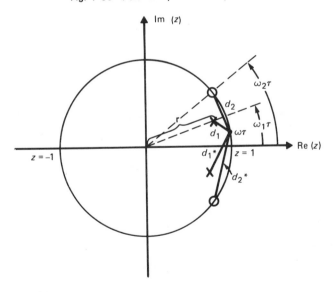

Here let us assume that the gain constant K in Eq. (1-113) is unity. $G(\omega)$ is then computed by measuring d_1, d_1^*, d_2, and d_2^* for each value of $\omega\tau$, as the point $\omega\tau$ travels around the unit circle from $\omega = 0$ (i.e., $z = 1$) to $\omega = \omega_s/2 = \pi/\tau$ (i.e., $z = -1$). Observe that when $\omega\tau = \omega_1\tau$, d_1 is minimum; hence, $G(\omega)$ has a peak. Also, when $\omega\tau = \omega_2\tau$, $G(\omega) = 0$ and we achieve a null at frequency ω_2. Since $\omega_2 > \omega_1$, the pole–zero plot in Fig. 1-36 is seen to describe a low-pass notch response. This low-pass notch response is sketched in Fig. 1-37. The

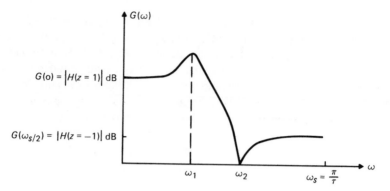

Fig. 1-37   Gain response $G(\omega)$ for the pole–zero plot in Fig. 1-36.

response shown in Fig. 1-37 is the base-band response. Therefore, to complete the sketch, the base-band response is repeated at integer multiples of ω_s. This repeated spectrum is shown in Fig. 1-38. It is noted that we may also write the z-domain transfer function, with the exception of the gain constant [i.e., K in Eq. (1-113)], from the pole–zero plot; that is, from Fig. 1-36,

$$H(z) = \frac{K(z - e^{j\omega_2\tau})(z - e^{-j\omega_2\tau})}{(z - re^{j\omega_1\tau})(z - re^{-j\omega_1\tau})} \tag{1-117a}$$

$$= \frac{K[z^2 - (2\cos\omega_2\tau)z + 1]}{z^2 - (2r\cos\omega_1\tau)z + r^2} \tag{1-117b}$$

or

$$H(z) = \frac{K[1 - (2\cos\omega_2\tau)z^{-1} + z^{-2}]}{1 - (2r\cos\omega_1\tau)z^{-1} + r^2z^{-2}} \tag{1-118}$$

Fig. 1-38   Repeated gain response for base band $G(\omega)$ in Fig. 1-37.

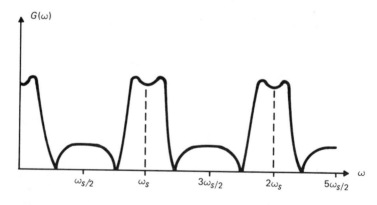

As seen in Chapter 6, $H(z)$ expressed in terms of inverse powers of z, as per Eq. (1-118), arises naturally in sampled-data switched-capacitor filters.

●

Filter Transmission and Related Topics

1.9.3 Analog Sampled-Data Systems

The z-transform, which we initially derived from consideration of an ideal impulse sampled signal $x(k\tau)$ in Eq. (1-98), is a convenient means for analyzing sampled systems on a sample-to-sample basis. In an analog sampled-data system, the input and output are analog or continuous signals. We discussed in Section 1.9.1 the need to band-limit the input spectrum to prevent aliasing, so that the base-band signal can be reconstructed without error. The anti-aliasing will also serve to band-limit high-frequency noise which would otherwise be aliased back into the base band. Inherent in all analog sampled-data systems is the means to provide some sort of analog reconstruction.

The simplest form of reconstruction is the *zero-order hold* or *sample-and-hold* (S/H). The impulse response of the zero-order hold is

$$h_0(t) = \begin{cases} 1/\tau & \text{for} \quad 0 \le t < \tau \\ 0 & \text{elsewhere} \end{cases} \tag{1-119}$$

The sample-and-hold impulse response is sketched in Fig. 1-39 and an S/H

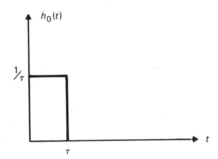

Fig. 1-39 Impulse response for sample-and-hold.

reconstructed signal $x_r(t)$ is shown in Fig. 1-40. The output of the sample-and-hold remains constant between sampling instants. If the sampling interval, τ, is small enough, the reconstructed signal closely approximates the original signal.

Fig. 1-40 Reconstruction with sample-and-hold.

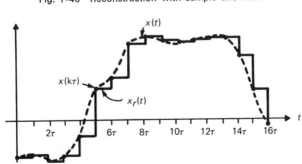

The transfer function for the sample-and-hold is

$$H_0(s) = \frac{1 - e^{-s\tau}}{s\tau}$$ (1-120)

and its frequency response is

$$H_0(j\omega) = \frac{1 - e^{-j\omega\tau}}{j\omega\tau} = e^{-j\omega\tau/2} \frac{\sin \omega\tau/2}{\omega\tau/2}$$ (1-121)

$H_0(j\omega)$ is seen to have the $\sin (x)/x$ gain response shown in Fig. 1-41a and the phase characteristic shown in Fig. 1-41b.

Fig. 1-41 (a) Gain and (b) phase responses for sample-and-hold.

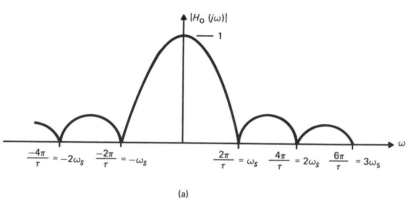

(a)

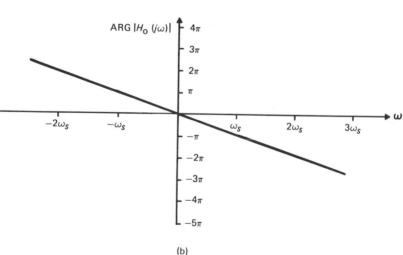

(b)

Let us now look at what happens to the impulse sampled spectrum in Fig. 1-32c when the sampled signal is reconstructed with a sample-and-hold. The spectrum of the reconstructed signal $X_r(j\omega)$ is simply the product of the

Filter Transmission and Related Topics

impulse sampled spectrum $X^{\#}(j\omega)$ and the sample-and-hold spectrum $H_0(j\omega)$:

$$X_r(j\omega) = X^{\#}(j\omega) \cdot H_0(j\omega) \qquad (1\text{-}122)$$

This spectrum is sketched in Fig. 1-42. Observe that the high-frequency content has been substantially attenuated in $X_r(j\omega)$. Note also that base-band spectrum is slightly altered near the band edge, by the attenuation in the main lobe of $H_0(j\omega)$. When $\omega_s/\omega_c \gg 1$, the base-band distortion can be made negligible.

Fig. 1-42   Spectrum of reconstructed sampled signal.

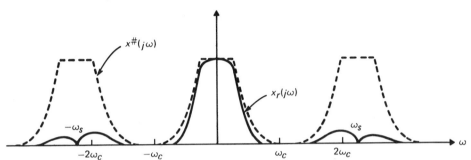

For lower sampling rates, the sampled data filter which operates on the signal X is usually designed with a peak to compensate for the sample-and-hold related band-edge droop. As we will see in Chapter 6, switched capacitor filters can be made to inherently provide sample-and-hold types of outputs. We note that when a smooth output is desired, the sample-and-hold can be followed by a continuous low-pass filter to attenuate the residual high-frequency components.

In Fig. 1-43 we show a complete analog input/analog output sampled-data system. This system demonstrates all the concepts discussed in this section. The relevant time-domain signals are shown below their respective points of observation.

1.10 D4-CHANNEL BANK

An interesting example that illustrates many of the concepts discussed in this chapter is the *D4-channel bank* [P5], which converts analog (voice-frequency) telephone signals into digital form and vice versa, using *pulse code modulation* (PCM). In PCM the analog signal is sampled at periodic intervals as in PAM, shown in Fig. 1-31d. However, for PCM, the samples are quantized into discrete steps (i.e., within a specified range of expected sample values, only discrete levels are allowed) and converted into a coded pattern of a series of equal-amplitude pulses. The D4-channel bank serves as a terminal for a PCM repeated line which carries 24 digitally encoded telephone conversations. Hence, each D4-channel bank is comprised of 24 channels. Systems of this type enjoy wide usage throughout the world-wide telephone industry.

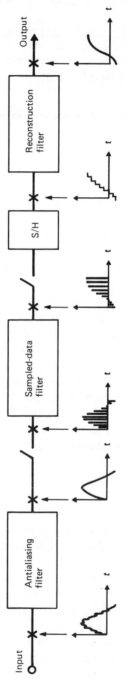

Fig. 1-43   Analog input/analog output sampled-data system.

A simplified schematic of the voice-frequency end of a D4-channel bank is shown in Fig. 1-44. At the extreme left of this schematic, we see that a voice-frequency analog signal from the switching equipment passes through a hybrid transformer. The hybrid transformer separates the common transmit and receive path into two separate paths. An incoming transmit signal is then amplified and band-limited by a transmitting band-pass filter. Subsequently, the filtered analog signal is sampled at an 8-kHz rate by a JFET (junction field-effect transistor), converting the signal into a PAM signal like that shown in Fig. 1-31c. Each channel's PAM signal is passed through the channel sampling gate and presented to the encoder at different times. The encoder (analog to digital) then sequentially encodes the PAM signals from the 24 channels into PCM form and sends the signals out on the PCM line. On the receiving side, the PCM signals are sequentially decoded (digital to analog) and converted to PAM signals. The analog signal for each individual channel is recovered by passing the PAM signal through a receiving low-pass filter. The analog signal then passes through the hybrid transformer to the switching equipment and

Fig. 1-44 Typical D-channel bank. Copyright 1975, American Telephone and Telegraph Company; Reprinted by permission.

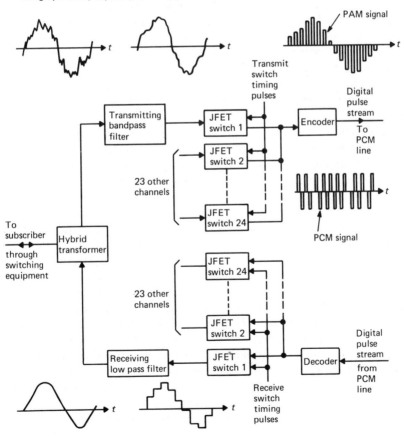

subsequently to the subscriber. The intermediate time-domain signals are shown as inserts in Fig. 1-44. Note that the transformers, transmitting filters, and receiving filters are replicated 24 times in each channel bank.

The main purpose of the transmitting bandpass filter is to perform the anti-aliasing for the 8-kHz sampling operation. It is noted that the maximum voice band frequency is 4-kHz; hence, the 8-kHz sampling frequency corresponds to the Nyquist rate. In addition, a low-frequency notch is placed at 60 Hz to reject power-line interference. The requirements for this filter, and a filter frequency response which meets these specifications, are shown in Fig. 1-45. The passband flatness and high-frequency rejection requirements can be met with a fifth-order elliptic transfer function of the form given in Eq. (1-57). The 60-Hz rejection specification can be met with either a second-order high-pass notch or a third-order high-pass transfer function. These types of functions were described in Section 1.3. Techniques for realizing this class of filter are given in Chapters 5 and 6.

The purpose of the receiving low-pass filter is to perform the smoothing of the reconstructed S/H signal provided by the decoder. Since the sample-and-hold operation is performed at an 8-kHz rate, its $(\sin x)/x$ response, with $x = \pi f/8000$, provides undesired attenuation to the higher-frequency content of the voice signal, as illustrated in Fig. 1-42. A peak is placed near the cutoff of the receiving low-pass filter to compensate for this attenuation. Thus, the filter and decoder, in combination, provide the same in-band flatness as the transmitting filter. The specifications for this filter, and a frequency response meeting these requirements, are shown in Fig. 1-46. The dashed line shows the filter response alone, and the solid line shows the filter response when multiplied by $(\sin x)/x$.

The filter functions shown in Figs. 1-45 and 1-46 are referred to as the *nominal* or *desired functions*. With pencil and paper or with a computer program, we can design a filter (i.e., evaluate the component values) to realize these frequency responses. A simple design exercise was performed in Example 1-4. However, when the filter is manufactured, the manufactured component values will deviate unintentionally from the design values, owing to imperfections in the manufacturing process. That is, if we were to measure the same component (e.g., an R or a C) in each of a large number of "identical" manufactured filters, very few would be identical in value or equal exactly to the design value. Similarly, if we were to measure the frequency responses of each of these filters, very few would look identical or be exactly the function we intended. Obviously, the variations in the component values and in the observed frequency responses are related. This relationship is referred to as *sensitivity*. The problem is further complicated by the fact of nature that values of physical components in any filter will vary as the ambient environment changes or with age during the life cycle of the filter. In the design of a filter intended for manufacture, or intended to function in a varying environment, or intended to function without replacement over a long period of time, these effects must be carefully analyzed. The bottom-line question with all these sources of variation is: Does the filter still meet requirements as per the specification windows shown in Figs. 1-45 and 1-46? The number of filters that continue to meet requirements versus the total

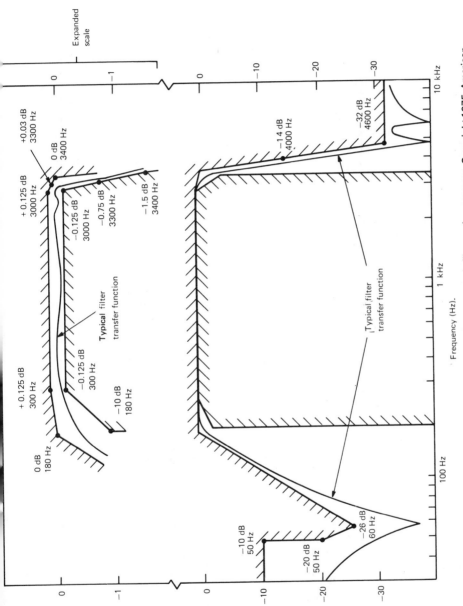

Fig. 1-45 D4-channel bank transmitting filter requirements and filter gain response. Copyright 1975, American Telephone and Telegraph Company; Reprinted by permission.

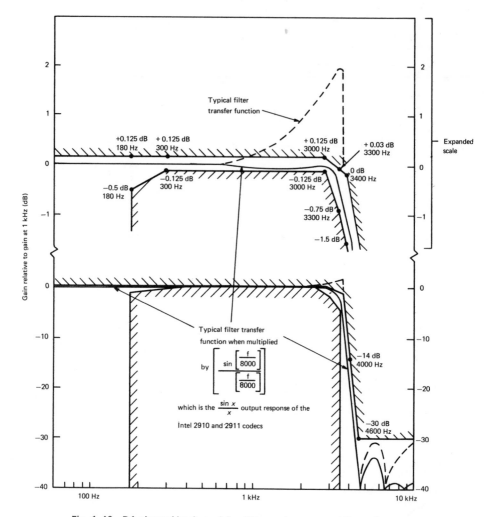

Fig. 1-46 D4-channel bank receiving filter requirements and filter gain response. Copyright 1975, American Telephone and Telegraph Company; Reprinted by permission.

population of "identical" filters is referred to as *yield*. These concepts are treated in detail in Chapter 3, but let us illustrate this phenomenon here with the frequency responses of Fig. 1-47. Shown in this figure is the nominal receiving filter [multiplied by (sin x)/x response], denoted by the solid line, and the worst-case deviations from nominal, predicted from a complete simulation of 1000 networks with component variations modeled to account for manufacturing tolerances, temperature variations (0°C to 60°C), and aging. The area between the dashed worst-case limits represents a continuum of responses for the 1000

Filter Transmission and Related Topics

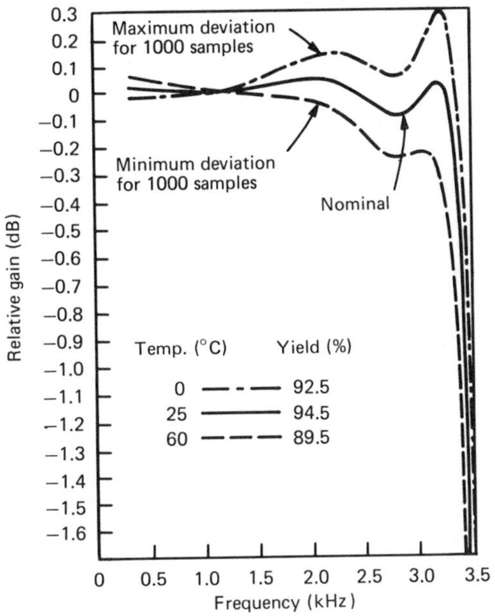

Fig. 1-47 Computer simulated worst case magnitude response deviations for 1000 samples of the receiving low-pass filter. The response deviations are due to variations in the component values from the combined effects of manufacturing tolerances, temperature variations, and aging. The dashed curves represent the worst-case response deviations and the solid curve represents the nominal response. Copyright 1975, American Telephone and Telegraph Company; Reprinted by permission.

networks. The yields at the temperatures 0°C, 25°C, and 60°C are also given in the figure. Thus, of the 1000 samples analyzed, at 0°C, 92.5% at 25°C, 94.5% and at 60°C, 89.9% of the filters were predicted to meet specifications.

1.11 SUMMARY

In this chapter we have reviewed the fundamental concepts relevant to continuous and sampled data systems. Although this chapter cannot possibly provide a complete treatment of these subjects, we have found this material to be a sufficient background for understanding and applying the tools given in the forthcoming chapters to the design of practical active filters.

In Chapter 2 we consider the basic principles that underlie operation of the operational amplifier, the workhorse of active filters, and its use in simple building block networks. The principal problem faced by the designer of a manufacturable filter is the variation of the filter's response due to unintended variations in the component values. Such variations occur because of manufacturing imperfections, changes in the ambient environment, and aging. Techniques for predicting the severity of this problem and for minimizing it are treated in Chapter 3. Chapter 4 is devoted entirely to the realization of continuous active networks for implementing second-order transfer functions. Such networks often serve as building blocks for higher-order active filters, as described in Chapter 5. Finally, in Chapter 6 we treat the analysis and design of active switched-capacitor, sampled-data networks.

REFERENCES

Books

B1. DARYANANI, G., *Principles of Active Network Synthesis and Design*. New York: Wiley, 1976.

B2. GHAUSI, M. S., *Principles and Design of Linear Active Circuits*. New York: McGraw-Hill, 1965, Chap. 4.

B3. WEINBERG, L., *Network Analysis and Synthesis*. New York: McGraw-Hill, 1968, Chap. 11.

B4. DANIELS, R. W., *Approximation Methods for Electronic Filter Design*. New York: McGraw-Hill, 1974.

B5. LINDQUIST, C., *Active Network Design with Signal Filtering Applications*. Long Beach, CA: Steward and Sons, 1977.

B6. CHRISTIAN, E., AND E. EISENMANN, *Filter Design Tables and Graphs*. New York: Wiley, 1966.

B7. SEDRA, A. S., AND P. O. BRACKETT, *Filter Theory and Design: Active and Passive*. Portland, Oreg.: Matrix, 1978.

B8. SU, K. L., *Time Domain Synthesis of Linear Networks*. Englewood Cliffs, N.J.: Prentice-Hall, 1971.

B9. RABINER, L. R., AND B. GOLD, *Theory and Application of Digital Signal Processing*. Englewood Cliffs, N.J.: Prentice-Hall, 1975.

B10. OPPENHEIM, A. V., AND R. W. SCHAFER, *Digital Signal Processing*. Englewood Cliffs, N.J.: Prentice-Hall, 1975.

B11. TRETTER, S. A., *Introduction to Discrete Time Signal Processing*. New York: Wiley, 1976.

Papers

P1. STICHT, D. S., AND L. P. HUELSMAN, "Direct Determination of Elliptic Network Functions," *Int. J. Comput. Elec. Eng.*, 1 (1973), 272–280.

P2. HENDERSON, K. W., AND W. H. KAUTZ, "Transient Response of Conventional Filters," *IRE Trans. Circuit Theory*, CT-5 (December 1958), 333–347.

P3. OLIVER, B. M., J. R. PIERCE, AND C. E. SHANNON, "The Philosophy of PCM," *Proc. IRE*, 36 (November 1948), 1324–1331.

P4. GAUNT, W. B., AND J. B. EVANS, "The D3 Channel Bank," *Bell Lab. Rec.*, 50 (August 1972), 229–233.

P5. FRIEDENSON, R. A., R. W. DANIELS, R. J. DOW, AND P. H. MCDONALD, "RC Active Filters for the D3 Channel Bank," *Bell Syst. Tech. J.*, 54, no. 3 (March 1975), 507–529.

PROBLEMS

1.1 A Butterworth low-pass filter is required in an application for which the magnitude response should be down at least 30 dB for $\omega = 1.5\omega_{3dB}$. What is the lowest-order Butterworth filter that can be used?

1.2 A low-pass filter is needed such that the allowable ripple in the magnitude response in the passband is less or equal to 0.5 dB. The magnitude must be down at least 30 dB at $\omega = 1.5\omega_c$. A Chebyshev filter of what order is required?

1.3   The magnitude response of a nominal filter must lie within the shaded area shown in Fig. P1-3. Determine the transfer function of a filter that would meet the given specifications, using:
   (a) A Chebyshev filter
   (b) A Cauer filter

Fig. P1-3

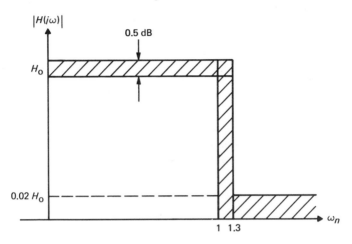

1.4   A filter with Chebyshev magnitude response is defined by $n = 5$ and $\epsilon = 0.1$. Determine the transfer function and the 3-dB cutoff frequency of the filter.

1.5   A transfer function is given by

$$H(s) = \frac{s^2 + sa_1 + a_1 b_1}{s^3 + s^2(1 + a_1) + sb_1 + a_1 b_1}$$

Determine the values of a_1 and b_1 if $|H(j\omega)|$ is specified to be a maximally flat magnitude response.

1.6   Given

$$|H(j\omega)|^2 = \frac{1 + a_2\omega^2 + a_4\omega^4 + \ldots + a_{2m}\omega^{2m}}{1 + b_2\omega^2 + b_4\omega^4 + \ldots + b_{2n}\omega^{2n}} \qquad m < n$$

show that if the magnitude is to be maximally flat,

$$b_{2i} = \begin{cases} a_{2i} & i = 1, \ldots, m \\ 0 & i = m+1, \ldots, n-1 \end{cases}$$

1.7   The argument of the transfer function for a general linear network can be written as

$$\phi = \arg H(j\omega) = \tan^{-1}\frac{\alpha_1\omega + \alpha_3\omega^3 + \alpha_5\omega^5 + \ldots}{1 + \beta_2\omega^2 + \beta_4\omega^4 + \beta_6\omega^6 + \ldots}$$

Show that for a linear phase response, the first two conditions are

$$\alpha_3 - \beta_2\alpha_1 = \tfrac{1}{3}\alpha_1^3$$

$$\alpha_5 - \beta_4\alpha_1 = \tfrac{1}{3}\alpha_1^3(\tfrac{2}{5}\alpha_1^2 + \beta_2)$$

1.8 Repeat Prob. 1.5 if $H(j\omega)$ is to have a linear phase function.

1.9 Use Table B-6 to determine the pole–zero location of an elliptic filter for the following specifications: $\omega_s/\omega_p = 1.5$, the minimum stop-band attenuation is ≥ 20 dB, and the ripple in the passband is ≤ 0.5 dB.

1.10 Consider the sixth-order all-pole transfer functions for Butterworth, Chebyshev ($\tfrac{1}{2}$-dB ripple), and Thomson filters. Determine the attenuation at $\omega = 2\omega_{3dB}$ in each case.

1.11 If the voltage ratio transfer function of an RC network, N_a, is given by $H_a(s)$, show that the voltage ratio of another network N_b obtained by $RC:CR$ transformation is given by

$$H_b(s) = H_a\left(\frac{1}{s}\right)$$

If $H_a(s)$ represents the transfer function corresponding to a third-order Butterworth filter, sketch the magnitude response of $H_b(s)$.

1.12 For the transfer function given by

$$H(s) = \frac{s + 2}{(s + 1)(s + 3)}$$

Find the step input response, the rise time, and the 3-dB bandwidth.

1.13 For the fifth-order Thomson filter, the transfer function is given in Appendix B.
(a) Determine the rise time and the delay time by using Elmore's definition and compare the results with those in Fig. 1.27c.
(b) Determine ω_{3dB} for the filter using the results in part (a) and Eq. (1-69).
(c) Determine ω_{3dB} by using Eq. (1-53).

1.14 Derive Eqs. (1-79) and (1-80).

1.15 A low-pass prototype is given by the transfer function

$$H(p) = \frac{1}{p^2 + \sqrt{2}\,p + 1}$$

Obtain the band-pass transfer function using Eq. (1-64) and sketch the response. Assume that $\omega_0 = 1$ and $Q = 10$. Show the pole–zero locations of the band-pass What is $\Delta\omega_{3dB}$ of the band-pass transfer function?

1.16 For the transfer functions
(a) $H(z) = 1 - z^{-1}$
(b) $H(z) = \dfrac{1}{z^2 - (2\alpha \cos \beta)z + \beta^2}$
sketch the magnitude of the frequency response $|H(j\omega)|$.

1.17 A low-pass sampled-data analog filter (Fig. P1-17) has the following measured gain response 0 dB at dc, 0.2 dB at 2.2 kHz, -50 dB at 4 kHz, -30 dB at 4.5 kHz, and -40 dB at 15 kHz. The output is a sampled-and-hold waveform with a hold period of $\tau = 1/0.32$ μsec. It is required to post-filter this sampled-data output

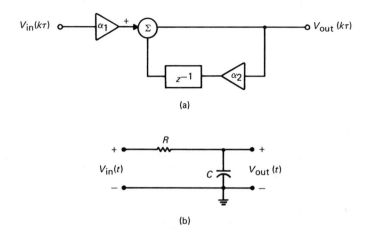

(a)

(b)

Fig. P1-17

with a continuous analog low-pass filter so that all out-of-band components, for $f > 50$ kHz, are attenuated by at least 50 dB.

(a) Give the out-of-band attenuation requirements for the post filter.
(b) If the passband of the sampled data filter is to be unaffected (error in passband ≤ 0.01 dB) by the post filter, give the order, approximation type, and cutoff frequency for a low-pass post filter that satisfies the attenuation requirements in part (a).

1.18 Repeat Prob. 1.17 for $\tau = 1/1.28$ μsec.

chapter two

....

Operational Amplifiers

2.1 INTRODUCTION

The *operational amplifier*, often referred to as the *op amp*, is one of the most versatile building blocks in linear circuits applications. The availability of high-performance, inexpensive op amps has had a dramatic influence in the electronic industry. Op amps are the workhorses in linear circuit applications and play a role comparable to that of microprocessors in digital circuitry. Integrated circuit op amps of relatively high performance cost less than $0.50 (1980) and are composed of small capacitors, resistors, and many transistors. In fact, in certain op amp packages one may find four identical op amps in one package (e.g., LM 348; see Appendix C). Their cheap price, off-the-shelf availability, and good performance justify their use in industry and emphasis in this book.

Basically, op amps are direct-coupled differential amplifiers with extremely high gain that are generally used with external feedback for gain–bandwidth control. Until recently, typical op amps were realized as silicon integrated circuits composed of bipolar transistors. These circuits are very high quality, small, and quite inexpensive. Bipolar op amps [P1] typically consume < 3 mm^2 of silicon area. To further reduce the cost of linear circuits and combined linear-digital subsystems, there has been considerable economic motivation to develop op amp technologies which are compatible with well-established large-scale integrated (LSI) circuit technologies, such as the N-channel metal-oxide-semi-conductor (NMOS) and complementary-metal-oxide-semiconductor (CMOS) technologies. Currently available NMOS operational amplifiers [P3, P4] do not compete well with the performance characteristics of their bipolar transistor counterparts; nevertheless, they require about one-fifth of the silicon area, with

performance acceptable for such linear circuit applications as switched capacitor, sampled-data filters. On the other hand, high performance CMOS op amps
are available. Moreover, given the same output drive, a CMOS op amp con-
sumes a fraction of the power consumed by an NMOS op amp.

In this chapter we consider the op amp as a circuit block defined only by
its behavior observed at its input–output ports. Some knowledge and familiarity
with the internal circuitry of the op amp is very useful; however, one does not
need to know the intricate details of its design and configuration to use it as a
component in active filters. Those readers interested in these details are referred
to the many good textbooks and tutorial papers on this subject [B1–B3, P1, P2].
In this chapter we present the fundamentals of op amp operation, terminology,
and circuit applications which are germane to the design of active filters.

2.2 OP AMP NOTATION AND CHARACTERISTICS

In practical circuits, the op amp is invariably used with some form of external
feedback. Op amp operation with feedback, referred to as closed-loop operation,
is treated in succeeding sections. In this section we consider the notation and
basic operation of the open-loop op amp (i.e., the op amp in the absence of
external feedback).

Typically an op amp has five terminals, two inputs, one output, and two
power supply terminals, as designated by the well-known circuit symbol shown
in Fig. 2-1a. Usually, the supply terminals are omitted in the symbol and only
the output and input terminals are shown as in Fig. 2-1b. The equivalent

Fig. 2-1   (a) Symbol of an op amp, showing the principal terminals. (b) Symbol.

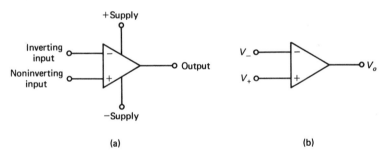

(a)   (b)

circuit corresponding to Fig. 2-1 is shown in Fig. 2-2. The transfer characteristics
of a typical op amp are shown in Fig. 2-3. Note that the range of input voltage
for linear operation is very small. When the input signals exceed this range, the
operation is nonlinear. This property may be used in nonlinear applications,
such as Schmitt trigger circuits and multivibrator circuits for digital electronics.
A thorough treatment of the various nonlinear applications is beyond the scope
of this book. The interested reader can find this subject matter in any one of
several books on digital electronics [B2, B8].

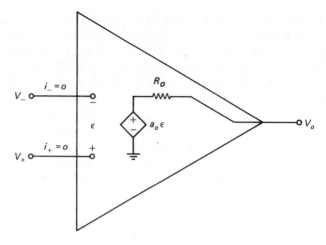

Fig. 2-2   Circuit model of an op amp.

Fig. 2-3   Typical op amp transfer characteristic.

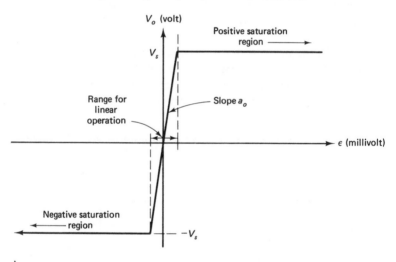

The perhaps most discriminating characteristic of the op amp is its extremely high open loop gain, which we denote a_0. Typical bipolar op amps achieve open-loop gains on the order of 10^5 (100 dB). NMOS and CMOS op amp gains are somewhat lower, on the order of 60 and 80 dB, respectively. Referring to Fig. 2-3, for linear operation the differential input voltage

$$\epsilon = v_+ - v_- < \frac{V_s}{a_0} \tag{2-1}$$

which implies that the differential input signal ϵ must be very small. Moreover, since the input impedance is very high, this implies that the input currents i_+ and i_- are also very small.

When analyzing linear circuits containing op amps, it is most conveient to consider the operation of the ideal op amp. The ideal op amp is characterized as follows: open-loop gain $a_0 = \infty$, input impedance $R_i = \infty$, and output impedance $R_o = 0$. These conditions imply that

$$i_+ = i_- = 0 \tag{2-2a}$$

and for a finite output v_o,

$$\epsilon = v_+ - v_- = 0 \tag{2-2b}$$

Often, one of the input terminals to the op amp is connected to ground. If, for example, the positive terminal is connected to ground, then by definition $v_+ = 0$ and according to Eq. (2-2) $v_- = 0$ and $i_- = 0$. The negative terminal is then said to be at virtual ground.

These ideal conditions, which fortunately represent a good first-order approximation to a real op amp, greatly simplify the analysis and synthesis of active filters. In a given application, the accuracy of this approximation depends on several factors. These factors will be identified, on an as needed basis, as the text progresses. Certainly, when appropriate computer aids are available, the op amp should be modeled more accurately. More will be said about practical op amp models in Section 2.5.

2.3 CLOSED-LOOP AMPLIFIER CONFIGURATIONS

To introduce the fundamentals of op amp circuit techniques, consider the simple amplifying stages shown in Figs. 2-4, 2-6, and 2-8. In each case the gain is controlled by placing some form of feedback around the op amp.

Fig. 2-4 (a) Inverting amplifier circuit. (b) Equivalent circuit.

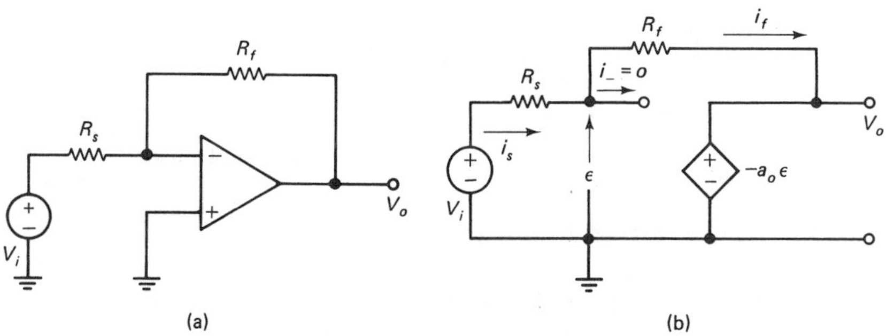

(a)   (b)

2.3.1 Inverting Amplifier

Consider first the simple inverting amplifier shown in Fig. 2-4. To determine the gain of the amplifier, the closed-loop gain $A = v_o/v_i$, we write the circuit v–i relations from Fig. 2-4.

$$v_i = R_s i_s + \epsilon \tag{2-3}$$

$$i_f = i_s \tag{2-4}$$

$$v_o = -i_f R_f + \epsilon \tag{2-5}$$

$$v_o = -a_0 \epsilon \tag{2-6}$$

From these equations, eliminating ϵ and solving for A, we obtain

$$A = \frac{v_o}{v_i} = -\left(\frac{R_f}{R_s}\right)\frac{1}{1 + 1/a_0 + R_f/(a_0 R_s)} \tag{2-7}$$

Note that since $|\epsilon| \ll v_i$ and a_0 is very high, we could set $\epsilon \simeq 0$ in Eqs. (2-3) and (2-5) or set $a_0 \rightarrow \infty$ in Eq. (2-7) to get

$$A = \frac{v_o}{v_i} \simeq -\left(\frac{R_f}{R_s}\right) \tag{2-8}$$

It should be noted, however, that from Eq. (2-7) our approximations imply that $R_f/R_s \ll a_0$. In practice, the closed-loop gain A is almost always much smaller than the open-loop gain a_o. A detailed discussion of the error associated with this approximation is reserved for Section 2.5. Note that since in Eq. (2-8) the polarity of output is the negative of the input, the amplifier is called an *inverting amplifier*.

In some cases it might be convenient to make the approximation at the outset and represent the op amp with the virtual ground model shown in Fig. 2-5. From this circuit model the relationship given in Eq. (2-8) can be written by inspection.

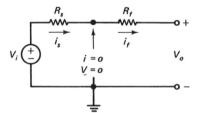

Fig. 2-5 Op amp circuit model with virtual ground.

2.3.2 Noninverting Amplifier

A noninverting amplifier using the op amp is shown in Fig. 2-6. From Fig. 2-6 we have the following relations (we are using the approximation $a_0 \rightarrow \infty$ as is almost always done in practice):

$$v_+ = v_i \tag{2-9}$$

$$v_- = \left(\frac{R_2}{R_1 + R_2}\right)v_o \tag{2-10}$$

Operational Amplifiers

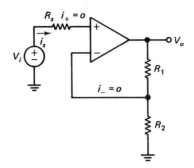

Fig. 2-6 Noninverting amplifier
circuit.

The approximation of $\epsilon = 0$ yields $v_+ = v_-$. Hence from Eqs. (2-9) and (2-10), we have

$$\frac{v_o}{v_i} = \frac{R_1 + R_2}{R_2} = 1 + \frac{R_1}{R_2} \tag{2-11}$$

If $R_1 = 0$, Eq. (2-11) becomes that of a "follower" circuit (i.e., output voltage follows the input). In this case, of course, R_2 is redundant and the follower in Fig. 2-7 is a closed-loop unity-gain circuit.

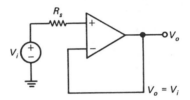

Fig. 2-7 Unity-gain amplifier.

2.3.3 Differential Input/Differential Output Amplifier

A closed-loop differential input/differential output amplifier is shown in Fig. 2-8. This amplifier requires a balanced output as well as input. These amplifiers are not as common as the single-ended output type but they are available and can be used to advantage in some applications (see Chapter 4).

The gains K_1 and K_2 are given by (see Prob. 2.2):

$$K_1 = \frac{v_o}{v_1}\bigg|_{v_2=0} = -\frac{K_o}{1 + K_o} \tag{2-12a}$$

$$K_2 = \frac{v_o}{v_2}\bigg|_{v_1=0} = \frac{1}{1 + K_o} \tag{2-12b}$$

where

$$K_o = \frac{R_b}{R_a} = \frac{R_d}{R_c} \tag{2-12c}$$

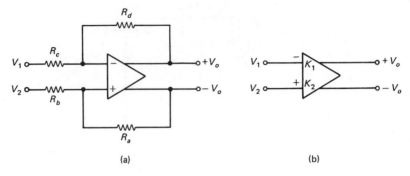

Fig. 2-8 (a) Differential output op amp. (b) Circuit symbol.

2.4 ANALOG COMPUTATION

Op amps are the basic elements of an analog computer. In fact, the terms "operational amplifier" originally came from analog computers, where the circuit was required to perform various mathematical operations (addition, subtraction, integration etc.). These same basic analog computer functions are used today to construct filters (see Chapters 4 and 5). For a complete treatment of this subject, the reader is referred to [B5].

The two basic operations in analog computers are addition and integration. The other operations can be obtained by using these basic blocks.

Addition Op amps may be used as summers to perform the addition operation as shown in Fig. 2-9. From Fig. 2-9 we have

$$i_1 = \frac{v_1 - v_-}{R_1} \simeq \frac{v_1}{R_1} \tag{2-13a}$$

$$i_2 = \frac{v_2 - v_-}{R_2} \simeq \frac{v_2}{R_2} \quad \text{and so on} \tag{2-13b}$$

and

$$i_f = -\frac{v_o}{R_f} = i_1 + i_2 + \ldots + i_n \tag{2-14}$$

From Eqs. (2-13) and (2-14), we get

$$v_o = -R_f \left(\frac{v_1}{R_1} + \frac{v_2}{R_2} + \ldots + \frac{v_n}{R_n} \right) \tag{2-15}$$

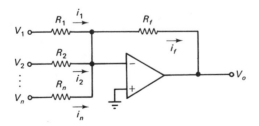

Fig. 2-9 Analog adder.

Note that in addition to summation, we can also do scale changing by properly adjusting $R_1, R_2, \ldots$.

Integration Consider the circuit shown in Fig. 2-10a. The circuit operation is described by

$$V_o = -\left(\frac{Z_f}{R_1}\right) V_i = -\left(\frac{1}{R_1 C_f}\right) \frac{1}{s} V_i \qquad (2\text{-}16)$$

Fig. 2-10 (a) Integrator with zero initial charge on the capacitor. (b) Integrator with initial condition applied.

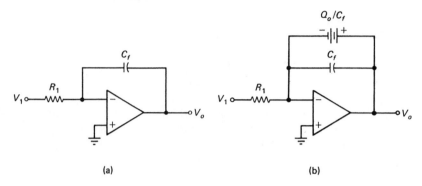

(a)   (b)

One readily recognizes Eq. (2-16) to be the integration operation. We may rewrite Eq. (2-16) in the time demain as

$$v_o = -\frac{1}{R_1 C_f} \int v_i \, dt \qquad (2\text{-}17)$$

Note that the initial charge on the capacitors [or initial conditions in Eq. (2-17)] can be provided by the additional battery of value Q_0/C_f as shown in Fig. 2-10b. It is to be noted that differentiation can be performed if C_f and R_1 are interchanged in Fig. 2-10a. However, integrators are preferred over the differentiators for the following practical reasons:

1. It is more convenient to introduce initial conditions in an integrator.
2. Saturation can be avoided since the gain of an integrator decreases with frequency.
3. Sudden changes can easily overload a differentiator.

Simulation of a differential equation for solution Using only integrators and adders we can simulate and solve a linear, time-invariant differential equation as follows:

Consider the ordinary differential equation

$$a_n \frac{d^n x}{dt^n} + a_{n-1} \frac{d^{n-1} x}{dt^{n-1}} + \ldots + a_1 \frac{dx}{dt} + a_0 x = f(t) \qquad (2\text{-}18)$$

We rewrite Eq. (2-18) as

$$\frac{d^n x}{dt^n} = -\frac{a_{n-1}}{a_n}\frac{d^{n-1}x}{dt^{n-1}} - \cdots - \frac{a_1}{a_n}\frac{dx}{dt} - \frac{a_0}{a_n}x + \frac{1}{a_n}f(t) \qquad (2\text{-}19)$$

A simulation of Eq. (2-19) is shown in Fig. 2-11.

Note that different systems may use different variables, such as temperature, pressure, and so on. The independent variable of the system will correspond to the time t in an analog computer, whereas the dependent variable, whatever it is, corresponds to the voltage in the analog computer. In the solution of some given problems, the "real" time scale may be either too fast or too slow for the computer, and the magnitude scale may be too large or too small. To handle this problem we use magnitude and time scaling:

$$\tau = \alpha t \qquad (2\text{-}20a)$$

$$y = \beta x \qquad (2\text{-}20b)$$

where τ is the "machine" time and t the "real" time, y is the "machine" variable in volts and x the "real" variable of the problem, whatever quantity it might be. Substitution of Eq. (2-20) in (2-18) will yield a scaled differential equation that can be solved via the simulation given in Fig. 2-11 and then rescaled to get the actual values.

●

EXAMPLE 2-1

Consider, for example, the simulation of a fourth-order Butterworth filter to determine its step response. The transfer function is

$$\frac{V_o}{V_i}(s) = H(s) = \frac{K}{s_n^4 + 2.613 s_n^3 + 3.414 s_n^2 + 2.613 s_n + 1} \qquad (2\text{-}21)$$

where $s_n = s/(\pi \times 10^4)$ (i.e., the 3-dB bandwidth of the filter $\omega_{3dB}/2\pi = 5$ kHz). The differential equation corresponding to transfer function (2-21) is

$$\frac{d^4 v_o}{dt^4} + 2.613\frac{d^3 v_o}{dt^3} + 3.414\frac{d^2 v_o}{dt^2} + 2.613\frac{dv_o}{dt} + v_o = Kv_i \qquad (2\text{-}22)$$

We can rewrite as

$$\frac{d^4 v_o}{dt^4} = -2.613\frac{d^3 v_o}{dt^3} - 3.414\frac{d^2 v_o}{dt^2} - 2.613\frac{dv_o}{dt} - v_o + Kv_i \qquad (2\text{-}23)$$

The simulation of Eq. (2-23) is shown in Fig. 2-12.

The normalized step response [i.e., $v_i(t) = u(t)$ and $K = 1$] is shown in Fig. 2-13. Compare this figure with the one shown in Fig. 1-27a. From Fig. 2-13 one readily finds that

$$(\tau_r)_n = 2.43 \qquad (\tau_d)_n = 2.82 \qquad \gamma = 10.9\%$$

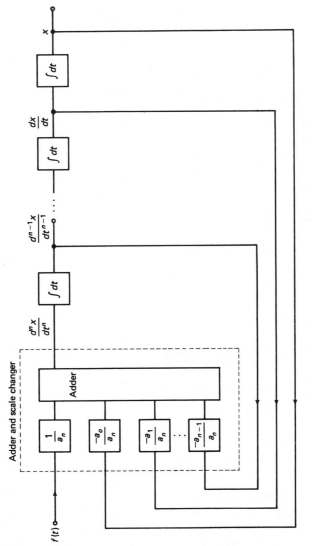

Fig. 2-11 Analog simulation for the solution of Eq. (2-19).

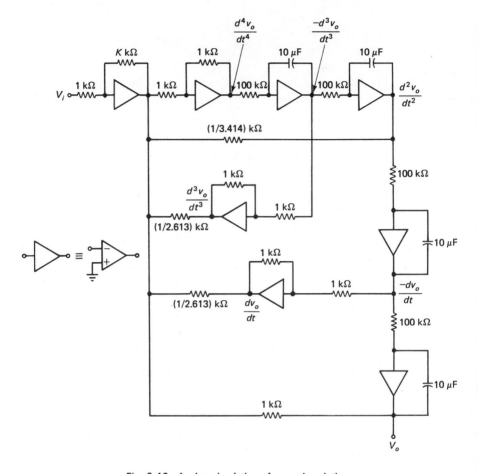

Fig. 2-12 Analog simulation of example solution.

The actual (i.e., denormalized) rise time and delay times are

$$\tau_r = 77.4 \ \mu\text{sec} \qquad \tau_d = 89.8 \ \mu\text{sec}$$

●

The network in Fig. 2-12 is not only an analog computer simulation, it is a fourth-order low-pass active-RC filter with transfer function given by Eq. (2-21). This circuit by no means represents an optimum structure; nevertheless, it does indeed demonstrate the principle of active filtering. One area in which this circuit can be improved is in the amount of hardware employed. In particular, op amps, although inexpensive, are not free. Perhaps, more important, op amps consume power and are the primary sources of noise. As shown in Chapters 4 and 5, the transfer function in Eq. (2-21) can be realized with two op amps, as compared to the eight used in Fig. 2-12. Another important consideration is the degree to which the output or frequency response drifts

Operational Amplifiers

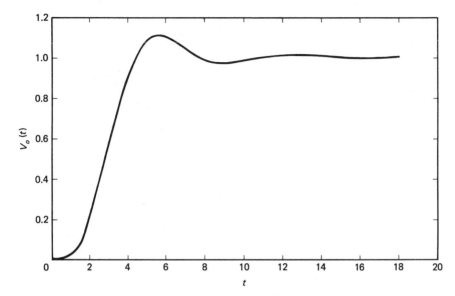

Fig. 2-13   Normalized step response of the analog simulation.

from the desired response [i.e., Eq. (2-21)] as the component values, namely R's, C's, and op amp a_0's, vary due to manufacturing tolerances, temperature, aging, and humidity. If this drift is too large, as it would be in the circuit in Fig. 2-12, the network is rendered useless. To some degree this statistical variation in the response can be predicted and minimized. This area of active-filter analysis and synthesis is treated in detail in Chapters 3, 4, and 5.

2.5 PRACTICAL OPERATIONAL AMPLIFIER CONSIDERATIONS

As mentioned earlier, an op amp is composed of many transistors, resistors, and capacitors. In integrated circuits, transistors are cheaper and require less chip area than do passive components; therefore, the large number of transistors required to realize an op amp is not a limiting factor as far as cost and size are concerned. A fundamental parameter of the op amp used in the analysis and design of active filters, as discussed later in Chapters 4 and 5, is the unity gain–bandwidth product, which we denote by GB. A practical, yet still approximate, circuit model for the op amp is shown in Fig. 2-14, where

$$a(s) \simeq \frac{a_0}{1 + s/p_0} \tag{2-24}$$

A Bode plot corresponding to Eq. (2-24) is given in Fig. 2-15. The unity gain–bandwidth product is seen to be $\mathrm{GB} = a_0 p_0$. In the model of Fig. 2-14 we include the finite input and output impedances.

It is noted that Eq. (2-24) represents a dominant pole approximation to

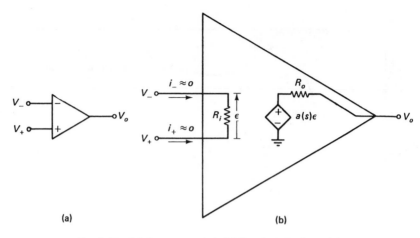

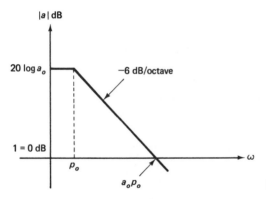

Fig. 2-14　(a) Op amp symbol. (b) Practical circuit model.

Fig. 2-15　Bode plot of Eq. (2-24).

the actual op amp frequency-dependent open-loop gain. Since most op amps are internally or externally compensated to achieve a single-pole roll-off, Eq. (2-24) represents a generally practical and accurate model for most active-filter applications. Typical parameters for the op amps used in active filters are shown in Table 2-1.

TABLE 2-1　Typical op amp parameters

$R_i \geq 10^6 \; \Omega$	$10 \text{ Hz} \leq p_0/2\pi \leq 100 \text{ Hz}$
$R_o < 1 \text{k} \; \Omega$	$\text{GB}/2\pi \geq 10^6 \text{ Hz}$
$a_0 > 10^4$	

It is noted that in most active filter applications, $\omega \gg p_0$; thus, Eq. (2-24) is often further approximated to be

$$a(s) \simeq \frac{a_0 p_0}{s} = \frac{\text{GB}}{s} \qquad (2\text{-}25)$$

(i.e., as an integrator).

　　　　　　　　　　　　　　　　　　　　　Operational Amplifiers

In uncompensated op amps, terminals are provided for adding a discrete compensation network. Typically, the single pole compensation is provided by a single small capacitor (~ 30 pF). The effect of the additional capacitance is to get a roll-off closer to the ideal 6 dB/octave for stability reasons, and the price paid is a reduction in the unity gain–bandwidth product. The reduction in unity gain–bandwidth product forces the operation of the device to drop to a lower frequency and reduces the excess phase shift due to nondominant poles. Excess phase shift, as will be shown later, can cause instability in feedback-connected op amps. Whenever a large amount of feedback is applied to an amplifier, such as is the case with op amp circuits, one must ascertain the stability for the closed-loop circuit. The stability of feedback systems is discussed in Section 2.7. We only mention here that for unconditional stability, it is necessary that the open-loop phase shift be less than 180° at the frequency for which the open-loop gain is unity. It is for this reason that a number of op amp manufacturers use internal compensation.

An equivalent circuit for the analysis of compensation is shown in Fig. 2-16. For simplicity we can assume that the first stage can be represented by

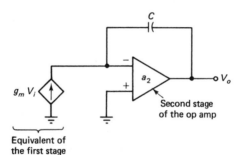

Fig. 2-16 Circuit for op amp compensation analysis.

Equivalent of the first stage

$g_m V_i$ and the second stage to be an ideal amp (i.e., $a_2 \rightarrow \infty$). The overall gain of the circuit is given by

$$\frac{V_o}{V_i} = \frac{-g_m}{sC} \qquad (2\text{-}26)$$

The gain of an op amp is unity at GB; hence, from Eq. (2-26), we have

$$\text{GB} = \frac{g_m}{C} \qquad (2\text{-}27)$$

It is noted that for a GB on the order of $2\pi \times 10^6$ with a compensation capacitor $C = 30$ pF, g_m must be about 200 μmhos.

The inclusion of the compensated op amp frequency response in the analysis and simulation of active-RC filters is often required. To get some feeling for the severity of the effect, let us return to the active-RC integrator in Fig.

2-10. The transfer function, with finite op-amp gain $a(s)$, is as follows:

$$H(s) = -\left(\frac{1}{sR_1C_f}\right)\frac{1}{1 + 1/a(s)(1 + 1/sR_1C_f)} \qquad (2\text{-}28)$$

Let us evaluate, for a frequency of 3 kHz, the integrator gain with $a = \infty$ and $a(j6\pi \times 10^3) = -j10^2$, a typical value for an op amp compensated for a single pole roll-off. From Eq. (2-28) we obtain the following values for $|H(j6\pi \times 10^3)|$ when $R_1C_f = (1/2\pi)10^{-4}$:

$$|H(j6\pi \times 10^3)|_{a=\infty} = \tfrac{10}{3} \longrightarrow 10.46 \text{ dB} \qquad (2\text{-}29a)$$

$$|H(j6\pi \times 10^3)|_{a=j10^2} = \tfrac{10}{3}(1 - 0.032) \longrightarrow 10.17 \text{ dB} \qquad (2\text{-}29b)$$

Comparing Eqs (2-29a) and (2-29b), the gain error introduced by the finite op amp gain is seen to be about 0.3 dB. Active-RC filters are typically comprised of several integrators; thus the accumulative effect of these errors is often unacceptable. It should be noted, however, that the effect of the op amp phase error is often more serious than its gain error. We will discuss this matter later.

One way to reduce this effect is to alter the form of compensation. A particularly useful compensation for precision audio-frequency active filters is a two-pole single-zero compensation, resulting in an open-loop gain of the form

$$a(s) = \frac{a_0(1 + s/z_1)}{(1 + s/p_0)(1 + s/p_1)} \qquad (2\text{-}30)$$

where $z_1 > p_1$. The actual values of the zero and pole (z_1 and p_1) may be determined from measurements. For example, a typical op amp gain function with one-pole compensations is compared to a two-pole single-zero compensation in Fig. 2-17. Note that the latter compensation provides at least an order-of-magnitude-greater op amp gain over the audio-frequency range. To appreciate the impact of this increased gain, let us compute the integrator 3-kHz gain using Eqs. (2-28) and (2-30). At 3 kHz the two-pole single-zero gain is $a(j6\pi \times 10^3) \simeq -j3000$. Evaluating this integrator gain, we obtain

$$|H(j6\pi \times 10^3)|_{a=-j3000} = \tfrac{10}{3}(1 - 0.001) \longrightarrow 10.45 \text{ dB} \qquad (2\text{-}31)$$

Comparing Eqs. (2-31) and (2-29a), we find that the error in integrator gain due to the finite op amp gain is a mere 0.01 dB. This compensation, for the most part, eliminates the need to consider the op amp gain in designing active filters for many applications. This is particularly true for voice-frequency applications, where the filtering is performed at frequencies of less than 5 kHz.

Another means for eliminating the error due to the finite op amp gain is simply to adjust the passive components in the network to compensate for the error at some critical frequency. This technique, in which the ideal transfer function is changed to compensate for the predicted op amp gain, is referred to as *predistortion*. We will have more to say about predistortion in Chapter 4.

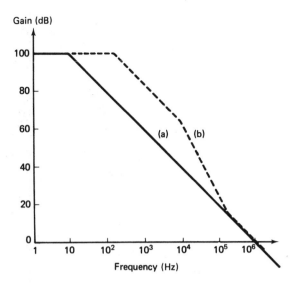

Fig. 2-17   Bode plot of an op amp modeled by a single-pole and a two-pole single-zero transfer function.

To point that we wish to make here is that once the circuit is predistorted, natural variations in the predicted gain, due to meanufacturing tolerances and environmental changes, will cause the transfer function to drift from its desired form. In general, the values of a_0 and p_0 individually vary considerably from unit to unit and with temperature. However, their values tend to vary in opposite directions, so that the variation in their product GB may be less than 20%. With the exception of the temperature-compensated op amp, such as the LM 324, variations in GB with temperature are about $\pm 10\%$ over the full 0 to 70°C range. With the LM 324, GB variations have been measured to be about $\pm 3\%$ over the same temperature range. Op amps with good unit-to-unit uniformity can be obtained. For example, the measure voltage gain versus frequency for four different op amps within a unit (LM 348N) selected at random is shown in Fig. 2-18a. The 6-dB/octave roll-off is also indicated by a heavy line. Note that when multiple (two to four) op amps are housed in one package (e.g., LM 348), one cannot differentiate one op amp from another within the package. This op amp matching phenomenon is used to good advantage in some active-filter structures.

Currently, the gain–bandwidth can be varied in the *programmable op amp* (e.g., Harris HA-2720/2725). For many applications, however, their output impedance ($R_o \simeq 5$ kΩ) is prohibitively high. Certainly, it is highly desirable that the load impedance $Z_L \gg R_o$ so that variations in R_o do not affect the circuit performance. To a limited extent, one can vary (i.e., decrease) the GB by properly selecting the values of external capacitors which are connected at the terminals marked for compensation purposes.

The problem of slew-rate limiting is also an important consideration in high-frequency filter design. This is discussed in some detail in Appendix C.

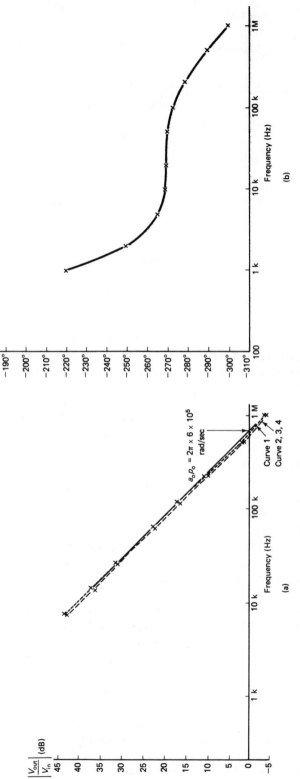

Fig. 2-18 (a) Measured voltage gain versus frequency for four different LM 348N op amp units (solid line is the −6-dB/octave roll-off). (b) Measured phase shift versus frequency for an LM 348N op amp test circuit ($R_f = 100$ kΩ, $R_s = 100\Omega$ in Fig. 2-4).

2.5.1 Second-Order Effects

In measuring the gain frequency response of an op amp, one often finds that the gain variation follows that of Fig. 2-15 pretty well over the frequency range $p_0 < \omega \leq a_0 p_0$, but the phase response does not, as can be seen in Fig. 2-18a and b. Moreover, for frequencies slightly larger than $a_0 p_0$, the gain is smaller and the phase shift larger than that predicted by a one-pole model. A typical measured phase response of the LM 348N is shown in Fig. 2-18b. Note that the phase exceeds $-270°$ in the vicinity and beyond $f \geq 100$ kHz, which is considerably below the unity gain–bandwidth product $a_0 f_0 = 600$ kHz. If the one-pole model were accurate, the phase response would not exceed $-270°$ at any frequency ($-90°$ for the one-pole and $-180°$ for the inverting configuration). The phase error can be reasonably accounted for by inserting another pole in the gain function $a(s)$ near (but larger than) $a_0 p_0$. In most applications these variations are insignificant. However, in some cases the excess phase shift might be enough to cause serious distortion in the response of an active filter and even instability. This problem can become particularly serious, as we see in Chapter 5, when negative feedback is placed around a high-order active-filter operating at frequencies above the audio band. To determine the impact of this second-order effect, we approximate the gain as follows:

$$a(s) \simeq \frac{a_0 e^{-s/m a_0 p_0}}{1 + s/p_0} \tag{2-32a}$$

or

$$a(s) \simeq \frac{a_0}{(1 + s/p_0)(1 + s/m a_0 p_0)} \tag{2-32b}$$

where, depending on the op amp, m is usually larger than 1, with a typical value of $m \simeq 2$ for LM 348N. With $s = j\omega$, the exponent $\omega/m a_0 p_0 \ll 1$ on the right-hand side of Eq. (2-32a) is referred to as the *excess phase*. Since $|e^{-j\omega/m a_0 p_0}| = 1$, the second (nondominant) pole is seen to negligibly alter the magnitude response. However, its effect on the phase response may result in instability when large amounts of negative feedback are applied to the amplifier. The effect of the excess phase as exhibited by Eqs. (2-32) must be included in applications such as active R and active C filters discussed in Section 4.6.

2.6 MONOLITHIC INTEGRATORS

The integrator is unparalleled in its importance in active filters. Most active filters use at least one integrator, with an nth-order filter requiring n integrators to derive the frequency dependence needed to realize a specified frequency response. A typical active-RC integrator is shown in Fig. 2-10. This integrator circuit, although the workhorse of audio-frequency active filters, is not suited for realization as a monolithic integrated circuit in any technology. for either audio or higher frequencies. For audio frequencies the feedback capacitor $C_f \simeq$ 5000 to 10,000 pF is orders of magnitude too large for integration. For high-frequency operation, the loss of gain due to the initial pole of $a(s)$ is sufficient to render the integrator useless.

2.6.1 Bipolar Integrators

For high frequencies the solution is in principle simple and, in fact, it has already been given in Eqs. (2-25) and (2-26). These equations describe the response for an op amp compensated for a single-pole roll-off. This circuit,which requires a small compensation capacitance of 30 pF, is easily integrated in any standard bipolar technology. With integrators of this type, as we shall see in Chapter 4, an active filter can be realized with only resistors and compensated op amps. These filters are typically referred to as *active-R filters* (Section 4.6). For op amps with GB $= 2\pi \times 10^6$ rad/sec, filters of this type are limited to operate in the frequency range 50 kHz to about 200 kHz. At frequencies below 50 kHz, as shown in Chapter 4, the resistors become large and negate much of the size advantage of the approach. At frequencies above 200 kHz, the performance is limited due to the nonlinearity of the op amp and may be difficult to stabilize due to the op amp excess phase. A more severe problem is the wide variability in GB, or g_m/C, in manufacture or with changes in the environment. These effects can be minimized by stabilizing the integrator GBs using well-known phase-lock-loop techniques [P5].

For the more common audio-frequency operation, compensated op amp integrators do not yield monolithic integrated active filters. It has been shown, however, that very small values of $g_m \simeq 0.1$ μmho can be achieved with a bipolar-compatible ion-implanted JFET process [P3]. According to Eq. (2-26), $g_m = 1 \times 10^{-6}$ mho and $C = 30$ pF yields GB $= 2\pi \times 5300$ rad/sec, that is a value quite suitable for audio-frequency applications. Like in compensated op amp integrators, GB may vary widely with manufacturing tolerances and changes in the environment. Again, phase-lock-loop techniques can be used to stabilize the GB of these integrators.

2.6.2 MOS Integrators

Although it has been demonstrated that precision monolithic filters can be realized in a bipolar technology, many integrated-circuit manufacturers are committed to metal-oxide-semiconductor (MOS) processes for the realization of such high-volume large-scale integrated (LSI) digital circuits and subsystems as memories and microprocessors. From a system point of view, it is most desirable to have a monolithic linear circuit technology which is compatible with the well-established digital technology. Only through this compatibility will analog and digital circuits be integrable economically on a single silicon chip.

As noted in the introductory section, MOS op amps are available which are suitable for audio-frequency filtering. In addition to op amps, standard MOS processes provide high-quality switches and capacitors of varying sizes from 0.1 pF to about 100 pF [P5]. Each of these components can be realized within very reasonable chip areas. Diffused MOS resistors, on the other hand, can have poor temperature and linearity characteristics and require large chip areas. The absolute value of a MOS capacitor is determined by the photolithographic definition of its area. Since the capacitance per unit area is uniform across an IC chip, it is possible for absolute capacitor values to vary by as much as ±10%, while ratios of capacitors track to achieve precision capacitor ratios of ±0.1 to ±1.0%.

Let us now examine an interesting MOS circuit [P5] which performs a function similar to that of a resistor, yet takes advantage of the components and precision available in MOS technologies. This circuit, shown in Fig. 2-19a, is comprised of a two-phase switch and a capacitor C. This type of circuit is referred to as a *switched capacitor* (SC) *circuit*. The operation of the circuit may be stated as follows. Initially, with the switch in the left-hand position, the capacitor charges to the voltage V_1. When the switch is thrown to the right,

Fig. 2-19 (a) Switched capacitor element. (b) MOS implementation. (c) Biphase nonoverlapping clock.

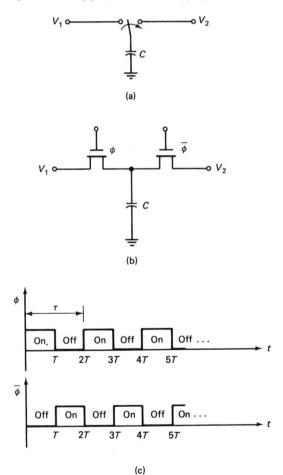

(a)

(b)

(c)

capacitor C discharges to the voltage V_2. The amount of charge that flows from port 1 to port 2 as the switch changes position from left to right is $Q = C(V_2 - V_1)$. If the switch is toggled back and forth every τ seconds, the current flow i, from port 1 to port 2, will be

$$i = C\frac{dv}{dt} \simeq C\left(\frac{V_2 - V_1}{\tau}\right) = \frac{V_2 - V_1}{R_{eq}} \qquad (2\text{-}33a)$$

where

$$R_{eq} = \frac{\tau}{C} \qquad (2\text{-}33b)$$

From Eq. (2-33) the switched capacitor circuit in Fig. 2-19a is seen to provide a function similar to that of a resistor of value $R = \tau/C$. The model described by Eq. (2-33) is somewhat simplistic, as shown in Chapter 6. Nevertheless, it very adequately demonstrates the point that resistor-like functions can be achieved in MOS integrated circuits with simple switched capacitor networks. It is to be noted that this switched-capacitor circuit is a simple sampled-data system with a sample rate of $f_s = 1/\tau$. As noted in Chapter 1, sampled-data systems, like any other discrete-time system, must see an input signal that is band-limited below $f_s/2$ to avoid distortion due to aliasing. Typically, switched capacitor filters operate with a sampling frequency f_s which is much higher than any signal frequency in the band of interest. In addition to the sampled-data nature of the circuit, there is a time delay as the signal is switched from port 1 to port 2. This time delay is not included in the simplistic model in Eq. (2-33). Since the purpose of this discussion is only to introduce the switched capacitor concept, these modeling concerns shall be reserved for a detailed discussion in Chapter 6.

The MOS realization for this simple switched capacitor circuit is shown schematically in Fig. 2-19b. Switches ϕ and $\bar{\phi}$ are ideally biphase MOS switches clocked according to the schedule given in Fig. 2-19c. The switches are shown to be ON and OFF for equal $\tau/2$ time intervals (i.e., a 50% duty cycle). In practice, due to the rise and fall times of the switches, the duty cycle is typically less than 50%. This less-than-50% duty cycle ensures that the ϕ and $\bar{\phi}$ on times do not overlap.

Let us now consider the operation of the MOS switched capacitor integrator shown in Fig. 2-20. This circuit is recognized to be very similar to the active-RC integrator shown previously in Fig. 2-10. As we might well expect from the previous discussion, the input resistor of the active-RC integrator is replaced by the switched capacitor circuit in Fig. 2-19. As an approximation, we can represent the switched capacitor integrator as an analog integrator, as per Eq. (2-16), with transfer function

$$\frac{V_o}{V_i} \simeq -\left(\frac{1C_1}{\tau C_F}\right)\frac{1}{s} = -f_s\left(\frac{C_1}{C_F}\right)\frac{1}{s} \qquad (2\text{-}34a)$$

where

$$f_s = \frac{1}{\tau} \qquad (2\text{-}34b)$$

Operational Amplifiers

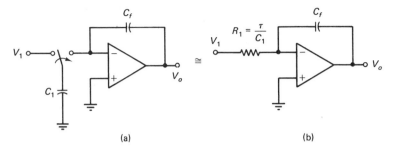

Fig. 2-20   (a) Switched capacitor integrator. (b) Equivalent analog integrator.

Equation (2-34) conveys a good deal of information about the variability of this integrator. As noted earlier, MOS technologies offer the ability to obtain precision capacitor ratios and essentially ideal switches. With this inherent property of the technology and a very stable frequency source or clock for f_s, the integrator gain is indeed a precision quantity. Typically, f_s is derived through simple logic circuitry from a very stable quartz crystal oscillator; thus, f_s is usually assumed invariant.

2.7 STABILITY CONSIDERATIONS

Let us now digress a bit and consider in more detail the important topic of stability in feedback systems. Any active network that employs feedback, whether it be analog or switched capacitor, can become unstable if the amount of feedback is sufficiently large and if there is inherent delay in the system. As shown in Chapter 6, due to the time delays associated with the switching arrangement, even apparently stable active switched capacitor networks can be made to be unstable.

In any event, invariably an op amp and its surrounding circuitry define a feedback system. One can define a feedback system as a system in which some portion of the transmitted signal is returned to the input via some intentional or unitentional path. The use of feedback provides several advantages to the circuit designer; however, extreme care should be exercised to ensure that it results in a stable system. In a linear system, the necessary and sufficient condition for stability is that all poles of the closed-loop transfer function lie in the left-half complex frequency plane. This simple condition is readily understood if we look at the time-domain response (Section 1.6). The output will grow exponentially with time due to any right-half-plane pole contribution and thus cause instability.

Before we consider stability tests, it is advisable to consider briefly some basic and commonly used terminologies in feedback systems [B1, B6].

The open-loop transfer function is denoted by $a(s)$; the feedback function is given by $f(s)$. The loop transmission function is denoted by $T(s)$, namely,

$$T(s) = -a(s)f(s) \qquad (2\text{-}35)$$

The closed-loop transfer function $A(s)$, as obtained from Fig. 2-21, is given by

$$A(s) = \frac{a(s)}{1 - a(s)f(s)} \qquad (2\text{-}36)$$

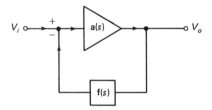

Fig. 2-21 Single-loop feedback system.

The poles of the closed-loop transfer function must be in the left-half s-plane for the system to be stable. These poles are also referred to as the *roots* of the characteristic equation. The characteristic equation is given by

$$F(s) = 1 - a(s)f(s) = 0 \qquad (2\text{-}37)$$

The quantity $20 \log_{10} |F(j\omega)|$ is called the *amount of feedback*, expressed in decibels.

As a specific example, consider the following system (frequency normalized):

$$a(s) = -\frac{a_0}{(s+1)^3} \qquad (2\text{-}38a)$$

$$f(s) = f_0 \qquad (2\text{-}38b)$$

From Eq. (2-36), we have

$$A(s) = \frac{-a_0}{(s+1)^3 + a_0 f_0} = \frac{-a_0}{s^3 + 3s^2 + 3s + 1 + a_0 f_0} \qquad (2\text{-}39)$$

An examination of the poles of Eq. (2-39) will show that the roots will be in the right half-plane for $a_0 f_0 \geq 8.0$. For $a_0 f_0 = 8$, one pair of roots will be on the $j\omega$-axis (we shall include $j\omega$-axis poles as unstable in this text).

Stability of a linear feedback system is readily determined by the following tests. For justification of these tests, the reader is referred to any text on feedback control systems (e.g., [B7]).

2.7.1 Routh–Hurwitz Test

The *Routh–Hurwitz test* is a mathematical method to check the existence of any root of the polynomial (characteristic equation) with positive real parts. Of course, if a root-finder subroutine is used in a computer, such a test is not needed and the stability problem can be resolved by an inspection of the roots. The absence of any right-half-plane roots of the characteristic equation guar-

Operational Amplifiers

antees system stability. It is, of course, assumed that the characteristic equation accurately describes the system (we shall consider this point later in the section).

The characteristic equation of a system may be written in general form as

$$F(s) = s^n + a_1 s^{n-1} + \ldots + a_{n-1}s + a_n = 0 \tag{2-40}$$

We rearrange Eq. (2-40) in two lines and compute the triangular array as follows:

$$
\begin{array}{c|cccc}
 & \downarrow & & & \\
s^n & 1 & a_2 & a_4 & a_6 & \cdots \\
s^{n-1} & a_1 & a_3 & a_5 & a_7 & \cdots \\
\hline
 & b_0 & b_2 & b_4 & & \cdots \\
 & c_1 & c_3 & c_5 & & \cdots \\
 & d_0 & \cdot\cdot & \cdot\cdot & & \cdots \\
\end{array}
$$

where

$$b_0 = \frac{a_1 a_2 - a_3}{a_1} \qquad b_2 = \frac{a_1 a_4 - a_5}{a_1} \qquad b_4 = \frac{a_1 a_6 - a_7}{a_1}$$

$$b_1 = \frac{b_0 a_3 - a_1 b_2}{b_0} \qquad c_3 = \frac{b_0 b_5 - a_1 b_4}{b_0}$$

$$d_0 = \frac{c_1 b_2 - b_0 c_3}{c_1} \qquad \text{etc.}$$

The test is simply to examine the elements of the entire column 1, a_1, b_0, c_1, d_0, ... (marked by an arrow). If any of the numbers is either a zero or negative, the system is unstable. For example, consider the system described by Eq. (2-39):

$$F(s) = s^3 + 3s^2 + 3s + 1 + a_0 f_0 = 0$$

We form the array

$$
\begin{array}{c|cc}
s^3 & 1 & 3 \\
s^2 & 3 & 1 + a_0 f_0 \\
\hline
 & (8 - a_0 f_0)/3 & 0 \\
 & 1 + a_0 f_0 & \\
\end{array}
$$

From the test column it is seen that for $a_0 f_0 \geq 8$, the system is unstable. In fact, it can readily be shown that for $a_0 f_0 = 8$, the roots are at $s = -3$ and $s = \pm j\sqrt{3}$ (i.e., one pair of poles are on the $j\omega$-axis as shown in Fig. 2-22).

2.7.2 Root-Locus Technique

The *root-locus technique* provides a graphical location of the roots of the characteristic equation as the feedback factor $a_0 f_0$ is varied. For example,

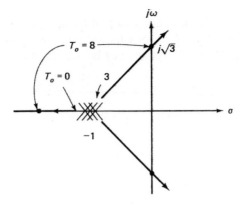

Fig. 2-22   Root locus for Eq. (2-41).

consider the system described by Eq. (2-38). The characteristic equation is given by

$$F(s) = 1 + T(s) = 1 + \frac{a_0 f_0}{(s + 1)^3} = 0 \qquad (2\text{-}41\text{a})$$

$$(s + 1)^3 + T_0 = 0 \qquad (2\text{-}41\text{b})$$

The root locus of $F(s)$, as $a_0 f_0$ $(= T_0)$ is varied, is shown in Fig. 2-22.

A root-finding program where T_0 is varied selectively from zero to a very large number in logarithmic manner (i.e., $T_0 = 0, 2, 5, 10, 20, 50, 100, 500, 1000, \ldots$) is most helpful. This can easily be implemented via a computer. Any region of interest in the design may be examined further by varying T_0 in a finer subdivision in that region. Root-locus construction rules, to sketch the loci, are also of considerable aid to the designer. Easy-to-follow rules can be found in many texts (see [B6] and [B7]) and will not be listed here.

In op amps where a large amount of feedback is employed, care should be exercised to describe accurately the op amp model. Consider, for example, the system shown schematically in Fig. 2-23. In this circuit two op amps are utilized. If the op amps are described by the dominant poles (assumed to be the same for convenience purposes only):

$$K_1(s) = \frac{K_{01}}{1 + s/p_1} \qquad (2\text{-}42\text{a})$$

$$K_2(s) = \frac{K_{02}}{1 + s/p_1} \qquad (2\text{-}42\text{b})$$

Fig. 2-23   Single-loop feedback system using two op amps.

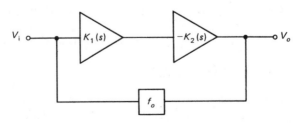

We have

$$a(s) = \frac{-K_{01}K_{02}}{(1 + s/p_1)^2} = \frac{-a_0}{(1 + s_n)^2} \qquad (2\text{-}43a)$$

$$f(s) = f_0 \qquad (2\text{-}43b)$$

where $a_0 = K_{01}K_{02}$ and $s_n = s/p_1$. From Eqs. (2-42), (2-43), and (2-36), we have

$$A(s) = \frac{-a_0}{(1 + s_n)^2 + a_0 f_0} = \frac{-a_0}{s_n^2 + 2s_n + 1 + a_0 f_0} \qquad (2\text{-}44)$$

The system is always stable, no matter what the value of $a_0 f_0$ is, as shown in the root-locus diagram in Fig. 2-24a.

Fig. 2-24 (a) Root locus when the op amp is modeled with one pole in Fig. 2-23. (b) Root locus when the op amp model includes the nondominant pole.

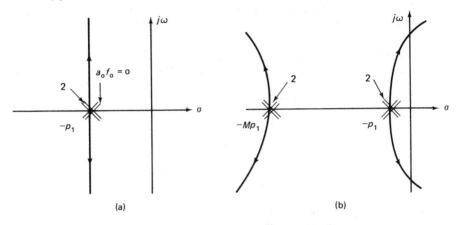

(a)

(b)

Now consider the inclusion of second-order effects (i.e., the nondominant poles, as discussed in Section 2.6). In this case we have

$$K_i = \frac{K_{0i}}{(1 + s/p_1)(1 + s/Mp_1)} \qquad i = 1, 2 \qquad (2\text{-}45)$$

where, as in Eq. (2-32),

$$M = ma_0, \qquad s_n = \frac{s}{p_1}$$

The closed-loop gain function becomes

$$A(s) = \frac{a_0}{(1 + s_n)^2(1 + s_n/M)^2 + a_0 f_0} \qquad (2\text{-}46)$$

The system described in Eq. (2-46) can clearly become unstable, as shown by the root-locus diagram in Fig. 2-24b. The significance of the nondominant poles in a stability check is further illustrated in conjunction with actual practical high-Q filters (see Chapter 4). Note that for a high Q, the dominant pole pair is very close to the $j\omega$-axis (e.g., for $Q = 20$, the imaginary part of the dominant pole is 40 times its real part.)

2.7.3 Nyquist Criterion

The *Nyquist criterion* is a graphical test based on the steady-state frequency response of the loop transmission function. It is an extremely useful and practical method, since measured data can be used to predict the stability of the system prior to closing the feedback loop. If measured data are used, one need not worry whether the model is realistic, because calculation is not based on a proposed model but the actual system.

It is assumed in the following that the open-loop transfer function $a(s)$ and the feedback function $f(s)$ have no pole in the right half-plane. Thus, the closed-loop gain function can have poles in the right half-plane (RHP) only if the characteristic equation has zeros in the RHP. The test is to find in a simple way whether there is any value of s with a nonnegative real part at which $1 + T(s) = 0$.

It can be shown [B7] that the Nyquist diagram maps the right half s-plane into the interior of a contour in the T-plane. If there is any root of

Fig. 2-25   Nyquist plot for Eq. (2-47).

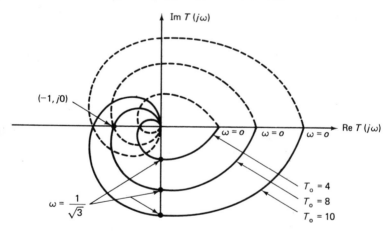

$1 + T(s) = 0$ is the RHP, the T-plane contour will enclose the point $(-1, j0)$, which is usually referred to as the *critical point*. Hence, encirclement of the critical point is the Nyquist criterion of instability.

To use the Nyquist criterion, we plot the imaginary part of $T(j\omega)$ versus the real part of $T(j\omega)$ for $0 \leq \omega < \infty$. A polar plot may be more convenient for these purposes. A Nyquist plot of the system described by Eq. (2-38), i.e.,

$$T(s) = \frac{a_0 f_0}{(s + 1)^3} = \frac{T_0}{(s + 1)^3} \tag{2-47}$$

Operational Amplifiers

is shown in Fig. 2-25. Note that for the Nyquist plot we need the entire $j\omega$-axis (i.e., $-\infty < \omega < \infty$). However, since, Re $T(j\omega)$ is even and Im $T(j\omega)$ odd, at any point ω_1 we have the mirror reflection point for $-\omega_1$ as shown by the dashed lines in Fig. 2-25. Note that for $T_0 = 8$, the system is on the verge of instability and hence unstable for $T_0 \geq 8$.

In brief, the Nyquist criterion states that the system is stable if the closed curve Im $T(j\omega)$ versus Re $T(j\omega)$ [or $|T(j\omega)|$ versus arg $T(j\omega)$] does not enclose or pass through the point $(-1, j0)$ and unstable otherwise. Two quantities of interest, in the design of feedback systems, are the gain margin and the phase margin. The *gain margin* G_m is defined as the value of $|T(j\omega)|^{-1}$ at the frequency at which arg $T(j\omega) = 180°$, and this frequency is referred to as the *phase cross-over frequency* ω_p. G_m is usually expressed in decibels. The *phase margin* Φ_m is defined as $180°$ plus arg $T(j\omega)$ at the frequency at which $|T(j\omega)| = 1$. The frequency at which $|T(j\omega)| = 1$ is called the *gain crossover frequency*, ω_g. These quantities are shown in a Nyquist plot in Fig. 2-26. In some cases it may be

Fig. 2-26   Gain and phase crossover frequencies in a Nyquist plot.

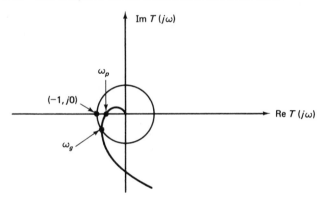

more convenient to indicate the stability margin (i.e., gain and phase margins) in a Bode plot as shown in Fig. 2-27. The values of G_m and Φ_m also give indication of frequency and transient response, but not in a direct manner. Usually, in a feedback amplifier, a gain margin of at least 10 dB and a phase margin of at least 60° are provided.

Analytically, the gain and phase margins are determined by their definitions as follows. From $|T(j\omega)| = 1$, we find ω_g, and from arg $T(j\omega) = 180°$, we find ω_p. Then:

$$G_m = 20 \log |T(j\omega_p)^{-1}| \quad \text{dB} \qquad (2\text{-}48)$$

$$\Phi_m = 180 + \arg T(j\omega_g) \quad \text{deg} \qquad (2\text{-}49)$$

For example, if $T(s) = 4/(1 + s)^3$, from Eqs. (2-48) and (2-49) we find that $\omega_g = 1.23$, $\omega_p = 1.732$, $G_m = 6$ dB, and $\Phi_m = 27$ deg.

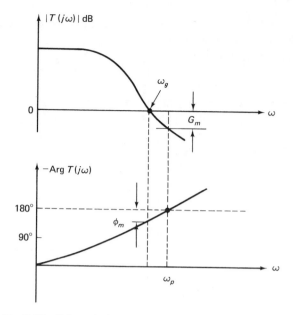

Fig. 2-27 Gain and phase crossover frequencies in a Bode plot, also showing the gain and phase margins.

Note that in a Nyquist or Bode plot, we examine $T(s)$ for $s = j\omega$ only (i.e., the steady-state response). Hence, the transient response is completely covered, and only in an indirect manner may information regarding the transient response be obtained by the gain and phase margins. In a root-locus approach we examine $T(s)$ for $s = \sigma + j\omega$; hence, the transient response is directly obtained by the closed-loop pole locations for a given T_0.

2.8 NEGATIVE-IMPEDANCE CONVERTER AND GYRATOR REALIZATIONS

Up to this point we have considered the realization of analog-computer-like blocks: the integrator, the summer, the inverting amplifier, and the noninverting amplifier, by embedding an op amp in an appropriate feedback network. The behavior of these blocks was uniquely defined by their voltage transfer functions. In this section we consider another class of op amp circuits whose behavior is defined by their terminal impedance or admittance functions.

2.8.1 Negative-Impedance Converter

Active-RC filters in their earlier stages of development used negative-impedance converters to realize filter transmission. The *negative-impedance converter* (NIC) is an active two-port circuit that converts an impedance termination Z_L at one port into a negative impedance at the other port, proportional to Z_L. We shall not consider NICs in detail, as they are not used in practice now. We have included this topic here only for the sake of completeness.

Operational Amplifiers

Consider the linear two-port shown schematically in Fig. 2-28. The input impedance (using the h-parameters) is given by

$$Z_i = h_{11} - \frac{h_{12}h_{21}}{h_{22} + Y_L} \qquad (2\text{-}50)$$

Fig. 2-28 Two-port NIC schematic.

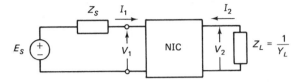

The output admittance is given by

$$Y_0 = h_{22} - \frac{h_{12}h_{21}}{h_{11} + Z_S} \qquad (2\text{-}51)$$

Necessary and sufficient conditions for the active two-port to be an ideal NIC are

$$h_{11} = h_{22} = 0 \qquad (2\text{-}52a)$$

$$h_{12}h_{21} = K \qquad (2\text{-}52b)$$

If we choose $h_{12} = h_{21} = \pm 1$, we have two types of NICs:

1. The current-inversion-type INIC which occurs for $h_{12} = h_{21} = 1$. (Note that the actual current through the load is $-I_2$.) In this case

$$V_1 = V_2, \qquad I_1 = I_2 \qquad (2\text{-}53)$$

2. The voltage-inversion-type VNIC which occurs for $h_{12} = h_{21} = -1$. In this case,

$$V_1 = -V_2, \qquad I_1 = -I_2 \qquad (2\text{-}54)$$

Op amp realization of these two types are shown in Fig. 2-29a and b, respectively.

Note that NICs are potentially unstable two-ports; hence, care should be exercised in choosing their termination. The open-circuit stable and short-circuit stable sides are indicated by OCS and SCS, respectively. In other words, the SCS port should be terminated by a small resistor and the OCS port should be terminated by a large resistor for stability reasons. Because of their potentially unstable character, NICs are not used in practical active filters.

2.8.2 Gyrator

An *ideal* gyrator is a passive, nonreciprocal two-port shown schematically in Fig. 2-30. An actual gyrator is realized with active elements such as op amps.

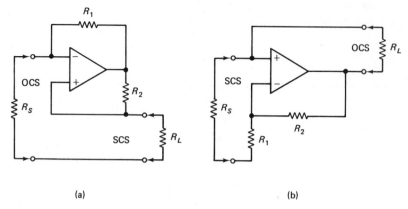

Fig. 2-29 Op amp realization of NIC circuits.

Fig. 2-30 Gyrator symbol.

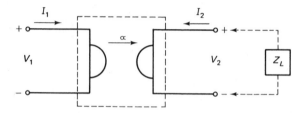

Nonidealness in the op amp therefore creates nonidealness in the gyrator characteristics. The defining voltage–current relationships of an ideal gyrator are given by

$$V_1 = rI_2 \qquad I_1 = -gV_2$$
$$V_2 = rI_1 \qquad \Longleftrightarrow \qquad I_2 = gV_1 \qquad (2\text{-}55)$$

The constant $r = 1/g$, which has the dimension of a resistance, is called the *gyration resistance*.

From Eq. (2-55), solving for the input impedance, we obtain

$$Z_{\text{in}} = \frac{V_1}{I_1} = \frac{r^2}{Z_L} \qquad (2\text{-}56a)$$

In Eq. (2-56a) it is readily seen that if Z_L is the impedance of a capacitor (i.e., $Z_L = 1/sC$), the input impedance of the entire circuit is equivalent to that of of an inductor:

$$Z_{\text{in}} = \frac{r^2}{Z_L} = (r^2C)s = L_{\text{eq}}s \qquad (2\text{-}56b)$$

where

$$L_{\text{eq}} = r^2C \qquad (2\text{-}56c)$$

This property of the gyrator is attractive in integrated circuits, as the designer can simulate inductors with capacitively terminated gyrators. A high-quality inductor simulation circuit that utilizes the circuit in Fig. 2-33 is discussed in Section 2.9.1.

A gyrator realization that uses op amps, with dc gains of a_0, is shown in Fig. 2-31. For this circuit the y-parameter matrix is

$$[y_{ij}] = \frac{1}{(1 + 1/a_0)R} \begin{bmatrix} \dfrac{1}{a_0} & -1 \\ \dfrac{1}{1 + 2/a_0} & \dfrac{1}{a_0} \end{bmatrix} \tag{2-57a}$$

$$\simeq \frac{1}{R} \begin{bmatrix} 0 & -1 \\ 1 & 0 \end{bmatrix} \quad \text{for } a_0 \longrightarrow \infty \tag{2-57b}$$

Fig. 2-31   Gyrator realization with two op amps.

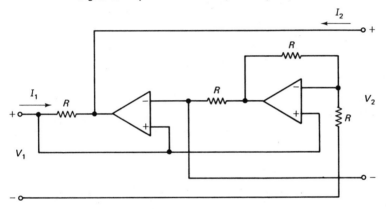

Fig. 2-32   (a) GIC circuit. (b) Schematic representation.

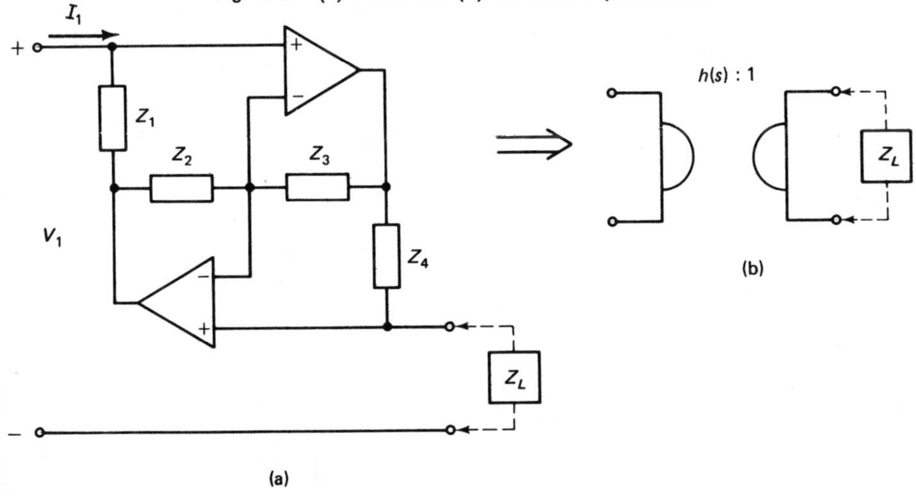

(b)

(a)

2.9 REALIZATION OF GENERALIZED-IMMITTANCE CONVERTER

For NIC we saw that $Z_i = -KZ_L$, where K is a positive constant nominally unity. If $-K$ is replaced by a general function $f(s)$, one obtains

$$Z_i = f(s)Z_L \qquad (2\text{-}58a)$$

$$Y_i = h(s)\,Y_L \qquad \text{where} \qquad h(s) = \frac{1}{f(s)} \qquad (2\text{-}58b)$$

A two-port described by Eq. (2-58) is called a generalized-immittance converter (GIC) and $f(s)$ is referred to as the impedance-conversion function. As for an NIC, two special types of GICs exist, a voltage-conversion type (VGIC), in which

$$V_1 = f(s)V_2 \qquad (2\text{-}59a)$$

$$I_1 = -I_2 \qquad (2\text{-}59b)$$

and a current-conversion type (IGIC), in which

$$V_1 = V_2 \qquad (2\text{-}60a)$$

$$I_1 = h(s)I_2 \qquad (2\text{-}60b)$$

One of the most useful GIC circuits is shown in Fig. 2-32 [P8]. For finite-gain operational amplifiers, straightforward analysis of the circuit in Fig. 2-32, yields

$$Y_{\text{in}} = \frac{I_1}{V_1} \simeq \frac{Y_1 Y_3}{Y_2 Y_4}\,Y_L\,\frac{1 + \dfrac{1}{a_1}\dfrac{Y_2}{Y_3}\left(1 + \dfrac{Y_4}{Y_L}\right) + \dfrac{1}{a_2}\left(1 + \dfrac{Y_4}{Y_L}\right)}{1 + \dfrac{1}{a_1}\left(1 + \dfrac{Y_L}{Y_4}\right) + \dfrac{1}{a_2}\dfrac{Y_3}{Y_2}\left(1 + \dfrac{Y_L}{Y_4}\right)} \qquad (2\text{-}61)$$

where $Y_i = 1/Z_i$. We have assumed in the derivation of Eq. (2-61) that $\omega/\text{GB}_i \ll 1$; thus, all terms containing the product of the inverse amplifier gains $(1/a_1 a_2)$ have been ignored. When $a_1 = a_2 = \infty$, the desired result is

$$Y_{\text{in}} = \frac{I_1}{V_1} = \frac{Y_1 Y_3}{Y_2 Y_4}\,Y_L = h(s)\,Y_L \qquad (2\text{-}62)$$

is obtained.

2.9.1 Inductor Simulation

One of the more important applications of the GIC in Fig. 2-32 is inductor simulation. As illustrated in Chapter 5, low-sensitivity active-RC filters can be derived directly from passive-RLC prototype networks by replacing the passive inductors with active, GIC-simulated inductors.

If we set $Y_1 = G_1$, $Y_2 = G_2$, $Y_3 = G_3$, $Y_4 = sC$, and $Y_L = G_L$, we obtain, assuming that $a_1 = a_2 = \infty$, for Y_{in},

$$Y_{in} = \frac{G_1 G_3}{s G_2 C} G_L = h(s) G_L = \frac{1}{s L_0} \qquad (2\text{-}63)$$

where

$$L_0 = \frac{G_2 C}{G_1 G_3 G_L} \qquad (2\text{-}64)$$

The circuit that realizes a grounded inductor according to Eqs. (2-63) and (2-64) is shown in Fig. 2-33.

Fig. 2-33 GIC, inductor simulation.

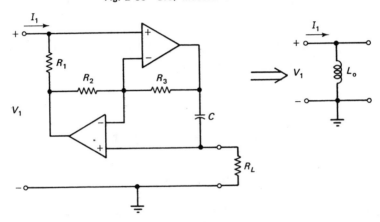

In practical circuits, op amp gains are finite with $a_i \simeq -(GB)_i/s$. Finite op amp gains result in a finite inductor quality factor and an error in the inductance value. To ascertain the magnitude of these two nonideal effects, let us compute the actual input admittance according to Eq. (2-61), with $s = j\omega$. We obtain

$$Y_{in} \simeq \frac{G_1 G_3 G_L}{j\omega C G_2} \frac{1 + \dfrac{j\omega}{GB_1} \dfrac{G_2}{G_3}\left(1 + \dfrac{j\omega C}{G_L}\right) + \dfrac{j\omega}{GB_2}\left(1 + \dfrac{j\omega C}{G_L}\right)}{1 + \dfrac{j\omega}{GB_1}\left(1 + \dfrac{G_L}{j\omega C}\right) + \dfrac{j\omega}{GB_2} \dfrac{G_3}{G_2}\left(1 + \dfrac{G_L}{j\omega C}\right)} \qquad (2\text{-}65)$$

Assuming that $\omega/(GB)_1 \ll 1$ and $\omega/(GB)_2 \ll 1$, we can approximate Eq. (2-65) as follows:

$$Y_{in} = \frac{1}{j\omega L_0}(\alpha + j\beta) = \operatorname{Re} Y_{in} + j \operatorname{Im} Y_{in} \qquad (2\text{-}66)$$

where

$$\alpha \simeq \frac{1 - \dfrac{\omega^2 C}{G_L}\left(\dfrac{G_2}{G_3}\dfrac{1}{GB_1} + \dfrac{1}{GB_2}\right) + \dfrac{G_L}{C}\left(\dfrac{G_3}{G_2}\dfrac{1}{GB_1} + \dfrac{1}{GB_2}\right)}{1 + \dfrac{2G_L}{C}\left(\dfrac{G_3}{G_2}\dfrac{1}{GB_2} + \dfrac{1}{GB_1}\right)} \qquad (2\text{-}67a)$$

and

$$\beta \simeq \frac{\omega\left(\dfrac{G_2}{G_3}\dfrac{1}{GB_1} + \dfrac{1}{GB_2}\right)\left(1 - \dfrac{G_3}{G_2}\right)}{1 + \dfrac{2G_L}{C}\left(\dfrac{G_3}{G_2}\dfrac{1}{GB_2} + \dfrac{1}{GB_1}\right)} \qquad (2\text{-}67b)$$

The quality factor for an inductor is defined as follows:

$$Q_L = -\frac{\text{Im } Y_{in}}{\text{Re } Y_{in}} = \frac{\alpha}{\beta} \qquad (2\text{-}68)$$

From Eq. (2-67b), Q_L can be made arbitrarily large, independent[1] of the mismatch in GB_1 and GB_2, by setting

$$G_2 = G_3 \qquad (2\text{-}69)$$

If we assume that $GB_1 = GB_2 = GB$ and $G_2 = G_3$, the fractional change in inductance value due to the nonideal op amp is given by

$$\frac{\Delta L}{L_0} = \frac{1}{\alpha} - 1 \simeq \frac{2}{GB}\left(\frac{\omega^2 C}{G_L} + \frac{G_L}{C}\right) = \frac{2\omega}{GB}\left(\frac{\omega C}{G_L} + \frac{G_L}{\omega C}\right) \qquad (2\text{-}70)$$

As noted in Section 2-5, $GB_1 = GB_2$ is an excellent assumption in dual op amp units.

The value of $\Delta L/L_0$ at some critical frequency ω_c can be minimized by setting

$$G_L = \omega_c C \qquad (2\text{-}71)$$

yielding the minimum inductance error of

$$\left(\frac{\Delta L}{L_0}\right)_{\min} = \frac{4\omega_c}{GB} \qquad (2\text{-}72)$$

Substituting Eqs. (2-69) and (2-71) into Eq. (2-64) yields the value for G_1:

$$G_1 = \frac{1}{L_0\omega_c} \qquad (2\text{-}73)$$

Equations (2-69), (2-71), and (2-73) in essence describe the design procedure for

[1] There exists another design which achieves high Q if the two GBs are matched. The design presented here, which is based on resistor matching, is superior [B9].

a minimum-inductance-error, maximum-Q, GIC-simulated inductor. The application of these GIC inductors to active filters is discussed in Chapter 5.

2.9.2 Frequency-Dependent Negative Resistor

Another important application of the GIC in Fig. 2-32 is the realization of frequency-dependent negative resistors (FDNRs). These elements arise in frequency-transformed passive RLC prototype networks [P10]. As shown in Chapter 5, the use of FDNRs rather than inductors can result in active simulations requiring fewer GICs (i.e., fewer op amps).

The FDNR is defined according to the following immittance functions:

$$Y(s) = K_1 s^2 \quad \text{or} \quad Z(s) = \frac{K_2}{s^2} \tag{2-74}$$

Clearly, for $s = j\omega$ we have a frequency-dependent negative resistance—hence the name FDNR.

If we set, in Fig. 2-32 and Eq. (2-61), $Y_1 = sC$, $Y_2 = G_2$, $Y_3 = G_3$, $Y_4 = G_4$, and $Y_L = sC$, then

$$Y_{\text{in}} = s^2 \frac{G_3 C^2}{G_2 G_4} = h(s)(sC) = s^2 D_0 \tag{2-75}$$

where

$$D_0 = \frac{G_3 C^2}{G_2 G_4} \tag{2-76}$$

Note that $a_1 = a_2 = \infty$ has been used.

The circuit realizations for a grounded FDNR is given in Fig. 2-34. The circuit symbol for the FDNR is also given in this figure and is used in the text.

Fig. 2-34 GIC, FDNR simulation.

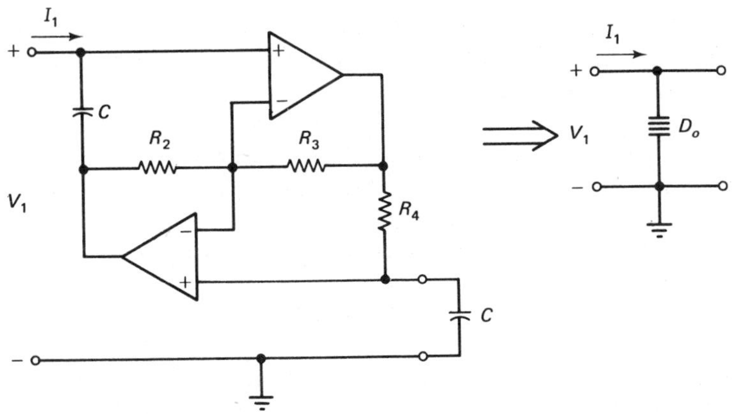

As in the GIC inductor, the nonideal op amp with gain–bandwidth products GB_1 and GB_2 will cause the FDNR D_0, value to be in error and the quality factor to be finite. Substituting $a_1 = GB_1/j\omega$, $a_2 = GB_2/j\omega$, $Y_1 = j\omega C$, $Y_2 = G_2$, $Y_3 = G_3$, $Y_4 = G_4$, and $Y_L = j\omega C$ into Eq. (2-61) results in the following nonideal admittance:

$$Y_{in} = -\omega^2 D_0 \frac{1 + \left(\dfrac{G_2}{G_3}\dfrac{j\omega}{GB_1} + \dfrac{j\omega}{GB_2}\right)\left(1 + \dfrac{G_4}{j\omega C}\right)}{1 + \left(\dfrac{j\omega}{GB_1} + \dfrac{G_3}{G_2}\dfrac{j\omega}{GB_2}\right)\left(1 + \dfrac{j\omega C}{G_4}\right)} \tag{2-77a}$$

$$= -\omega^2 D_0(\delta + j\gamma) = \operatorname{Re} Y_{in} + j \operatorname{Im} Y_{in} \tag{2-77b}$$

where

$$\delta = \frac{1 + \dfrac{G_4}{C}\left(\dfrac{G_2}{G_3}\dfrac{1}{GB_1} + \dfrac{1}{GB_2}\right) - \dfrac{\omega^2 C}{G_4}\left(\dfrac{G_3}{G_2}\dfrac{1}{GB_2} + \dfrac{1}{GB_1}\right)}{1 - \dfrac{2\omega^2 C}{G_4}\left(\dfrac{G_3}{G_2}\dfrac{1}{GB_2} + \dfrac{1}{GB_1}\right)} \tag{2-78a}$$

and

$$\gamma = \frac{\omega\left(\dfrac{G_2}{G_3}\dfrac{1}{GB_1} + \dfrac{1}{GB_2}\right)\left(1 - \dfrac{G_3}{G_2}\right)}{1 - \dfrac{2\omega^2 C}{G_4}\left(\dfrac{G_3}{G_2}\dfrac{1}{GB_2} + \dfrac{1}{GB_1}\right)} \tag{2-78b}$$

The quality factor for an FDNR is given by the relation

$$Q_D = \frac{\operatorname{Re} Y_{in}}{\operatorname{Im} Y_{in}} = \frac{\delta}{\gamma} \tag{2-79}$$

From Eq. (2-71b) we see that Q_D can be made arbitrarily large by setting

$$G_2 = G_3 \tag{2-80}$$

This high-Q condition is analogous to the high-inductor-Q condition given in Eq. (2-68). The fractional error in D_0 for a high-Q FDNR with matched operational amplifiers ($GB_1 = GB_2 = GB$) is

$$\frac{\Delta D}{D_0} \simeq \alpha - 1 \simeq \frac{2}{GB}\left(\frac{G_4}{C} + \frac{\omega^2 C}{G_4}\right) \tag{2-81}$$

Error $\Delta D/D_0$ can be minimized at some critical frequency ω_C by setting

$$G_4 = \omega_c C \tag{2-82}$$

This minimum error condition is analogous to the inductance minimum error condition given by Eq. (2-70). Substituting Eqs. (2-80) and (2-82) into Eq. (2-75), we find that the desired value for D_0 is obtained when

$$C = \omega_c D_0 \qquad (2\text{-}83)$$

Equations (2-80), (2-82), and (2-83) serve as the design equations for the non-ideal FDNR. The application of FDNR to active filter design is discussed in Chapter 5.

2.10 SUMMARY

In this chapter we have shown how the operational amplifier is used to realize simple functions used in active filters, such as summing amplifiers, integrators, inductor simulations, and FDNRs. Although in our pencil-and-paper analyses we typically, for simplicity, assume ideal op amp with infinite gain, the omission of their finite gain–bandwidth products results in some degree of error. Depending on the frequency of operation and the form of compensation used, this error may or may not be negligible. When the error is nonignorable, designs can frequently be nominally predistorted to minimize the error. Since op amp gain–bandwidth products (GBs) can vary by as much as $\pm 20\%$ from unit to unit or over a 0°C to 70°C temperature range, there is some treachery in applying this predistortion. Fortunately, for voice-frequency applications, the double-pole single-zero compensation of Eq. (2-31) has rendered active filters essentially independent of the op amp.

Circuits such as inductor simulations and FDNRs which rely on either accurate gain–bandwidth matching or resistor matching can achieve good performance with single-pole compensation. This is primarily due to the general availability of accurately matched dual and quad op amps as demonstrated in Fig. 2-18a.

The concept of a monolithic integrated-circuit active filter is now becoming a practical reality. Monolithic integrators were shown to be realizable in both bipolar and MOS-LSI technologies. Since the MOS technologies are dominant in the realization of digital circuits and subsystems, there has been considerable motivation to develop a compatible linear circuit technology. As noted in Section 2.6.2, switched capacitor networks take full advantage of the capabilities of the MOS technologies: precision capacitor ratios and nearly ideal switches. It is, however, no coincidence, that switched capacitor active filters followed an extensive development of good-quality, low-power NMOS and CMOS op amps. Although the op amps currently available are adequate for most applications, it is expected that we will see continued improvements in their size and performance as the technology matures.

In spite of the rapid development of monolithic active-filter technologies, the design of active-RC filters is by no means a dying art. It is expected that thick- and thin-film active-RC filters, which can be automatically tuned, will provide a most economical alternative for the wide variety of low-volume to moderate-volume applications. It is the high-volume applications, which can afford the startup costs of layout and mask making, which will benefit most from monolithic active filters.

REFERENCES

Books

B1. ROBERGE, J. K., *Operational Amplifier Theory and Practice*. New York: Wiley, 1975.

B2. WAIT, F., L. P. HUELSMAN, AND G. KORN, *Introduction to Operational Amplifiers: Theory and Applications*. New York: McGraw-Hill, 1975.

B3. GRAEME, J. G., G. E. TOBEY, AND L. P. HUELSMEN, *Operational Amplifiers*. New York: McGraw-Hill, 1971.

B4. HAMILTON, D., AND W. HOWARD, *Basic Integrated Circuits Engineering*. New York: McGraw-Hill, 1975, Chap. 10.

B5. KORN, G. A., AND T. M. KORN, *Electronic Analog and Hybrid Computers*, 2nd ed., New York: McGraw-Hill, 1972.

B6. GHAUSI, M., *Electronic Circuits*. New York: Van Nostrand Reinhold, 1971.

B7. KUO, B., *Automatic Control Systems*. Englewood Cliffs, N.J.: Prentice-Hall, 1962.

B8. TAUB, H., AND D. SCHILLING, *Digital Integrated Electronics*. New York: McGraw-Hill, 1972, Chap. 2.

B9. SEDRA, A., AND P. O. BRACKETT, *Filter Theory and Design: Active and Passive*. Portland, Oreg.: Matrix, 1978.

B10. MEYER, R.G., ed., *Integrated-Circuit Operational Amplifiers*. New York: IEEE Press, 1978.

Papers

P1. SOLOMON, J. E., "The Monolithic Op. Amp.: A Tutorial Survey," *IEEE J. Solid-State Circuits*, SC-9 (December 1974), 314–332.

P2. WIDLAR, J., "Design Techniques for Monolithic Operational Amplifiers," *IEEE J. Solid-State Circuits*, SC-4 (1969), 184–191.

P3. TAN, K. S., AND P. R. GRAY, "Fully Integrated Analog Filters Using Bipolar JFET Technology," *IEEE J. Solid-State Circuits*, SC-13, no. 6 (December 1978), 814–821.

P4. TSIVIDIS, Y. P., AND R. R. GRAY, "An Integrated NMOS Operational Amplifier with Internal Compensation," *IEEE J. Solid-State Circuits*, SC-11 (December 1976), 748–753.

P5. HODGES, D. A., P. R. GRAY, AND R. W. BRODERSEN, "Potential of MOS Technologies for Analog Integrated Circuits," *IEEE J. Solid-State Circuits*, SC-13 (June 1978), 285–294.

P6. BRAND, J. R., AND R. SCHAUMANN, "Active-R Filters: Review of Theory and Practice," *IEE J. Electron. Circuits Syst.*, 2, no. 4 (July 1978), 81–101.

P7. HOSTICKA, B. J., R. W. BRODERSEN, AND P. R. GRAY, "MOS Sampled Date Recursive Filters Using Switched Capacitor Integrators," *IEEE J. Solid-State Circuits*, SC-12 (December 1977), 600–608.

P8.	ANTONIOU, A., "Realization of Gyrators Using Operational Amplifiers and Their Use in *RC*-Active Network Synthesis," *Proc. IEE* (Lond.), November 1969, 1838–1850.

P9.	RIORDAN, H., "Simulated Inductors Using Differential Amplifiers," *Electron. Lett.*, 3 (February 1967), 50–51.

P10.	BRUTON, L. T., "Network Transfer Functions Using the Concept of Frequency-Dependent Negative Resistance," *IEEE Trans. Circuit Theory*, CT-16 (August 1969), 406–408.

PROBLEMS

2.1	For the inverting and noninverting op amp circuits shown in Fig. 2-4 and 2-6, derive the expressions for the voltage gain if the op amp is nonideal and characterized by $a(s) = GB/s$.

2.2	Derive Eq. (2-12).

2.3	Show an analog-computer simulation of a third-order Bessel filter to determine the step-input response of the filter.

2.4	Show an analog computation setup for the solution of a second-order state equations

$$\frac{d}{dt}\begin{bmatrix} x_1 \\ x_2 \end{bmatrix} = \begin{bmatrix} a_{11} & a_{12} \\ a_{21} & a_{22} \end{bmatrix}\begin{bmatrix} x_1 \\ x_2 \end{bmatrix} + \begin{bmatrix} b_1 \\ b_2 \end{bmatrix}u(t) \qquad \begin{bmatrix} x_1(0) \\ x_2(0) \end{bmatrix} = \begin{bmatrix} 0 \\ 0 \end{bmatrix}$$

2.5	Show an analog simulation for the following second-order elliptic filter:

$$\frac{V_o}{V_i}(s) = \frac{s^2 + a_1}{s^2 + b_1 s + b_0}$$

2.6	Using the practical circuit model of Fig. 2-14 for the op amp, show that the transfer function of the integrator of Fig. 2-10a is given by

$$T(s) = \frac{-\dfrac{1}{sR_1 C_f}\left[1 - \dfrac{sC_f R_0}{a(s)}\right]}{1 + \dfrac{1}{a(s)}\left[1 + \left(\dfrac{1}{sC_f} + R_0\right)\left(\dfrac{1}{R_1} + \dfrac{1}{R_i}\right)\right]}$$

Let $R_1 = 1\,k\Omega$, $C_1 = 0.1\,\mu F$ and use the Fairchild $\mu A741$ parameters:

$$R_i = 3\,M\Omega \qquad a(s) = \frac{3 \times 10^5}{\left(1 + \dfrac{s}{30}\right)\left(1 + \dfrac{s}{7 \times 10^6}\right)}$$
$$R_0 = 50\,\Omega$$

2.7	A compensated op amp has two negative real open-loop poles with magnitudes 100 Hz and 5×10^4 Hz and a negative real zero with magnitude 100 kHz. The dc gain of the op amp is 100 dB. If negative feedback is applied around the amplifier, sketch the root locus as *f* varies from 0 to 1. For what value of *f* is the poles at 45° with the negative real axis. Can this amplifier become unstable? At what value of *f* will the circuit become unstable if the open loop has another real pole with magnitude 10^6 Hz.

2.8 Determine the stability of the following characteristic equations:
(a) $s^4 + 2s^3 + 2s^2 + 3s + 10 = 0$
(b) $s^5 + 3s^4 + 2s^3 + 5s^2 + s + 1 = 0$

2.9 Sketch the root-locus for the following systems:

(a) $T(s) = \dfrac{K}{s(s + 2)^2}$

(b) $T(s) = \dfrac{K}{s(s + 2)^2(s + 10)}$

(c) $T(s) = \dfrac{K(s^2 + 1)}{(s + 1)(s + 1)(s + 2)}$

2.10 The following measured data were obtained for a system with a schematic shown in Fig. 2-21. Assume that $f(s) = f_0$ is a constant in the frequency range of interest.

Frequency	0.01	0.1	1	5	10	60
$\lvert a \rvert \angle \arg a(j\omega)$	1000	800	200	90	40	1
	$\angle -5°$	$\angle -20°$	$\angle -80°$	$\angle -120°$	$\angle -200°$	$\angle -260°$

Determine the value of f_0 for a gain margin of 10 dB. What is the corresponding phase margin? For what value of f_0 will the system become unstable?

2.11 For the system shown in Fig. P2-11, determine the maximum value of K as a function of n for which the system is stable.

Fig. P2-11

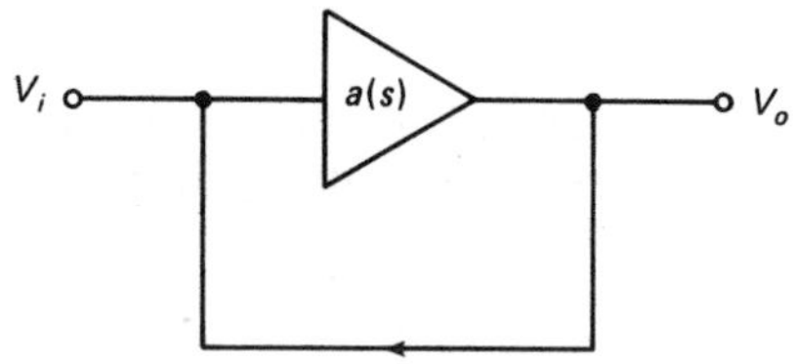

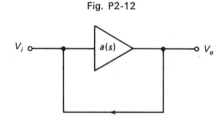

$$a(s) = -\dfrac{K}{(s + 1)^n}$$

2.12 For the system in Fig. P2-12, determine the value of K for stable operation if $a(s) = Ke^{-0.1s}/(s + 1)$.

Fig. P2-12

Operational Amplifiers

2.13 For the circuit in Fig. P2-13, show that the expression for the voltage transfer is given by

$$\frac{V_o}{V_i} = \frac{Y_1^b - Y_1^a}{(Y_1^b - Y_1^a) + (Y_2^b - Y_2^a)}$$

Fig. P2-13

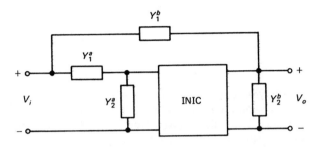

2.14 The circuit shown in Fig. P2-14 utilizes an op amp and two grounded RC two-port networks. If the op amp is assumed ideal, show that

$$\frac{V_o}{V_i} = -\frac{y_{12}^b}{y_{12}^a}$$

where y_{12} is the short-circuit admittance parameter.

Fig. P2-14

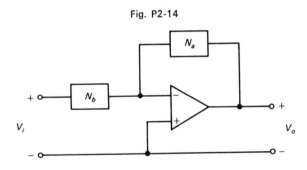

2.15 Show that the circuit in Fig. 2-31 realizes a gyrator; assume ideal op amps. In other words, derive Eq. (2-57b).

2.16 Derive the expression for the input impedance in Fig. 2-32 and verify the inductor simulation and FDNR realizations given in Figs. 2-33 and 2-34, respectively.

2.17 A single op amp (unity-gain) realization of a lossy grounded inductor is shown in Fig. P2-17. Determine the relations between L_{eq} and R_{eq} in terms of R's and C's assuming an ideal op amp.

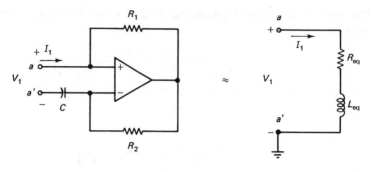

Fig. P2-17

2.18 An alternative gyrator realization circuit of Fig. 2-33 is shown in Fig. P2-18. Derive the expression for Z_{in} and show that this circuit performs best if the op amps are matched (i.e., $GB_1 = GB_2$). [Note that in Fig. 2-33, the corresponding condition was resistor matching given by Eq. (2-69).]

Fig. P2-18

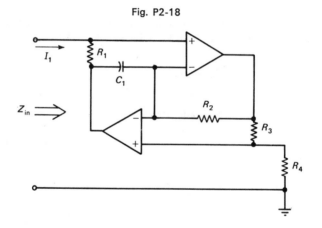

Operational Amplifiers

chapter three

....

Sensitivity

3.1 INTRODUCTION

In Chapter 1 we illustrated the means for obtaining filter transfer functions that meet given frequency or time-domain requirements. Given an appropriate circuit configuration, one then evaluates the circuit element values that yield the desired transfer function. This operation, referred to as *network synthesis,* implies two operations: (1) the selection of an appropriate active network configuration, and (2) the calculation of the element values. In Chapter 2 we described a number of simple operational amplifier building blocks for active filters. As shown in Chapters 4, 5, and 6, combinations of these active building blocks and passive components can be interconnected to provide a vast number of equivalent circuit configurations which are capable of realizing a given transfer function.

Given this difficult task of selecting a circuit configuration, it is valuable to identify criteria that can be used as a figure of merit for determining the goodness of a circuit configuration. One of the most important criteria for comparing equivalent circuit configurations and for establishing their practical utility in meeting desired requirements is sensitivity. In practice, real circuit components will deviate from their nominal design values (i.e., those computed in step 2 of the synthesis operation) due to manufacturing tolerances, environmental changes such as in temperature and humidity, and chemical changes which occur as the circuit ages. Furthemore, modeling inaccuracies such as the non ideal op amp characteristics discussed in Chapter 2 and such parasitics as capacitor losses can also cause a practical active and passive component to deviate from its ideal behavior. Component deviations of these types cause the circuit transfer function or response to drift away from the specified function. The cause-and-effect relationship between the circuit element variations and the resulting changes in the response or some other network function is referred to as *sensitivity*. Mathematically, we can relate changes in the response (Δr) to variations in the elements (Δx) in the following linear manner:

$$\frac{\Delta r}{r} = S \frac{\Delta x}{x} \tag{3-1}$$

we define S as the sensitivity of r to variations in x.

Ideally, we would like $\Delta r/r = 0$ or at least as small as possible. Small $\Delta r/r$ implies from Eq. (3-1) that either S, $\Delta x/x$, or both are small. In practice, precision elements (i.e., small $\Delta x/x$) imply an expensive network, with cost inversely proportional to $\Delta x/x$. Sensitivity S, which is a function of only the circuit configuration, and the element values, can often be reduced so that $\Delta r/r$ is acceptable, at no cost penalty. In fact, the smaller we make S, the cheaper the circuit becomes to manufacture; thus, the cost is reduced. A simple yet pointed example of minimzing sensitivity in an active network is the application of two-pole single-zero compensation, Eq. (2-30), in op amps. Comparing Eqs. (2-29) and (2-31), we see that this form of compensation reduces considerably the sensitivity of the integrator response $|H|$ to variations (from ∞) in the op amp gain a. Also in Chapter 2, we minimized the sensitivity of the inductance value for the GIC inductor simulation to GB by setting $G_L = \omega_c C$ in Eq. (2-71).

In the first case the integrator sensitivity was reduced by altering the circuit configuration: namely, the compensation. In the second case, the GIC inductor sensitivity was minimized by appropriately exercising our freedom in choosing some element values. As we shall demonstrate in Chapters 4, 5, and 6, it is these very operations that we will use to reduce sensitivities in more complex active filters.

In active filters, large S and/or large $\Delta x/x$ can alter the response of a filter beyond recognition and may even in some cases cause instability. By computing sensitivity, such problems can be identified before the network is constructed. To further appreciate the problem, consider the network shown in Fig. 1-15, repeated in Fig. 3-1 in the following example.

●

EXAMPLE 3-1

Consider the design of the circuit in Fig. 3-1 for a Butterworth filter response with $\omega_{3dB} = 2\pi(10^4)$ rad/sec. For convenience, let $R_1 = R_2 = R$, $C_1 = C_2 = C$. The voltage-gain function is given by

$$\frac{V_o}{V_i} = \frac{K/R^2 C^2}{s^2 + s\left(\dfrac{3-K}{RC}\right) + \dfrac{1}{R^2 C^2}} \tag{3-2}$$

For a second-order Butterworth filter design with specified bandwidth ω_{3dB}, the

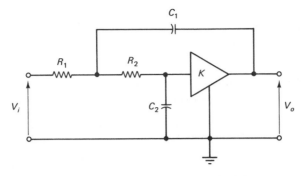

Fig. 3-1 Active-RC circuit for Example 3-1.

design equations are:

$$\frac{1}{R^2 C^2} = (\omega_{3dB})^2 \tag{3-3a}$$

$$\frac{3 - K}{RC} = \sqrt{2}\,\omega_{3dB} \tag{3-3b}$$

Now if we arbitrarily choose $R = 1$ kΩ, the values of C and K are

$$C = 0.0159\ \mu\text{F} \quad \text{and} \quad K = 1.586$$

The locus of the pole locations corresponding to Eq. (3-2) is given in Fig. 3-2.

Fig. 3-2 Locus of pole locations for Eq. (3-2).

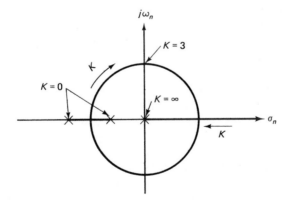

Now suppose that K varies by 20% (increased) due to any of a number of factors and that the other elements do not vary. The new pole locations, as a result of the increase in K, are shown in Table 3-1. Table 3-1 also shows the changes in the location of the pole pair if the R's changed by 10% or the C's changed by 10%. For a high-Q band-pass response, where the poles are located near the $j\omega$-axis, sometimes such changes can cause instability. For example, for a design

TABLE 3-1

Designed Nominal Value of the Pole × 10⁴	Poles due to +20% Increase in K	Poles If R's Increased by 10%	Poles If C's Increased by 10%
$-4.44654 \pm j4.44788$	$-3.44905 \pm j5.25922$	$-4.04231 \pm j4.04353$	$-4.04231 \pm j4.04353$

with $Q = 10$ and $\omega_0 = 1$ in Eq. (3-2), the value of K is 2.9. In this case even an increase of 5% in K will cause instability.

•

The purpose of this introduction has been to convey the concept of sensitivity and its importance to active filter designers. It should be clear at this point that a good understanding of sensitivity is essential to the successful realization of practical active filters. In the succeeding sections of this chapter, we turn our attention to the definitions and computation of various kinds of sensitivities. We begin by examining the most simplistic deterministic definitions of sensitivity and conclude with a comprehensive treatment of the more practical statistical multiparameter definitions of sensitivity.

3.2 SINGLE-PARAMETER SENSITIVITY DEFINITIONS

The most simplistic of sensitivity definitions are those which relate the change of some network function to the variation in a single circuit element or component. Since in a practical circuit many components are varying simultaneously, single-element sensitivities are of limited value. They do, however, provide an excellent vehicle for introducing the various forms of sensitivity. Often of practical importance are the largest and smallest single-parameter sensitivities, which define the most- and least-sensitive components. Clearly, we would like to use our most precise components where the sensitivities are large in magnitude and we can afford to use our cheaper components where the sensitivities are small. Finally, as we shall see in later sections, the more practical multiparameter definitions of sensitivity can be conveniently expressed in terms of single-parameter sensitivities. Let us now consider the definitions for the various types of single-parameter sensitivity.

3.2.1 Transfer Function Sensitivity

Let $H(s, x)$ be the gain function of a network and x any of the element values R_i, C_i, K_i, GB_i, and so on. The sensitivity of H with respect to x is given by

$$S_x^H = \frac{\Delta H/H}{\Delta x/x} = \frac{x}{H}\frac{\Delta H}{\Delta x} \qquad (3-4)$$

where

$$\Delta H = H(s, x + \Delta x) - H(s, x)$$

Note that S_x^H in Eq. (3-4) is of the same form as S in Eq. (3-1). Note that S_x^H gives the percentage change in H, due to a percentage change in x. Actually, if we expand ΔH in a Taylor series and, for small Δx, truncate the series after the linear term we obtain

$$\Delta H \simeq \frac{\partial H}{\partial x} \Delta x = x \frac{\partial H}{\partial x} \frac{\Delta x}{x} \qquad (3\text{-}5)$$

Normalizing Eq. (3-5) (i.e., dividing both sides by H) and utilizing Eq. (3-4), we obtain

$$S_x^H = \frac{\partial H/H}{\partial x/x} = \frac{\partial(\ln H)}{\partial(\ln x)} \simeq \frac{x}{H} \frac{\Delta H}{\Delta x} \qquad (3\text{-}6)$$

This is the classical definition of sensitivity, sometimes also referred to as the Bode sensitivity. The definition given in Eq. (3-6) is valid for small changes due to the linear approximation in Eq. (3-5). Theoretically, the change is differential, but in practice for changes up to 5%, the use of the definition above, Eq. (3-6), will give a meaningful result when interpreted properly. If $H = f(b)$ and $b = f(x)$, then from Eq. (3-6), we have

$$\frac{\partial H}{\partial x} = \frac{\partial H}{\partial b} \frac{\partial b}{\partial x} \qquad (3\text{-}7a)$$

and

$$\frac{x}{H} \frac{\partial H}{\partial x} = \left(\frac{\partial H}{\partial b} \frac{b}{H}\right)\left(\frac{\partial b}{\partial x} \frac{x}{b}\right) \qquad (3\text{-}7b)$$

or

$$S_x^H = S_b^H S_x^b \qquad (3\text{-}7c)$$

Some simple relations can be readily derived by using the definition given in Eq. (3-6). Let

$$H(s, x) = \frac{N(s, x)}{D(s, x)} \qquad (3\text{-}8)$$

From Eqs. (3-6) and (3-8), we obtain

$$S_x^H = \frac{\partial H}{\partial x}\left(\frac{x}{H}\right) = \frac{DN' - ND'}{D^2}\left(x\frac{D}{N}\right) = x\left(\frac{N'}{N} - \frac{D'}{D}\right) \qquad (3\text{-}9a)$$

where

$$N' = \frac{\partial N}{\partial x} \quad \text{and} \quad D' = \frac{\partial D}{\partial x}$$

In other words,

$$S_x^H = S_x^N - S_x^D \qquad (3\text{-}9b)$$

We may also express the network function as

$$H(j\omega, x) = |H(j\omega, x)|^{J\phi(\omega, x)} \qquad (3\text{-}10)$$

Then from Eqs. (3-6) and (3-10), we obtain

$$\ln H(j\omega, x) = \ln |H(j\omega, x)| + j\phi(\omega, x)$$

$$S_x^H = \frac{\partial \ln H(j\omega, x)}{\partial(\ln x)} = \frac{\partial \ln |H(j\omega, x)|}{\partial x/x} + j\frac{\partial \phi(\omega, x)}{\partial x/x} \qquad (3\text{-}11)$$

In other words,

$$\text{Magnitude sensitivity } S_x^{|H|} = \text{Re } S_x^H = \frac{\partial \ln |H|}{\partial \ln x} \qquad (3\text{-}12a)$$

$$\text{phase sensitivity } S_x^\phi = \text{Im } S_x^H = \frac{\partial \phi}{\partial \ln x} \qquad (3\text{-}12b)$$

The reader should note that in Eq. (3-12b) the phase sensitivity does not have $\partial\phi/\phi$. Consider now a general second-order biquadratic function

$$H(s) = \frac{b_2 s^2 + b_1 s + b_0}{a_2 s^2 + a_1 s + a_0} \qquad (3\text{-}13)$$

From Eq. (3-9b) we have $S_x^H = S_x^N - S_x^D$, and with (3-7c) and (3-13),

$$S_x^{|H|} = \sum_{i=0}^{2} S_{b_i}^{|N|} S_x^{b_i} - \sum_{i=0}^{2} S_{a_i}^{|D|} S_x^{a_i} \qquad (3\text{-}14)$$

It can be shown (Prob. 3.3) that

$$\sum_{i=0}^{2} S_{b_i}^{|N|} = \sum_{i=0}^{2} S_{a_i}^{|D|} = 1 \qquad (3\text{-}15)$$

●

EXAMPLE 3-2

Consider the derivation of S_K^H, $S_K^{|H|}$, and S_K^ϕ for the network given in Fig. 3-1 with H expressed in Eq. (3-2). First, let us derive N' and D' with respect to gain K, i.e.,

$$N'(j\omega) = \frac{1}{R^2 C^2} = \frac{1}{a_0} \qquad (3\text{-}16)$$

$$D'(j\omega) = \frac{-j\omega}{RC} = \frac{j\omega}{\sqrt{a_0}} \qquad (3\text{-}17)$$

where we define $a_0 = -1/R^2 C^2$. Using Eqs. (3-2), (3-9a), (3-16) and (3-17) we can write the following expression for S_K^H:

$$S_K^H = 1 + \frac{\sqrt{a_0}\, j\omega K}{(a_0 - \omega^2 + j\omega a_1)} = \left[1 + \frac{\omega^2 a_1 K(\sqrt{a_0})}{(a_0 - \omega^2)^2 + \omega^2 a_1^2}\right]$$
$$+ j\left[\frac{\omega K\sqrt{a_0}\,(a_0 - \omega^2)}{(a_0 - \omega^2)^2 + \omega^2 a_1^2}\right] \qquad (3\text{-}18a)$$

$$= \text{Re}\,(S_K^H) + j\,\text{Im}\,(S_K^H) \qquad (3\text{-}18b)$$

Then using Eqs. (3-12), we can write

$$S_K^{|H|} = 1 + \frac{\omega^2 a_1 K(\sqrt{a_0})}{(a_0 - \omega^2)^2 + \omega^2 a_1^2} \qquad (3\text{-}19a)$$

$$S_K^\phi = \frac{\omega K \sqrt{a_0}\,(a_0 - \omega^2)}{(a_0 - \omega^2)^2 + \omega^2 a_1^2} \qquad (3\text{-}19b)$$

Note that $S_K^{|H|} \geq 1$ and for all ω and the equality sign holds at $\omega = 0$. Also, $S_K^\phi \geq 0$ for $\omega < \sqrt{a_0}$ and $S_K^\phi \leq 0$ for $\omega > \sqrt{a_0}$, and it is zero at $\omega = 0$ and $\omega = \sqrt{a_0}$.

•

3.2.2 Root Sensitivity

In some cases we might be interested in the sensitivity of the roots of a polynomial (e.g., the sensitivity of the natural frequencies). For such calculations, we define the sensitivity of a root p_i as

$$S_x^{p_i} = \frac{dp_i}{dx/x} \qquad (3\text{-}20)$$

$S_x^{p_i}$ in general is a complex number and the parameter x is real. If $p_i = \sigma_i + j\omega_i$, then we can write

$$S_x^{p_i} = S_x^{\sigma_i} + jS_x^{\omega_i} \qquad (3\text{-}21)$$

Relations between pole variation and the transfer function variation are given later in this section.

•

EXAMPLE 3-3

Let us calculate the root sensitivity for the poles of the network shown in Fig. 3-1. The transfer function for this network with $R_1 = R_2 = R$ and $C_1 = C_2 = C$ was given in Eq. (3-2). Let us assume that the desired pole locations are complex, which implies for stability that $1 < K < 3$. For K within this range, the poles of Eq. (3-2) can be written in the form

$$p_i = -\frac{3 - K}{2RC} + j\frac{\sqrt{4 - (3 - K)^2}}{2RC} = \sigma_i + j\omega_i \qquad (3\text{-}22)$$

Let us evaluate the root sensitivity, Eq. (3-21), of complex root p_i in Eq. (3-22) with respect to gain K. Differentiating Eq. (3-22) with respect to K, and multiplying the result by K, yields

$$S_K^{p_i} = K\frac{\partial p_i}{\partial K} = \frac{K}{2RC}\left[1 + j\frac{3 - K}{\sqrt{4 - (3 - K)^2}}\right] \qquad (3\text{-}23)$$

For $R = 1\,\text{k}\Omega$, $C = 0.0159\,\mu\text{F}$, and $K = 1.586$, as in Example 3-1, we obtain the following numerical value for $S_K^{p_i}$:

$$S_K^{p_i} = 4.9874 \times 10^4 + j4.9859 \times 10^4 \tag{3.24}$$

In other words, if K changes by 20% (i.e., $\Delta K/K = 0.2$), the corresponding pole shift is approximately $\Delta p_i = 9.975 \times 10^3 + j9.972 \times 10^3$. This pole shift is seen to be a good approximation to the exact pole shift $9.97 \times 10^3 + j8.81 \times 10^3$, computed from the entries in Table 3-1, even though the percentage change is not small. For a smaller percentage change, the approximation will, of course, be better.

●

3.2.3 ω_0 and Q Sensitivities

In band-pass filters, two of the important parameters are the center frequency ω_0 and the quality factor Q, where $Q = \omega_0/\Delta\omega_{3\text{dB}}$. We are therefore interested in the sensitivities of these two parameters as they relate to the sensitivity of the magnitude function.

A gain-scaled second-order band-pass filter can be written as

$$H(s) = \frac{s}{s^2 + (\omega_0/Q)s + \omega_0^2} \tag{3-25}$$

The magnitude function, corresponding to $H(s)$ in Eq. (3-25), is

$$|H(j\omega)|^2 = \frac{\omega^2}{(\omega_0^2 - \omega^2)^2 + [(\omega_0/Q)\omega]^2} \tag{3-26}$$

From Eqs. (3-26) and (3-6), we obtain the following results (Prob. 3.4):
At $\omega = \omega_0$:

$$S_x^{|H(j\omega_0)|} = S_x^Q - S_x^{\omega_0} \tag{3-27}$$

At the 3-dB cutoff frequencies:

$$S_x^{|H(j\omega_{3\text{dB}})|} \simeq -S_x^{\omega_0}(Q + \tfrac{1}{2}) + \tfrac{1}{2}S_x^Q \tag{3-28}$$

In band-pass filters one is usually interested in high-Qs. For high-Q filters, Eq. (3-28) may be approximated as

$$S_x^{|H(j\omega_{3\text{dB}})|} \simeq -QS_x^{\omega_0} + \tfrac{1}{2}S_x^Q \tag{3-29}$$

From Eq. (3-29), we note that the center frequency sensitivity is multiplied by Q, which means that in a high-Q filter, the ω_0 sensitivity is much more important than Q sensitivity at the edges of the pass band.

An alternative way to obtain similar results for high Q is to consider a Taylor series expansion of the transfer function with respect to ω_0 and Q. Assuming that the percentage changes in ω_0, Q, and ω_{3dB} are small (i.e., $\Delta\omega_0/\omega_0$, $\Delta Q/Q$, and $Q\Delta\omega_0/\omega_0$ are small enough so that the terms beyond the linear term can be ignored),[1] we have

$$\frac{\Delta H(s)}{H(s)} \simeq S_{\omega_0}^H \frac{\Delta\omega_0}{\omega_0} + S_Q^H \frac{\Delta Q}{Q} \tag{3-30}$$

where

$$H(s) = \frac{H_0 s}{s^2 + (\omega_0/Q)s + \omega_0^2} \tag{3-31}$$

and we assume $Q \gg 1$. From Eqs. (3-27) and (3-31), we obtain

$$S_{\omega_0}^H(j\omega_0) = -1 + j2Q, \qquad S_Q^H(j\omega_0) = 1 \tag{3-32}$$

Hence, for a high Q:

$$\left|\frac{\Delta H}{H}\right|_{s=j\omega_0} \simeq \sqrt{\left(\frac{\Delta Q}{Q}\right)^2 + \left(2Q\frac{\Delta\omega_0}{\omega_0}\right)^2} \tag{3-33}$$

Note that in Eq. (3-33) the fractional change in ω_0 is weighted by Q and thus, in a high-Q circuit, the frequency response of a second-order network is $2Q$ times more sensitive to variations in center frequency that it is to variations in Q. Thus, in minimizing sensitivity, ω_0 sensitivity is more important.

It is interesting to note the relationship between ω_0, Q, and the pole variations. The pole variation dp_i/p_i, for high Q, is related to the transfer function sensitivity by

$$\left|\frac{\Delta H(s)}{H(s)}\right|_{s=j\omega_0} \simeq 2Q\left|\frac{\Delta p_i}{p_i}\right| \tag{3-34}$$

Thus, the variation of the transfer function is directly related to variations of the poles, as one would expect.

●

EXAMPLE 3-4

Let us derive the ω_0 and Q sensitivities for the network in Fig. 3-1. Strictly speaking, the circuit in Fig. 3-1 is a low-pass circuit. This circuit was designed for a low-pass Butterworth response in Example 3-1. However, for high values of Q, a frequency-peaked response can exhibit band-pass response characteristics within the pass band. The complete transfer function for this circuit, including the finite gain of the operational amplifier, can be determined as

[1]Note that in a practical situation it is the change in center frequency as a fraction of the bandwidth that is important [i.e., $(\Delta\omega_0/\omega_{3dB}) = Q(\Delta\omega_0/\omega_0)$]. For example, if the center frequency changes by 1% and $Q = 50$, the center frequency will change over 50% of the bandwidth. This consideration will limit the maximum allowable Q in a high-Q active filter.

$$H(s) = \frac{K\left(\frac{1}{1 + K/a_0}\right)\Big/ R_1 R_2 C_1 C_2}{s^2 + s\left[\frac{1}{R_1 C_1} + \frac{1}{R_2 C_1} + \frac{1}{R_2 C_2} - \frac{K}{R_2 C_2(1 + K/a_0)}\right] + \frac{1}{R_1 R_2 C_1 C_2}}$$

(3-35)

$$= \frac{G}{s^2 + (\omega_0/Q)s + \omega_0^2}$$

Note that when $a_0 \rightarrow \infty$ and $R_1 = R_2 = R$ and $C_1 = C_2 = C$, $H(s)$ reduces to Eq. (3-2). Also, gain K is realized using the noninverting amplifier configuration given in Fig. 2-6 and according to Eq. (2-11), $K = 1 + R_a/R_b$. From $H(s)$ we can write the following expressions for ω_0 and Q:

$$\omega_0 = \sqrt{\frac{1}{R_1 R_2 C_1 C_2}} \tag{3-36a}$$

$$Q = \sqrt{\frac{1}{R_1 R_2 C_1 C_2}}\left[\frac{1}{R_1 C_1} + \frac{1}{R_2 C_1} + \frac{1}{R_2 C_2} - \frac{K}{R_2 C_2(1 + K/a_0)}\right]^{-1} \tag{3-36b}$$

From the definition of $S_x^{\omega_0}$ we can write

$$S_{R_1}^{\omega_0} = S_{R_2}^{\omega_0} = S_{C_1}^{\omega_0} = S_{C_2}^{\omega_0} = -\tfrac{1}{2} \tag{3-37a}$$

$$S_K^{\omega_0} = S_{a_0}^{\omega_0} = 0 \tag{3-37b}$$

●

From the definition of S_x^Q, after some algebraic manipulation in Example 3-4, we may write

$$S_{R_1}^Q = -\frac{1}{2} + \frac{1}{R_1 C_1}\frac{Q}{\omega_0} \tag{3-37c}$$

$$S_{C_1}^Q = -\frac{1}{2} + \left(\frac{1}{R_1} + \frac{1}{R_2}\right)\frac{Q}{\omega_0 C_1} \tag{3-37d}$$

$$S_{R_2}^Q = -\frac{1}{2} + \left[\frac{1}{R_2 C_1} + \frac{1}{R_2 C_2}\left(1 - \frac{K}{1 + K/a_0}\right)\right]\frac{Q}{\omega_0}$$

$$\simeq -\frac{1}{2} + \left[\frac{1}{C_1} + \frac{1}{C_2}(1 - K)\right]\frac{Q}{\omega_0 R_2} \tag{3-37e}$$

$$S_{C_2}^Q = -\frac{1}{2} + \frac{1}{R_2 C_2}\left(1 - \frac{K}{1 + K/a_0}\right)\frac{Q}{\omega_0} \simeq -\frac{1}{2} + \frac{Q}{R_2 C_2 \omega_0}(1 - K) \tag{3-37f}$$

$$S_K^Q = +\frac{K}{R_2 C_2}\left(\frac{1}{1 + K/a_0}\right)^2\frac{Q}{\omega_0} \simeq +\frac{K}{R_2 C_2}\frac{Q}{\omega_0} \tag{3-37g}$$

$$S_{a_0}^Q = +\frac{1}{R_2 C_2}\left[\frac{K^2/a_0}{(1 + K/a_0)^2}\right]\frac{Q}{\omega_0} \simeq +\frac{K^2}{R_2 C_2}\frac{Q}{\omega_0 a_0} \tag{3-37h}$$

From Eqs. (3-37), the following are noted:

1. As $a_0 \rightarrow \infty$, $S_{a_0}^Q \rightarrow 0$. Also, $S_{a_0}^Q$ increases as K increases.
2. $S_x^{\omega_0}$ are at their theoretical minima.

3. S_x^Q are functions of R_1, R_2, C_1, C_2, and K and vary dramatically with the values of these parameters.

4. Note that $S_{R_1}^{\omega_0} + S_{R_2}^{\omega_0} = S_{C_1}^{\omega_0} + S_{C_2}^{\omega_0} = -1$ and $S_{R_1}^Q + S_{R_2}^Q = S_{C_1}^Q + S_{C_2}^Q = 0$. These results are not coincidences, but are directly related to the dimensions of ω_0 and Q in R and C. That is ω_0, which is of the form $1/RC$, is of dimension -1 in R and C and Q, which is a dimensionless quantity, is of dimension 0 in R and C. Hence, in general

$$\sum_{i=1}^{R} S_{R_i}^{\omega_0} = \sum_{j=1}^{C} S_{C_j}^{\omega_0} = -1 \quad \text{and} \quad \sum_{i=1}^{R} S_{R_i}^Q = \sum_{j=1}^{C} S_{C_j}^Q = 0.$$

In practice, it is usually desirable to minimize the total capacitance. One can easily verify that the minimum $C_{\text{tot}} = C_1 + C_2$ is obtained, in Example 3-4, when $C_1 = C_2 = C$. Setting $C_1 = C_2 = C$ in the expressions for ω_0 and Q result in the following formulas ($a_0 \rightarrow \infty$):

$$\omega_0 = \frac{1}{C}\sqrt{\frac{1}{R_1 R_2}} = \frac{1}{R_1 C}\sqrt{\frac{R_1}{R_2}} = \frac{\alpha}{R_1 C} \tag{3-39a}$$

$$Q = \omega_0 \left(\frac{1}{R_1 C} + \frac{2}{R_2 C} - \frac{K}{R_2 C}\right)^{-1} = \frac{\alpha}{1 + \alpha^2(2 - K)} \tag{3-39b}$$

where

$$\alpha = \sqrt{\frac{R_1}{R_2}}$$

or more appropriately for design, Eqs. (3-39) maybe rewritten as

$$R_1 C = \frac{\alpha}{\omega_0} \tag{3-40a}$$

$$K = \frac{1}{\alpha^2}\left(1 + 2\alpha^2 - \frac{\alpha}{Q}\right) \tag{3-40b}$$

where $K > 0$ implies that $2\alpha^2 - \alpha/Q > -1$. The sensitivities S_x^Q may also be written in terms of α as follows:

$$S_{R_1}^Q = -\frac{1}{2} + \frac{Q}{\alpha} \tag{3-41a}$$

$$S_{C_1}^Q = -\frac{1}{2} + Q\left(\frac{\alpha^2 + 1}{\alpha}\right) \tag{3-41b}$$

$$S_{R_2}^Q \simeq \frac{1}{2} - \frac{Q}{\alpha} \tag{3-41c}$$

$$S_{C_2}^Q \simeq +\frac{1}{2} - Q\left(\frac{\alpha^2 + 1}{\alpha}\right) \tag{3-41d}$$

$$S_K^Q \simeq -1 + Q\left(\frac{1 + 2\alpha^2}{\alpha}\right) \tag{3-41e}$$

$$S_{a_0}^Q \simeq \frac{Q}{\alpha^3 a_0}\left(1 + 2\alpha^2 - \frac{\alpha}{Q}\right)^2 \tag{3-41f}$$

Note that any S^Q_x can be minimized by appropriately choosing α where, for realizability, α must lie in the region $2\alpha^2 - \alpha/Q > -1$. For example, $\alpha = 2Q$ implies that $S^Q_{R_1} = S^Q_{R_s} = 0$. Another, perhaps more useful quantity to minimize would be the $\sum |S^Q_{x_i}|^2$. Of course, ideally, we would like a measure (S) of sensitivity that relates the multiple parameter variations to expected variations in the transfer function, magnitude response, or phase response. The derivation of such sensitivity measures will be treated in later sections of this chapter.

3.2.4 Gain Sensitivity Product [P13]

Consider the ω_0 and Q variations due to changes in the closed-loop amplifier gain K:

$$\frac{\Delta\omega_0}{\omega_0} \simeq S^{\omega_0}_K \frac{\Delta K}{K} \tag{3-42a}$$

$$\frac{\Delta Q}{Q} \simeq S^Q_K \frac{\Delta K}{K} \tag{3-42b}$$

The closed-loop gain for an inverting amplifier can be written from Eq. (2-7), namely,

$$K = K_0\left(\frac{1}{1 + 1/a_0 - K_0/a_0}\right) \tag{3-43}$$

where

$$K_0 = -\frac{R_f}{R_s}$$

We may now express the normalized variation in the closed-loop gain K in terms of the normalized variation in the open-loop gain a_0:

$$\frac{\Delta K}{K} \simeq S^K_{a_0} \frac{\Delta a_0}{a_0} \tag{3-44}$$

where from Eq. (3-43),

$$S^K_{a_0} = \left(\frac{1 - K_0}{K_0}\right)\bigg/\left(\frac{a_0}{K_0} + \frac{1 - K_0}{K_0}\right) \simeq \frac{-K_0}{a_0} \quad \text{for } a_0 \gg K_0 \gg 1 \tag{3-45}$$

Then substituting Eq. (3-45) into Eq. (3-44) and substituting the result into Eq. (3-42) yields

$$\frac{\Delta\omega_0}{\omega_0} \simeq S^{\omega_0}_K S^K_{a_0} \frac{\Delta a_0}{a_0} \simeq -K_0 S^{\omega_0}_K \frac{\Delta a_0}{a_0^2} \tag{3-46a}$$

$$\frac{\Delta Q}{Q} \simeq S^Q_K S^K_{a_0} \frac{\Delta a_0}{a_0} \simeq -K_0 S^Q_K \frac{\Delta a_0}{a_0^2} \tag{3-46b}$$

We may now define what is referred to as the gain sensitivity product (Γ_K) for the center frequency and Q as

$$\Gamma^{\omega_0}_K = -K_0 S^{\omega_0}_K \tag{3-47a}$$

$$\Gamma^Q_K = -K_0 S^Q_K \tag{3-47b}$$

Note that $\Delta a_0 / a_0^2$ depends only upon the op amp used, and Γ_K depends only on the circuitry surrounding the op amp. The gain sensitivity product has been used as a measure for evaluating sensitivity to variations of the open-loop gain either from its ideal infinite value or from some finite nominal value. The open-loop gain for typical op amps can vary by more than 50%; however, as is evident from Eq. (3-44), the variation in closed-loop gain K_0 can be considerably less. The important fact to note here is that $S_K^{\omega_0}$ and S_K^Q values alone are not in general adequate for estimating active filter sensitivity to op amp variations. In fact, their use can indeed be misleading and result in erroneous conclusions. However, one cannot go wrong if Eqs. (3-46) are used in their entirety. In most applications, active-RC filters are operated at sufficiently low frequencies to render the effect of the op amp variations in a closed loop smaller than that due to the passive component variations. We shall now conclude our discussion of gain sensitivity product. In Chapter 4 we shall again raise the issue of op amp variations when we discuss active-RC, active-R and second order sections.

3.3 MULTIPARAMETER SENSITIVITY—
DETERMINISTIC CASE

In any network design the network function is a function of several parameters. For example, in the circuit shown in Fig. 3-1, $H(s)$ depends on R_1, R_2, C_1, C_2, K_0, and a_0 and the variation of each component will have an effect on the magnitude and phase of $H(j\omega)$. In this section we consider the deterministic case (i.e., a case where the percentage of variations of the elements can be specified or known).

Let $H(s, \mathbf{x})$ be the transfer function that is a function of k parameters, $x_1, x_2, \ldots, x_k$ (i.e., the $\mathbf{x}$ vector is given by $[x_1, x_2, \ldots x_k]^t$). The linear term of the Taylor series expansion of $H(s, \mathbf{x})$ yields

$$dH = \sum_{i=1}^{k} \frac{\partial H}{\partial x_i} dx_i = H \sum_{i=1}^{k} \left(\frac{x_i}{H} \frac{\partial H}{\partial x_i} \right) \frac{dx_i}{x_i} \qquad (3\text{-}48)$$

or

$$\frac{dH}{H} = \sum_{i=1}^{k} S_{x_i}^H \frac{dx_i}{x_i} \qquad (3\text{-}49)$$

The incremental change corresponding to Eq. (3-49) may be written as

$$\frac{\Delta H}{H} \simeq \sum_{i=1}^{k} S_{x_i}^H \frac{\Delta x_i}{x_i} \qquad (3\text{-}50)$$

where $S_{x_i}^H$ is the sensitivity due to the individual parameters x_i. Equation (3-50) may be conveniently written as

$$\frac{\Delta H}{H} \simeq \mathbf{d}_T^t \, \Delta \hat{\mathbf{x}} \qquad (3\text{-}51)$$

where

$$\mathbf{d}_T^t = [S_{x_1}^H \quad S_{x_2}^H \quad \ldots \quad S_{x_k}^H] \tag{3-52}$$

$$\Delta\hat{\mathbf{x}} = \left[\frac{\Delta x_1}{x_1} \quad \frac{\Delta x_2}{x_2} \quad \ldots \quad \frac{\Delta x_k}{x_k}\right]^t \tag{3-53}$$

The boldface letters indicate vectors and the superscript t denotes transpose.

Equation (3-52) defines a multiparameter sensitivity vector $\mathbf{d}_T$ of order $k \times 1$. For sensitivity calculation comparisons and optimization, it is preferable to have a scalar quantity. In the sensitivity literature one finds a definition of multiparameter sensitivity given by Schoeffler [P2]:

$$S = \sum_{i=1}^{k} |S_{x_i}^H|^2 \tag{3-54a}$$

The square root of Eq. (3-54a) is also suggested by some authors. Note that Eq. (3-54a) is closely related to Eq. (3-52), i.e.,

$$S = (\mathbf{d}_T^*)(\mathbf{d}_T) \tag{3-54b}$$

The sensitivity definitions above are somewhat simplistic and do not take into account the random element variations (e.g., tolerance and correlations between the element values, etc.). In integrated circuits there is a definite correlation in parameter variations; for example, a change in temperature will change all resistors by some percentage, all capacitances by another percentage. Similarly, in production, mask errors and deviations in material parameters tend to cause errors in like elements which are statistically interdependent. Thus, ratios of like elements tend to be more tightly controlled than do absolute element values. These phenomena should be reflected in any meaningful formulation.

3.4 MULTIPARAMETER SENSITIVITY—
STATISTICAL CASE

Since element tolerances cannot be treated in a deterministic case, we must resort to a statistical or stochastic sensitivity measure to achieve a realistic and accurate representation. A statistical multiparameter sensitivity measure of the transfer function proposed [P4] is

$$E \int_{\omega_1}^{\omega_2} \left|\frac{\Delta H}{H}\right|^2 d\omega = \int_{\omega_1}^{\omega_2} S_T \, d\omega \tag{3-55}$$

where E denotes the expected value operation[2] and ω_1 and ω_2 are the designer-specified frequency band (e.g., the cutoff frequencies in a band-pass response) and S_T will be defined shortly in Eq. (3-60b). Note that the other sensitivity measures could be considered special cases of Eq. (3-55), as we show later.

[2]For the definition and interpretation of various statistical terms such as expected value, variance, and correlation coefficient, see the references listed in [B3] and [B4].

The advantages of the measure in Eq. (3-55) are the following:

1. Multiparameter element variations are included.
2. Correlated element variations are readily considered.
3. Sensitivity may be averaged over a frequency range (integral) and/or evaluated at each frequency (integrand).
4. The measure can be readily optimized (i.e., sensitivity minimization can be incorporated in the design).
5. Gives good agreement with Monte Carlo analysis (see Section 3.10).

Assuming that element variations are small enough[3] to utilize a linear approximation to the Taylor series expansions, we have from Eq. (3-51)

$$\frac{\Delta H}{H} \simeq \mathbf{d}_T^t \, \Delta \hat{\mathbf{x}} \tag{3-56}$$

where $\Delta \hat{x}_i = \Delta x_i / x_i$ denotes percent variations (i.e., parameter variations normalized with respect to nominal design values x_i). From Eqs. (3-55) and (3-56) the measure is

$$E \int_{\omega_1}^{\omega_2} \left| \frac{\Delta H}{H} \right|^2 d\omega \simeq E \int_{\omega_1}^{\omega_2} (\mathbf{d}_T^t \, \Delta \hat{\mathbf{x}})^* (\Delta \hat{\mathbf{x}}^t \mathbf{d}_T) \, d\omega \tag{3-57a}$$

$$= \int_{\omega_1}^{\omega_2} (\mathbf{d}_T^{t*} \mathbf{P} \mathbf{d}_T) \, d\omega \tag{3-57b}$$

where $\mathbf{P} = E[\Delta \hat{\mathbf{x}} \Delta \hat{\mathbf{x}}^t]$ is the covariance matrix of order $k \times k$ and * denotes a complex conjugate. Note also that a typical component of $\mathbf{d}_T$ from Eq. (3-52) is

$$(d_T)_{x_i} = \frac{x_i}{H} \frac{\partial H}{\partial x_i} = S_{x_i}^H \tag{3-58}$$

and if the covariance matrix is assumed diagonal (i.e., zero cross-correlation term), we have

$$\mathbf{d}_T^{t*} \mathbf{P} \mathbf{d}_T = \sum_{t=1}^{k} |S_{x_i}^H|^2 \sigma_{x_i}^2 \tag{3-59}$$

where

$$\sigma_{x_i}^2 = E\left[\left(\frac{\Delta x_i}{x_i} \right)^2 \right] \qquad i = 1, 2, \ldots, k$$

is the variance. One further notes that with $\sigma_{x_i}^2 = 1$, Eqs. (3-54a) and (3-59) are equivalent [i.e., Schoeffler's definition Eq. (3-54) is a special case of Eq. (3-59)].

Let us now define the integral and integrand in Eq. (3-57b) as the integral

---

[3]The reader is cautioned that even for small element variations, the series may not converge for a high Q. In that case, refer to Section 3.7 and use large change sensitivity.

sensitivity measure M_T and the frequency-dependent integrand sensitivity measure S_T:

$$M_T = \int_{\omega_1}^{\omega_2} (\mathbf{d}_T^{t*}\mathbf{Pd}_T)\, d\omega \simeq E \int_{\omega_1}^{\omega_2} \left| \frac{\Delta H}{H} \right|^2 d\omega \qquad (3\text{-}60a)$$

$$S_T = \mathbf{d}_T^{t*}\mathbf{Pd}_T \simeq E \left| \frac{\Delta H}{H} \right|^2 d\omega \qquad (3\text{-}60b)$$

We shall now consider in some detail the computation of Eq. (3-55) and its relation to the magnitude and phase sensitivities. Since $\mathbf{d}_T$ is in general complex quantity, we may write

$$\mathbf{d}_T = \mathbf{d}_\alpha + j\mathbf{d}_\beta \qquad (3\text{-}61)$$

where

$$\mathbf{d}_\alpha = \left[\frac{\partial \ln \alpha}{\partial \ln x_1} \quad \frac{\partial \ln \alpha}{\partial \ln x_2} \quad \cdots \quad \frac{\partial \ln \alpha}{\partial \ln x_k} \right]^t \qquad (3\text{-}62)$$

$$\mathbf{d}_\beta = \left[\frac{\partial \beta}{\partial \ln x_1} \quad \frac{\partial \beta}{\partial \ln x_2} \quad \frac{\partial \beta}{\partial \ln x_k} \right]^t \qquad (3\text{-}63)$$

Note that we have defined for convenience $\alpha(\omega) = |H(j\omega)|$, and it should not be confused with the attenuation α given in Chapter 1, and $\beta = \arg H(j\omega)$. From Eqs. (3-50) and (3-61), we obtain

$$\frac{\Delta H}{H} \simeq \mathbf{d}_\alpha^t \, \Delta\hat{\mathbf{x}} + j\mathbf{d}_\beta^t \, \Delta\hat{\mathbf{x}} \qquad (3\text{-}64)$$

$$\left| \frac{\Delta H}{H} \right|^2 \simeq \mathbf{d}_\alpha^t \, \Delta\hat{\mathbf{x}} \, \Delta\hat{\mathbf{x}}^t \, \mathbf{d}_\alpha + \mathbf{d}_\beta^t \, \Delta\hat{\mathbf{x}} \, \Delta\hat{\mathbf{x}}^t \, \mathbf{d}_\beta \qquad (3\text{-}65)$$

From Eq. (3-65) we get

$$E\left[\left| \frac{\Delta H}{H} \right|^2 \right] \simeq \mathbf{d}_T^t E[\Delta\hat{\mathbf{x}}\Delta\hat{\mathbf{x}}^t]\mathbf{d}_T = \mathbf{d}_\alpha^t \mathbf{Pd}_\alpha + \mathbf{d}_\beta^t \mathbf{Pd}_\beta \qquad (3\text{-}66)$$

where $\mathbf{P}$ is the covariance matrix of order $k \times k$:

$$\mathbf{P} = E[\Delta\hat{\mathbf{x}}\Delta\hat{\mathbf{x}}^t] \qquad k \times k \qquad (3\text{-}67)$$

Substitution of Eq. (3-66) into (3-60) yields

$$M_T = \int_{\omega_1}^{\omega_2} [\mathbf{d}_\alpha^t \mathbf{Pd}_\alpha + \mathbf{d}_\beta^t \mathbf{Pd}_\beta]\, d\omega = M_\alpha + M_\beta = \int_{\omega_1}^{\omega_2} S_T\, d\omega \qquad (3\text{-}68)$$

where M_α and M_β are the sensitivity measures of the magnitude and phase functions, respectively:

$$M_\alpha = \int_{\omega_1}^{\omega_2} \mathbf{d}_\alpha^t \mathbf{Pd}_\alpha\, d\omega = \int_{\omega_1}^{\omega_2} S_\alpha\, d\omega \simeq E \int_{\omega_1}^{\omega_2} \left| \frac{\Delta\alpha}{\alpha} \right|^2 d\omega \qquad (3\text{-}69a)$$

$$M_\beta = \int_{\omega_1}^{\omega_2} \mathbf{d}_\beta^t \mathbf{Pd}_\beta\, d\omega = \int_{\omega_1}^{\omega_2} S_\beta\, d\omega \simeq E \int_{\omega_1}^{\omega_2} |\Delta\beta|^2\, d\omega \qquad (3\text{-}69b)$$

It is important to note the physical meaning of sensitivity measures M_α, M_β, and M_T. These quantities, in general, describe the weighted average (since in general the component variations are not equal) over the frequency band $\omega_1 \leq \omega \leq \omega_2$, of the second statistical moment [B4] of the respective real functions, $\Delta\alpha/\alpha$, $\Delta\beta$, and the complex function $\Delta H/H$. For zero mean element variations ($\Delta\hat{x}$), they approximately equal the weighted average of the variance for these functions.

3.5 COMPUTATION OF STATISTICAL SENSITIVITY MEASURES [P6]

The stochastic sensitivity measure of the gain function given in Eq. (3-69a) can be written explicitly as

$$M_\alpha = \int_{\omega_1}^{\omega_2} \sum_{i=1}^{k} \sum_{j=1}^{k} (d_\alpha)_{x_i}(d_\alpha)_{x_j} p_{ij}\, d\omega \qquad (3\text{-}70)$$

Note that in Eq. (3-70) p_{ij} is the entry in the ith row and jth column of the covariance matrix $\mathbf{P}$ in (3-67). Also, $(d_\alpha)_{x_i}$ denotes the ith element of vector $\mathbf{d}_\alpha$; see Eq. (3-58).

If there is no correlation between the random variables, M_α takes the following simple form

$$M_\alpha = \int_{\omega_1}^{\omega_2} \sum_{i=1}^{k} (d_\alpha)_{x_i}^2 \sigma_{x_i}^2\, d\omega \qquad (3\text{-}71)$$

where $\sigma_{x_i}^2$ is the variance of the random variable $\Delta x_i/x_i$ (note that it is not the variance of elements); hence, M_α can be computed, once $\mathbf{d}_\alpha$ and p_{ij} are known. Similarly,

$$M_\beta = \int_{\omega_1}^{\omega_2} \sum_{i=1}^{k} \sum_{j=1}^{k} (d_\beta)_{x_i}(d_\beta)_{x_j} p_{ij}\, d\omega \qquad (3\text{-}72)$$

and, for no correlation,

$$M_\beta = \int_{\omega_1}^{\omega_2} \sum_{i=1}^{k} (d_\beta)_{x_i}^2 \sigma_{x_i}^2\, d\omega \qquad (3\text{-}73)$$

Note that for the no-correlation case, we have

$$M_T = M_\alpha + M_\beta = \int_{\omega_1}^{\omega_2} \sum_{i=1}^{k} [(d_\alpha)_{x_i}^2 + (d_\beta)_{x_i}^2]\sigma_{x_i}^2\, d\omega$$
$$= \int_{\omega_1}^{\omega_2} \sum_{i=1}^{k} |(d_T)_{x_i}|^2 \sigma_{x_i}^2\, d\omega \qquad (3\text{-}74)$$

A comparison of Eqs. (3-74) and (3-54) reveals that the deterministic case (3-54) defined at single frequency is a special case of Eq. (3-74), that is the integrand with $\sigma_{x_i}^2 = 1$ for $i = 1, \ldots, k$.

Consider now the general transfer function of a linear active network:

$$H(j\omega, \mathbf{x}) = \frac{N(j\omega, \mathbf{x})}{D(j\omega, \mathbf{x})} = \frac{b_m s^m + b_{m-1}s^{m-1} + \cdots + b_1 s + b_0}{s^n + a_{n-1}s^{n-1} + \cdots + a_1 s + a_0}\bigg|_{s=j\omega} \qquad (3\text{-}75)$$

We can express $(d_T)_{x_i}$ as

$$(d_T)_{x_i} = \frac{\partial \ln H}{\partial \ln x_i} = \frac{\partial \ln N}{\partial \ln x_i} - \frac{\partial \ln D}{\partial \ln x_i} \tag{3-76a}$$

$$= \frac{x_i}{N}\frac{\partial N}{\partial x_i} - \frac{x_i}{D}\frac{\partial D}{\partial x_i} \tag{3-76b}$$

$$= x_i \frac{\partial \mathbf{b}}{\partial x_i}\mathbf{B} - x_i \frac{\partial \mathbf{a}}{\partial x_i}\mathbf{A} \tag{3-76c}$$

where

$$\mathbf{b} = [b_m \quad b_{m-1} \quad \ldots \quad b_1 \quad b_0]^t \qquad (m+1) \times 1 \tag{3-77a}$$

$$\mathbf{a} = [a_{n-1} \quad a_{n-2} \quad \ldots \quad a_1 \quad a_0]^t \qquad n \times 1 \tag{3-77b}$$

$$\mathbf{B} = \frac{\nabla_b N}{N} = \frac{1}{N}\left[\frac{\partial N}{\partial b_m} \quad \frac{\partial N}{\partial b_{m-1}} \quad \ldots \quad \frac{\partial N}{\partial b_1} \quad \frac{\partial N}{\partial b_0}\right]^t \qquad (m+1) \times 1 \tag{3-77c}$$

$$\mathbf{A} = \frac{\nabla_a D}{D} = \frac{1}{D}\left[\frac{\partial D}{\partial a_{n-1}} \quad \frac{\partial D}{\partial a_{n-2}} \quad \ldots \quad \frac{\partial D}{\partial a_1} \quad \frac{\partial D}{\partial a_0}\right]^t \qquad n \times 1 \tag{3-77d}$$

Since $\mathbf{B}$ and $\mathbf{A}$ are complex vectors, we have

$$\mathbf{B} = \mathbf{B}_{re} + j\mathbf{B}_{im} = \operatorname{Re}\left(\frac{\nabla_b N}{N}\right) + j\operatorname{Im}\left(\frac{\nabla_b N}{N}\right) \tag{3-78}$$

$$\mathbf{A} = \mathbf{A}_{re} + j\mathbf{A}_{im} = \operatorname{Re}\left(\frac{\nabla_a D}{D}\right) + j\operatorname{Im}\left(\frac{\nabla_a D}{D}\right) \tag{3-79}$$

Substitution of Eqs. (3-78) and (3-79) into (3-76) yields

$$(d_T)_{x_i} = x_i \frac{\partial \mathbf{b}^t}{\partial x_i}\mathbf{B}_{re} - x_i \frac{\partial \mathbf{a}^t}{\partial x_i}\mathbf{A}_{re} + j\left(x_i \frac{\partial \mathbf{b}^t}{\partial x_i}\mathbf{B}_{im} - x_i \frac{\partial \mathbf{a}^t}{\partial x_i}\mathbf{A}_{im}\right) \tag{3-80}$$

We note that

$$(d_\alpha)_{x_i} = x_i \frac{\partial \mathbf{b}^t}{\partial x_i}\mathbf{B}_{re} - x_i \frac{\partial \mathbf{a}^t}{\partial x_i}\mathbf{A}_{re} = \mathbf{C}_i^t \begin{bmatrix} \mathbf{B}_{re} \\ -\mathbf{A}_{re} \end{bmatrix} \tag{3-81a}$$

$$(d_\beta)x_i = x_i \frac{\partial \mathbf{b}^t}{\partial x_i}\mathbf{B}_{im} - x_i \frac{\partial \mathbf{a}^t}{\partial x_i}\mathbf{A}_{im} = \mathbf{C}_i^t \begin{bmatrix} \mathbf{B}_{im} \\ -\mathbf{A}_{im} \end{bmatrix} \tag{3-81b}$$

where

$$\mathbf{C}_i^t = x_i \frac{\partial \mathbf{Y}^t}{\partial x_i} \quad (m+n+1) \times 1 \tag{3-82}$$

and

$$\mathbf{Y} = \begin{bmatrix} \mathbf{b} \\ \mathbf{a} \end{bmatrix} \quad (m+n+1) \times 1 \tag{3-83}$$

Clearly, since $(d_\alpha)_{x_i}$ is a scalar

$$(d_\alpha)_{x_j} = (d_\alpha)_{x_j}^t$$
$$= [\mathbf{B}_{re}^t - \mathbf{A}_{re}^t][\mathbf{C}_j] \tag{3-84}$$

Hence, the product of $(d_\alpha)_{x_i}$ and $(d_\alpha)_{x_j}$ from Eq. (3-81) and (3-84) can be written as

$$(d_\alpha)_{x_i}(d_\alpha)_{x_j} = \mathbf{C}_i^t \mathbf{E}_\alpha \mathbf{C}_j \qquad (3\text{-}85)$$

where $\mathbf{E}_\alpha$ is defined as

$$\mathbf{E}_\alpha = \begin{bmatrix} \mathbf{B}_r \mathbf{B}_r^t & -\mathbf{B}_r \mathbf{A}_r^t \\ -\mathbf{A}_r \mathbf{B}_r^t & \mathbf{A}_r \mathbf{A}_r^t \end{bmatrix} \qquad (m+n+1) \times (m+n+1) \qquad (3\text{-}86)$$

Similarly, by taking the imaginary part of $(\mathbf{S}_T)_{x_i}$, we obtain

$$\mathbf{E}_\beta = \begin{bmatrix} \mathbf{B}_i \mathbf{B}_i^t & -\mathbf{B}_i \mathbf{A}_i^t \\ -\mathbf{A}_i \mathbf{B}_i^t & \mathbf{A}_i \mathbf{A}_i^t \end{bmatrix} \qquad (m+n+1) \times (m+n+1) \qquad (3\text{-}87)$$

Note that both $\mathbf{E}_\alpha$ and $\mathbf{E}_\beta$ are symmetric matrices and $\mathbf{E}_T = \mathbf{E}_\alpha + \mathbf{E}_\beta$. Substitution of Eq. (3-86) into (3-70) yields

$$M_\alpha = \sum_{i=1}^{k} \sum_{j=1}^{k} \mathbf{C}_i^t \mathbf{K}_\alpha \mathbf{C}_j \rho_{ij} \qquad (3\text{-}88)$$

where

$$\mathbf{K}_\alpha = \int_{\omega_1}^{\omega_2} \mathbf{E}_\alpha \, d\omega$$

If there is no correlation between the random variables,

$$M_\alpha = \sum_{i=1}^{k} \mathbf{C}_i^t \mathbf{K}_\alpha \mathbf{C}_i \sigma_{x_i}^2 \qquad (3\text{-}89)$$

Similarly, from Eqs. (3-87) and (3-72),

$$M_\beta = \sum_{i=1}^{k} \sum_{j=1}^{k} \mathbf{C}_i^t \mathbf{K}_\beta \mathbf{C}_j \rho_{ij} \qquad (3\text{-}90)$$

where

$$\mathbf{K}_\beta = \int_{\omega_1}^{\omega_2} \mathbf{E}_\beta \, d\omega$$

For no correlation Eq. (3-90) reduces to

$$M_\beta = \sum_{i=1}^{k} \mathbf{C}_i^t \mathbf{K}_\beta \mathbf{C}_i \sigma_{x_i}^2 \qquad (3\text{-}91)$$

Finally,

$$M_T = \sum_{i=1}^{k} \sum_{j=1}^{k} \mathbf{C}_i^t \mathbf{K}_T \mathbf{C}_j \rho_{ij} \qquad (3\text{-}92)$$

where

$$\mathbf{K}_T = \int_{\omega_1}^{\omega_2} \mathbf{E}_T \, d\omega$$

and for no correlation,

$$M_T = \sum_{i=1}^{k} \mathbf{C}_i^t \mathbf{K}_T \mathbf{C}_i \sigma_{x_i}^2 \qquad (3\text{-}93)$$

Note that the vectors $\mathbf{C}_i$ are, in general, functions of the network topology and parameters. The values of the matrices $\mathbf{K}_\alpha$, $\mathbf{K}_\beta$, and $\mathbf{K}_T$ can be calculated directly from the given transfer function, independent of the network topology. Hence, for comparison of the magnitude sensitivity of various candidate networks for realizing a given transfer function, we need calculate $\mathbf{K}_\alpha$ only once. Then $\mathbf{C}_i$ are obtained for each individual network. Similarly, for the phase sensitivity, we need to calculate $\mathbf{K}_\beta$ only once. Note also that for comparison purposes at a particular frequency, one need only compare the integrands in Eq. (3-74) evaluated at the desired frequency.

•

EXAMPLE 3-5

To illustrate the computation of sensitivity described in this section, we shall consider an extremely simple example, shown in Fig. 3-3. The transfer function of the circuit is given by

$$\frac{V_o}{V_i} = \frac{1/RC}{s + 1/RC} \tag{3-94a}$$

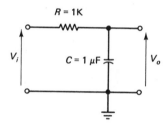

Fig. 3-3 Simple *RC* circuit for Example 3-5.

Assume that the design values are $R = 1 \text{ k}\Omega$, $C = 1 \text{ }\mu\text{F}$. We normalize the transfer function such that

$$\frac{V_o}{V_i} = \frac{10^3}{s + 10^3} = \frac{1}{s_n + 1} \tag{3-94b}$$

From Eq. (3-83), which is a vector containing the numerator and denominator coefficients, we have

$$\mathbf{Y} = [b_0 \quad a_0]^t = \left[\frac{1}{RC} \quad \frac{1}{RC}\right]^t \tag{3-95a}$$

The *k* parameters here are two: the resistor and the capacitor:

$$\mathbf{x} = [R \quad C]^t \tag{3-95b}$$

The matrices $\mathbf{A}_{re}$, $\mathbf{A}_{im}$, $\mathbf{B}_{re}$, and $\mathbf{B}_{im}$ are all scalar in this simple case:

$$\mathbf{A}_{re} = \frac{1}{1 + \omega^2} \qquad \mathbf{A}_{im} = \frac{-\omega}{1 + \omega^2}$$
$$\mathbf{B}_{re} = 1 \qquad \mathbf{B}_{im} = 0 \tag{3-96}$$

From Eqs. (3-86) and (3-87), the $\mathbf{E}_\alpha$ and $\mathbf{E}_\beta$ matrices are

$$\mathbf{E}_\alpha = \begin{bmatrix} 1 & -\dfrac{1}{1+\omega^2} \\ -\dfrac{1}{1+\omega^2} & \dfrac{1}{(1+\omega^2)^2} \end{bmatrix}, \quad \mathbf{E}_\beta = \begin{bmatrix} 0 & 0 \\ 0 & \dfrac{\omega^2}{(1+\omega^2)^2} \end{bmatrix} \quad (3\text{-}97\text{a})$$

Integrating $\mathbf{E}_\alpha$ and $\mathbf{E}_\beta$ over the passband $(0, 1)$, we get

$$\mathbf{K}_\alpha = \begin{bmatrix} 1 & -0.7854 \\ -0.7854 & 0.6427 \end{bmatrix}, \quad \mathbf{K}_\beta = \begin{bmatrix} 0 & 0 \\ 0 & 0.1427 \end{bmatrix} \quad (3\text{-}97\text{b})$$

The vectors $\mathbf{C}_1$ and $\mathbf{C}_2$ from (3-82) are given by

$$\mathbf{C}_1 = \mathbf{C}_2 = \begin{bmatrix} -\dfrac{1}{RC} & -\dfrac{1}{RC} \end{bmatrix}^t = [-1 \quad -1]^t \quad (3\text{-}98)$$

If we assume no correlation and variance $\sigma_{x_i}^2 = 10^{-4}$ (i.e., 1% variation in each element), the statistical sensitivity measures M_α, M_β from Eqs. (3-89) and (3-91) are given by

$$M_\alpha = \sum_{i=1}^{2} \mathbf{C}_i^t \mathbf{K}_\alpha \mathbf{C}_i \sigma_{x_i}^2 = 0.1438 \times 10^{-4}$$

$$M_\beta = \sum_{i=1}^{2} \mathbf{C}_i^t \mathbf{K}_\beta \mathbf{C}_i \sigma_{x_i}^2 = 0.2854 \times 10^{-4}$$

The transfer function sensitivity $M_T = M_\alpha + M_\beta = 0.4292 \times 10^{-4}$. Note that the statistical sensitivity is of the order of magnitude of the variance. Convenient numbers for comparison purposes may be normalized value with respect to a variance (i.e., $M_T/\sigma_{x_i}^2$), which in this case is 0.4292.

EXAMPLE 3-6

To further illustrate the computation of sensitivity described in this section, let us consider the network shown in Fig. 3-4. The transfer function of the network, assuming ideal finite-gain amplifiers (i.e., ideal op amps used in a noninverting mode), is

$$H(s) = \frac{K_1 K_2 G_1 S_1 G_2 S_2}{s^2 + [G_1 S_1 + G_2 S_2(1 - K_2 K_1)]s + G_1 S_1 G_2 S_2} \quad (3\text{-}99)$$

where $G_i = 1/R_i$, $S_i = 1/C_i$, $i = 1, 2$.

Consider the design of a second-order Butterworth filter with ω_{3dB} normalized to unity. The design equations are

$$\begin{aligned} b_0 &= K_1 K_2 G_1 S_1 G_2 S_2 = 1 \\ a_1 &= G_1 S_1 + G_2 S_2(1 - K_1 K_2) = \sqrt{2} \\ a_0 &= G_1 S_1 G_2 S_2 = 1 \end{aligned} \quad (3\text{-}100)$$

Let

$$K_1 = K_2 = 1$$

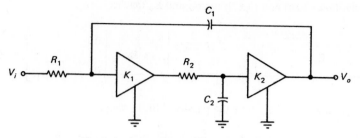

Fig. 3-4   Active-RC circuit for Example 3-6.

The vectors $\mathbf{Y}$ and $\mathbf{x}$ are

$$\mathbf{Y} = [b_0 \quad a_1 \quad a_0]^t = [1 \quad \sqrt{2} \quad 1]^t \tag{3-101a}$$

$$\mathbf{x} = [G_1 \quad S_1 \quad G_2 \quad S_2 \quad K_1 \quad K_2]^t \tag{3-101b}$$

The matrices $\mathbf{A}_{re}$ and $\mathbf{B}_{re}$ and $\mathbf{E}_\alpha$ are

$$\mathbf{A}_{re} = \left[\frac{\sqrt{2}\,\omega^2}{1+\omega^4} \quad \frac{1-\omega^2}{1+\omega^4}\right]^t \tag{3-102a}$$

$$\mathbf{B}_{re} = 1 \tag{3-102b}$$

$$\mathbf{E}_\alpha = \frac{1}{1+\omega^4}\begin{bmatrix} (1+\omega^4) & -\sqrt{2}\,\omega^2 & -(1-\omega^2) \\[2mm] \sqrt{2}\,\omega^2(1-\omega^2) & \dfrac{2\omega^4}{1+\omega^4} & \dfrac{\sqrt{2}\,\omega^2(1-\omega^2)}{1+\omega^4} \\[2mm] -(1-\omega^2) & \dfrac{\sqrt{2}\,\omega^2(1-\omega^2)}{1+\omega^4} & \dfrac{(1-\omega^2)^2}{1+\omega^4} \end{bmatrix}$$

$$\tag{3-103a}$$

The matrix $\mathbf{K}_\alpha$ is calculated from the matrix $\mathbf{E}_\alpha$ for the passband region $[0, 1]$ using, for example, the trapezoidal rule with increments of 0.1 rad/sec.

$$\mathbf{K}_\alpha = \begin{bmatrix} 1 & -0.345 & -0.623 \\ -0.345 & 0.183 & 0.133 \\ -0.623 & 0.133 & 0.495 \end{bmatrix} \tag{3-103b}$$

The vectors $\mathbf{C}_i$, $i = 1, \ldots, 6$, are evaluated from (3-82):

$$\mathbf{C}_1 = \mathbf{C}_2 = [1 \quad \sqrt{2} \quad 1]^t$$
$$\mathbf{C}_3 = \mathbf{C}_4 = [1 \quad 0 \quad 1]^t \tag{3-104}$$
$$\mathbf{C}_6 = \mathbf{C}_6 = \left[1 \quad -\frac{1}{\sqrt{2}} \quad 0\right]^t$$

The statistical magnitude sensitivity measure M_α, assuming no correlation and variances $\sigma_{x_i}^2 = 10^{-4}$, is determined from Eq. (3-89).

$$M_\alpha = \sum_{i=1}^{6} \mathbf{C}_i^t \mathbf{K}_\alpha \mathbf{C}_i \sigma_{x_i}^2 = 3.685 \times 10^{-4}$$

Similarly, we determine

$$\mathbf{K}_\beta = \begin{bmatrix} 0 & 0 & 0 \\ 0 & +0.060 & -0.133 \\ 0 & -0.133 & 0.372 \end{bmatrix}$$

$$M_\beta = \sum_{i=1}^{6} \mathbf{C}_i^t \mathbf{K}_\beta \mathbf{C}_i \sigma_{x_i}^2 = 1.036 \times 10^{-4}$$

and the total sensitivity measure

$$M_T = M_\alpha + M_\beta = 4.724 \times 10^{-4}$$

As can be seen from this example, the statistical sensitivity computation can be best done via computer.

●

3.6 STATISTICAL MULTIPARAMETER SENSITIVITY OF HIGH-Q ACTIVE FILTERS

In most band-pass applications we are interested in a narrow-band (i.e., a high-Q), frequency response. A second-order high-Q network has a transfer function of the form

$$H(s) = \frac{N(s)}{s^2 + (\omega_0/Q)s + \omega_0^2} = \frac{N(s)}{D(s)} \tag{3-105}$$

where Q is high (i.e., $Q \geq 10$) and the roots of $N(s)$ are assumed not close to those of $D(s)$. Under this approximation, $N(s)$ may be considered constant within the passband. Let us express for convenience the denominator $D(s)$ as

$$D(s) = s^2 + \frac{\omega_0}{Q}s + \omega_0^2 \tag{3-106a}$$

$$= s^2 + a_1 s + a_0 \tag{3-106b}$$

From Eq. (3-106), we have

$$\mathbf{a} = [a_1 \quad a_0]^t = [\omega_0/Q \quad \omega_0^2]^t \tag{3-107}$$

$$(d_T)_{x_i} = x_i \frac{\partial \mathbf{b}^t}{\partial x_i} \mathbf{B} - x_i \frac{\partial \mathbf{a}^t}{\partial x_i} \mathbf{A} \simeq -x_i \frac{\partial \mathbf{a}^t}{\partial x_i} \mathbf{A} \tag{3-108}$$

since we have assumed that the numerator is relatively constant in the passband. But

$$x_i \frac{\partial \mathbf{a}^t}{\partial x_i} = \left[\frac{\omega_0}{Q} (S_{x_i}^{\omega_0} - S_{x_i}^{Q}) \quad 2\omega_0 S_{x_i}^{\omega_0} \right] \tag{3-109}$$

where

$$S_{x_i}^{Q} = \frac{x_i}{Q} \frac{\partial Q}{\partial x_i} \quad \text{and} \quad S_{x_i}^{\omega_0} = \frac{x_i}{\omega_0} \frac{\partial \omega_0}{\partial x_i} \tag{3-110}$$

and from Eq. (3-77d),

$$\mathbf{A} = \frac{1}{D}[j\omega, 1] \tag{3-111}$$

Substitution of Eq. (3-111) into (3-108) yields

$$(d_\alpha)_{x_i} = -\mathrm{Re}\left(\frac{1}{D}\right)2\omega_0^2 S_{x_i}^{\omega_0} - \mathrm{Re}\left(\frac{j\omega}{D}\right)\frac{\omega_0}{Q}(S_{x_i}^{\omega_0} - S_{x_i}^{Q}) \tag{3-112}$$

$$(d_\beta)_{x_i} = -\mathrm{Im}\left(\frac{1}{D}\right)2\omega_0^2 S_{x_i}^{\omega_0} - \mathrm{Im}\left(\frac{j\omega}{D}\right)\frac{\omega_0}{Q}(S_{x_i}^{\omega_0} - S_{x_i}^{Q}) \tag{3-113}$$

In the passband region, Eqs. (3-112) and (3-113) can be written as the following approximate expressions

$$(d_\alpha)_{x_i} \simeq -2\omega_0^2 \, \mathrm{Re}\left(\frac{1}{D}\right)S_{x_i}^{\omega_0} \tag{3-114}$$

$$(d_\beta)_{x_i} \simeq -2\omega_0^2 \, \mathrm{Im}\left(\frac{1}{D}\right)S_{x_i}^{\omega_0} \tag{3-115}$$

Also,

$$(d_T)_{x_i} \simeq -2\omega_0^2 \left(\frac{1}{D}\right)S_{x_i}^{\omega_0} \tag{3-116}$$

Of course, for Eqs. (3-114) to (3-116) to provide good approximations, the Q sensitivities have to be sufficiently low (i.e., $|S_{x_i}^{Q}| \ll Q|S_{x_i}^{\omega_0}|$). Note that there are some cases where this inequality is not valid, but it is generally true for all useful circuits. Thus, we may write the following approximate expressions for the integrands of M_α, M_β, and M_T.

$$S_\alpha \simeq 4\omega_0^4\left[\mathrm{Re}\left(\frac{1}{D}\right)\right]^2 \sum_{i=1}^{k}\sum_{j=1}^{k} S_{x_i}^{\omega_0}S_{x_j}^{\omega_0}\rho_{ij} = \kappa\left[\mathrm{Re}\left(\frac{1}{D}\right)\right]^2 \tag{3-117a}$$

$$S_\beta \simeq 4\omega_0^4\left[\mathrm{Im}\left(\frac{1}{D}\right)\right]^2 \sum_{i=1}^{k}\sum_{j=1}^{k} S_{x_i}^{\omega_0}S_{x_j}^{\omega_0}\rho_{ij} = \kappa\left[\mathrm{Im}\left(\frac{1}{D}\right)\right]^2 \tag{3-117b}$$

$$S_T = 4\omega_0^4\left|\frac{1}{D}\right|^2 \sum_{i=1}^{k}\sum_{j=1}^{k} S_{x_i}^{\omega_0}S_{x_j}^{\omega_0}\rho_{ij} = \kappa\left|\frac{1}{D}\right|^2 \tag{3-117c}$$

where

$$\kappa = 4\omega_0^4 \sum_{i=1}^{k}\sum_{j=1}^{k} S_{x_i}^{\omega_0}S_{x_j}^{\omega_0}\rho_{ij} \tag{3-117d}$$

These approximate integrands will be found to be particularly convenient in Chapter 5 when we consider the computation of M_α, M_β, and M_T for high-order filters in which second-order, high-Q networks are cascaded together or interconnected in one of several multiple-loop feedback topologies.

Another useful form for $(d_\alpha)_{x_i}$ and $(d_\beta)_{x_i}$ can be obtained by substituting the following expressions:

$$\text{Re}\left(\frac{1}{D}\right) = \frac{\omega_0^2 - \omega^2}{|D|^2} \simeq \frac{-2Q^2 u}{\omega_0^2 \Delta} \tag{3-118a}$$

$$\text{Re}\left(\frac{j\omega}{D}\right) = \frac{\omega^2 \omega_0/Q}{|D|^2} \simeq \frac{Q}{\omega_0 \Delta} \tag{3-118b}$$

$$\text{Im}\left(\frac{1}{D}\right) = \frac{-\omega\omega_0/Q}{|D|^2} \simeq -\frac{Q}{\omega_0^2 \Delta} \tag{3-118c}$$

$$\text{Im}\left(\frac{j\omega}{D}\right) = \frac{\omega(\omega_0^2 - \omega^2)}{|D|^2} \simeq -\frac{2Q^2 u}{\omega_0 \Delta} \tag{3-118d}$$

namely, Eq. (3-118) into (3-112) and (3-113), respectively, where in the passband $\omega^2 - \omega_0^2 \simeq 2\omega_0(\omega - \omega_0)$, $\Delta = |D|^2/[\text{Im}(D)]^2 \simeq 1 + 4Q^2 u^2$ and $u = (\omega - \omega_0)/\omega_0$. Performing this substitution, $(d_\alpha)_{x_i}$ and $(d_\beta)_{x_i}$ may be written in the following form:

$$(d_\alpha)_{x_i} \simeq \mathbf{C}_i^t \begin{bmatrix} \dfrac{4Q^2 u - 1}{\Delta} \\ \dfrac{1}{\Delta} \end{bmatrix} \tag{3-119a}$$

$$(d_\beta)_{x_i} \simeq \mathbf{C}_i^t \begin{bmatrix} \dfrac{2Q(1 + u)}{\Delta} \\ \dfrac{-2Qu}{\Delta} \end{bmatrix} \tag{3-119b}$$

where

$$\mathbf{C}_i^t = [S_{x_i}^{\omega_0} \quad S_{x_i}^{Q}] \quad \text{a } 1 \times 2 \text{ vector}$$

We may now write the following expressions for S_α, S_β, and S_T:

$$S_\alpha \simeq \sum_{i=1}^{k} \sum_{j=1}^{k} \mathbf{C}_i^t \mathbf{E}_\alpha \mathbf{C}_j \rho_{ij} \tag{3-120a}$$

$$S_\beta \simeq \sum_{i=1}^{k} \sum_{j=1}^{k} \mathbf{C}_i^t \mathbf{E}_\beta \mathbf{C}_j \rho_{ij} \tag{3-120b}$$

$$S_T \simeq \sum_{i=1}^{k} \sum_{j=1}^{k} \mathbf{C}_i^t \mathbf{E}_T \mathbf{C}_j \rho_{ij} \tag{3-120c}$$

where

$$\mathbf{E}_\alpha = \frac{1}{\Delta^2} \begin{bmatrix} (4Q^2 u - 1)^2 & 4Q^2 u - 1 \\ 4Q^2 u - 1 & 1 \end{bmatrix}$$

$$\mathbf{E}_\beta = \frac{1}{\Delta^2} \begin{bmatrix} 4Q^2(1 + u)^2 & -4Q^2 u(1 + u) \\ -4Q^2 u(1 + u) & 4Q^2 u^2 \end{bmatrix}$$

$$\mathbf{E}_T = \mathbf{E}_\alpha + \mathbf{E}_\beta \simeq \frac{1}{\Delta} \begin{bmatrix} 4Q^2 + 1 & -1 \\ -1 & 1 \end{bmatrix}$$

Finally, we may write the integral sensitivity measures M_α, M_β, and M_T, for $\omega_1 = \omega_0(1 - 1/Q) \le \omega \le \omega_2 = \omega_0(1 + 1/Q)$, according to Eqs. (3-88), (3-90), and (3-92) as follows:

$$M_\alpha \simeq \sum_{i=1}^{k} \sum_{j=1}^{k} \mathbf{C}_i^t \mathbf{K}_\alpha \mathbf{C}_j \rho_{ij} \qquad (3\text{-}121a)$$

with

$$\mathbf{K}_\alpha = \int_{\omega_1}^{\omega_2} \mathbf{E}_\alpha \, d\omega = \omega_0 \int_{-1/Q}^{1/Q} \mathbf{E}_\alpha \, du$$

and

$$M_\beta \simeq \sum_{i=1}^{k} \sum_{j=1}^{k} \mathbf{C}_i^t \mathbf{K}_\beta \mathbf{C}_j \rho_{ij} \qquad (3\text{-}121b)$$

with

$$\mathbf{K}_\beta = \int_{\omega_1}^{\omega_2} \mathbf{E}_\beta \, d\omega = \omega_0 \int_{-1/Q}^{1/Q} \mathbf{E}_\beta \, du$$

Hence

$$M_T = \sum_{i=1}^{k} \sum_{j=1}^{k} \mathbf{C}_i^t \mathbf{K}_T \mathbf{C}_j \rho_{ij} \qquad (3\text{-}121c)$$

with

$$\mathbf{K}_T = \int_{\omega_1}^{\omega_2} \mathbf{E}_T \, d\omega = \omega_0 \int_{-1/Q}^{1/Q} \mathbf{E}_T \, du$$

The values of $\mathbf{K}_\alpha$ and $\mathbf{K}_\beta$ (hence $\mathbf{K}_T$) can be calculated directly from the given transfer function independent of the network topology. For the case under consideration[4] (i.e., $-1/Q \le u \le 1/Q$) $\mathbf{K}_\alpha$, $\mathbf{K}_\beta$, and $\mathbf{K}_T$ are computed to be

$$\mathbf{K}_\alpha = \omega_0 \begin{bmatrix} k_{1\alpha}Q + \dfrac{k_{2\alpha}}{Q} & \dfrac{-k_{2\alpha}}{Q} \\[2ex] \dfrac{-k_{2\alpha}}{Q} & \dfrac{k_{2\alpha}}{Q} \end{bmatrix} \qquad (3\text{-}122a)$$

$$\mathbf{K}_\beta = \omega_0 \begin{bmatrix} k_{1\beta}Q + \dfrac{k_{2\beta}}{Q} & \dfrac{-k_{2\beta}}{Q} \\[2ex] \dfrac{-k_{2\beta}}{Q} & \dfrac{k_{2\beta}}{Q} \end{bmatrix} \qquad (3\text{-}122b)$$

$$\mathbf{K}_T = \omega_0 \begin{bmatrix} k_T\left(\dfrac{4Q^2 + 1}{Q}\right) & \dfrac{-k_T}{Q} \\[2ex] \dfrac{-k_T}{Q} & \dfrac{k_T}{Q} \end{bmatrix} \qquad (3\text{-}122c)$$

where $k_{1\alpha} = 1.414$, $k_{2\alpha} = 0.754$, $k_{1\beta} = 3.014$, $k_{2\beta} = 0.354$, and $k_T = 1.108$.

[4]For $-1/2Q \le u \le 1/2Q$, $k_{1\alpha} = 0.571$, $k_{2\alpha} = 0.643$, $k_{1\beta} = 2.571$, $k_{2\beta} = 0.143$, and $k_T = 0.785$.

Substituting Eqs. (3-122) into Eqs. (3-121) yields the following statistical sensitivity formulas:

$$M_\alpha = \omega_0 \left(k_{1\alpha} Q + \frac{k_{2\alpha}}{Q} \right) T^{\omega_0, \omega_0}_{x_i x_j} + \omega_0 \frac{k_{2\alpha}}{Q} (T^{Q,Q}_{x_i x_j} - T^{\omega_0, Q}_{x_i x_j}) \tag{3-123a}$$

$$\simeq \omega_0 k_{1\alpha} Q T^{\omega_0, \omega_0}_{x_i x_j} \quad \text{for high } Q \tag{3-123b}$$

$$M_\beta = \omega_0 \left(k_{1\beta} Q + \frac{k_{2\beta}}{Q} \right) T^{\omega_0, \omega_0}_{x_i x_j} + \omega_0 \frac{k_{2\beta}}{Q} (T^{Q,Q}_{x_i x_j} - T^{\omega_0, Q}_{x_i x_j}) \tag{3-124a}$$

$$\simeq \omega_0 k_{1\beta} Q T^{\omega_0, \omega_0}_{x_i x_j} \quad \text{for high } Q \tag{3-124b}$$

$$M_T = \omega_0 k_T \left(\frac{4Q^2 + 1}{Q} \right) T^{\omega_0, \omega_0}_{x_i x_j} + \omega_0 \frac{k_T}{Q} (T^{Q,Q}_{x_i x_j} - T^{\omega_0, Q}_{x_i x_j}) \tag{3-125a}$$

$$\simeq 4\omega_0 k_T T^{\omega_0, \omega_0}_{x_i x_j} \quad \text{for high } Q \tag{3-125b}$$

where

$$T^{\omega_0, \omega_0}_{x_i x_j} = \sum_{i=1}^{k} \sum_{j=1}^{k} S^{\omega_0}_{x_i} S^{\omega_0}_{x_j} \rho_{ij} \tag{3-126a}$$

$$T^{Q,Q}_{x_i x_j} = \sum_{i=1}^{k} \sum_{j=1}^{k} S^{Q}_{x_i} S^{Q}_{x_j} \rho_{ij} \tag{3-126b}$$

$$T^{\omega_0, Q}_{x_i x_j} = \sum_{i=1}^{k} \sum_{j=1}^{k} S^{\omega_0}_{x_i} S^{Q}_{x_j} \rho_{ij} \tag{3-126c}$$

Note that for high-Q networks, $M_\beta > M_\alpha$:

$$M_\beta \simeq \frac{k_{1\beta}}{k_{1\alpha}} M_\alpha \tag{3-127}$$

where

$$\frac{k_{1\beta}}{k_{1\alpha}} = \begin{cases} 2.13 & \text{for } \dfrac{-1}{Q} \leq u \leq \dfrac{1}{Q} \\[2ex] 4.50 & \text{for } \dfrac{-1}{2Q} \leq u \leq \dfrac{1}{2Q} \end{cases}$$

●

EXAMPLE 3-7

As an example of the calculation of sensitivity for high-Q networks, consider the simple circuit shown in Fig. 3-5. This circuit is not very suitable for very high Q applications, for reasons discussed in Chapter 4. The circuit is considered here just to illustrate the use of the equations developed in Section 3.6. The voltage transfer function of the circuit, assuming ideal amplifiers, is given by

$$H(s) = \frac{V_o}{V_i} = \frac{K_1 K_2 G_1 S_2 s}{(1 - K_1 K_2)s^2 + (G_1 S_2 + G_2 S_1)s + G_1 S_1 G_2 S_2}$$

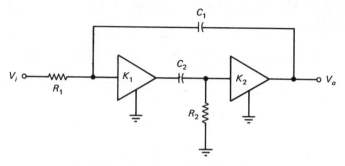

Fig. 3-5 Active-RC circuit for Example 3-7.

Let us design the circuit for a normalized center frequency of unity and $Q = 10$:

$$(\omega_0)_n = \sqrt{\frac{1}{R_1 R_2 C_1 C_2 (1 - K_1 K_2)}} = 1$$

$$Q = \sqrt{\frac{R_1 R_2 C_1 C_2 (1 - K_1 K_2)}{R_1 C_2 + R_2 C_1}} = 10$$

Note that $G_i = 1/R_i$ and $S_i = 1/C_i$.

The normalized element values for the design are chosen as

$$R_1 = R_2 = 1, \qquad C_1 = C_2 = \frac{1}{2Q} = 0.05$$

$$K_1 = -K_2 = \sqrt{4Q^2 - 1} = 19.975$$

The ω_0 sensitivities $S_{x_i}^{\omega_0}$, as determined from above, are

$$S_{R_1}^{\omega_0} = S_{R_2}^{\omega_0} = S_{C_1}^{\omega_0} = S_{C_2}^{\omega_0} = -\tfrac{1}{2}$$

$$S_{K_1}^{\omega_0} = S_{K_2}^{\omega_0} \simeq \tfrac{1}{2}$$

Note that for high-Q circuits, the Q sensitivities $S_{x_i}^{Q}$ need not be calculated, as these quantities do not appear in the approximate expressions (3-123b) and (3-124b).

For the calculation of sensitivity, let us assume the following:

1. Uncorrelated element variations (i.e., $\rho_{ij} = 0$, $i \neq j$) and with variance $\sigma_i^2 = 10^{-4}$.

2. The sensitivity measure is to be determined over a band of frequencies, $\omega_1, \omega_2 = \omega_0 \pm \omega_0/Q$ (i.e., $-1/Q \leq u \leq 1/Q$).

For high Q we need only calculate $T_{x_i x_j}^{\omega_0, \omega_0}$; thus, from Eq. (3-126a), we have

$$T_{x_i x_j}^{\omega_0, \omega_0} = \sum_{i=1}^{6} (S_{x_i}^{\omega_0})^2 \times 10^{-4} = 1.5 \times 10^{-4}$$

From Eqs. (3-123b) and (3-124b), we obtain

$$M_\alpha = (1.414)(10)(1.5 \times 10^{-4}) = 2.121 \times 10^{-3}$$

$$M_\beta = (3.014)(10)(1.5 \times 10^{-4}) = 4.521 \times 10^{-3}$$

$$M_T = M_\alpha + M_\beta = 6.642 \times 10^{-3}$$

Note that the sensitivity of the circuit is higher by an order of magnitude than the low-Q example of Eq. (3-6).

●

3.7 LARGE-PARAMETER-CHANGE STATISTICAL SENSITIVITY [P8]

When the parameter variations are not small, the nonlinear term in the Taylor series expansion must be included for an accurate measure of sensitivity. It should also be noted, however, that for very high Q networks and incremental changes in the parameters x, the series expansion of $\Delta H/H$ in Eq. (3-50) may not converge; hence, the Taylor series expansion in such cases is not valid.

To appreciate the differences between $\Delta H/H$ and $\Delta H/H'$ (where $H' = H + \Delta H$), consider again the Sallen and Key circuit shown in Fig. 3-1. The transfer function, given in Eq. (1-50), is rewritten for convenience, i.e.,

$$H(s) = \frac{V_o}{V_i} = \frac{KG_1 S_1 G_2 S_2}{s^2 + s[G_1 S_1 + G_2 S_1 + G_2 S_2(1 - K)] + G_1 S_1 G_2 S_2} \tag{3-128}$$

where

$$G_i = \frac{1}{R_i} \quad \text{and} \quad S_i = \frac{1}{C_i}$$

For normalized values $G_1 S_1 = G_2 S_2 = 1$, $G_2 S_1 = 8.1$ and $K = 10$, the circuit has a Q of 10 and $\omega_0 = 1$:

$$H(s) = \frac{10}{s^2 + 0.1s + 1} \tag{3-129}$$

Note that $|H(j1)| = 100$.

Now suppose that the gain changes by 1% from the design value of 10 to 9.9. The actual transfer function can be calculated to give

$$|H'(j1)| = 49.5 \tag{3-130}$$

By simple calculation at $s = j1$, we have

$$\left| \frac{\Delta H}{H} \right| = \frac{H' - H}{H} = 51\% \tag{3-131}$$

Evaluation of $|\Delta H/H|$ by the linear term of the series approximation from Eq. (3-51) is

$$\left|\frac{\Delta H}{H}\right| = \left|S_K^H \frac{\Delta K}{K}\right| = 101\%$$ (3-132)

A comparison of Eqs. (3-131) and (3-132) clearly shows the inaccuracy of the approximation. Fortunately, this problem can be readily resolved by deriving the Taylor series expansion for the following quantity:

$$\int_{\omega_1}^{\omega_2} \left|\frac{\Delta H}{H'}\right|^2 d\omega$$ (3-133)

where we recall that

$$\Delta H = H(j\omega, \mathbf{x} + \Delta\mathbf{x}) - H(j\omega, \mathbf{x}) = H' - H$$ (3-134)

Let us now examine the accuracy of the linear term of the Taylor series for $\Delta H/H'$, by comparing this approximation with actual variation obtained by directly calculating H and H'. The actual change is

$$\left|\frac{\Delta H}{H(K + \Delta K)}\right| = \left|\frac{H' - H}{H'}\right| = 102.02\%$$ (3-135)

To find the percentage change using a linear approximation, let us first rewrite the expression for $\Delta H/H'$ in the following form.

$$\frac{H' - H}{H'} = 1 - \frac{H}{H'}$$ (3-136)

The Taylor series expansion retaining only the linear term of the series approximation is[5]

$$\frac{1}{H'} \simeq \frac{1}{H} + \left[\mathbf{\nabla}\left(\frac{1}{H}\right)^t\right]\Delta\mathbf{x}$$ (3-137)

where

$$\mathbf{\nabla}\left(\frac{1}{H}\right) = \left[\frac{\partial(1/H)}{\partial x_i}\right] \quad \text{a } k \times 1 \text{ vector}$$ (3-138)

and t denotes transpose. Substitution of N/D for H in Eq. (3-138) yields

$$\frac{1}{H'} \simeq \frac{D}{N} + \mathbf{\nabla}_x\left(\frac{D}{N}\right)\mathbf{x}\,\Delta\hat{\mathbf{x}}$$ (3-139)

where

$$\mathbf{x} = [\text{diag. } x_i] \quad \text{a diagonal } k \times k \text{ matrix}$$

[5]It should be noted that the Taylor series in this formulation possesses a region of convergence whose definition is independent of Q. This is not true for $\Delta H/H$ expansion (Prob. 3.13).

A typical component of the vector $\nabla_x (D/N)$ is

$$\nabla\left(\frac{D}{N}\right)_i = \frac{\partial(D/N)}{\partial x_i} = \frac{1}{N}\frac{\partial D}{\partial x_i} - \frac{D}{N^2}\frac{\partial N}{\partial x_i} \qquad (3\text{-}140)$$

Hence, in general

$$\nabla(D/N) = \frac{1}{N}\,\nabla D - \frac{D}{N^2}\,\nabla N \qquad (3\text{-}141)$$

From Eqs. (3-141), (3-140), and (3-51), we obtain

$$\frac{\Delta H}{H'} \simeq [\mathbf{d}_T^t\, \Delta \hat{\mathbf{x}}] \qquad (3\text{-}142)$$

where

$$\mathbf{d}_T = \mathbf{x}\left[\frac{\nabla N}{N} - \frac{\nabla D}{D}\right] \qquad \text{a } k \times 1 \text{ vector} \qquad (3\text{-}143a)$$

$$= \mathbf{S}_x^N - \mathbf{S}_x^D \qquad (3\text{-}143b)$$

For the narrow-band band-pass case (i.e., $Q \geq 10$) we have $|D| \ll N$ in the bandpass region and (3-143) reduces to

$$\mathbf{d}_T \simeq -\mathbf{x}\frac{\nabla D}{D} = -\mathbf{S}_x^D \qquad \text{for high } Q \qquad (3\text{-}144)$$

Noting that

$$\left|\frac{\Delta H}{H'}\right|^2 = \left|\frac{\Delta H}{H'}\right|^{t^*}\left|\frac{\Delta H}{H'}\right| \qquad (3\text{-}145)$$

from Eqs. (3-145) and (3-142), we obtain

$$\left|\frac{\Delta H}{H'}\right|^2 \simeq [\mathbf{d}_T^* \,\Delta\hat{\mathbf{x}}\,\hat{\mathbf{x}}'\mathbf{d}_T] \qquad (3\text{-}146)$$

Substitution of Eqs. (3-146) into (3-55) yields

$$E\int_{\omega_1}^{\omega_2}\left|\frac{\Delta H}{H'}\right|^2 d\omega \simeq \int_{\omega_1}^{\omega_2}(\mathbf{d}_T^*\mathbf{P}\mathbf{d}_T)\,d\omega = M_T \qquad (3\text{-}147a)$$

Note carefully that the Taylor series expansions for $|\Delta H/H'|^2$ and $|\Delta H/H|^2$ have identical linear terms. In other words, if linear approximations are valid for both expansion, we have

$$M_T = E\int_{\omega_1}^{\omega_2}\left|\frac{\Delta H}{H}\right|^2 d\omega \simeq E\int_{\omega_1}^{\omega_2}\left|\frac{\Delta H}{H'}\right|^2 d\omega \qquad (3\text{-}147b)$$

To say it differently, for Eq. (3-147b) to be valid not only must $\Delta x_i/x_i \ll 1$, but we must also have $Q|\Delta x_i/x_i| \ll 1$. For high Q and $\Delta\mathbf{x}$ sufficiently large, the Taylor series from which M_T was obtained ceases to converge. Under these severe conditions, it can be shown [P8] that the Taylor series for $|\Delta H/H'|^2$ is still convergent, with the linear term M_T still an acceptable approximation. Thus, the sensitivity measure M_T retains its validity; however, when comparing M_T or its integrand S_T with an exact calculation or with experimental data, it is most important to apply the appropriate normalized difference equation to interpret the meaning for M_T.

Going back to our example, we calculate, from Eq. (3-146),

$$\frac{H' - H}{H'} = \frac{\Delta H}{H'} = 101\% \tag{3-148}$$

From a comparison of Eqs. (3-135) and (3-148), we see that the approximation in this formulation is very good.

One may also derive similar results for gain and phase functions, when Q and $\Delta \hat{x}$ are large:

$$E \int_{\omega_1}^{\omega_2} \left| \frac{\Delta\alpha}{\alpha'} \right|^2 d\omega \simeq \int_{\omega_1}^{\omega_2} \mathbf{S}_\alpha^t \mathbf{PS}_\alpha \, d\omega \tag{3-149}$$

$$E \int_{\omega_1}^{\omega_2} |\Delta\beta|^2 \, d\omega \simeq \int_{\omega_1}^{\omega_2} \mathbf{S}_\beta^t \mathbf{PS}_\beta \, d\omega \tag{3-150}$$

Again, for only incremental component variations (i.e., $Q\,|\Delta x_i/x_i| \ll 1$),

$$E \int_{\omega_1}^{\omega_2} \left| \frac{\Delta\alpha}{\alpha'} \right|^2 d\omega \simeq E \int_{\omega_1}^{\omega_2} \left| \frac{\Delta\alpha}{\alpha} \right|^2 d\omega = M_\alpha \tag{3-151}$$

For large component variations and high Q,

$$M_\alpha \simeq E \int_{\omega_1}^{\omega_2} \left| \frac{\Delta\alpha}{\alpha'} \right|^2 d\omega \neq E \int_{\omega_1}^{\omega_2} \left| \frac{\Delta\alpha}{\alpha} \right|^2 d\omega \tag{3-152}$$

Note that the interpretation for M_β does not change with $\Delta \mathbf{x}$ or Q.

For finite attenuation band-reject transfer functions [i.e., $N(s)$ with high Q], one can readily verify (Prob. 3.13) that Q-independent convergence occurs for the Taylor series expansion of $\Delta H/H$ and the Taylor series expansion for $\Delta H/H'$ diverges with Q. Thus, for this class of transfer functions, the measures are:

$$M_T \simeq E \int_{\omega_1}^{\omega_2} \left| \frac{\Delta H}{H} \right|^2 d\omega \tag{3-153a}$$

$$M_\alpha \simeq E \int_{\omega_1}^{\omega_2} \left| \frac{\Delta\alpha}{\alpha} \right| d\omega \tag{3-153b}$$

$$M_\beta \simeq E \int_{\omega_1}^{\omega_2} |\Delta\beta|^2 \, d\omega \tag{3-153c}$$

Again the linear approximate measures retain their validity, but care must be taken to provide the correct interpretations for them.

One class of filters for which measures M_T, M_α and M_β are not applicable is the notch filter. For the purposes of the present discussion, a notch filter is one in which the transfer function H becomes zero at one or more finite frequencies. The $j\omega$-axis zeros of H determine the presence of frequency domain nulls (traps or notches); that is, a filter with $H = N/D$ and $N = \prod_{j=1}^{k} (s^2 + \omega_j^2)$ will have nulls at frequencies ω_j, for $j = 1, \ldots, k$, which are often referred to as *notch frequencies*. In filters, notches appear in several types of transfer functions: for example, for the purpose of increasing stop-band attenuation in a band-pass filter or to remove specific, unwanted frequency components from a signal spectrum, e.g. a band-reject filter. If the function of the notch(es) is only to increase stopband attenuation, sensitivities about the notch frequency(s) are usually unimportant, with the filter sensitivity within the passband predominating. However, if the notch(es) are used to remove specific spectral components, then the sensitivities at and about the notch frequency(s) are crucial to acceptable filter performance. It is this latter class of applications with which we are now concerned.

As noted previously, the aforementioned sensitivity measures are not applicable for determining sensitivity at or about notch frequencies where $N = 0$ and $Q_N = \infty$. This is simply due to the fact that both $\Delta H/H$ and $\Delta H/H'$ require the calculation of N^{-1}, which implies an undefined sensitivity at any notch frequency. The difficulty is quickly removed by simply evaluating a measure based on the complex function ΔH which does not require an N^{-1} evaluation. Let us now consider the following sensitivity measure.

$$E \int_{\omega_1}^{\omega_2} |\Delta H|^2 \, d\omega \tag{3-154}$$

It is noted that the Taylor series expansion of ΔH yields (Prob. 3.13) the same convergence criteria as the $\Delta H/H$ series. Using a linear approximation for ΔH, Eq. (3-154) can be placed in the following familiar form:

$$(M_T)_N = \int_{\omega_1}^{\omega_2} (\mathbf{d}^{t*}\mathbf{P}\mathbf{d}) \, d\omega = \int_{\omega_1}^{\omega_2} (S_T)_N \, d\omega \simeq E \int_{\omega_1}^{\omega_2} |\Delta H|^2 \, d\omega \tag{3-155}$$

where

$$\mathbf{d} = \mathbf{x}\left[\frac{1}{D} \nabla N - \frac{N}{D^2} \nabla D\right] \quad (\textit{note: } \mathbf{d} \neq \mathbf{d}_T) \tag{3-156}$$

The derivation of the higher-order terms of the series expansion for ΔH and the expression for $(M_T)_N$, retaining the second-order term, is reserved as exercises for the reader.

Let us now consider in detail the computation of $(M_T)_N$ for a second-order notch filter, with a transfer function of the following form:

$$H(s) = \frac{G[s^2 + (\omega_z/Q_z)s + \omega_z^2]}{s^2 + (\omega_0/Q_0)s + \omega_0^2} = G\frac{N(s)}{D(s)} \tag{3-157}$$

Consider specifically the case with $Q_0 \leq 1$, and $\omega_z/Q_z = 0$, implying that $Q_z = \infty$. These conditions apply in the case of band-reject or "trap" filter.

One may evaluate the sensitivity for this filter using a procedure similar to that described in Section 3.5. This procedure remains as stated except that vectors **A** and **B** in Eqs. (3-78) and (3-79) must each be premultiplied by H in Eq. (3-157). Another useful form is to express **d** in terms of its G, ω_z, Q_0, $1/Q_z$, and ω_0 sensitivities. Thus, for Eqs. (3-156) and (3-148) we obtain

$$\mathbf{d} = H\mathbf{S}_x^G + \frac{G}{D}(j\omega\omega_z\mathbf{S}_x^{1/Q_z} + 2\omega_z^2\mathbf{S}_x^{\omega_z})$$
$$- \frac{H}{D}\left[j\omega\frac{\omega_0}{Q_0}(\mathbf{S}_x^{\omega_0} - \mathbf{S}_x^{Q_0}) + 2\omega_0^2\mathbf{S}_x^{\omega_0} \right] \tag{3-158}$$

Here, the sensitivity vectors $\mathbf{S}_x^G$, $\mathbf{S}_x^{\omega_z}$, $\mathbf{S}_x^{Q_0}$, and $\mathbf{S}_x^{\omega_0}$ are defined according to

$$\mathbf{S}_x^\theta = \left[\frac{x_1}{\theta}\frac{\partial\theta}{\partial x_1} \quad \frac{x_2}{\theta}\frac{\partial\theta}{\partial x_2} \cdots \frac{x_k}{\theta}\frac{\partial\theta}{\partial x_k}\right]^t \quad \text{for } \theta = G, \omega_z, Q_0, \omega_0$$

and

$$\mathbf{S}_x^{1/Q_z} = \left[x_1\frac{\partial(1/Q_z)}{\partial x_1} \quad x_2\frac{\partial(1/Q_z)}{\partial x_2} \cdots x_k\frac{\partial(1/Q_z)}{\partial x_k}\right]^t \tag{3-159}$$

Sensitivities $\mathbf{S}_{x_i}^{1/Q_z}$ are defined in the nonstandard manner given in Eq. (3-159) in order to avoid the undefined value for Q_z.

Thus, given any second-order notch filter transfer function, written in the form of Eq. (3-157), the parameter sensitivities (which are readily evaluated from the transfer function) and the sensitivity [using Eqs. (3-155) and (3-158)] can be determined.

●

EXAMPLE 3-8

Let us now consider a numerical example based upon a high-pass notch (HPN) filter design using the network [P9] shown in Fig. 3-6. The transfer function for this circuit is as follows:

$$H(s) = \frac{G\left[s^2 + s\left(\dfrac{1}{R_1C_1} + \dfrac{1}{R_{eq2}C_2} - \dfrac{k_2}{R_{eq1}C_1}\right) + \dfrac{1}{R_1R_2C_1C_2}\right]}{s^2 + s\left[\dfrac{1}{R_{eq1}C_1}\left(1 - \dfrac{R_{eq1}R_{eq2}}{R_1R_2}\right)\right] + \dfrac{1}{R_{eq1}R_{eq2}C_1C_2}} \tag{3-160a}$$

$$= G\frac{s^2 + \omega_z/Q_z s + \omega_z^2}{s^2 + \omega_0/Q_0 s + \omega_0^2} \tag{3-160b}$$

where
$$G = 1 + R_5/R_4$$
$$R_{eq1} = R_1 \| R_3 \| R_6$$
$$R_{eq2} = R_2 + R_4 \| R_5$$
$$k_2 = (R_4 \| R_5)/R_{eq2}$$

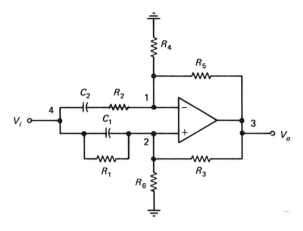

Fig. 3-6   Active-RC high-pass notch filter for Example 3-8.

In order for $|H(j\omega)|$ to achieve a null or notch at frequency ω_z, the term that defines ω_z/Q_z must be set to zero; that is, from Eq. (3-159a),

$$\frac{\omega_z}{Q_z} = \frac{1}{R_1 C_1} + \frac{1}{R_{eq2} C_2} - \frac{k_2}{R_{eq1} C_1} = 0 \qquad (3\text{-}161a)$$

or solving for k_2, the notch condition can be rewritten in the form

$$k_2 = \frac{R_{eq1}}{R_1} + \frac{R_{eq1} C_1}{R_{eq2} C_2} \qquad (3\text{-}161b)$$

The component sensitivities required to calculate $(M_T)_N$ given in Eq. (3-155) are listed in Table 3-2. The parameters α, β, and γ in Table 3-2 are defined as follows:

$$\alpha = G\frac{R_2}{R_1}, \qquad \beta = \frac{\omega_z}{\omega_0}, \qquad \gamma = \sqrt{\frac{R_{eq2} C_2}{R_{eq1} C_1}} \qquad (3\text{-}162)$$

The reader should not confuse the parameters α and β, defined in this example, with the gain and phase notations introduced earlier.

Normalizing frequency (see Section 1.6) such that the notch frequency $\omega_z = 1$ (i.e., defining the normalized radian frequency variable $\omega_n = \omega/\omega_z$), and setting the components to the normalized values (see Section 1.6) $R_1 = 2$, $R_2 = 1$, $R_3 = R_5 = 4$, $R_4 = \frac{4}{3}$, $R_6 = 0.8$, and $C_1 = C_2 = 1/\sqrt{2}$, results in the HPN transfer function

$$H(j\omega_n) = \frac{4(1 - \omega_n^2)}{2 - \omega_n^2 + j\omega_n\sqrt{2}} \qquad (3\text{-}163)$$

For $\omega_z = 2\pi \times 60$ Hz, the response described by Eq. (3-163) is given in Fig. 3-7, where the appropriate values for the component sensitivities are computed and listed in Table 3-2.

Substituting the appropriate entries in Table 3-2, with the element values assigned above, into Eqs. (3-155) and (3-158), the integrand of the sensitivity measure $(M_T)_N$ can be evaluated as a function of normalized radian frequency

TABLE 3-2

x	$S_x^{Q_2}$	$S_x^{\omega_0}$	$S_x^{Q_0}$	S_x^{1/Q_r}	S_x^G
C_1	$-\dfrac{1}{2}$	$-\dfrac{1}{2}$	$\dfrac{1}{2} - \dfrac{Q_0}{\gamma}\left(1 + \dfrac{1}{\alpha}\right) = -\dfrac{1}{2}$	$\dfrac{\alpha+1}{\alpha\beta\gamma} = \sqrt{2}$	0
C_2	$-\dfrac{1}{2}$	$-\dfrac{1}{2}$	$-\dfrac{1}{2} + \dfrac{Q_0}{\gamma}\left(1 + \dfrac{1}{\alpha}\right) = \dfrac{1}{2}$	$-\dfrac{\alpha+1}{\alpha\beta\gamma} = -\sqrt{2}$	0
R_1	$-\dfrac{1}{2}$	$-\dfrac{1}{2}\beta^2 = -\dfrac{1}{4}$	$-\dfrac{1}{2}\beta^2 + Q_0\gamma\beta^2\left(\dfrac{\alpha}{1+\alpha}\right) = \dfrac{1}{4}$	$-\dfrac{\alpha\beta\gamma}{1+\alpha} = -\dfrac{1}{\sqrt{2}}$	0
R_2	$-\dfrac{1}{2}$	$-\dfrac{1}{2}$	$\dfrac{1}{2} - Q_0\gamma\left(\dfrac{\alpha}{1+\alpha}\right) = -\dfrac{1}{2}$	$\dfrac{\alpha\beta\gamma}{1+\alpha} = \dfrac{1}{\sqrt{2}}$	0
R_3	0	$\dfrac{1}{2}\left(\dfrac{G-1}{G}\right)\dfrac{1}{\alpha} = \dfrac{3}{8}$	$\dfrac{G-1}{2\alpha G} - \dfrac{Q_0\gamma}{1+\alpha}\left(\dfrac{1}{\alpha} + \dfrac{G-1}{G}\right) = -\dfrac{11}{8}$	$\dfrac{\gamma}{\alpha\beta G(1+\alpha)} = \dfrac{1}{2\sqrt{2}}$	0
R_4	0	$\dfrac{1}{2}\left(\dfrac{G-1}{G}\right)\dfrac{1}{\alpha} = \dfrac{3}{8}$	$\dfrac{G-1}{2\alpha G} - \dfrac{Q_0\gamma}{1+\alpha}\dfrac{G-1}{G} = -\dfrac{3}{8}$	$-\dfrac{\gamma(G-1)}{\alpha\beta G}\left(1 - \dfrac{\alpha\beta^2}{1+\alpha}\right)$ $= -\dfrac{4}{8}\sqrt{2}$	$-\dfrac{G-1}{G} = -\dfrac{3}{4}$
R_5	0	$-\dfrac{1}{2}\left(\dfrac{G-1}{G}\right)\dfrac{1}{\alpha} = -\dfrac{3}{8}$	$-\dfrac{G-1}{2\alpha G} + \dfrac{Q_0\gamma}{1+\alpha}\left(\dfrac{1}{\alpha} + \dfrac{G-1}{G}\right) = \dfrac{11}{8}$	$-\dfrac{\gamma}{\alpha\beta G}\left(1 - \dfrac{\alpha\beta^2}{1+\alpha}\right)$ $= -\dfrac{3}{8}\sqrt{2}$	$\dfrac{G-1}{G} = \dfrac{3}{4}$
R_6	0	$-\dfrac{1}{2}\left(1 - \beta^2 + \dfrac{G-1}{G}\dfrac{1}{\alpha}\right)$ $= -\dfrac{5}{8}$	$-\dfrac{1}{2}\left(1 - \beta^2 + \dfrac{G-1}{\alpha G}\right)$ $+ Q_0\gamma\left[1 - \dfrac{\alpha\beta^2}{1+\alpha} - \dfrac{1}{G(1+\alpha)}\right] = \dfrac{5}{8}$	$\dfrac{\gamma}{\alpha\beta}\left(1 - \dfrac{1}{G(1+\alpha)} - \dfrac{\alpha\beta^2}{1+\alpha}\right)$ $= \dfrac{5}{4}\sqrt{2}$	0

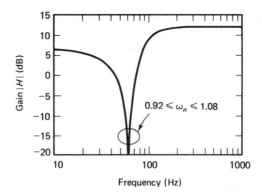

Fig. 3-7 Gain response for $H(j\omega_n)$ in Eq. (3-163) with $\omega_n = \omega/(2\pi \times 60 \text{ Hz})$.

ω_n. A plot of this integrand over the frequency range $0.92 \leq \omega_n \leq 1.08$, with all element variations $\Delta x_i/x_i$ assumed to be statistically independent zero-mean Gaussian random variables with $\sigma = 1\%$, is given in Fig. 3-8. The frequency band defined in this analysis includes only the notch, as shown in Fig. 3-7. Any attempt to use the other sensitivity measures (i.e., M_T, M_α, M_β) described in this chapter would result in a very erroneous analysis.

Finally, integrating $(S_T)_N$, given in Fig. 3-8, over the frequency band defined above (i.e., $\omega_1/\omega_z = 0.92$ and $\omega_2/\omega_z = 1.08$) results in the following

Fig. 3-8 Plot of $(S_T)_N$ versus frequency for a 60-Hz HPN filter realized with the circuit in Fig. 3-6.

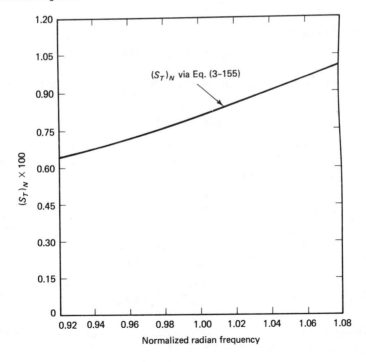

value for $(M_T)_N$:

$$(M_T)_N = 0.00130 \qquad \text{according to Eq. (3-155)}$$

[The integration in Eq. (3-155) was performed on a computer using the trapezoidal integration algorithm over a field of 50 points.]

•

Thus far we have only considered the transfer function sensitivity; however, one may also calculate the gain and phase function sensitivities. For most notch filter applications, for obvious reasons, the magnitude response sensitivity at and about the notch frequency has the most practical importance. Let us now write the transfer function H in the following familiar form: $H = |H|e^{j\phi} = \alpha e^{j\beta}$. As in previous measures, we must assume that the higher-order terms in the Taylor series expansion of ΔH are negligible if a simple relation among the gain, phase, and transfer function sensitivities is to be obtained. From Eq. (3-65) we can write the following relation:

$$\left|\frac{\Delta H}{H}\right|^2 \simeq \left|\frac{\Delta \alpha}{\alpha}\right|^2 + |\Delta\beta|^2 \qquad (3\text{-}164)$$

Then by simply premultiplying Eq. (3-164) by $|H|^2 = \alpha^2$, we obtain

$$|\Delta H|^2 = |\Delta\alpha|^2 + \alpha^2|\Delta\beta|^2 \qquad (3\text{-}165)$$

which implies that

$$(M_T)_N \simeq (M_\alpha)_N + \alpha^2(M_\beta)_N \qquad (3\text{-}166)$$

Therefore, at or in the vicinity of the notch frequency ω_z, $\alpha^2(M_\beta)_N \ll (M_\alpha)_N$ and

$$(M_\alpha)_N \simeq (M_T)_N \qquad (3\text{-}167)$$

as one might expect.

Let us now summarize the sensitivity measure formulation for large parameter changes as follows:

$$E \int_{\omega_1}^{\omega_2} \left|\frac{\Delta H}{H'}\right|^2 d\omega \simeq M_T \qquad \text{for high-}Q\text{ band-pass,} \atop \text{low-pass, and high-pass} \qquad (3\text{-}168a)$$

$$E \int_{\omega_1}^{\omega_2} \left|\frac{\Delta H}{H}\right|^2 d\omega \simeq M_T \qquad \text{for finite-attenuation} \atop \text{band-reject filters} \qquad (3\text{-}168b)$$

$$E \int_{\omega_1}^{\omega_2} |\Delta H|^2 d\omega \simeq (M_T)_N \qquad \text{for notch and finite-attenuation} \atop \text{band-reject filters} \qquad (3\text{-}168c)$$

For low-Q filters, either of these interpretations are equally valid, in the passband, for M_T. Also, $(M_T)_N$ may be used for these low-Q cases.

3.8 SENSITIVITY MEASURES FOR HIGH-ORDER FILTERS REALIZED AS A CASCADE OF BIQUADRATIC SECTIONS

There are several methods available for realizing high-order filters with active networks, whereby "high order" we mean order greater than 2. Perhaps the most widely used method is to realize a $2N$th-order filter as a cascade of N biquadratic active networks. For $(2N + 1)$-order filters, one can cascade N biquadratic sections plus one first-order section or $N - 1$ biquadratic sections plus one third-order section. The schematic for a N biquadratic-section cascade is shown symbolically in Fig. 3-9. Each box, denoted T_j with $j = 1, \ldots, N$, represents an active biquadratic network.

Fig. 3-9 Schematic for an N-section cascade configuration.

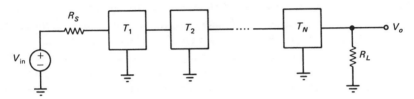

If the active networks T_j are noninteracting, a property that can be guaranteed when the active elements are op amps and the section output is an op amp input (the property is discussed in greater detail in Chapters 4 and 5), the transfer function for the N-section cascade may be written as follows:

$$H(s) = \frac{V_0}{V_{in}} = T_1(s)T_2(s)\ldots T_N(s) = \prod_{j=1}^{N} T_j(s) \qquad (3\text{-}169)$$

Using Eqs. (3-58), we can write the following expression for an arbitrary component of vector $\mathbf{d}_T$:

$$(d_T)_{x_i} = \frac{x_i}{H}\frac{\partial H}{\partial x_i} = \frac{x_i}{H}\left(\sum_{k=1}^{N} \frac{\partial H}{\partial T_k}\frac{\partial T_k}{\partial x_i}\right) \qquad (3\text{-}170)$$

However, from Eq. (3-169), we can write

$$\frac{\partial H}{\partial T_k} = \frac{\prod_{j=1}^{N} T_j}{T_k} = \frac{H}{T_k} \qquad (3\text{-}171)$$

Substituting Eq. (3-171) into Eq. (3-170) yields

$$(d_T)_{x_i} = \frac{x_i}{H}\sum_{k=1}^{N} \frac{H}{T_k}\frac{\partial T_k}{\partial x_i} = x_i\sum_{k=1}^{N} \frac{1}{T_k}\frac{\partial T_k}{\partial x_i} = \sum_{k=1}^{N} S_{x_i}^{T_k} \qquad (3\text{-}172)$$

Since sections T_i do not share common components, the vector $\mathbf{d}_T$ will have the following form:

$$\mathbf{d}_T^t = [S_{x_1}^{T_1} \ldots S_{x_m}^{T_1} \quad S_{x_{m+1}}^{T_2} \ldots S_{x_{2m}}^{T_2} \ldots S_{x_{(N+1)m+1}}^{T_N} \ldots S_{x_{Nm}}^{T_N}]$$

or

$$\mathbf{d}_T^t = [(\mathbf{S}_{x_1}^{T_1})^t \quad (\mathbf{S}_{x_2}^{T_2})^t \quad \ldots \quad (\mathbf{S}_{x_N}^{T_N})^t] \tag{3-173}$$

Typically, active-RC filters are realized in hybrid form, with each section fabricated on a separate chip. For this type of realization the biquadratic sections are statistically independent. That is, components that comprise a given section may be correlated, or statistically track each other, but components comprising different sections are uncorrelated. In a monolithic filter the statistical behavior can arise due to the physical separation of components comprising one section from those of another. For example, random oxide gradients along the surface of a silicon chip will cause capacitors that are physically separated by a sufficiently large area to vary in value in an uncorrelated manner. Thus, sections of a monolithic switched capacitor filter, which are widely separated on the chip, can be uncorrelated. In practice, it is then a good assumption, with the exception of densely packed monolithic filters, that only components within a section are correlated. Under this condition the covariance matrix $\mathbf{P}$, for the N-section cascade, can be expressed in the following diagonalized form:

$$\mathbf{P} = \begin{bmatrix} \mathbf{P}_1 & 0 & \ldots & 0 \\ 0 & \mathbf{P}_2 & & \vdots \\ \vdots & & \ddots & \vdots \\ \vdots & & & \ddots & 0 \\ 0 & & \ldots & 0 & \mathbf{P}_N \end{bmatrix} \tag{3-174}$$

where $\mathbf{P}_i$ for $i = 1, \ldots, N$ are the covariance matrices for the N biquadratic sections.

From Eqs. (3-60b), (3-173), and (3-174), the integrand sensitivity measure S_T can be written in the form

$$S_T = \mathbf{d}_T^{t*}\mathbf{P}\mathbf{d}_T = \sum_{k=1}^{N} \mathbf{d}_{Tk}^{t*}\mathbf{P}_k\mathbf{d}_{Tk} = \sum_{k=1}^{N} S_{Tk} \quad \text{where } \mathbf{d}_{Tk} = \mathbf{S}_{x_k}^{T_k} \tag{3-175}$$

Therefore, from Eqs. (3-60a) and (3-175) we can express the integral measure M_T as follows:

$$M_T = \int_{\omega_1}^{\omega_2} (\mathbf{d}_T^{t*}\mathbf{P}\mathbf{d}_T)\, d\omega = \int_{\omega_1}^{\omega_2} \left(\sum_{k=1}^{N} \mathbf{d}_{Tk}^{t*}\mathbf{P}_k\mathbf{d}_{Tk} \right) d\omega$$

Exchanging the order in which the summation and integration operations are performed, we obtain

$$M_T = \sum_{k=1}^{N} \int_{\omega_1}^{\omega_2} (\mathbf{d}_{Tk}^{t*}\mathbf{P}_k\mathbf{d}_{Tk})\, d\omega = \sum_{k=1}^{N} M_{Tk} \tag{3-176a}$$

In deriving Eq. (3-176), we have obtained the important result that the sensitivity for a cascade of N noninteracting biquadratic sections is simply the sum of the sensitivities for each of the sections. Of course, Eq. (3-176) is valid regardless of the order of the transfer function realized by T_k. However, the noninteraction, both electrical and statistical, of the sections is a crucial assumption in the derivation of this result.

It is noted that similar expressions can be derived for M_α and M_β, i.e.,

$$M_\alpha = \sum_{k=1}^{N} M_{\alpha k} \qquad (3\text{-}176b)$$

$$M_\beta = \sum_{k=1}^{N} M_{\beta k} \qquad (3\text{-}176c)$$

Note that for narrow-band band-pass filters where $T_k = N_k/D_k$ are high-Q second-order sections, we can combine Eqs. (3-117c) and (3-175) to obtain

$$S_{Tk} = \sum_{k=1}^{N} \kappa_k \left| \frac{1}{D_k} \right|^2 \qquad (3\text{-}177a)$$

where

$$\kappa_k = 4\omega_{0k}^4 \sum_{i=1}^{M} \sum_{j=1}^{M} S_{x_{ki}}^{\omega_{0k}} S_{x_{kj}}^{\omega_{0k}} (\rho_{ij})_k \qquad (3\text{-}177b)$$

$$D_k = D_k(j\omega) = \omega_{0k}^2 - \omega^2 + \frac{j\omega\omega_{0k}}{Q_k} \qquad (3\text{-}177c)$$

and M denotes the total number of elements that comprise section T_k.

3.9 SENSITIVITY MEASURES FOR HIGH-ORDER FILTERS REALIZED WITH BIQUADRATIC SECTIONS IN A FOLLOW-THE-LEADER (FLF) MULTIPLE-LOOP FEEDBACK CONFIGURATION

For high-order, narrow-band band-pass filters, cascade realizations can yield intolerably large sensitivities. One design method that has proven valuable in the design of low-sensitivity narrow-band filters has been to force controlled intrasection interaction via external feedback paths. One such multiple-loop feedback topology is the FLF topology, shown symbolically in Fig. 3-10 for a three-section design. There are several other multiple-loop feedback topologies and design methods for realizing low-sensitivity, high-order active filters. A more detailed treatment of this subject is reserved for Chapter 5.

To derive the transfer function for the three-section FLF topology in Fig. 3-10, we assume that the sections T_1, T_2 and T_3 are noninteracting. With this assumption, the derivation proceeds as follows:

$$V_1 = -T_1(V_{\text{in}} + F_{12}V_2 + F_{13}V_0) \qquad (3\text{-}178)$$

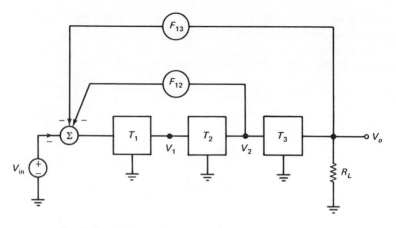

Fig. 3-10 Schematic for a three-section FLF configuration.

with

$$V_1 = \frac{1}{T_2 T_3} V_0 \qquad (3\text{-}179\text{a})$$

$$V_2 = \frac{1}{T_3} V_0 \qquad (3\text{-}179\text{b})$$

Substituting Eqs. (3-179) into Eq. (3-178) we can arrive at the following expression for the transfer function:

$$H = \frac{V_o}{V_i} = \frac{-T_1 T_2 T_3}{1 + F_{12} T_1 T_2 + F_{13} T_1 T_2 T_3} \qquad (3\text{-}180)$$

It is noted that the summing function, including gains F_{12} and F_{13}, in Fig. 3-10, can be realized with the op amp summing network shown in Fig. 2-9. Comparing Eqs. (3-178) and (2-15) we observe that F_{12} and F_{13} each represent a pair of ratioed resistors.

To derive the sensitivity measure for this network, we start, as we did for the cascade topology, with the derivation for an arbitrary component of vector $\mathbf{d}_T$, i.e.,

$$\{d_T\}_{x_i} = -H\left[\frac{1}{T_1 T_2 T_3}(S_{x_i}^{T_1} + S_{x_i}^{T_2}) + \frac{1 + F_{12} T_1 T_2}{T_1 T_2 T_3} S_{x_i}^{T_3} + \frac{F_{12}}{T_3} S_{x_i}^{F_{12}} + F_{13} S_{x_i}^{F_{13}}\right]$$

$$(3\text{-}181)$$

For narrow-band H, when T_i are high-Q band-pass second-order sections,

$$|S_{x_i}^{T_i}| \gg |S_{x_i}^{F_{12}}|, |S_{x_i}^{F_{13}}|$$

Therefore,

$$\{d_T\}_{x_i} \simeq -H\left[\frac{1}{T_1 T_2 T_3}(S_{x_i}^{T_1} + S_{x_i}^{T_2}) + \frac{1 + F_{12} T_1 T_2}{T_1 T_2 T_3} S_{x_i}^{T_3}\right] \qquad (3\text{-}182)$$

Since sections H_i do not share common elements, vector $\mathbf{d}_T$ may be written in the form

$$\mathbf{d}_T^t \simeq -H\left(\frac{1}{T_1 T_2 T_3} S_{x_1}^{T_1} \quad \frac{1}{T_1 T_2 T_3} S_{x_2}^{T_2} \quad \frac{1 + F_{12} T_1 T_2}{T_1 T_2 T_3} S_{x_3}^{T_3}\right) \qquad (3\text{-}183)$$

Assuming the same fabrication conditions as for the cascade configuration in Section 3.8, and substituting Eqs. (3-183) and (3-174) into Eq. (3-60b), yields the following expression for S_T:

$$S_T \simeq |H|^2 \left(\frac{1}{|T_1 T_2 T_2|^2} \mathbf{d}_{T_1}^{t*} \mathbf{P}_1 \mathbf{d}_{T_1} + \frac{1}{|T_1 T_2 T_3|^2} \mathbf{d}_{T_2}^{t*} \mathbf{P}_2 \mathbf{d}_{T_2}\right.$$
$$\left. + \frac{|1 + F_{12} T_1 T_2|^2}{|T_1 T_2 T_3|^2} \mathbf{d}_{T_3}^{t*} \mathbf{P}_3 \mathbf{d}_{T_3}\right) \qquad (3\text{-}184)$$
$$= \sum_{i=1}^{3} S_{T_i} \qquad (3\text{-}185)$$

Note that the approximate equality is due to the deletion of the small sensitivity contributions from the components that comprise the external multiple-loop feedback network.

For narrow-band band-pass filters where the T_i are high-Q band-pass sections, we can use Eqs. (3-177) to rewrite Eq. (3-185) into the following useful form:

$$S_T \simeq \frac{1}{|D|^2}\left(|D_1 D_2 D_3|^2 \frac{1}{|D_1|^2} \kappa_1 + |D_1 D_2 D_3|^2 \frac{1}{|D_2|^2} \kappa_2\right.$$
$$\left. + |D_1 D_2 D_3 + F_{12} N_1 N_2 D_3|^2 \frac{1}{|D_3|^2} \kappa_3\right) \qquad (3\text{-}186a)$$

where

$$D = D_1 D_2 D_3 + F_{12} N_1 N_2 D_3 + F_{13} N_1 N_2 N_3$$

Similarly, it can be shown that

$$S_\alpha \simeq \left[\operatorname{Re}\left(\frac{D_1 D_2 D_3}{D_1 D}\right)\right]^2 \kappa_1 + \left[\operatorname{Re}\left(\frac{D_1 D_2 D_3}{D_2 D}\right)\right]^2 \kappa_2$$
$$+ \left[\operatorname{Re}\left(\frac{D_1 D_2 D_3 + F_{12} N_1 N_2 D_3}{D D_3}\right)\right]^2 \kappa_3 \qquad (3\text{-}186b)$$
$$= \sum_{i=1}^{3} S_{\alpha_i}$$

$$S_\beta \simeq \left[\operatorname{Im}\left(\frac{D_1 D_2 D_3}{D_1 D}\right)\right]^2 \kappa_1 + \left[\operatorname{Im}\left(\frac{D_1 D_2 D_3}{D_2 D}\right)\right]^2 \kappa_2$$
$$+ \left[\operatorname{Im}\left(\frac{D_1 D_2 D_3 + F_{12} N_1 N_2 D_3}{D D_3}\right)\right]^2 \kappa_3 \qquad (3\text{-}186c)$$
$$= \sum_{i=1}^{3} S_{\beta_i}$$

Comparing Eq. (3-186) with Eq. (3-177a) with $N = 3$, we see that the FLF realization will have lower transfer function sensitivity at frequencies where

$$(S_T)_{FLF} < (S_T)_{cascade} \qquad (3\text{-}187a)$$

or

$$\left| \frac{D_2 D_3}{D} \right|^2 + \left| \frac{D_1 D_3}{D} \right|^2 + \left| \frac{D_1 D_2 + F_{12} N_1 N_2}{D} \right|^2 < \frac{1}{|\hat{D}_1|^2} + \frac{1}{|\hat{D}_2|^2} + \frac{1}{|\hat{D}_3|^2}$$

$$(3\text{-}187b)$$

The use of the caret over the D_i's for the cascade design is to emphasize that they will differ in both ω_0 and Q from the D_i's required for the FLF design. The design of FLF active filters is treated in detail in Chapter 5. It should be noted that an improperly designed FLF filter can have poorer sensitivity than an equivalent cascade design. This is demonstrated in Chapter 5.

3.10 MONTE CARLO SIMULATION

Thus far we have discussed approximate methods for evaluating the statistical quantities

$$E\left(\int_{\omega_1}^{\omega_2} \left| \frac{\Delta H}{H'} \right|^2 d\omega \right), \qquad E\left(\int_{\omega_1}^{\omega_2} \left| \frac{\Delta H}{H} \right|^2 d\omega \right), \qquad E\left(\int_{\omega_1}^{\omega_2} |\Delta H|^2 d\omega \right)$$

The primary advantage of the approximate methods is that they permit statistical network variations to be evaluated with modest amounts of computer time. They are reasonably accurate if they are carefully used according to the guidelines established in Section 3.10. This computational efficiency is particularly important when sensitivity is to be minimized (see the next section) and when sensitivity[6] is evaluated using a hand calculator or computer.

However, when accuracies, beyond the capabilities of approximate methods, are desired, one may evaluate sensitivity using a brute-force method known as Monte Carlo simulation. The price paid for this accuracy is a significant increase in computing time. For those readers with ready access to a large digital computer, this capability may already exist in a circuit analysis package on the library in your computing facility. In this case the computing costs associated with Monte Carlo simulations may be offset by the engineering time required to write special computer/calculator programs for, or hand compute, the approximate sensitivity measures. Thus it is clear that both forms of sensitivity analysis have their place, with required accuracy and computer tool availability by-and-large dictating the appropriate method to be used.

To implement a Monte Carlo simulation, one simply analyzes the circuit a statistically significant number of times, usually 100 to 10,000 times. The integrity and repeatability of the resulting statistics increases with the number

---

[6]The integration can be performed using any one of several numerical algorithms, such as the trapezoidal rule.

of analyses. In each analysis the component value(s) are changed, using a random generator, according to prespecified probability distribution(s). Thus, in the simulation, each random circuit component or parameter in the active filter is replaced, in principle, by a random-number generator that generates component values according to known probability distributions [B3, B4]

$$x_i = x_{0_i}(1 + \eta_i) \tag{3-188}$$

where x_{0_i} is the nominal value for component x_i and η_i is a random variable of known probability distribution, whose value is obtained from a random-number generator. Typically, the η_i are assumed to be independent random variables. Correlation (i.e., statistical dependence) is sometimes taken into account by artificial means. Correlation is, however, an important statistical mechanism in integrated-circuit filters, such as switched capacitor filters (see Chapter 6), where like elements are subjected to errors of the same sign and nearly equal value. For example, MOS capacitors in switched capacitor filters may vary significantly in value ($\Delta C_i/C_i \simeq \pm 20\%$) from one unit to another, but they are highly correlated such that these deviations tend to cancel when the capacitors are ratioed [i.e., $\Delta(C_i/C_j)/(C_i/C_j) < 0.5\%$]. One means for accounting for this correlation phenomenon is to identify capacitor ratios as the random variables (x_i) rather than the individual capacitors. Here manufacturing data would be maintained on the values of ratioed capacitors, from which appropriate probability distributions for generating the η_i can be determined. It is noted that for symmetric probability distributions, such as the uniform and Gaussian distributions shown in Fig. 3-11, x_{0_i} is the mean value and η_i is a zero mean sym-

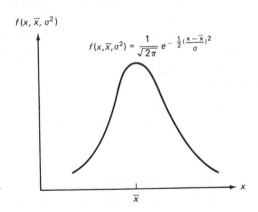

$$f(x, \bar{x}, \sigma^2)$$

$$f(x, \bar{x}, \sigma^2) = \frac{1}{\sqrt{2\pi}} e^{-\frac{1}{2}(\frac{x - \bar{x}}{\sigma})^2}$$

Fig. 3-11 Gaussian (normal) distribution.

metrically distributed random variable. For jointly Gaussian η_i, correlation is readily taken into account via the covariance matrix. Extensive use of the covariance matrix was made in Sections 3.6 and 3.7.

The transfer function variations $|\Delta H/H'|$, $|\Delta H/H|$, and $|\Delta H|$ are also random variables with unknown distributions. It is, of course, the purpose of this exercise to determine one or more of these distributions, or at least a selected number of their statistical moments. The random variables $|\Delta H/H'|$,

$|\Delta H/H|$ and $|\Delta H|$ are somewhat different from random variables η_i or x_i in that first they are functions of random variables x_i, which may be represented as a random vector $\mathbf{x}$, but they are also a function of the deterministic frequency variable ω. Thus these variations can be considered random processes, the value at each frequency being a random variable. With these concepts in mind, we shall now define

$$\left|\frac{\Delta H}{H'}\right|_i = \left|\frac{\Delta H}{H'}\right|_{\omega=\omega_i} \quad \text{with } \omega_1 \leq \omega_i \leq \omega_2 \qquad (3\text{-}189)$$

Similar definitions are assumed for random variables $|\Delta H|_i$ and $|\Delta H/H|_i$.

The Monte Carlo analyses would now proceed by evaluating at each random vector $\mathbf{x}$ values of, for instance, $|\Delta H/H'|_i$ for each frequency point within the frequency range $\omega_1 \leq \omega_i \leq \omega_2$. From these data one may construct, for each frequency point, a histogram or probability distribution for the appropriate random transfer function distribution. This is usually much more information than is required and perhaps too much to be readily interpreted. It is of more practical value to compute a selected number of the statistical moments, such as the first moment (m_1) or mean and the second moment (m_2), from which we may calculate the dispersion or variance $\sigma^2 = m_2 - m_1^2$ or the standard deviation σ. Moments m_1 and m_2 are readily computed, at each frequency ω_i according to the following relations:

$$m_1(\omega_i) = E(|Z|_i) = \frac{1}{\text{NSAMP}} \sum_{j=1}^{\text{NSAMP}} |Z_j|_i \qquad (3\text{-}190\text{a})$$

$$m_2(\omega_i) = E(|Z|_i^2) = \frac{1}{\text{NSAMP}} \sum_{j=1}^{\text{NSAMP}} |Z_j|_i^2 \qquad (3\text{-}190\text{b})$$

with Z_j being the jth sample of $\Delta H/H'$ and NSAMP being the total number of samples or network analyses. Then, using numerical integration, their integrals over the limits ω_1, ω_2 can be readily evaluated. For example, using trapezoidal rule integration,

$$E\left(\int_{\omega_1}^{\omega_2} \left|\frac{\Delta H}{H'}\right|^2 d\omega\right) = \left\{\tfrac{1}{2}[m_2(\omega_1) + m_2(\omega_{N_f})] + \sum_{i=2}^{N_f-1} m_2(\omega_i)\right\} \Delta\omega \qquad (3\text{-}191)$$

where N_f is the number of discrete frequency points and $\Delta\omega$ is the frequency step size, with $\Delta\omega = (\omega_2 - \omega_1)/N_f$. Thus, $E\left(\int_{\omega_1}^{\omega_2} |\Delta H/H'|^2 d\omega\right)$ may be calculated "exactly" using Eq. (3-191) or approximately using Eq. (3-147). For incremental variations one may also determine the gain and phase response sensitivities by merely keeping track of both the real and imaginary parts of random complex variable Z_j:

$$[m_1(\omega_i)]_\alpha = E\left(\left|\frac{\Delta\alpha}{\alpha'}\right|_i\right) = \frac{1}{\text{NSAMP}} \sum_{j=1}^{\text{NSAMP}} \text{Re}\,(Z_j)_i \qquad (3\text{-}192\text{a})$$

$$[m_2(\omega_i)]_\alpha = E\left(\left|\frac{\Delta\alpha}{\alpha'}\right|_i^2\right) = \frac{1}{\text{NSAMP}} \sum_{j=1}^{\text{NSAMP}} [\text{Re}\,(Z_j)_i]^2 \qquad (3\text{-}192\text{b})$$

Similarly, for phase function sensitivity,

$$[m_1(\omega_i)]_\beta = E(|\Delta\beta|_i) = \frac{1}{\text{NSAMP}} \sum_{j=1}^{\text{NSAMP}} \text{Im}\,(Z_j)_i \qquad (3\text{-}193\text{a})$$

$$[m_2(\omega_i)]_\beta = E(|\Delta\beta|_i^2) = \frac{1}{\text{NSAMP}} \sum_{j=1}^{\text{NSAMP}} [\text{Im}\,(Z_j)_i]^2 \qquad (3\text{-}193\text{b})$$

Equivalent expressions for $E|\Delta H/H|$, $E|\Delta H/H|^2$, $E|\Delta\alpha/\alpha|$, $E|\Delta\alpha/\alpha|^2$, $E|\Delta\beta|_i$, and $E|\Delta\beta|_i^2$ can be obtained with $Z = \Delta H/H$.

For large variations (i.e., large σ_{x_i}), $m_1(\omega_i) \neq [m_1(\omega_i)]_\alpha + [m_1(\omega_i)]_\beta$ and $m_2(\omega_i) \neq [m_2(\omega_i)]_\alpha + [m_2(\omega_i)]_\beta$. For these cases $[m_1(\omega_i)]_{\alpha,\beta}$ and $[m_2(\omega_i)]_{\alpha,\beta}$, in Eqs. (3-190), must be computed independently with $Z = |\Delta\alpha/\alpha'|, |\Delta\beta|$.

Since Monte Carlo analysis does not involve a series expansion and convergence is not an issue, it would be perfectly appropriate for the reader to question the value of retaining all three of the statistical response variation forms. Since $\Delta H/H$ and $\Delta H/H'$ are undefined when $N(j\omega) = 0$, $E|\Delta H|^2$ retains its utility. The usefulness of $E|\Delta H/H'|^2$ for calculating the statistical response variance for band-pass filters is appreciated when $|\Delta H/H'|$ and $|\Delta H/H|$ are expressed as follows:

$$\left|\frac{\Delta H}{H'}\right| = \left|\frac{H' - H}{H'}\right| = \left|1 - \frac{H}{H'}\right| = \left|1 - \frac{N}{N'}\frac{D'}{D}\right| \qquad (3\text{-}194)$$

It is noted for band-pass transfer functions that H is far more sensitive to changes in D than to changes in N, which results in many of the simplifying approximations made in Section 3.6. Also, it was shown in that section that center frequency shifts were more important than Q shifts. Thus, as $|\Delta x/x| \to$ large, this implies that $|\Delta\omega_0| \to$ large and one can further show, for $\Delta\omega_0 \neq 0$, that $|D'| > |D|$ with $|D'|/|D| \to$ large as $|\Delta\omega_0| \to$ large. As a result $|\Delta H/H'|$ becomes large with large component variations $|\Delta x|$. One may say that $|\Delta H/H'|$, for band-pass H, is very sensitive to large $|\Delta x|$. Now let us look at $|\Delta H/H|$ when

$$\left|\frac{\Delta H}{H}\right| = \left|\frac{H' - H}{H}\right| = \frac{N'/D' - N/D}{N/D} \left|\frac{N'}{N}\frac{D}{D'} - 1\right| \qquad (3\text{-}195)$$

Using these same arguments, one can establish that as $|D'| \ll |D|$, $|\Delta H/H| \to 1$ with $|\Delta x| \to$ large. Therefore, one can say that $|\Delta H/H|$, for band-pass H, is relatively insensitive to large $|\Delta x|$. One may argue similarly that for band-reject H, both $|\Delta H/H|$ and $|\Delta H|$ are very sensitive to large $|\Delta x|$.

It is appropriate at this point to briefly discuss random-number generation, particularly the generation of random numbers according to prespecified distributions. Since many of scientific subroutine libraries available in most, if not all, computer centers contain at least one random-number-generation routine, this book will assume that availability. Furthermore, statistical function libraries for several programmable calculators and microcomputers contain random-number-generation programs. Most random-number generators will generate random numbers over an interval from 0 to 1 according to a uniform probability distribution. The accuracy of the Monte Carlo analysis, which is in principle exact, is totally determined by the degree to which we have modeled the true manufacturing distributions and the degree of "randomness" of the random-number generator used. Since random-number generators are determined in a deterministic manner using a computer program, the "randomness" of the generated numbers should be scrutinized. A sequence of random numbers, generated by a particular random-number generator to obey a known probability distribution, can be tested in several ways, such as:

1. Calculating the mean and variance and cross-checking the results with their specified values.
2. Testing pairs of random numbers to ensure that they are uncorrelated. This can be done by merely adding pairwise products of two successive random-number generations for NSAMP cases. If the numbers are truly uncorrelated, then $\sum_{j=1}^{NSAMP} (x'_j - \bar{x}')(x''_j - \bar{x}'') = 0$ (or a very small number), where x' and x'' are two random variables and $\bar{x}'$ and $\bar{x}''$ are their respective means.
3. Reconstructing the probability distribution as a histogram and comparing shapes.

Once the uniformly distributed random number generator has been determined to be sufficiently random, one may then use it as a basis for generating random numbers to most all other distributions. For example, from the central limit theorem, the distribution of the random variable formed from the sum of n uniformly distributed independent variables tends to approach the Gaussian distribution as $n \rightarrow \infty$. For a uniform basis distribution, which generates numbers over the interval $(0, 1)$, $n = 12$ is typically used to yield a Gaussian distribution having a mean $\bar{x} = \frac{1}{2}$ and a variance $\sigma^2 = 1$. Other mean and variance values can be accommodated by modifying the random numbers x, with Gaussian distribution and $\sigma^2 = 1$, $\bar{x} = \frac{1}{2}$, according to the following algorithm:

$$x = \sigma \left(x - \frac{1}{2} + \frac{\bar{x}}{\sigma} \right) \tag{3-196}$$

●

EXAMPLE 3-9

Consider the network shown in Fig. 3-1, with transfer function as given in Eq. (3-2). Let the design equations be

$$\frac{1}{R^2 C^2} = 1 \tag{3-197a}$$

$$\frac{3 - K}{RC} = 0.1 \tag{3-197b}$$

To realize this design, let the normalized element values be $K = 2.9$, $R = C = 1.0$.

Let us now assume that only R and C are varying in a random fashion about their nominal values. For simplicity we shall let NSAMP $= 12$, and R, C, $\Delta R/R$, and $\Delta C/C$ be the values given in Table 3-3. It is noted that for an actual

TABLE 3-3[a]

R	C	$\dfrac{\Delta R}{R}$	$\dfrac{\Delta C}{C}$	$\left(\dfrac{\Delta R}{R}\right)^2$	$\left(\dfrac{\Delta C}{C}\right)^2$	$\left(\dfrac{\Delta R}{R}\right)\left(\dfrac{\Delta C}{C}\right)$	$\left\|\dfrac{\Delta H}{H'}\right\|^2_{\omega=1}$
1.010	1.005	0.010	0.005	1.00×10^{-4}	2.50×10^{-5}	5.00×10^{-5}	0.09220
0.997	0.998	-0.003	-0.002	9.00×10^{-6}	4.00×10^{-6}	6.00×10^{-6}	0.00995
1.003	0.990	0.003	-0.010	9.00×10^{-6}	1.00×10^{-4}	-3.00×10^{-5}	0.01968
0.990	1.005	-0.010	0.005	1.00×10^{-4}	2.50×10^{-5}	-5.00×10^{-5}	0.01018
1.003	1.002	0.003	0.002	9.00×10^{-6}	4.00×10^{-6}	6.00×10^{-6}	0.01010
0.997	1.010	-0.003	0.010	9.00×10^{-6}	1.00×10^{-4}	-3.00×10^{-5}	0.01962
1.010	0.995	0.010	-0.005	1.00×10^{-4}	2.50×10^{-5}	-5.00×10^{-5}	0.00987
0.997	1.002	-0.003	0.002	9.00×10^{-6}	4.00×10^{-6}	-6.00×10^{-6}	0.00041
0.997	0.990	-0.003	-0.010	9.00×10^{-6}	1.00×10^{-4}	3.00×10^{-5}	0.06659
1.003	1.010	0.003	0.010	9.00×10^{-6}	1.00×10^{-4}	3.00×10^{-5}	0.06897
0.990	0.995	-0.010	-0.005	1.00×10^{-4}	2.50×10^{-5}	5.00×10^{-5}	0.08829
1.003	0.998	0.003	-0.002	9.00×10^{-6}	4.00×10^{-6}	-6.00×10^{-6}	0.00040
Σ		0.000	0.000	5.16×10^{-4}	4.72×10^{-4}	0.00	0.39624

[a]These numbers for $\Delta R/R$ and $\Delta C/C$ were chosen intentionally to give $\sum \Delta R/R = \sum \Delta C/C = 0$ to have them independent. In a truly random generator with very large samples $> 10,000$, the sums generally come to $10^{-12} \simeq 0$.

$$\sigma_R = \frac{\overset{N}{\sum} \left(\dfrac{\Delta R}{R}\right)^2}{\text{NSAMP}} = \frac{5.16 \times 10^{-4}}{12} = 0.00656$$

$$\sigma_C = \frac{\overset{N}{\sum} \left(\dfrac{\Delta C}{C}\right)^2}{\text{NSAMP}} = \frac{4.72 \times 10^{-4}}{12} = 0.00627$$

$$\rho_{RC} = \frac{\overset{N}{\sum} \left(\dfrac{\Delta R}{R}\right)\left(\dfrac{\Delta C}{C}\right)}{\text{NAMP } \sigma_R \sigma_C} = \frac{0.000}{12(0.00656)(0.00627)} = 0.000$$

From Eq. (3–190 b),

$$m_2(\omega_0) = m_2(1) = E\left(\left\|\frac{\Delta H}{H'}\right\|^2_{\omega=1}\right) = \frac{\sum |\Delta H/H'|^2_{\omega=1}}{\text{NSAMP}} = \frac{0.39624}{12} = 0.03302$$

Monte Carlo simulation, usually NSAMP ≥ 100. Furthermore, we shall assume that $\omega = \omega_0 = 1$ is the only frequency of interest. Computing the means, standard deviation, and correlation coefficient for the R and C variations reveals that $\Delta R/R$ and $\Delta C/C$ are zero-mean uncorrelated random variables with $\sigma_R = 0.656\%$ and $\sigma_C = 0.627\%$. The resulting second moment, computed according to Eq. (3-193b) at $\omega = 1$ with $Z = \Delta H/H$, is

$$m_2(1) = 0.03302$$

●

3.11 COMPARISON OF MONTE CARLO AND APPROXIMATE SENSITIVITY ANALYSIS

To illustrate the accuracy of the approximate measures given in this chapter and to illustrate the difference between $E|\Delta H/H|^2$ and $|\Delta H/H'|^2$, let us consider the following two examples.

●

EXAMPLE 3-10

To evaluate the accuracy of $(S_T)_N$ and $(M_T)_N$, let us first compare the integrand measure $(S_T)_N$ plotted in Fig. 3-8 with $E|\Delta H|^2$ obtained exactly via a Monte Carlo simulation. Computation of Eq. (3-190b) with NSAMP $= 100$ and $\Delta x_i/x_i$ independent Gaussian random variables with $\sigma = 1\%$ yields the curve shown in Fig. 3-12. Superimposing the previously calculated approximate $(S_T)_N$ shown in Fig. 3-8 onto the exact $E|\Delta H|^2$ curve, one observes the good accuracy of the approximate analysis. Finally, integrating the curve for the Monte Carlo-derived $E|\Delta H|^2$ curve, in the same manner indicated previously for $(M_T)_N$, yields

$$E \int_{\omega_1}^{\omega_2} |\Delta H|^2 \, d\omega = 0.00127$$

Fig. 3-12 Comparison of $(S_T)_N$ versus frequency computed via Eq. (3-155) with $E|\Delta H|^2$ versus frequency obtained from Monte Carlo analysis. The random component (R's and C's) variations are assumed independent, Gaussian distributed with 1.0% standard deviations.

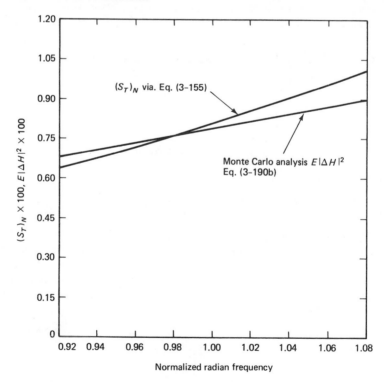

Comparing this value to $(M_T)_N = 0.00130$, we see that the agreement is to better than 2%.

EXAMPLE 3-11

Consider now the sensitivity analysis of a three-section (sixth-order) narrow-band Butterworth bandpass filter. Let us initially assume that the normalized component variations are independent, normally distributed random variables with variance $\sigma = 1\%$. Furthermore, the filter is designed to have an overall $f_0/\omega_{3dB} \equiv \hat{Q}$ ratio of 25 and the biquadratic sections T_i are high-Q bandpass sections of the form

$$H_i = \frac{H_{\beta i} s_n / Q_i}{s_n^2 + s_n / Q_i + 1} = \frac{N_i}{D_i}$$

where $s_n = s/\omega_0$. The $H_{\beta i}$, Q_i, F_{12}, and F_{13} values used to realize the FLF filter are given in Table 3-4. These parameters were computed using techniques

TABLE 3-4   FLF filter parameters for Example 3-9

i	$QH_{\beta i}$	Q_i	F_{1i}	κ_i
1	2.0983	44.154	—	$4 \times \sigma^2$
2	1.8220	44.154	0.5595	$4 \times \sigma^2$
3	1.227	28.816	0.0943	$4 \times \sigma^2$

discussed in Chapter 5. Finally, it is assumed that a low-Q sensitivity active-RC implementation is used to realize each T_i, where $\kappa_i \simeq 4 \times \sigma^2$ as defined in Eq. (3-177b). Using Eq. (3-117c) or Eq. (3-177b), it can be readily verified that this expression for κ_i results from a low-Q sensitivity network (i.e., $|S_X^Q| \leq 1$) where the center frequency is defined by four component values [i.e., $\omega_0 = R_1 C_1 R_2 C_2)^{-1/2}$].

Computing S_T over the frequency band $0.96 = \omega_1/\omega_0 \leq \omega_n \leq \omega_2/\omega_0 = 1.04$, which corresponds to about twice the filter bandwidth, results in the $*$ curve shown in Fig. 3-13. Let us now compare S_T with corresponding Monte Carlo simulations of $E|\Delta H/H'|^2$ and $E|\Delta H/H|^2$. These curves are superimposed on to Fig. 3-13 for direct comparison with S_T. Note that:

1. $E|\Delta H/H'|^2 \neq E|\Delta H/H|^2$, except at two crossover frequencies.
2. $M_T \simeq E|\Delta H/H'|^2$ according to Eq. (3-147).
3. $\sigma = 1\%$ corresponds to a large component variation for this network.

Let us now recompute the curves S_T, $E|\Delta H/H'|^2$, and $E|\Delta H/H|^2$ for $\sigma = 0.1\%$. These curves are shown in Fig. 3-14. Note that:

1. $M_T \simeq E|\Delta H/H'|^2 \simeq E|\Delta H/H|^2$.
2. At the band edges, where the curves peak, M_T is seen to more closely approximate the Monte Carlo simulation for $E|\Delta H/H'|^2$ than that for $E|\Delta H/H|^2$.
3. $\sigma = 0.1\%$ corresponds to a small or incremental variation for this network.

●

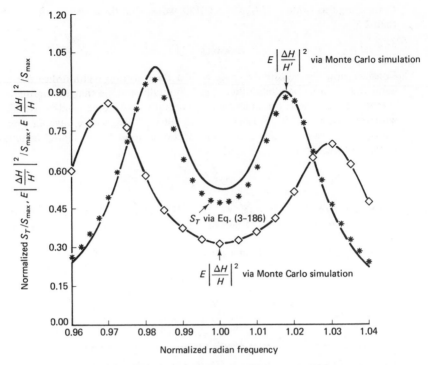

Fig. 3-13 Comparison of S_T via Eq. (3-186), $E|\Delta H/H'|^2$ via Monte Carlo simulation, and $E|\Delta H/H|^2$ via Monte Carlo simulation versus frequency for a high-Q sixth-order FLF band-pass filter ($f_0/BW = 25$). The passive-component variations are assumed to be statistically independent, Gaussian distributed with 1% standard deviations ($S_{max} = 0.949$).

3.12 SENSITIVITY OPTIMIZATION

Active filters are not efficient in their use of components; that is, there are more components (both active and passive) than are needed to completely specify the filter. For example, consider the second-order filter of Fig. 3-1. This filter realizes a transfer function of the form

$$\frac{V_o}{V_i} = \frac{a_0}{s^2 + b_1 s + b_0} \qquad (3\text{-}198)$$

From Eq. (3-165) it is obvious that only *three* components are required to specify the transfer function: one for each coefficient a_0, b_1, and b_0. However, the active filter in Fig. 3-1 contains *five* specifiable components: R_1, R_2, C_1, C_2, and K. Thus, to arrive at the transfer function Eq. (3-1) we arbitrarily set $R_1 = R_2 = R$ and $C_1 = C_2 = C$. This choice reduces the number of independent variables to three: R, C, and K. However, in this exercise we have thrown away two potentially valuable degrees of freedom or independent variables.

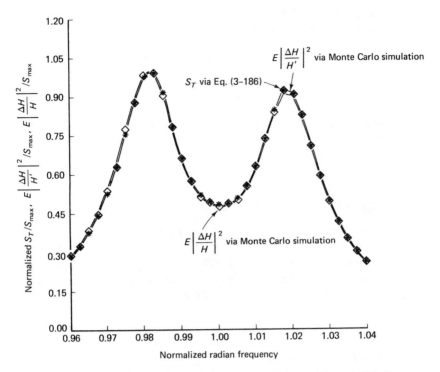

Fig. 3-14 Same as Fig. 3-13, but passive components have 0.1% standard deviations ($S_{max} = 0.0078$).

Rather than discarding this freedom, let us attempt to reduce the filter sensitivity, by properly adjusting these free parameters.

●

EXAMPLE 3-12

To illustrate the manner in which sensitivity can be minimized in an actual filter, let us consider the following simple example, the low-pass active filter shown in Fig. 3-1. The complete transfer function for this second-order network is given in Eq. (1-50) and repeated below for convenience:

$$H(s) = \frac{V_o}{V_i} = \frac{K/R_1R_2C_1C_2}{s^2 + s\left(\dfrac{1}{R_1C_1} + \dfrac{1}{R_2C_1} + \dfrac{1-K}{R_1C_2}\right) + \dfrac{1}{R_1C_1R_2C_2}} \quad (3\text{-}199)$$

We shall assume that the desired filter response has the following denominator coefficients:

$$b_1 = \frac{1}{R_1C_1} + \frac{1}{R_2C_1} + \frac{1-K}{R_2C_2} = 0.1 \quad (3\text{-}200a)$$

$$b_0 = \frac{1}{R_1C_1R_2C_2} = 1 \quad (3\text{-}200b)$$

For simplicity and ease of fabrication, let $C_1 = C_2$. With this assumption, Eqs. (3-200) reduce to

$$b_1 = \frac{1}{R_1 C} + \frac{1}{R_2 C}(2 - K) = 0.1 \qquad (3\text{-}201a)$$

$$b_0 = \frac{1}{R_1 R_2 C^2} = 1 \qquad (3\text{-}201b)$$

Let us now express the dependent parameters $R_1 C$ and K in terms of the independent parameter $R_2 C$. From Eq. (3-201), with $\omega_{3dB} = 1$,

$$R_1 C = \frac{1}{R_2 C} \qquad (3\text{-}202)$$

Substituting Eq. (3-202) into Eq. (3-201a) and solving for K yields

$$K = (R_2 C)^2 - 0.1 R_2 C + 2 \qquad (3\text{-}203)$$

It is noted that the desired response, being a high Q peaking low-pass, has characteristics similar to those of a high-Q band-pass. Thus, according to the guidelines provided in Section 3.7, we shall use the following sensitivity measure:

$$M_T = \int_{\omega_1}^{\omega_2} (\mathbf{d}_T^{t*} \mathbf{P} \mathbf{d}_T)\, d\omega \qquad (3\text{-}204)$$

as the measure to minimize.

Let us make a further simplifying assumption that only the resistor R_2 varies from unit to unit or with changes in the ambient environment. Thus, from Eq. (3-144),

$$\mathbf{d} = \frac{-1}{D}\left(j\omega\, R_2 \frac{\partial b_1}{\partial R_2} + R_2 \frac{\partial b_0}{\partial R_2} \right) \frac{\Delta R_2}{R_2} \qquad (3\text{-}205a)$$

where

$$R_2 \frac{\partial b_1}{\partial R_2} = -\frac{1}{R_2 C}(2 - K) \quad \text{and} \quad R_2 \frac{\partial b_0}{\partial R_2} = -1 \qquad (3\text{-}205b)$$

The integrand of Eq. (3-204) can now be expressed in terms of the independent parameter $R_2 C$. From Eq. (3-59), we have

$$\mathbf{d}_T^{t*} \mathbf{P} \mathbf{d}_T = \frac{1}{|D|^2}\left[1 + \omega^2 \frac{(2 - K)^2}{(R_2 C)^2} \right] \sigma_{R_2}^2 \qquad (3\text{-}206)$$

Substituting Eq. (3-203) into Eq. (3-206) gives $\mathbf{d}_T^{t*} \mathbf{P} \mathbf{d}_T$ as a function of only $R_2 C$:

$$\mathbf{d}_T^{t*} \mathbf{P} \mathbf{d}_T = \frac{1}{|D|^2}\left[1 + \omega^2 \frac{(0.1 - R_2 C)^2}{R_2 C} \right] \sigma_{R_2}^2 \qquad (3\text{-}207)$$

To further simplify the exercise, consider the integral in Eq. (3-204) evaluated at a single frequency $\omega = 1$ and let $x = R_2 C$ in Eq. (3-207), which reduces M_T to

$$M_T = 100\left[1 + \frac{(0.1 - x)^2}{x} \right] \sigma_{R_2}^2 \qquad (3\text{-}208)$$

From Eq. (3-208) the optimum value of x is found by setting $dM_T/dx = 0$:

$$\frac{dM_T}{dx} = 100\sigma_{R_2}^2\frac{x^2 - (0.1)^2}{x^2} = 0 \tag{3-209}$$

Hence, the minimum value of M_T is at $x = R_2C = 0.1$. To check whether this value of R_2C yields the minimum value of M_T, we must show that d^2M_T/dx^2 is positive: (In this case the answer is obvious by inspection.)

$$\frac{d^2M_T}{dx^2} = 100\sigma_{R_2}^2\frac{2}{0.1} > 0 \tag{3-210}$$

Thus, the optimum value is $(R_2C)_0 = 0.1$. Finally, we can determine $(R_1C)_0$ and K_0 from Eqs. (3-202) and (3-203):

$$(R_1C)_0 = 10 \quad \text{and} \quad K_0 = 2 \tag{3-211}$$

●

Although this example nicely illustrates the sensitivity minimization process, it is much oversimplified. In general, all element values vary simultaneously and the optimization is performed over a range of frequencies rather than at a single frequency. Furthermore, fabrication limitations and/or other performance specifications force the designer to place limitations on the values that each component can assume. These limitations, often referred to as *constraints*, further complicate the optimization process. Thus, in general, numerical methods and a digital computer are required to perform an optimization. Optimization methods have been, and still are, a subject of considerable study. Because of the importance of optimization and its applications to a variety of disciplines, many papers and books have been devoted to the subject. An overview of selected numerical optimization methods, particularly those which are most easily applied to sensitivity minimization, can be found in [B2] and [P10–P12]. The minimization of sensitivity in second- and higher-order active-*RC* networks is discussed further in Chapters 4 and 5.

REFERENCES

Books

B1. HEINLEIN, W., AND H. HOLMES, *Active Filters for Integrated Circuits.* Vienna: Oldenbourg Verlag, 1974, Chap. 4.

B2. PIERRE, D. A., *Optimization Theory with Applications.* New York: Wiley, 1969.

B3. PAPOULIS, A., *Probability Random Variables and Stochastic Processes.* New York: McGraw-Hill, 1965.

B4. COOPER, G. R., AND G. D. MCGILLEM, *Probablistic Methods of Signals and System.* New York: Holt, Rinehart and Winston, 1971.

Papers

P1. GOLDSTEIN, A. J., AND F. F. KUO, "Multiparameter Sensitivity," *IRE Trans. Circuit Theory,* CT-18 (1961), 177–178.

P2. SCHOEFFLER, J. D., "The Synthesis of Minimum Sensitivity Networks," *IEEE Trans. Circuit Theory*, CT-11, (1964), 271–276.

P3. HAYKIN, S. S., AND W. J. BUTLER, "Multiparameter Sensitivity Indexes of Performance for Linear Time-Invariant Networks," *Proc. IEE (Lond.)*, 117 (1970), 1239–1247.

P4. ROSENBLUM, A. L., AND M. S. GHAUSI, "Multiparameter Sensitivity in Active *RC* Networks," *IEEE Trans. Circuit Theory* (Special Issue on Active and Digital Networks), CT-18 (1971), 592–599.

P5. BISWAS, R. N., AND E. S. KUH, "A Multiparameter Sensitivity Measure for Linear Systems," *IEEE Trans. Circuit Theory*, CT-18 (1971), 718–719.

P6. ACAR, C., AND M. GHAUSI, "Statistical Multiparameter Sensitivity Measure of Gain and Phase Functions," *Int. J. Circuit Theory Appl.*, 5 (1977), 13–22.

P7. ACAR, A., K. LAKER, AND M. S. GHAUSI, "Statistical Multiparameter Sensitivity Measure in High *Q* Networks," *J. Franklin Inst.*, 280 (October 1975), 281–297.

P8. LAKER, K. R., AND M. S. GHAUSI, "A Large Change Multiparameter Sensitivity," *J. Franklin Inst.*, 298 (December 1974), 395–414.

P9. BOCTOR, S. A., "A Novel Second Order Canonical *RC*-Active Realization of a High Pass Notch Filter," *IEEE Trans. Circuits Syst.*, CAS-22 (May 1975), 397–404.

P10. ROSENBLUM, A. L., AND M. S. GHAUSI, "Sensitivity Minimization in Active *RC* Networks," *J. Franklin Inst.*, 294 (August 1972), 95–111.

P11. FLEISCHER, P. E., "Sensitivity Minimization in a Single Amplifier Biquad Circuit," *IEEE Trans. Circuits Syst.*, CAS-23 (January 1976), 45–55.

P12. LAKER, K. R., AND M. S. GHAUSI, "A Comparison of Active Multiple-Loop Feedback Techniques for Realizing High Order Bandpass Filters," *IEEE Trans. Circuits Syst.*, CAS-21 (November 1974), 774–783.

P13. MOSCHYTZ, G. S., "Gain-Sensitivity Product—A Figure of Merit for Hybrid-Integrated Filters Using Single Operational Amplifier," *IEEE J. Solid-State Circuits*, SC-6 (June 1971), 103–110.

PROBLEMS

3.1 If y, u, and v are single-valued differentiable functions of x, use the definition of classical sensitivity to derive the following properties:

(a) $S_x^{y(u)} = S_u^y S_x^u$

(b) $S_x^{1/y} = -S_x^y$

(c) $S_x^{uv} = S_x^u + S_x^v$

3.2 Under the same assumptions as in Prob. 3.1, derive the following if k is a constant:

(a) $S_x^{ky} = S_x^y$

(b) $S_x^{y+k} = \dfrac{y}{y+k} S_x^y$

(c) $S_x^{u+v} = \dfrac{1}{u+v}(u S_x^u + v S_x^v)$

3.3 Derive Eq. (3-15).

3.4 Derive Eqs. (3-27) and (3-28).

3.5 For the circuit shown in Fig. P3-5 (assume an ideal op amp):

(a) Show that $\dfrac{V_o}{V_i} = -\dfrac{1}{RC}\,\dfrac{s}{s^2 + \left(\dfrac{1}{R_1 C} + \dfrac{2}{R_2 C}\right)s + \dfrac{1}{R_1 R_2 C^2}}$.

Fig. P3-5

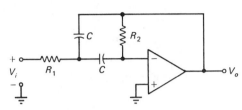

(b) Determine the various sensitivity functions $S^{\omega_0}_{x_i}$ and $S^{Q}_{x_i}$, where x_i are the R's and the C's.

(c) Obtain the expression for the deterministic multiparameter sensitivity given by (3-54).

3.6 For the circuit shown in Fig. P3-6, calculate the following:

Fig. P3-6

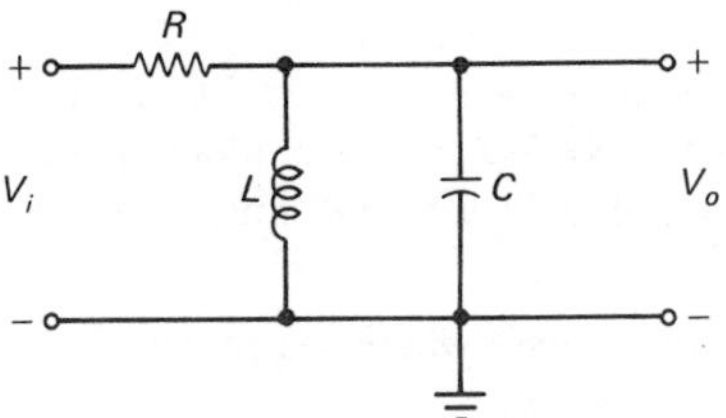

(a) $\sum S^{H}_{x_i}$, where x_i are the RLC components.

(b) The sensitivities of ω_0 and Q relative to $R_1 C$ and L.

(c) For a design value of $\omega_0 = 2\pi(10^5)$ rad/sec and $Q = 10$, determine the changes in ω_0 and ω_{3dB} if:

(1) C changes by $+5\%$.

(2) R changes by $+5\%$.

(d) Find the deterministic multiparameter sensitivity Eq. (3-54) of the circuit and evaluate it at:

(1) $\omega = \omega_0$

(2) $\omega = \omega_{3dB}$

3.7 Determine the multiparameter sensitivity measures M_α, M_β, and M_T over a frequency range $\omega_1 = 2\pi(0.95 \times 10^5)$ to $\omega_2 = 2\pi(1.05) \times 10^5$ rad/sec for the circuit in Fig. P3-6 with the design values $\omega_0 = 2\pi(10^5)$ rad/sec and $Q = 10$. Assume no correlation and a uniform distribution with variance $\sigma^2_{x_i} = 10^{-4}$.

3.8 Determine M_α, M_β, and M_T for Example 3-5 if the correlation between the elements are $\rho_{RR} = \rho_{CC} = +0.8$ and $\rho_{RC} = -0.8$.

3.9 Consider the circuit shown in Fig. P3-9.

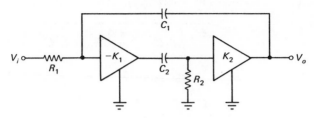

Fig. P3-9

(a) Show that the voltage transfer function is given by

$$\frac{V_o}{V_i} = \frac{-(K_1 K_2 G_1 C_2)s}{s^2[(1 + K_1 K_2)C_1 C_2] + s(G_1 C_2 + G_2 C_1) + G_1 G_2}$$

(b) For $G_1 = G_2 = 1$, $C_1 = C_2 = 1/2Q$, show that $|S^{\omega_0}_{x_i}| = \frac{1}{2}$ and $S^Q_{x_i} = 0$, where x_i are the passive circuit components.

(c) Derive the corresponding expressions for the active sensitivities (i.e., $|S^{\omega_0}_{K_i}|$ and $|S^Q_{K_i}|$). What are the bounds on these for high values of Q (i.e., $Q \geq 10$)?

3.10 (a) For the circuit shown in Fig. P3-9, derive the expression for the voltage transfer function. Include the op amp finite gain–bandwidth product in the derivation; that is, assume that the inverting and noninverting op amps are used to realize $-K_1$ and K_2 and that $a_i = GB/s$.

(b) Determine the circuit element values for a design of $\omega_0 = 10^5$ rad/sec, $Q = 20$, and $K_1 K_2 = 4Q^2$. Assume an ideal op amp and $R_1 = R_2$, $C_1 = C_2$.

(c) Determine the actual values of ω_0 and Q if the op amp GB $= 5 \times 10^5$ rad/sec.

3.11 Repeat Example 3-7 for the case where the correlation coefficients are as follows:

$$\sigma_R^2 = \sigma_C^2 = 10^{-4}, \qquad \rho_{R_1 R_2} = \rho_{C_1 C_2} = 0.8, \qquad \rho_{RC} = 0$$

3.12 Repeat Prob. 3.11 but assume a nonideal op amp with GB $= 10\omega_0$.

3.13 Examine the convergence properties of $\Delta H/H$ and $\Delta H/H'$ where $\Delta H = H' - H$, $H' = H(j\omega, x + \Delta x)$, and $H(j\omega) = N(j\omega, x)/D(j\omega, x)$. For simplicity assume a single-element variation.

Next, consider a special case of

$$H(s, x) = \frac{k(x)s}{s^2 + [\omega_0(x)/Q(x)]s + \omega_0^2(x)}$$

for a nominal value of $\omega_0 = 1$ and $Q = 10$. For what percent of element variations does each series expansion converge?

3.14 For the band-pass circuit shown in Fig. P3-14, the transfer function is given by

$$\frac{V_o}{V_i}(s) = \frac{KG_1 S_1 s}{s^2 + s[S_1(G_1 + G_2 + G_3 - KG_2) + S_2 G_3] + S_1 S_2 G_3(G_1 + G_2)}$$

where $S_i = 1/C_i$ and $G_i = 1/R_i$. Determine the statistical sensitivity measures

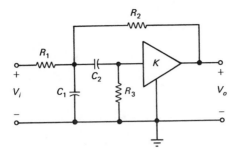

Fig. P3-14

M_α, M_β, and M_T for the following design specifications:

$$\omega_0 = 10^4 \text{ rad/sec}, \qquad Q = 10$$

Assume no correlation between elements and $\sigma_{x_i}^2 = 10^{-4}$.

3.15 Compute M_α, M_β, and M_T given by Eqs. (3-123) to (3-126) for the circuit in Fig. P3-9, taking $\omega_0 = 1$, $Q = 20$, $\omega_1 = \omega_0(1 - 1/2Q)$, $\omega_2 = \omega_0 (1 + 1/2Q)$, and $GB_1 = GB_2 = GB = 1000$. Assume 0.2% tolerances for the R's and C's and a 10% tolerance for the gain–bandwidth products, and no correlation between elements. Also assume that $-K_1$ and K_2 are realized as shown in Figs. 2-4a and 2-6.

3.16 Repeat Prob. 3.15 for $GB = 50$. Note that the ω_0 and Q sensitivities of the circuit are given in Eq. (4-76).

3.17 For the GIC-derived circuit shown in Fig. P3-17:
(a) Show that

$$T(s) = \frac{sC_2G_1(G_4 + G_5) + (1/a_1)G_{11}(G_4 + G_5)(sC_2 + G_3)}{D(s)}$$

where

$$D(s) = s^2C_1C_2G_4\left[1 + \frac{1}{a_2}\left(1 + \frac{G_5}{G_4}\right) + \frac{1}{a_1a_2}\left(1 + \frac{G_5}{G_4}\right)\right]$$

$$+ sC_2G_1G_4\left\{1 + \frac{1}{a_1}\frac{C_1G_3(G_4 + G_5)}{C_2G_1G_4} + \frac{1}{a_2}\frac{(G_1 + G_2)(G_4 + G_5)}{G_1G_4}\right.$$

$$+ \frac{1}{a_1a_2}\left[\frac{G_1G_3(G_4 + G_5)}{C_2G_1G_4} + \frac{(G_1 + G_3)(G_4 + G_5)}{G_1G_4}\right]\right\}$$

$$+ G_2G_3G_5\left[1 + \frac{1}{a_1}\frac{(G_1 + G_2)(G_4 + G_5)}{G_2G_5}\right.$$

$$+ \frac{1}{a_1a_2}\frac{(G_1 + G_2)(G_4 + G_5)}{G_2G_5}\right]$$

(b) Calculate the sensitivities of ω_0 and Q to various passive elements assuming ideal op amp (i.e., $a_1 = a_2 = \infty$).
(c) Assuming that $a_i(s) = GB_i/s$, show that

$$S_{GB_1}^{\omega_0} = \frac{\omega_0}{GB_1} \qquad S_{GB_2}^{\omega_0} = \frac{\omega_0}{GB_2}$$

$$S_{GB_1}^{Q} = \frac{2\omega_0Q}{GB_1} \qquad S_{GB_2}^{Q} = \frac{2\omega_0Q}{GB_2}$$

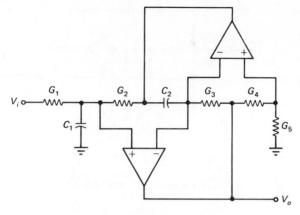

Fig. P3-17

3.18 Using the sensitivities derived in Prob. 3.16(b) and (c), compute M_α, M_β, and M_T for $Q = 50$, $\omega_0 = 1$, $\omega_1 = \omega_0(1 - 1/2Q)$, and $\omega_2 = \omega_0 (1 + 1/2Q)$, assuming $GB_1 = GB_2 = 100$, $\sigma_R = \sigma_C = 0.2\%$, $\sigma_{GB} = 10\%$, and no correlation between elements.

chapter four

....

Continuous-Time Active
Filters—Biquadratic
Realizations

4.1 INTRODUCTION

Active filters provide compact, lightweight signal-processing components for a wide variety of applications, such as voice and video signal processing for the communications industry, telemetry, biomedical electronics, radar, sonar, and seismic analysis, to name a few. Much of the motivation for considering active filters stems from the progress that has occurred in integrated-circuit technology. Such developments as the monolithic operational amplifier and the thick- and thin-film technologies enable one to realize high-quality, miniature, hybrid integrated-circuit active-RC filters at low cost. These filters are usually designed in the frequency range under 20 kHz, with much of the volume applications at voice frequencies [i.e., 4 kHz and below, e.g., D4-channel bank active-RC filters (see Fig. 1-45)]. For active filters, the upper frequency limit is imposed primarily by the cutoff frequency, or the unity gain–bandwidth product, of available op amps. Active-R and active-C filters, which derive their frequency dependence from the one-pole roll-off of the op amp, serve as alternative means for implementation of active-filters. For this class of filters, the upper limit in Q, and the center frequency is also determined by the stability conditions imposed by the phase response characteristic of the op amp and dynamic range limitations which stem from the noise levels and slew rates (Appendix C) of practical op amps.

This chapter and Chapter 5 are devoted to the analysis, design, and performance of continuous-time active filters. This includes traditional active-RC-type filters and, for high-frequency operation, active-R- and active-C-type filters. It has been widely demonstrated that monolithic filters can be achieved at voice-band frequencies (i.e., 4 kHz and below) using active switched capacitor (SC) filters. Active-SC filters, which are sampled data in nature, will be treated in detail in Chapter 6.

Basically, the design of active filters, whether they are the active-RC, active-R, or active-C type, can be achieved using one of four general approaches. As we show in Chapter 6, these same approaches are applicable to the design of active-SC filters.

Direct form   In the direct form a specified transfer function is realized directly by a combination of active elements (usually op amps) and passive components (R's and C's or only R's or only C's). In such a synthesis, the high-order filter is directly realized without resorting to biquadratic realizations. One type of direct synthesis uses only one or two active elements and RC networks to derive the desired transfer function [B1]. One example of this type of synthesis is shown in Fig. 4-1. The transfer function for this network is given by the expression

$$H(s) = \frac{V_o}{V_i} = \frac{Y_A - K_1 Y_B}{Y_A + Y_B + Y_1 - (K_2 - 1)Y_2} \qquad (4\text{-}1)$$

Fig. 4-1   Circuit configuration for direct realization.

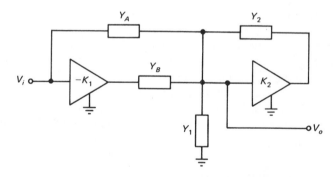

where Y_A, Y_B, Y_1, and Y_2 are driving-point RC admittance functions and K_1 and K_2 are the voltage gains of the active blocks. Some of the early active filters were synthesized in this manner; however, they are not particularly practical due to their poor sensitivity performance (Prob. 4.1). Thus, they will receive little treatment, beyond the aforementioned reference, except as exercises at the end of the chapter to illustrate this point.

Simulation   A second and more practical form of direct synthesis is to simulate inductors using gyrators, as shown in Fig. 4-2b. Alternatively, one may use scaling and the frequency-dependent negative-resistance (FDNR) approach, as in Fig. 4-2c. Another approach is to simulate L's and C's with active-R networks (i.e., op amps and resistors) for an active-R realization [P24, P25]. The disadvantage of the simulation approach is that the dynamic range critically depends on the choice of a passive prototype. However, the

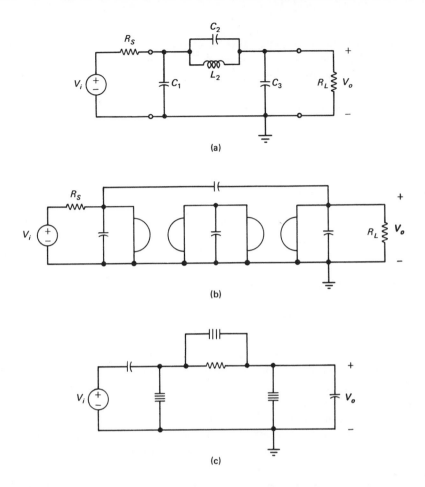

Fig. 4-2 (a) Passive-RLC filter. (b) Inductor simulation of part (a). (c) FDNR simulation of part (a).

sensitivity performance of these filters can be quite good and can be used in practical circuits. The simulation approach is discussed in more detail in Chapter 5, since one deals directly with the realization of high-order filters without resorting to biquadratic sections.

Cascade realization Cascade realization uses active second-order filter sections (active-RC, active-R, or active-C) and passive (or active) first-order sections. The second-order filter sections generally use one or more op amps as either positive- or negative-gain blocks (see Section 2.3). For active-R synthesis, the minimum number of op amps required to realize a second-order transfer function is two (see Section 4.6). In this approach the designer realizes an appropriate number of biquadratic (second-order numerator and second-order denominator) active-filter sections with transfer functions of the form

$$T_i(s) = \frac{b_{i2}s^2 + b_{i1}s + b_{i0}}{a_{i2}s^2 + a_{i1}s + a_{i0}} \tag{4-2}$$

where a_{1i} and b_{1i} are real numbers and $a_{1i} > 0$, and the subscript l denotes the lth section in the cascade. The transfer function for the overall filter, shown in Fig. 4-3 for an Nth-order filter, is

$$H(s) = \prod_{l=1}^{N/2} T_l(s) \qquad (4\text{-}3a)$$

Fig. 4-3 Cascade realization.

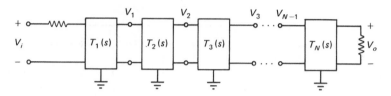

and for a $(N + 1)$th-order filter is

$$H(s) = \prod_{l=1}^{N/2} T_l(s) \times \frac{b_1 s + b_0}{a_1 s + a_0} \qquad (4\text{-}3b)$$

where N or $N + 1$ is the order of the desired filters as determined by the highest power of the complex frequency variable s in the denominator of $H(s)$.

It is noted that each $T_l(s)$ section is assumed to have a very high input impedance and a very low output impedance, so that the interaction between adjacent sections is negligible. Thus, adjustments to alter the coefficient(s) of one section do not alter the coefficients of any other section(s), so that this form of realization is attractive because of its simplicity and ease of design and tuning, and each section can be designed and tuned independently.

The sensitivity performance of cascade realizations, although acceptable for many applications, is typically the sum of the individual section sensitivities, as derived in Section 3.8 for statistically independent, biquadratic sections.

Clearly, to reduce the overall sensitivity M_α, M_β, or M_T for a cascade of biquadratic sections, one must reduce the sensitivities of the individual sections. Thus, low-sensitivity biquadratic sections play an important role in the design of high-order active filters.

A large number of biquadratic realizations are available in the literature. Most of the simple one-op amp filter structures use either positive or negative feedback, such as the Sallen and Key networks [P1]. These circuits are not generally suitable for high-Q realizations, as shown later in the chapter. Low-sensitivity biquadratic sections suitable for high-Q realizations generally use either multiple-loop feedback and/or more than one op amp. Much of this chapter will be devoted to those more useful low-sensitivity structures.

Multiple-loop feedback systems   As mentioned in Section 3.9 and as we demonstrate in Chapter 5, one may substantially reduce the sensitivity of a cascade of noninteracting biquad sections by forcing the sections to interact in a controlled manner. This can be done by placing multiple feedback loops

Continuous-Time Active Filters—Biquadratic Realizations

around various combinations of sections within the cascade. One such example is the follow-the-leader feedback structure shown in Fig. 4-4 for N even. Note that the cascade configuration can be considered a special, limiting case of a more general class of multiple-loop feedback configurations. With this class of realizations, one can simultaneously achieve the low (and perhaps lower) sensitivities of direct RLC-simulated synthesis and the advantages of modular design obtained with cascade synthesis.

Fig. 4-4   Multiple-loop feedback realization (FLF scheme).

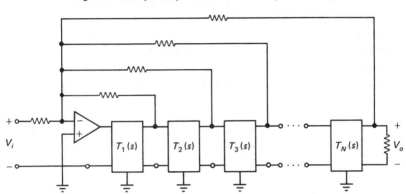

The primary advantage of the multiple-loop feedback approach is that intersectional coupling arises only due to the multiple-loop feedback topology. Thus, the coupling can be readily controlled by merely altering the multiple-loop feedback topology and/or the feedback resistors. As a result, one can achieve low sensitivity with little sacrifice in ease of design and tuning. Also, any of the low-sensitivity active-RC, active-R, or active-C biquadratic structures, to be discussed in this chapter, can be used, as in cascade synthesis. A detailed treatment of the design and sensitivity performance of multiple-loop feedback, high-order (order > 2) active filters is reserved until Chapter 5. Of course, the sensitivity of the multiple-loop feedback realization will always decrease as the individual section sensitivities decrease.

At this point it should be pointed out that the biquadratic active filter is second only to the op amp in importance as a building block for general active-filter synthesis. For this reason, we devote this entire chapter to the development of these key networks.

4.2 SINGLE-OP AMP OR SINGLE-AMPLIFIER BIQUAD (SAB) CONFIGURATIONS

We consider first the biquadratic active-RC sections that use a single op amp, configured in either an inverting (negative-feedback) or noninverting (positive-feedback) mode, in a single-loop feedback configuration. The sections are subdivided into low-pass, band-pass, high-pass, and notch filters.

4.2.1 Low-Pass Filter Sections with Positive Feedback

A low-pass biquadratic filter is characterized by a transfer function of the form

$$H(s) = \frac{b_0}{s^2 + a_1 s + a_0} \tag{4-4}$$

In searching the literature one can find a number of configurations capable of realizing Eq. (4-4). Two such representative configurations, one using positive feedback around an op amp configured in a noninverting mode and a second using negative feedback around an op amp configured in an inverting mode, are discussed here. Other interesting configurations are reserved for the exercises at the end of the chapter.

The first circuit, using positive feedback, is shown in Fig. 4-5. This circuit

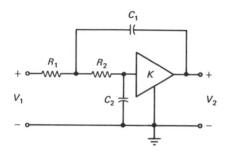

Fig. 4-5 Sallen and Key positive-feedback circuit.

is one of many developed by Sallen and Key. The voltage transfer function assuming an ideal op amp is given by

$$H(s) = \frac{V_2}{V_1} = \frac{KG_1 G_2 S_1 S_2}{s^2 + s(G_1 S_1 + G_2 S_1 + G_2 S_2 - KG_2 S_2) + G_1 S_1 G_2 S_2} \tag{4-5}$$

where K is the closed-loop gain of a noninverting amplifier [Eq. (2-11)],

$$G_i = \frac{1}{R_i} \quad \text{and} \quad S_i = \frac{1}{C_i} \quad \text{for } i = 1, 2$$

Note that the coefficient $a_1/\sqrt{a_0}$ in Eq. (4-5) involves a cancellation of terms. This cancellation mechanism, which is characteristic of positive-feedback biquadratic sections, will be shown to render large sensitivities to the elements that enter into the calculation, particularly when $a_1 \ll 1$, (i.e., for large Q).

In a cascade design, the section described above realizes one of the N pole pairs of the overall realization. However, in a multiple-loop feedback design, this section realizes one of the N pole pairs of the open-loop response (i.e., the response with all feedback paths external to the sections open-circuited). Thus, the closing of the multiple feedback loops shifts the poles to their desired closed-loop positions.

To design the ideal circuit, one need only match the coefficients of like powers of s in Eqs. (4-4) and (4-5):

$$b_0 = \frac{K}{R_1 R_2 C_1 C_2} \tag{4-6a}$$

$$a_1 = \frac{1}{R_1 C_1} + \frac{1}{R_2 C_1} + \frac{1}{R_2 C_2}(1 - K) \tag{4-6b}$$

$$a_0 = \frac{1}{R_1 R_2 C_1 C_2} \tag{4-6c}$$

Neglecting the gain constant b_0, it is noted that only two independent element values are required to specify the response of this circuit [i.e., to satisfy Eqs. (4-6b) and (4-6c)]. However, in the circuit of Fig. 4-5 there are five elements; thus, five degrees of freedom are available to the designer. The solution to the synthesis equations is therefore *not* unique. Hence, some elements, in this case three, may be preselected to make maximum use of available, off-the-shelf element values. Alternatively, they may be selected to optimize some other performance criteria, such as sensitivity (see, e.g., Examples 3-4 and 3-10). Note that in the case of the circuit of Fig. 4-5 there are two element values that can be selected independent of the synthesis equations; thus, these elements are referred to as *independent elements*. The remaining three element values depend upon both the synthesis equations and the values selected for the independent element values; thus, these elements are referred to as *dependent elements*. The choice as to whether a specific element is to be designated as independent or dependent is somewhat arbitrary, with the choice usually determined by convenience. For example, it is typically desirable to minimize the "spread in capacitance" by setting $C_1 = C_2$.

Let us for convenience define $C_1 = C$, $C_2 = \alpha C$; thus, $C_2/C_1 = \alpha$. With these definitions, α determines the capacitance spread and C is a scale factor that determines the actual capacitance levels. By solving Eqs. (4-6) for R_1, R_2 and K in terms of the specified transfer function coefficients a_0, a_1, and b_0 and capacitor parameters α and C, we obtain the following ideal design equations:

$$K = \frac{b_0}{a_0} \tag{4-7a}$$

$$R_2 = \frac{1}{2a_0 \alpha C}\{a_1 + \sqrt{a_1^2 + 4[b_0 - a_0(1 + \alpha)]}\} \tag{4-7b}$$

$$R_1 = \frac{1}{a_0 \alpha C^2 R_2} \tag{4-7c}$$

Note that for $K \geq 1$, $b_0 \geq a_0$ and R_2 is real as long as $b_0 \geq a_0(1 + \alpha) - a_1^2/4$. Thus, for a given pole location, the range of realizable gain constant b_0 is limited. There are several ways of altering the gain constant: (1) attenuating the input signal with an input voltage dividing network, (2) amplifying the output signal with either an inverting or a noninverting op amp amplifier, or (3) resistive gain enhancement [P2]. For now, let us not unnecessarily restrict or constrain the synthesis by requiring a specific value of b_0. Removing this constraint, we may rewrite design Eqs. (4-7) as follows:

$$R_2 = \frac{1}{2a_0\alpha C}\{a_1 + \sqrt{a_1^2 + 4a_0(K - 1 - \alpha)}\} \qquad (4\text{-}8a)$$

$$R_1 = \frac{1}{a_0\alpha C^2 R_2} \qquad (4\text{-}8b)$$

where R_2 is real for $K \geq 1 + \alpha - a_1^2/4a_0$. To evaluate R_1 and R_2, values for $K, C,$ and α must be chosen. Typically, C is chosen to accommodate conveniently obtained discrete capacitors or for economical hybrid integrated-circuit realization. Gain K may then be chosen arbitrarily, or to minimize the total resistance, or to minimize sensitivity. In practice, the value of K chosen represents an engineering compromise between minimum total resistance and minimum sensitivity.

To illustrate the sensitivity problem, let us plot the loci of the pole positions as a function of gain K. The poles, shown in Fig. 4-6, are given by the roots of

$$s^2 + 2\gamma s + (\gamma^2 + \beta^2) = 0 \qquad (4\text{-}9)$$

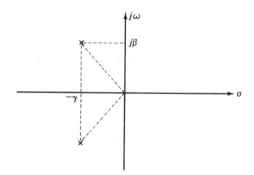

Fig. 4-6 Pole locations of Fig. 4-5 for a specific value of K, with K real.

The design equations are then obtained by simply substituting $a_1 = 2\gamma$ and $a_0 = \gamma^2 + \beta^2$ in Eqs. (4-6) and (4-8)

Let us consider the normalized frequency case:

$$H(s_n) = \frac{K}{s_n^2 + \sigma s_n + 1} = \frac{K}{s_n^2 + (1/Q)s_n + 1} \qquad (4\text{-}10)$$

where

$$s_n = \frac{s}{\omega_0} = \frac{s}{\sqrt{\gamma^2 + \beta^2}} \quad \text{and} \quad \sigma = \frac{1}{Q} = \frac{2\gamma}{\sqrt{\gamma^2 + \beta^2}}$$

If we select $G_1 = G_2 = S_1 = S_2 = 1$, then Eq. (4-5) becomes

$$H(s_n) = \frac{K}{s_n^2 + (3 - K)s_n + 1} \qquad (4\text{-}11)$$

Continuous-Time Active Filters—Biquadratic Realizations

The root locus of the poles as K is varied as shown in Fig. 4-7. Note from Eqs. (4-10) and (4-11) that

$$Q = \frac{1}{3 - K} \qquad \omega_0 = 1 \qquad (4\text{-}12)$$

Fig. 4-7   Root locus of Fig. 4-5 as K is varied (K real).

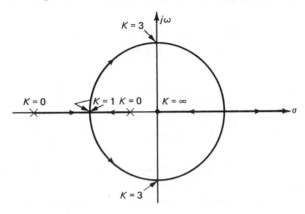

and that the circuit will be unstable for $K \geq 3$. Thus, for high Q in this design, small variations in K will result in large shifts in the pole locations (i.e., the pole positions and transfer function are very sensitive to gain K).

Let us now compute the ω_0 and Q sensitivities to both the passive and active components. The passive component sensitivities can be readily derived from Eq. (4-5), where

$$\omega_0 = \sqrt{\frac{1}{R_1 R_2 C_1 C_2}} \qquad (4\text{-}13a)$$

$$Q = \frac{\sqrt{\dfrac{1}{R_1 R_2 C_1 C_2}}}{\dfrac{1}{R_1 C_1} + \dfrac{1}{R_2 C_1} + \dfrac{1}{R_2 C_2}(1 - K)} \qquad (4\text{-}13b)$$

and $K = 1 + R_a/R_b$ according to Eq. (2-11). Using the ω_0 and Q sensitivity relationships given in (3-38), we have

$$S_{R_1}^{\omega_0} = S_{R_2}^{\omega_0} = S_{C_1}^{\omega_0} = S_{C_2}^{\omega_0} = -\tfrac{1}{2} \qquad (4\text{-}14a)$$

$$S_K^{\omega_0} = 0 \qquad (4\text{-}14b)$$

$$S_{R_1}^{Q} = -S_{R_2}^{Q} = -\tfrac{1}{2} + Q\sqrt{\frac{R_2}{R_1 \alpha}} \qquad (4\text{-}14c)$$

$$S_{C_2}^{Q} = -S_{C_1}^{Q} = -\tfrac{1}{2} + (1 - K)Q\sqrt{\frac{R_1}{R_2 \alpha}} \qquad (4\text{-}14d)$$

$$S_{R_a}^{Q} = -S_{R_b}^{Q} = (1 - K)Q\sqrt{\frac{R_1}{R_2 \alpha}} \qquad (4\text{-}14e)$$

$$S_K^{Q} = QK\sqrt{\frac{R_1}{R_2 \alpha}} \qquad (4\text{-}14f)$$

The following should be noted:

1. In an active-RC filter the most efficient realization of the ω_0^2 term requires only two R's and two C's; thus, the center frequency is minimally sensitive to the passive components.

2. The passive component Q sensitivities are all directly proportional to Q in this circuit.

3. S_{R_a/R_b}^Q and $S_{C_2}^Q$ are directly proportional to $(1 - K)Q$.

4. The Q sensitivities can become intolerably large for $Q > 5$, unless α is selected very large.

5. The passive component Q sensitivities are minimized when $K = 1$, $R_1 = R_2 = 1$, and $\alpha = 1/4Q^2$; however, the realization of this capacitance ratio is unrealistic for $Q > 5$. Since $K \geq 1$, $K = 1$ represents a minimum positive-feedback design. This, of course, is just one limiting case of a continuum of various values of parameter α. The value of α can be used as a parameter to control the passive sensitivities with implications on active sensitivities to be pointed out later.

6. The component spreads can be minimized by applying more positive feedback. In this design, which corresponds to a maximum positive feedback design, $R_1 = R_2 = 1$, $C_1 = C_2 = 1$, and $K = 3 - 1/Q$. Substituting these values into the sensitivities listed in Eqs. (4-14), it can be readily verified that the passive Q sensitivities are maximized with

$$S_{R_1}^Q = -S_{R_2}^Q = -\tfrac{1}{2} + Q \qquad (4\text{-}14\text{g})$$

$$S_{C_2}^Q = -S_{C_1}^Q = \tfrac{1}{2} - 2Q \qquad (4\text{-}14\text{h})$$

$$S_{R_a/R_b}^Q = -2Q \qquad (4\text{-}14\text{i})$$

7. Positive feedback is seen to reduce component spreads, thus reduce the size of an integrated circuit implementation of the circuit, at the expense of increased passive component sensitivities.

Let us now determine the effect the finite op amp gain has on the definition of ω_0 and Q. For a real op amp of gain $a(s)$, the actual gain K, which we denote as $\hat{K}$, is a function of $a(s)$, as described earlier in Chapter 2.

$$\hat{K} = \frac{K}{1 + \dfrac{K}{a(s)}} \qquad (4\text{-}15)$$

Note that as $|a(s)| \to \infty$, $\hat{K} \to K$. Substituting Eq. (4-15) into Eq. (4-5) yields

$$H = \frac{V_2}{V_1} = \frac{K\left[1 + \dfrac{K}{a(s)}\right]^{-1} G_1 G_2 S_1 S_2}{s^2 + s\left\{G_1 S_1 + G_2 S_1 + G_2 S_2\left[1 - K\left(1 + \dfrac{K}{a(s)}\right)^{-1}\right]\right\} + G_1 G_2 S_1 S_2} \qquad (4\text{-}16)$$

For $|K/a(s)| \ll 1$,

$$\left[1 + \frac{K}{a(s)}\right]^{-1} \simeq 1 - \frac{K}{a(s)}$$

Making this substitution in Eq. (4-16) results in the following approximate expression for H:

$$H \simeq \frac{K\left[1 - \dfrac{K}{a(s)}\right]G_1G_2S_1S_2}{s^2 + s\left\{G_1S_1 + G_2S_1 + G_2S_2\left[1 - K + \dfrac{K^2}{a(s)}\right]\right\} + G_1G_2S_1S_2} \qquad (4\text{-}17)$$

For op amps compensated for single-pole roll-off, $a(s) \simeq a_0 p_0/s = GB/s$; thus,

$$H \simeq \frac{K\left(1 - \dfrac{K}{GB}s\right)G_1G_2S_1S_2}{s^2\left(1 + G_2S_2\dfrac{K^2}{GB}\right) + s[G_1S_1 + G_2S_1 + G_2S_2(1 - K)] + G_1G_2S_1S_2}$$

$$\simeq \frac{\left(\dfrac{K}{1 + G_2S_2K^2/GB}\right)\left(1 - \dfrac{K}{GB}s\right)G_1G_2S_1S_2}{s^2 + s\left[\dfrac{G_1S_1 + G_2S_1 + G_2S_2(1 - K)}{1 + G_2S_2K^2/GB}\right] + \dfrac{G_1G_2S_1S_2}{1 + G_2S_2K^2/GB}}$$

Again for $G_2S_2K^2/GB \ll 1$, we can make the substitution

$$\left(1 + \frac{G_2S_2K^2}{GB}\right)^{-1} \simeq 1 - \frac{G_2S_2K^2}{GB}$$

yielding for H:

$$H \simeq \frac{K\left(1 - \dfrac{G_2S_2K^2}{GB}\right)\left(1 - \dfrac{K}{GB}s\right)G_1G_2S_1S_2}{s^2 + s\left[\left(G_1S_1 + G_2S_1 + G_2S_2(1-K)\right)\left(1 - \dfrac{G_2S_2K^2}{GB}\right)\right] + G_1G_2S_1S_2\left(1 - \dfrac{G_2S_2K^2}{GB}\right)}$$

$$\hspace{12cm} (4\text{-}18a)$$

$$= \frac{N(s)}{s^2 + a_1 s + a_0} \qquad (4\text{-}18b)$$

Solving for R_1 and R_2 in terms of the specified transfer function coefficients a_1 and a_0, capacitor parameters α and C, gain K, and the nominal op amp GB, we obtain the following design equations:

$$a_1 = [G_1S_1 + G_2S_1 + G_2S_2(1 - K)]\left(1 - \frac{G_2S_2K^2}{GB}\right)^2 \qquad (4\text{-}19a)$$

$$a_0 = G_1G_2S_1S_2\left(1 - \frac{G_2S_2K^2}{GB}\right) \qquad (4\text{-}19b)$$

From (4-19a) and (4-19b), we obtain

$$R_2^3 - \frac{a_1}{\gamma C a_0}R_2^2 + \frac{\gamma + 1 - K}{\gamma^2 C^2 a_0}R_2 - \frac{K^2(\gamma + 1 - K)}{\gamma^3 C^3 a_0 GB} = 0 \qquad (4\text{-}19c)$$

Equations (4-19), which reduce to Eqs. (4-8) when $GB \rightarrow \infty$, can be used to compensate the design for the frequency response of the actual op amp.

●

EXAMPLE 4-1

Let us design the positive-feedback network in Fig. 4-5 for $\omega_0 = 2\pi \times 2000$ rad/sec (i.e., $a_0 = 16\pi^2 \times 10^6$) and $Q = 10$ (i.e., $a_1 = 4\pi \times 100$) with $C = 2/a_1$, $\alpha = a_1^2/4a_0$, $K = 1$, and $GB = \infty$. From Eqs. (4-8), we determine

$$R_1 = R_2 = 1$$

Let us now redesign this network to operate with a μA741 type op amp with a nominal $GB = 2\pi \times 10^6$ rad/sec. From Eq. (4-19C), we obtain the following:

$$R_2^3 - 2R_2^2 + R_2 - 0.04 = 0$$

The two practical solutions are $R_2 = 1.18382$ and 0.77244. The corresponding values of R_1 are 0.81618 and 1.22756. The third value of R_2 leads to a very high R_1/R_2 ratio (i.e., 44.6) and hence is impractical. It should be pointed out that the effect of the op amp can be significantly decreased by using the two-pole one-zero op amp compensation given in Eq. (2-30) and discussed in Section 2.5.

●

To derive the ω_0 and Q sensitivities to the op amp GB, let us first write the following expressions, from (4-18a):

$$\tilde{\omega}_0 \simeq \sqrt{\frac{1}{R_1 R_2 C_1 C_2}\left(1 - \frac{K^2}{R_2 C_2 GB}\right)} = \omega_0 \sqrt{1 - \frac{K^2}{R_2 C_2 GB}} \quad (4\text{-}20a)$$

$$\tilde{Q} \simeq \frac{\sqrt{\dfrac{1}{R_1 R_2 C_1 C_2}}}{\left(\dfrac{1}{R_1 C_1} + \dfrac{1}{R_2 C_1} + \dfrac{1}{R_2 C_2}(1 - K)\right)} \sqrt{1 + \frac{K^2}{R_2 C_2 GB}} \quad (4\text{-}20b)$$

$$= Q \sqrt{1 + \frac{K^2}{R_2 C_2 GB}}$$

where ω_0 and Q denote ideal values for $GB = \infty$ and $\tilde{\omega}_0$ and $\tilde{Q}$ denote actual values obtained with a single-pole compensated op amp. Note that as K increases and/or GB decreases, the center frequency $\tilde{\omega}_0$ becomes smaller and $\tilde{Q}$ is enhanced.

The sensitivities of ω_0 and Q to the op amp gain–bandwidth product can be written as follows, $\alpha = C_2/C_1$

$$S_{GB}^{\tilde{\omega}_0} \simeq \frac{K^2}{2}\frac{\omega_0}{GB}\sqrt{\frac{R_1}{R_2 \alpha}} \quad (4\text{-}21a)$$

$$S_{GB}^{\tilde{Q}} \simeq -\frac{K^2}{2}\frac{\omega_0}{GB}\sqrt{\frac{R_1}{R_2 \alpha}} \quad (4\text{-}21b)$$

From (4-21a) and (4-21b), the following observations are to be noted:

1. $S_{GB}^{\tilde{\omega}_0}$ and $S_{GB}^{\tilde{Q}}$ increase with the square of gain K and with the ratio ω_0/GB. For an op amp with a GB of $2\pi \times 1$ MHz, such as the μA741 type, these sensitivities will become intolerably large for $f_0 > 10$ kHz.

2. For $K = 1$, $R_1 = R_2 = 1$, and $\alpha = 1/4Q^2$, which minimize the passive component sensitivities, the GB sensitivities are

$$S_{GB}^{\tilde{\omega}_0} \simeq Q\frac{\omega_0}{GB} \quad \text{and} \quad S_{GB}^{\tilde{Q}} \simeq -Q\frac{\omega_0}{GB} \qquad (4\text{-}21c)$$

These sensitivities again suggest a $Q < 5$ limit for practical realizations.

3. For $K = 3 - 1/Q$, $R_1 = R_2 = 1$, and $C_1 = C_2 = 1$ (i.e., $\alpha = 1$), which minimize the component spreads while maximizing the passive Q sensitivities, the GB sensitivities are

$$S_{GB}^{\tilde{\omega}_0} \simeq \frac{9}{2}\frac{\omega_0}{GB} \quad \text{and} \quad S_{GB}^{\tilde{Q}} \simeq -\frac{9}{2}\frac{\omega_0}{GB} \qquad (4\text{-}21d)$$

In contrast to the passive sensitivities, increasing the positive feedback serves to decrease the op amp GB sensitivity.

4. Again, it should be noted that the designs in cases 2 and 3 represent extreme cases. The fact that α can be used to optimize sensitivities has already been mentioned. For example [P3], if

$$R_1 = R_2 \qquad \alpha = \frac{1}{4Q_0^2}$$

(with Q_0 a design parameter $Q_0 < Q$), then

$$S_{GB}^{\tilde{\omega}_0} \simeq Q\frac{\omega_0}{GB}$$

These vastly differing sensitivity properties with component design values suggest that an optimum design which minimizes the overall sensitivity, such as M_T, M_α, or M_β discussed in Chapter 3, lies somewhere between the minimum and maximum positive-feedback designs. Sensitivity minimization in active biquad filters is treated in Section 4.2.

Although the ω_0 and Q sensitivities provide intuition as to the relative roles played by the various individual components, a more complete prediction of the circuit sensitivity can be made by computing the expected variation in the circuits transfer function (M_T), magnitude response (M_α), or phase response (M_β), as appropriate. The computation of statistical measures for these variations, M_T, M_α, and M_β, respectively, was treated in detail in Chapter 3. These sensitivity measures are functions of both the passive and active ω_0 and Q

sensitivities, and the derivation of ω_0 and Q sensitivities serves as an intermediate step in the computation of M_T, M_α, and M_β as discussed in Section 3.6.

●

EXAMPLE 4-2

Compute M_T, M_α, and M_β for the design given in Example 4-1 using a nine-point trapezoidal rule integration over the frequency range $\omega_0(1 - 1/Q)$ $\leq \omega \leq \omega_0(1 + 1/Q)$. To compute these sensitivity measures, let us use the following practical component standard deviations: $\sigma_{R_1} = \sigma_{R_2} = 0.1\%$, σ_{C_1} $= \sigma_{C_2} = 0.3\%$, and $\sigma_{\mathrm{GB}} = 15\%$. All components are uncorrelated. Using $S_{x_i}^{\omega_0}$ and $S_{x_i}^{Q}$ given in Eqs. (4-14) and (4-21), we compute the following sensitivity measures from (Eqs. 3-117 a–d) and for $R_2 = 1.18382$, $R_1 = 0.81618$:

$$M_\alpha \simeq \kappa \int_{\omega_1}^{\omega_2} \left[\mathrm{Re}\!\left(\frac{1}{D}\right) \right]^2 d\omega = 1.60 \times 10^{-4} \tag{4-22a}$$

$$M_\beta \simeq \kappa \int_{\omega_1}^{\omega_2} \left[\mathrm{Im}\!\left(\frac{1}{D}\right) \right]^2 d\omega = 3.41 \times 10^{-4} \tag{4-22b}$$

$$M_T \simeq \kappa \int_{\omega_1}^{\omega_2} \left| \frac{1}{D} \right|^2 d\omega = 5.01 \times 10^{-4} \tag{4-22c}$$

where $D = \omega_0^2 - \omega^2 + j\omega(\omega_0/Q)$, Re denotes real part of, Im denotes imaginary part of, and

$$\kappa \simeq 4\left\{ \left[\sum_R (S_R^{\omega_0})^2 + \frac{1}{Q^2} \sum_R (S_R^Q)^2 \right] \sigma_R^2 + \left[\sum_C (S_C^{\omega_0})^2 + \frac{1}{Q^2} \sum_C (S_C^Q)^2 \right] \sigma_C^2 \right. \tag{4-22d}$$

$$\left. + \left[(S_{\mathrm{GB}}^{\omega_0})^2 + \frac{1}{Q^2} (S_{\mathrm{GB}}^Q)^2 \right] \sigma_{\mathrm{GB}}^2 \right\}$$

From (4-22), the following are noted:

1. Since $S_{\mathrm{GB}}^{\omega_0}$ and S_{GB}^Q are dependent on the center frequency ω_0, κ is in general a function of Q and ω_0.

2. In performing the integration to compute M_α, M_β, and M_T, it is convenient to normalize ω_0 to unity so that $D = 1 - \omega^2 + j\omega(1/Q)$ and the limits of integration become $1 - 1/Q \leq \omega \leq 1 + 1/Q$.

3. The integrals $\int_{\omega_1}^{\omega_2} [\mathrm{Re}\,(1/D)]^2\, d\omega$, $\int_{\omega_1}^{\omega_2} [\mathrm{Im}\,(1/D)]^2\, d\omega$, and $\int_{\omega_1}^{\omega_2} |1/D|^2\, d\omega$ depend only on the limits of integration and Q. These quantities are independent of the network topology and the component statistics. Thus, for commonly specified ω_0, Q and the band of interest (ω_1, ω_2), κ serves as an easily computed figure of merit for comparing different biquad realizations. As we discuss in Chapter 5, to obtain a meaningful estimate of sensitivity for higher-order filters, we must compute M_α, M_β, M_T, or their integrands.

●

Continuous-Time Active Filters—Biquadratic Realizations

4.2.2 Low-Pass Filters with Negative Feedback

Let us now look at another simple low-pass second-order section, shown in Fig. 4-8.[1] In contrast to the circuit in Fig. 4-5, which uses positive feedback,

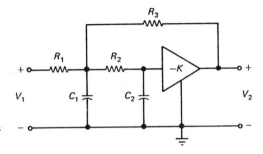

Fig. 4-8 Low-pass second-order negative-feedback circuit.

the circuit employs negative feedback. Analysis of the circuit, assuming an ideal op amp, yields the following voltage transfer function:

$$H(s) = \frac{V_2}{V_1} = \frac{-KG_1G_2S_1S_2}{s^2 + (G_1S_1 + G_2S_1 + G_3S_1 + G_2S_2)s + S_1S_2G_2(G_1 + G_3 + KG_3)}$$

(4-23a)

$$= \frac{b_0}{s^2 + a_1s + a_0}$$

(4-23b)

In contrast to the positive-feedback circuit, the transfer function coefficients are not derived by a cancellation of terms. This absence of canceling terms is characteristic of negative-feedback active biquadratic networks. Letting $C_1 = C$, $C_2 = \gamma C$, and $R_3 = \delta R_1$, the ideal design equations can be derived to be

$$R_2 = \frac{a_1(1 + \delta + K)}{2a_0\gamma C(1 + \delta)}\left[1 + \sqrt{1 - \frac{4a_0(1 + \delta)(1 + \gamma)}{a_1^2(1 + \delta + K)}}\right]$$

(4-24a)

$$R_1 = \frac{1 + \delta + K}{a_0\delta\gamma C^2R_2}, \quad \delta = \frac{R_3}{R_1}, \quad \gamma = \frac{C_2}{C_1}$$

(4-24b)

For realizability, we find from Eq. (4-24a) that

$$K \geq (1 + \delta)\left[\frac{4a_0}{a_1^2}(1 + \gamma) - 1\right]$$

(4-24c)

To illustrate the sensitivity and stability of this network, let us plot the loci of

[1]The input resistance of the $(-K)$ finite-gain amplifier is low and can cause difficulty in a practical realization using a single op amp.

the pole positions as a function gain (with the normalized element values $G_1 = G_2 = G_3 = 1$, $S_1 = S_2 = 1$). Substituting these values into Eq. (4-23) yields

$$H(s_n) = \frac{-K}{s_n^2 + 4s_n + 2 + K} \qquad (4\text{-}25)$$

The root locus as K is varied is shown in Fig. 4-9. From Fig. 4-9 it is quite apparent that the circuit is stable for all values of $K > 0$. However, for a high

Fig. 4-9   Root locus of Fig. 4-8 as K is varied (K real).

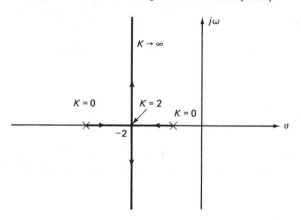

value of Q, which implies a large value of K, the cutoff frequency $\omega_p = \sqrt{2 + K}$ becomes large, approaching the cutoff frequency of the gain block, which becomes lower as K increases. Thus, the pole of the op amp under these conditions cannot be ignored and the circuit can become quite unstable. One would expect this circuit to be very sensitive, for high Q, to the finite gain–bandwidth of the op amp (Prob. 4.6). Thus, this circuit is limited to low-Q, low-frequency realizations. Also unlike the positive-feedback circuit, the negative-feedback circuit does afford the designer the flexibility of passive versus active sensitivity trade-off.

●

EXAMPLE 4-3

Let us consider the design of the negative-feedback circuit in Fig. 4-8 to realize the following normalized transfer function:

$$H(s_n) = \frac{b_0}{s_n^2 + (1/Q)s_n + 1}$$

If we arbitrarily set $\gamma = \delta = 1$, $C = 1$, and $K = (1 + \delta)[(4a_0/a_1^2)(1 + \gamma) - 1]$, we can compute the following component values:

$$C_1 = C_2 = 1 \quad \text{and} \quad \frac{R_a}{R_b} = K = 16Q^2 - 2$$

and from Eqs. (4-24a) and (4-24b),

$$R_1 = R_3 = 2Q$$
$$R_2 = 8Q$$

The following should be noted:

1. In this negative-feedback circuit, high Q implies large component spreads (i.e., here the ratio of gain determining resistors $\simeq 16Q^2$. This is characteristic of negative-feedback single-amplifier biquad networks.
2. Owing to the large gain K, we should expect large GB sensitivities.

Since $\omega_0 = \sqrt{a_0}$ and $Q = \sqrt{a_0}/a_1$, we can write the following expressions for ω_0 and Q:

$$\omega_0 = \sqrt{\frac{1 + (R_1/R_3)(1 + K)}{R_1 R_2 C_1 C_2}} \tag{4-26a}$$

$$Q = \frac{\sqrt{\dfrac{1 + (R_1/R_3)(1 + K)}{R_1 R_2 C_1 C_2}}}{\dfrac{1}{R_1 C_1} + \dfrac{1}{R_2 C_1} + \dfrac{1}{R_3 C_1} + \dfrac{1}{R_2 C_2}} \tag{4-26b}$$

From Eqs. (4-26a), we can derive the following ω_0 sensitivities:

$$S_{R_1}^{\omega_0} = -\frac{1}{2}\left(\frac{\delta}{1 + \delta + K}\right) \tag{4-27a}$$

$$S_{R_2}^{\omega_0} = S_{C_1}^{\omega_0} = S_{C_2}^{\omega_0} = -\frac{1}{2} \tag{4-27b}$$

$$S_{R_3}^{\omega_0} = -\frac{1}{2}\left(\frac{1 + K}{1 + \delta + K}\right) \tag{4-27c}$$

$$S_{R_a}^{\omega_0} = -S_{R_b}^{\omega_0} = \frac{1}{2}\left(\frac{K}{1 + \delta + K}\right) \tag{4-27d}$$

where from Eq. (2-8), $K = R_a/R_b$. Using Eq. (4-26b), we can derive the Q sensitivities. This derivation is facilitated if the following identity is used:

$$S_x^Q = S_x^{\omega_0} - S_x^{a_1} \tag{4-28}$$

with

$$a_1 = \frac{1}{R_1 C_1} + \frac{1}{R_2 C_1} + \frac{1}{R_3 C_1} + \frac{1}{R_2 C_2}$$

From Eq. (4-28) and Eqs. (4-27), the following Q sensitivities are derived:

$$S_{R_1}^Q = \frac{\delta}{1 + \delta + \dfrac{R_1}{R_2}\dfrac{\delta}{\gamma}(1 + \gamma)} - \frac{1}{2}\frac{\delta}{1 + \delta + K} \qquad (4\text{-}29\text{a})$$

$$S_{R_2}^Q = -\left[\frac{1}{2} - \frac{\dfrac{R_1}{R_2}\dfrac{\delta}{\gamma}(1 + \gamma)}{1 + \delta + \dfrac{R_1}{R_2}\dfrac{\delta}{\gamma}(1 + \gamma)}\right] \qquad (4\text{-}29\text{b})$$

$$S_{R_3}^Q = \frac{1}{1 + \delta + \dfrac{R_1}{R_2}\dfrac{\delta}{\gamma}(1 + \gamma)} - \frac{1}{2}\frac{1 + K}{1 + \delta + K} \qquad (4\text{-}29\text{c})$$

$$S_{C_1}^Q = -\left[\frac{1}{2} - \frac{\delta\dfrac{R_1}{R_2} + 1 + \delta}{1 + \delta + \dfrac{R_1}{R_2}\dfrac{\delta}{\gamma}(1 + \gamma)}\right] \qquad (4\text{-}29\text{d})$$

$$S_{C_2}^Q = -\left[\frac{1}{2} - \frac{\dfrac{\delta}{\gamma}\dfrac{R_1}{R_2}}{1 + \delta + \dfrac{R_1}{R_2}\dfrac{\delta}{\gamma}(1 + \gamma)}\right] \qquad (4\text{-}29\text{e})$$

$$S_{R_a}^Q = -S_{R_b}^Q = \frac{1}{2}\left(\frac{K}{1 + \delta + K}\right) \qquad (4\text{-}29\text{f})$$

Note that $|S_{x_i}^{\omega_0}| \le \frac{1}{2}$ and $|S_{x_i}^Q| < 1$. In contrast to positive-feedback circuits, negative-feedback circuits are quite insensitive to the passive components.

To observe the effect of the finite op amp GB, let us substitute for gain K in Eq. (4-23) the actual gain $\hat{K}$, where

$$\hat{K} = \frac{K}{1 + s\left(\dfrac{1 + K}{\mathrm{GB}}\right)}$$

Performing this substitution, we can derive the following approximate expression for the actual transfer function:

$$H \simeq$$
$$\frac{-K\left(1 - s\dfrac{1 + K}{\mathrm{GB}}\right)G_1 G_2 S_1 S_2}{s^2 + \left[G_1 S_1 + G_2 S_1 + G_3 S_1 + G_2 S_2 - \dfrac{K(1+K)G_3 G_2 S_1 S_2}{\mathrm{GB}}\right]s + S_1 S_2 G_2(G_1 + G_3 + KG_3)} \qquad (4\text{-}30)$$

where it is assumed that

$$\omega\frac{1 + K}{\mathrm{GB}} \ll 1$$

From Eq. (4-30) it is observed that the s-dependent gain of the op amp has caused a subtraction in the a_1 coefficient. This of course implies that the op amp will tend to enhance the Q and that instability is possible in a high-Q design. One could compensate for the nominal op amp GB as we did earlier for the positive-feedback circuit.

The GB sensitivities can now be determined:

$$S_{GB}^{\tilde{\omega}_0} = 0 \tag{4-31a}$$

$$S_{GB}^{\tilde{Q}} = \frac{K(1 + K)C}{GB} \left[\frac{1}{1 + \delta + \frac{R_1}{R_2}\frac{\delta}{\gamma}(1 + \gamma)} \right] \tag{4-31b}$$

●

Note that, for Example 4-3, S_{GB}^{Q} is prohibitively large. This behavior can be significantly improved using the op amp compensation in Eq. (2-31).

It should be evident at this point that single-amplifier biquads using either negative or positive feedback do not render practical high-Q circuits. However, they do offer simple, efficient, low-Q realizations. Thus, they are suited to the low-Q realization of such second-order low-pass functions as Butterworth and Bessel. Also, the design of positive-feedback circuits can be optimized, thus enabling their use for Q values up to about 30, as will be explained later.

In addition to low-pass transfer functions, one can realize band-pass, high-pass, and notch functions with single-amplifier biquads. Positive- and negative-feedback circuits for these applications are discussed in the forthcoming subsections. However, for high-Q design, they will perform similarly to the low-pass positive- and negative-feedback circuits. For this reason these circuits will not be given the detailed treatment which the low-pass circuits received.

4.2.3 Band-Pass Filter Sections

A band-pass biquadratic filter is characterized by the relation

$$H(s) = \frac{H_2 s}{s^2 + a_1 s + a_0} = \frac{H_2 s}{s^2 + (\omega_0/Q)s + \omega_0^2} \tag{4-32}$$

The realization of this transfer function can be achieved with any one of several circuit configurations [P3]. One of the best negative-feedback circuits using a single op amp is shown in Fig. 4-10. The voltage ratio transfer function of the circuit, assuming an ideal op amp, is

$$H(s) = \frac{V_2}{V_1} = \frac{-G_1 S_2 s}{s^2 + sG_2(S_1 + S_2) + G_1 S_1 G_2 S_2} \tag{4-33}$$

where the reader is reminded that $G_i = 1/R_i$ and $S_i = 1/C_i$ for $i = 1, 2$. Here there are four degrees of freedom, two more than that required to specify the pole-pair positions. As an example, we may narrow the choice by setting

$$S_1 = S_2 = 1$$

$$G_2 = 1 \quad \text{and} \quad G_1 = \frac{1}{\gamma} \tag{4-34}$$

Note that if we set $\gamma = 1$, then $Q = 1/2$; thus, the design has been overly constrained. If we substitute the normalized specifications of Eq. (4-34) into Eq. (4-33), we obtain

$$H(s_n) = \frac{-s_n(1/\gamma)}{s_n^2 + 2s_n + 1/\gamma} \tag{4-35a}$$

From (4-35a), we see that

$$\omega_0 = \frac{1}{\sqrt{\gamma}}$$

$$Q = \frac{1}{2\sqrt{\gamma}} \tag{4-35b}$$

Thus, given a value for Q, we obtain γ as

$$\gamma = \frac{1}{4Q^2} \tag{4-35c}$$

and then denormalize the circuit to obtain the desired value of ω_0. The sensitivity performance of this negative-feedback bandpass is quite similar to the negative-feedback low-pass circuit discussed previously—the passive sensitivities are quite low but the active sensitivities are very high. As will be seen later, this circuit can be made to have the sensitivity trade-off characteristics of positive-feedback realizations by the addition of some resistive positive feedback. Without this Q-enhancing positive feedback, however, the circuit in Fig. 4-10 is limited to a maximum Q of about 5.

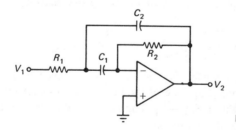

Fig. 4-10 Good negative-feedback band-pass circuit.

As another example, consider the circuit shown in Fig. 4-11, in which positive feedback is placed around a noninverting, ideal op amp with a resulting gain of K. Analysis of this circuit yields

$$H(s) = \frac{V_2}{V_1} = \frac{KG_1 S_1 s}{s^2 + [S_1(G_1 + G_2 + G_3 - KG_2) + S_2 G_3]s + S_1 S_2 G_3(G_1 + G_2)} \tag{4-36}$$

where, of course, $G_i = 1/R_i$ and $S_i = 1/C_i$ for $i = 1, 2, 3$.

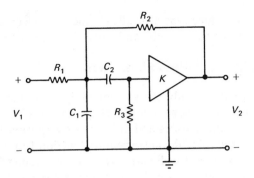

Fig. 4-11 Positive-feedback
band-pass circuit.

If we set the center frequency ω_0 to a normalized value of unity, then one possible set of element value assignments is

$$S_1 = S_2 = 1$$
$$G_1 = G_2 = G_3 = 1 \tag{4-37}$$

If we substitute the specifications Eq. (4-37) into the transfer function of Eq. (4-36), we obtain

$$H(s_n) = \frac{Ks_n}{s_n^2 + (4 - K)s_n + 1} \tag{4-38}$$

where, it is noted, $s_n = s/\omega_0$.

A sketch of the root locus for the pole positions of Eq. (4-38) is shown in Fig. 4-12. Note that the sensitivity properties of this circuit are similar to those

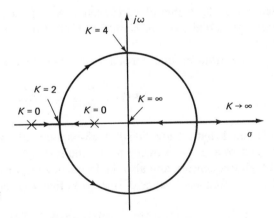

Fig. 4-12 Root locus of Fig. 4-11
as K is varied (K real).

of low-pass circuit of Fig. 4-5. The transmission zeros, at 0 and ∞, are well away from the filter passband, and have negligible effect on the filter sensitivity. Thus, filters, whether they be low-pass, band-pass, or high-pass biquadratic realizations, will exhibit similar sensitivity performance if they are configured similarily (i.e., use the same type of feedback, the same op amp closed-loop mode, and an equal number of elements) and suffer from comparable element deviations.

4.2.4 High-Pass Filter Sections

We have already shown that a high-pass inductorless active network is readily obtained from a low-pass prototype circuit merely by replacing all capacitors with resistors and all resistors with capacitors. This comes about as a result of the $RC:CR$ transformation (Section 1.6) of the low-pass filter section.

For example, two high-pass circuits, corresponding to the low-pass circuits in Figs. 4-5 and 4-8, are shown in Fig. 4-13a and b, respectively. As noted in

Fig. 4-13   (a) Positive-feedback high-pass circuit. (b) Negative-feedback high-pass circuit.

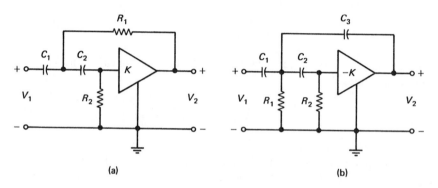

(a)   (b)

Section 4.2.3, it should be expected that the sensitivity performance for the high-pass circuits will be similar to that of their corresponding low-pass prototypes.

The transfer function of a high-pass biquadratic filter section is of the form

$$H(s) = \frac{V_2}{V_1} = \frac{H_3 s^2}{s^2 + a_1 s + a_0} \tag{4-39}$$

It is noted that a truly "ideal" high-pass circuit does not exist, since $H(s)$ must go to zero as $s \to \infty$ owing to the shunt stray capacitances which are apparent at high frequencies, and also the frequency response of the op amp. It is only in the finite frequency range of operation that $H(s)$ behaves as a high-pass filter.

4.2.5 Filter Sections with jω-Axis Zeros; Elliptic and Notch Filters

The elliptic approximation (also referred to as the Cauer approximation) is one of the most commonly used functions in filter design. As discussed in Section 1.3, the elliptic approximation is able to achieve a given stop-band requirement with a lower-order function than either the Butterworth or Chebyshev approximations. Thus, for high-order filter synthesis, the elliptic approximation will lead to a less expensive filter realization.

A biquadratic filter section, which can be used to realize an elliptic or notch filter response, has a transfer function of the form

$$H(s) = \frac{H_4(s^2 + \omega_z^2)}{s^2 + \omega_p s/Q + \omega_p^2} \tag{4-40}$$

One realization for the response in Eq. (4-40) is shown in Fig. 4-14, where $C_3 = C_1 + C_2$ and $G_3 = G_1 + G_2$. Like many of the other sections discussed thus far in this chapter, the section shown in Fig. 4-14 uses positive feedback around an noninverting operational amplifier with finite closed-loop gain K. The transfer function of the circuit, assuming an ideal op amp, is given by

$$H(s) = \frac{V_2}{V_1} = \frac{K(C_1 C_2 s^2 + G_1 G_2)}{C_1 C_2 s^2 + [(C_1 + C_2)G_2 + C_2(G_1 + G_2)(1 - K)]s + G_1 G_2} \tag{4-41}$$

where, as mentioned previously, $C_3 = C_1 + C_2$, $G_3 = G_1 + G_2$ and $G_i = 1/R_i$. Comparing Eqs. (4-40) and (4-41), one notes that in Eq. (4-41) that $\omega_z = \omega_p$. In general, however, one usually requires that ω_z be different from ω_p. One way of separating ω_p and ω_z is demonstrated in Prob. 4.12. Also, the reader should note that the network in Fig. 4-14 contains three capacitors; thus, in general, it

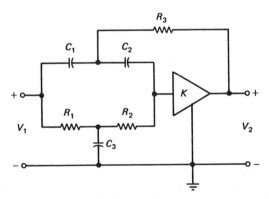

Fig. 4-14 Notch circuit.

will have a third-order response. But by choosing C_3 and G_3 as mentioned above, a zero is forced to cancel the real pole, resulting in the biquadratic transfer function in Eq. (4-41). It should be noted, however, that even though the canceled pole and zero do not appear in (4-41), their degradation effect on sensitivity will be present and should be accounted for. This is readily done by writing the complete transfer function, including the canceled pole and zero. Because of the tolerance and variations of the components, exact cancellation does not occur. Statistical sensitivity measures, discussed in Chapter 3, take this fact into account.

With this filter section we have seven specifiable components and only two independent design constraints, with $\omega_p = \omega_z$, imposed by the pole–zero pairing. For example, if we assume a normalized frequency such that $\omega_p = \omega_z = 1$, then one possible set of design equations is

$$K = 1$$
$$C_1 = C_2 = 1 \tag{4-42}$$
$$G_1 = \frac{1}{G_2} = \gamma$$

The substitution of Eq. (4-42) into the transfer function Eq. (4-41) yields

$$H(s_n) = \frac{s_n^2 + 1}{s_n^2 + (2/\gamma)s_n + 1} \tag{4-43}$$

Now, if we let $G_1 = G_2 = 1$, $C_1 = C_2 = 1$ and let K vary, the root locus of the pole positions as K varies is sketched in Fig. 4-15. Note that the behavior of this circuit is similar to the behavior sketched in Fig. 4-7, for the low-pass, positive (single-loop)-feedback circuit of Fig. 4-5.

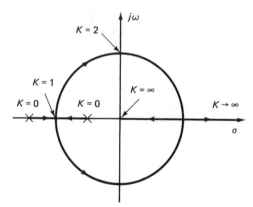

Fig. 4-15 Root locus of Fig. 4-14 as K is varied (K real).

One application where it is desirable for $\omega_p = \omega_z$ is in the realization of a band-reject or notch filter. For this type of filter, the gain at high and low frequencies approaches some specified value, say K, and the attenuation at $s = j\omega_z$, the notch frequency, is infinite. The primary application for this type of filter is to remove undesired tones from the frequency spectrum of a signal, while (ideally) not altering the remaining signal characteristics. An alternative circuit to that of Fig. 4-14, for realizing a band-reject response, is shown in Fig. 4-16. The transfer function of this circuit is given by

$$\frac{V_2}{V_1} = \frac{K(s^2 + G^2 S^2)}{s^2 + 2(2 - K)GSs + G^2 S^2} \tag{4-44}$$

where $G = 1/R$ and $S = 1/C$.

Continuous-Time Active Filters—Biquadratic Realizations

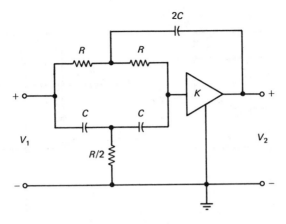

Fig. 4-16   Notch circuit.

4.2.6 Single-Op Amp Biquadratic Realizations Using Both Negative and Positive Feedback

Let us briefly summarize, in somewhat general terms, what we have observed thus far regarding the use of negative and positive feedback. For the negative-feedback circuits, we found that the magnitude of Q sensitivities to variations in the passive components were quite small. The Q sensitivities relative to the amplifier gain and bandwidth, however, were quite large, of the order of Q. This restricts the negative-feedback circuits to low-Q ($Q \le 5$) applications. The positive-feedback circuit, on the other hand, offers the ability of trade-off between active and passive sensitivities. At one extreme, if we select $K = 1$, thus minimizing the amount of positive feedback, we obtain a realization with low passive sensitivities and very high active sensitivities, in fact identical to the negative feedback case. At the other extreme, by increasing the positive feedback one can considerably decrease the active sensitivities at the expense of increased passive sensitivities. It should be apparent, then, that an optimum exists by using the "proper" amount of positive feedback.

In the following we study a circuit in which this sensitivity optimization process can be carried out and clearly illustrated. The circuit to be considered is similar to the negative feedback circuit of Fig. 4-8 except for the addition of some resistive positive feedback. The results obtained can be directly applied to the design of positive-feedback circuits, since the two families of circuits are related by the complementary transformation [P3b].

Consider the circuit shown in Fig. 4-17, which was originally introduced by Delyiannis [P4]. The transfer function for this circuit, assuming that GB $= \infty$, is as follows:

$$\frac{V_2}{V_1} = \frac{\dfrac{-G_1 S_2}{(1 - 1/K)} s}{s^2 + s\left(G_2 S_1 + G_2 S_2 - \dfrac{G_1 S_2}{K - 1}\right) + G_1 G_2 S_1 S_2} \tag{4-45a}$$

$$= \frac{b_0 s}{s^2 + a_1 s + a_0} \tag{4-45b}$$

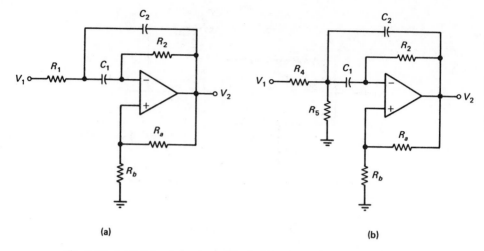

Fig. 4-17 (a) Delyiannis band-pass circuit. (b) Delyiannis circuit with provisions for gain adjustment.

where, of course, $G_i = 1/R_i$, $S_i = 1/C_i$ for $i = 1, 2$ and

$$K = 1 + \frac{R_a}{R_b} > 1$$

Note the subtractive term $G_1 S_1/(K - 1)$ in Eq. (4-45), which occurs due to positive feedback. For $C_1 = C$, $C_2 = \gamma C$, and $R_2 = \beta R_1$, the ideal design equations to determine the values for R_2 and K can be written as follows:

$$R_1 = \frac{1}{C}\sqrt{\frac{1}{a_0 \gamma \beta}} \tag{4-46a}$$

$$K = \frac{\beta + \gamma + 1 - a_1\sqrt{\beta\gamma/a_0}}{1 + \gamma - a_1\sqrt{\beta\gamma/a_0}} \tag{4-46b}$$

or substituting $a_0 = \omega_0^2$ and $a_1 = \omega_0/Q$, Eqs. (4-46) can be rewritten in the following convenient forms:

$$R_1 = \frac{1}{C\omega_0}\sqrt{\frac{1}{\gamma\beta}} \tag{4-47a}$$

$$K = \frac{Q(\beta + \gamma + 1) - \sqrt{\beta\gamma}}{Q(1 + \gamma) - \sqrt{\beta\gamma}} \tag{4-47b}$$

It should be noted that the circuit in Fig. 4-17a does not provide sufficient freedom to independently specify the gain constant b_0. To provide this freedom, all that is required is to replace R_1 in Fig. 4-17a with the voltage divider formed by R_4 and R_5 in Fig. 4-17b. This circuit is a special case of the general Friend SAB discussed in Section 4.3 (see Prob. 4.13).

Noting that

$$\omega_0 = \sqrt{\frac{1}{R_1 R_2 C_1 C_2}} \qquad (4\text{-}48a)$$

$$Q = \frac{1/\sqrt{R_1 R_2 C_1 C_2}}{\dfrac{1}{R_2 C_1} + \dfrac{1}{R_2 C_2} - \dfrac{1}{K-1}\dfrac{1}{R_1 C_2}}$$

$$= \frac{\sqrt{R_2/R_1}}{\sqrt{\dfrac{C_2}{C_1}} + \sqrt{\dfrac{C_1}{C_2}} - \dfrac{1}{K-1}\dfrac{R_2}{R_1}\sqrt{\dfrac{C_1}{C_2}}} \qquad (4\text{-}48b)$$

The sensitivities of ω_0 and Q to component can be written as follows:

$$S_{R_1}^{\omega_0} = S_{R_2}^{\omega_0} = S_{C_1}^{\omega_0} = S_{C_2}^{\omega_0} = -\frac{1}{2} \qquad (4\text{-}49a)$$

$$S_{R_2}^{Q} = -S_{R_1}^{Q} = -\frac{1}{2} + \frac{Q}{\sqrt{\beta}}\left(\frac{\gamma+1}{\sqrt{\gamma}}\right) \qquad (4\text{-}49b)$$

$$S_{C_1}^{Q} = -S_{C_2}^{Q} = -\frac{1}{2} + \frac{Q\sqrt{\gamma}}{\sqrt{\beta}} \qquad (4\text{-}49c)$$

$$S_{R_a}^{Q} = -S_{R_b}^{Q} = -\left(\frac{2Q}{\sqrt{\beta}} - 1\right)\sqrt{\frac{1}{\gamma}} \qquad (4\text{-}49d)$$

The following observations are noteworthy:

1. The center frequency ω_0 is minimally sensitive to the passive components.
2. Characteristic of positive-feedback circuits, the passive component Q sensitivities are directly proportional to Q.
3. The passive component Q sensitivities can be reduced by increasing β. Thus, low passive component sensitivities can be traded for large resistance spreads.
4. As the capacitance ratio γ increases, some Q sensitivities decrease while others increase. As far as the passive component sensitivities are concerned, little design freedom is lost by setting $\gamma = 1$.

To determine the effect of the finite op amp GB, let us reanalyze the Delyiannis circuit with a single-pole compensated op amp of gain $a(s) \simeq GB/s$. This analysis results in the following approximate transfer function:

$$\frac{V_2}{V_1} \simeq \frac{-sG_1 S_2 \dfrac{K}{K-1}\left[1 - \dfrac{s}{GB}\dfrac{K}{K-1} - \dfrac{K^2 G_1 S_2}{GB(K-1)^2}\right]}{s^2 + s\left(G_2 S_1 + G_2 S_2 - \dfrac{G_1 S_2}{K-1}\right)\left[1 - \dfrac{K^2 G_1 S_2}{GB(K-1)^2}\right] + G_1 G_2 S_1 S_2\left[1 - \dfrac{K^2 G_1 S_2}{GB(K-1)^2}\right]} \qquad (4\text{-}50a)$$

$$= \frac{N(s)}{s^2 + a_1 s + a_0} \qquad (4\text{-}50b)$$

It is noted that the derivation of Eq. (4-50) proceeds in much the same manner as the derivations given previously for Eqs. (4-18) and (4-30).

Comparing the coefficients of like powers of s in Eqs. (4-50a) and (4-50b), the following design equations, which compensate for the nominal op amp GB, can be derived. Let us assume that K is computed according to the ideal expression given in Eq. (4-46b). Given a computed value for K, γ, and C, we can compensate for the nominal GB by adjusting R_1 and R_2, determined from the following equations:

$$a_1 = \left(G_2 S_1 + G_2 S_2 - \frac{G_1 S_2}{K-1} \right)(1 - \eta G_1 S_2) \qquad (4\text{-}51a)$$

$$a_0 = G_1 G_2 S_1 S_2 (1 - \eta G_1 S_2) \qquad (4\text{-}51b)$$

where

$$\eta = \frac{K^2}{(K-1)^2 \text{GB}}$$

Finally, the ω_0 and Q sensitivities to the op amp GB can be determined from the transfer function Eq. (4-50a). Approximate expressions for these sensitivities (Prob. 4.10) are as follows:

$$S_{\text{GB}}^{\omega_0} \simeq -S_{GB}^{\mathring{Q}} \simeq \frac{1}{2} \frac{\omega_0}{\text{GB}} \frac{K^2}{(K-1)^2} \sqrt{\frac{\beta}{\gamma}} \qquad (4\text{-}52)$$

Note the following:

1. The ω_0 and Q sensitivities increase as the desired center frequency increases.

2. These sensitivities increase with increasing β. In contrast, the passive component Q sensitivities decrease with increasing β. By minimizing either of the statistical multiparameter sensitivity measures M_α, M_β, or M_T over some predetermined frequency range, an optimum value for β can be determined. Fleischer [P2] has shown that the optimum value of β is approximately

$$\beta_{\text{opt}} \simeq 4 \frac{\text{GB}}{\omega_0} \sqrt{\frac{8\sigma_R^2 + \sigma_C^2}{8\sigma_{\text{GB}}^2}} \qquad (4\text{-}53)$$

for $\sigma_{R_1}^2 = \sigma_{R_2}^2 = \sigma_{R_a}^2 = \sigma_{R_b}^2 = \sigma_R^2$ and $\sigma_{C_1}^2 = \sigma_{C_2}^2 = \sigma_C^2$.

3. Using the more complex double-pole single-zero op amp compensation given in Eq. (2-31), the sensitivities to GB can be considerably decreased. For voice-frequency applications where $f_0 < 4\,\text{kHz}$, the op amp GB sensitivities can be shown to be negligible compared to the passive component sensitivities. For these applications, β can be made as large as is practical to minimize the passive component Q sensitivities. In this case the sensitivity is dominated by the passive ω_0 sensitivities and the overall

Continuous-Time Active Filters—Biquadratic Realizations

sensitivities can be written approximately using Eqs. (3-123), namely

$$M_\alpha \simeq \omega_0 k_{1\alpha} Q T_{x_i x_j}^{\omega_0, \omega_0} \tag{4-54a}$$

$$M_\beta \simeq \omega_0 k_{1\beta} Q T_{x_i x_j}^{\omega_0, \omega_0} \tag{4-54b}$$

$$M_T \simeq 4\omega_0 k_T T_{x_i x_j}^{\omega_0, \omega_0} \tag{4-54c}$$

or the equivalent relations given in Eqs. (4-22).

●

EXAMPLE 4-4

Design a second-order band-pass active-RC filter with $Q = 20$ and $\omega_0 = 2\pi \times 2 \times 10^3$ rad/sec, using the Delyiannis circuit in Fig. 4-17 for minimum overall sensitivity. The op amp $GB = 2\pi \times 10^6$ rad/sec and the component standard deviations are

$$\sigma_{GB} = 0.20$$
$$\sigma_R = 0.001 \quad \text{for } R = R_1 = R_2 = R_a = R_b$$
$$\sigma_C = 0.001 \quad \text{for } C = C_1 = C_2$$

Furthermore, to minimize the capacitance spread, let $\gamma = 1$. The design is as follows:

$$C = 0.01 \ \mu F$$
$$\beta_{opt} = 10.607$$
$$R_1 = 2.43 \ k\Omega$$
$$R_2 = 25.8 \ k\Omega$$
$$K = 6.774$$

Computing κ, M_α, M_β, and M_T from Eqs. (4-22) over the frequency $\omega_0(1 - 1/Q) \leq \omega \leq \omega_0 (1 + 1/Q)$ and plotting, as a function of frequency, the integrand measures yield

$$\kappa = 0.132 \times 10^{-4}$$
$$M_\alpha = 0.9318 \times 10^{-4}$$
$$M_\beta = 1.9875 \times 10^{-4}$$
$$M_T = 2.9193 \times 10^{-4}$$

The plots are shown in Fig. 4-18. It should be noted that these numbers are relative and that any comparison between circuits should be made on the same basis. However, note that from the example in Fig. 4-18a, the figure shows that the deviation at the center frequency is minimal while it increases as we deviate from the center frequency.

●

Another interesting circuit that employs both positive and negative feedback is the high-pass notch circuit proposed by Boctor [P21], shown in Fig. 4-19.

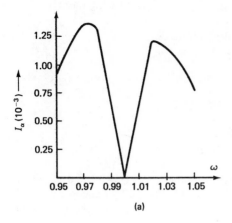

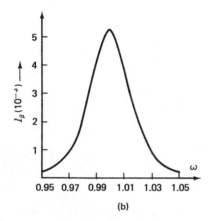

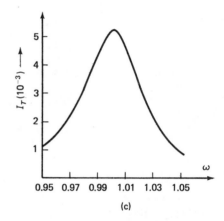

Fig. 4-18 Sensitivity integrands of the example (a) gain, (b) phase, and (c) transfer function.

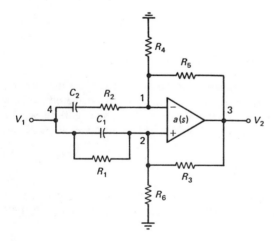

Fig. 4-19 High-pass notch network.

202

This circuit was discussed briefly in Example 3-6. It is noted that this high-pass notch circuit requires only two capacitors, as compared to the three-capacitor circuits shown in Figs. 4-14 and 4-16.

Analyzing Boctor's network, assuming an op amp with finite gain $a(s)$, the transfer function is determined to be the following:

$$H(s) = \frac{s^2(1 - k_2) + s(G_1 S_1 + G_{eq_2} S_2 - k_2 G_{eq_1} S_1) + G_1 G_{eq_2} S_1 S_2}{s^2\left[\dfrac{1}{a(s)} + k_1\right] + s\left[\dfrac{1}{a(s)}(G_{eq_1} S_1 + G_{eq_2} S_2) + k_1(G_2 S_2 + G_{eq_1} S_1) - G_3 C_1\right]}$$
$$+ G_{eq_1} G_{eq_2} S_1 S_2\left[\dfrac{1}{a(s)} + k_1\left(\dfrac{G_2}{G_{eq_2}} - \dfrac{G_3}{k_1 G_{eq_1}}\right)\right]$$

(4-55a)

$$= \frac{b_2 s^2 + b_1 s + b_0}{a_2 s^2 + a_1 s + a_0}$$

(4-55b)

where

$$G_{eq_1} = G_1 + G_3 + G_6, \qquad \frac{1}{G_{eq_2}} = \frac{1}{G_2} + \frac{1}{G_{45}},$$

$$G_{45} = G_4 + G_5, \qquad k_1 = \frac{G_5}{G_2} k_2, \qquad k_2 = \frac{G_{eq_2}}{G_{45}}$$

Recall that $G_i = 1/R_i$ and $S_i = 1/C_i$.

As indicated in Eq. (4-55b), to obtain the desired notch the b_1 coefficient must be set to zero. Making this assignment in Eq. (4-55a) results in the following notch condition for k_2:

$$k_2 = \frac{R_{eq_1}}{R_1} + \frac{R_{eq_1}}{R_{eq_2}}\left(\frac{C_1}{C_2}\right)$$

(4-56)

Furthermore, we can minimize the dependence of the pole frequency on the op amp gain by setting $G_2/G_{eq_2} - G_3/k_1 G_{eq_1} = 1$ in coefficient a_0 of Eq. (4-55a). This leads to a constraint on k_1:

$$k_1 = \frac{R_{eq_1}}{R_3}\left(\frac{R_2}{R_{45}}\right)$$

(4-57)

Substituting Eqs. (4-56) and (4-57) into Eq. (4-55a) yields the following transfer function:

$$H(s) = \left(\frac{1 - k_2}{\dfrac{1}{a(s)} + k_1}\right)\frac{s^2 + G_1 G_2 S_1 S_2}{\left\{s^2 + \dfrac{s}{1 + \dfrac{1}{k_1 a(s)}}\left[\dfrac{1}{k_1 a(s)}(G_{eq_1} S_1 + G_{eq_2} S_2) + G_{eq_1} S_1\left(1 - \dfrac{G_1 G_2}{G_{eq_1} G_{eq_2}}\right)\right]\right.}$$
$$\left. + G_{eq_1} G_{eq_2} S_1 S_2\right\}$$

(4-58)

It is noted when $|a(s)| \to \infty$, the ideal transfer function is

$$H(s) = \frac{1 - k_2}{k_1} \left\{ \frac{s^2 + G_1 G_2 S_1 S_2}{s_2 + s G_{eq_1} S_1 \left[\left(1 - \frac{G_1 G_2}{G_{eq_1} G_{eq_2}} \right) \right] + G_{eq_1} G_{eq_2} S_1 S_2} \right\} \quad (4\text{-}59)$$

The sensitivities for this circuit can be obtained from the more general transfer function given in Eq. (4-55a). In comparison with the Delyiannis circuit, the pole frequency of the Boctor circuit is seen to depend on eight passive components (six resistors and two capacitors) as opposed to four. On the other hand, the zero or notch frequency depends on only four passive components. Thus, the notch frequency for the Boctor circuit is minimally sensitive. As in the Delyiannis circuit, the pole Q sensitivities to the passive components are directly proportional to Q. From this comparison it can be reasonably estimated that the Boctor circuit will be more sensitive to passive component variations than will Delyiannis-type circuits. Notch and more general types of circuits, derived by extension of the Delyiannis concept, are treated in the next section.

To assess the effect of the finite op amp gain and to facilitate the comparison with other circuits, let us assume a one-pole compensated op amp with $a(s) \simeq GB/s$. Making this substitution into Eq. (4-58), we can derive the following approximate transfer function (for $|s/k_1 GB| \ll 1$):

$$H(s) \simeq \frac{1 - k_2}{k_1 \left(1 + \frac{\alpha}{k_1 GB} \right)} \left[\frac{s^2 + \omega_z^2}{s^2 + \left(\frac{\omega_0}{Q} + \frac{\omega_0^2}{k_1 GB} \right) \left(1 - \frac{\alpha}{k_1 GB} \right) s + \omega_0^2 \left(1 - \frac{\alpha}{k_1 GB} \right)} \right]$$

$$(4\text{-}60)$$

where $\omega_0^2 = G_{eq_1} G_{eq_2} S_1 S_2$

$$\frac{\omega_0}{Q} = G_{eq_1} S_1 \left(1 - \frac{G_1 G_2}{G_{eq_1} G_{eq_2}} \right)$$

$$\alpha = G_{eq_1} S_1 + G_{eq_2} S_2$$

$$\omega_z^2 = G_1 G_2 S_1 S_2$$

Thus, the actual pole frequency and Q are given by

$$\tilde{\omega}_0 \simeq \omega_0 \sqrt{1 - \frac{\alpha}{k_1 GB}} \simeq \omega_0 \left(1 - \frac{\alpha}{2 k_1 GB} \right) \quad (4\text{-}61a)$$

$$\tilde{Q} = \frac{\omega_0}{\left(\frac{\omega_0}{Q} + \frac{\omega_0^2}{k_1 GB} \right) \left(1 - \frac{\alpha}{k_1 GB} \right)} \simeq \frac{Q \left(1 + \frac{\alpha}{2 k_1 GB} \right)}{\left(1 + \frac{\omega_0 Q}{k_1 GB} \right)} \quad (4\text{-}61b)$$

The $\tilde{\omega}_0$ and $\tilde{Q}$ sensitivities to the op amp GB are then

$$S_{GB}^{\tilde{\omega}_0} \simeq \frac{\alpha}{2 k_1 GB} \quad (4\text{-}62a)$$

$$S_{GB}^{\tilde{Q}} \simeq -Q \frac{\omega_0}{k_1 GB} \quad (4\text{-}62b)$$

Note that $S_{GB}^{\omega_0}$ is independent of ω_0 and negligible for $\alpha/2k_1 GB \ll 1$. The Q sensitivity to GB is on the same order as that for the Delyiannis circuit.

4.3 GENERAL BIQUAD CIRCUIT USING A SINGLE OP AMP

One circuit that may be used to realize a general biquadratic transfer function, and of course can be used to realize special-case low-pass, high-pass, band-pass, and band-reject functions, is shown in Fig. 4-20. This circuit is strikingly similar

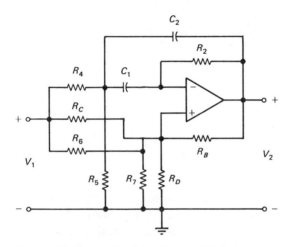

Fig. 4-20 Friend general biquad circuit.

to the positive feedback/negative feedback circuit of Delyiannis, shown in Fig. 4-17. The circuit has, in essence, two feedforward paths added to synthesize the zeros [P6]. Henceforth, we shall refer to these circuits as the Delyiannis–Friend (DF) circuits. The configuration in Fig. 4-20 is designed for and is highly suited to thin-film, hybrid integrated-circuit fabrication. In the literature this circuit is either referred to as the single amplifier biquad (SAB) or as the *standard tantalum active resonator* (STAR). It is currently being used extensively in the Bell Telephone System for audio- and voice-frequency signal-processing applications.

The transfer function of this circuit, assuming an infinite-gain op amp, is given by

$$H(s) = \frac{V_2}{V_1} = \frac{b_2 s^2 + b_1 s + b_0}{s^2 + a_1 s + a_0} \tag{4-63}$$

The coefficients of Eq. (4-63) are related to the circuit element values, for GB =

∞, according to the following relations:

$$b_2 = K_2 \tag{4-64a}$$

$$b_1 = \frac{K_2}{C_2}(G_1 + G_2 + G_3) + \frac{K_2}{C_1}(G_2 + G_3)$$
$$- \left(1 + \frac{G_B}{G_A}\right)\left[\frac{K_1}{C_2}G_1 + K_3G_3\left(\frac{1}{C_1} + \frac{1}{C_2}\right)\right] \tag{4-64b}$$

$$b_0 = \frac{1}{C_1C_2}\left[K_2G_1(G_2 + G_3) - K_3G_1G_3\left(1 + \frac{G_B}{G_A}\right)\right] \tag{4-64c}$$

$$a_1 = \frac{C_1 + C_2}{C_1C_2}\left(G_2 - \frac{G_3G_B}{G_A}\right) - \frac{G_3G_B}{C_1G_A} \tag{4-64d}$$

$$a_0 = \frac{G_1}{C_1C_2}\left(G_2 - G_3\frac{G_B}{G_A}\right) \tag{4-64e}$$

where

$$R_1 = R_4 \| R_5, \qquad R_A = R_C \| R_D, \qquad R_3 = R_6 \| R_7,$$
$$K_1 = \frac{R_5}{R_4 + R_5}, \qquad K_2 = \frac{R_D}{R_C + R_D}, \qquad K_3 = \frac{R_7}{R_6 + R_7} \quad \text{and} \quad G_i = \frac{1}{R_i} \tag{4-65}$$

It is noted that there are nine variables $(C_1, C_2, R_1, R_2, R_3, K_1, K_2, K_3, R_A/R_B)$ and only five specifiable coefficients in Eq. (4-63). Therefore, four of these variables and either R_A or R_B can be independently prespecified by the designer. Usually, the capacitances are set to equal values, such as the normalized values $C_1 = C_2 = 1$.

To synthesize this network, we choose C_1, C_2, R_A, R_B, and K_3 and compute R_1, K_1, R_3, and R_2, respectively. It is noted that the optimum value for β [Eq. (4-53)] found by Fleischer for the Delyiannis circuit is applicable here (i.e., approximately the same value). An optimum $R_A/R_B = 1/(K - 1)$ is related to the optimum β for $\gamma = 1$, according to Eq. (4-46b). After some algebraic manipulation, Eqs. (4-64) and (4-65) lead to the following ideal design equations:

$$R_1 = \frac{2}{\gamma C}\frac{R_A}{R_B}\left[-a_1 + \sqrt{a_1^2 + 4\left(1 + \frac{1}{\gamma}\right)a_0\frac{R_A}{R_B}}\right]^{-1} \tag{4-66a}$$

$$K_1 = \left(1 + \frac{R_A}{R_B}\right)^{-1}[b_2 + \gamma(1 + \gamma)b_0C^2R_1^2 - b_1\gamma CR_1] \tag{4-66b}$$

$$R_3 = \left(1 + \frac{R_A}{R_B}\right)(b_2 - K_3)\left[R_1\gamma C^2a_0\left(\frac{b_0}{a_0} - b_2\right)\right]^{-1} \tag{4-66c}$$

$$R_2 = R_3\left(\gamma C^2a_0R_1R_3 + \frac{R_A}{R_B}\right)^{-1} \tag{4-66d}$$

where $0 \le K_3 \le 1$ and $\gamma = C_2/C_1 = 1$. Also,

$$R_4 = \frac{R_1}{K_1} \quad \text{thus} \quad R_5 = \frac{R_4 R_1}{R_4 - R_1} \qquad (4\text{-}66\text{e})$$

$$R_6 = \frac{R_3}{K_3} \quad \text{thus} \quad R_7 = \frac{R_6 R_3}{R_6 - R_3} \qquad (4\text{-}66\text{f})$$

$$R_C = \frac{R_A}{b_2} \quad \text{thus} \quad R_D = \frac{R_C R_A}{R_C - R_A} \qquad (4\text{-}66\text{g})$$

K_3 is selected such that the resulting value for R_3 is nonnegative.

For band-pass realizations, the SAB reduces directly to the Delyiannis circuit of Fig. 4-17. In addition, the low-pass notch, high-pass notch and all-pass response cases of the SAB are shown in Figs. 4-21, 4-22, and 4-23, respectively.

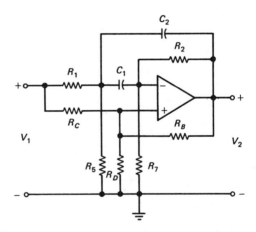

Fig. 4-21 Low-pass notch network.

The low-pass notch configuration, $b_1 = 0$ and $b_0 > a_0$ in Eq. (4-63), is obtained by setting G_6 to zero in Fig. 4-20. The low-pass notch circuit shown in Fig. 4-21 has a transfer function (with GB $= \infty$) given by

$$H(s) = \frac{\dfrac{R_A}{R_C}(s^2 + \omega_z^2)}{s^2 + \left[\dfrac{1}{R_2 C_2} + \dfrac{1}{R_2 C_1} - \dfrac{R_A}{R_B}\left(\dfrac{1}{R_3 C_1} + \dfrac{1}{R_3 R_2} + \dfrac{1}{R_1 C_2}\right)\right]s + \omega_p^2} \qquad (4\text{-}67)$$

where

$$\omega_z = \sqrt{b_0} = \sqrt{\frac{1 + \dfrac{R_2}{R_3}}{R_1 R_2 C_1 C_2}}, \qquad \omega_p = \sqrt{a_0} = \sqrt{\frac{1 - \dfrac{R_A R_2}{R_B R_3}}{R_1 R_2 C_1 C_2}}$$

and for $b_1 = 0$,

$$R_4 = \frac{\dfrac{R_C R_2}{R_A}\left(1 + \dfrac{R_A}{R_B}\right)}{\dfrac{R_2}{R_1} + \left(1 + \dfrac{C_2}{C_1}\right)\left(1 + \dfrac{R_2}{R_3}\right)} \qquad (4\text{-}68)$$

Note that for realizability $(R_A/R_B)(R_2/R_3) < 1$; also, as desired, $\omega_p < \omega_z$. It is further noted that the low-pass notch circuit in Fig. 4-21 contains only two capacitors, in contrast to the circuits in Figs. 4-14 and 4-16. Thus, no pole–zero cancellation is needed in the SAB to obtain a second-order transfer function.

One can similarly realize a high-pass notch response, where $\omega_p > \omega_z$, by setting G_5 and G_7 to zero in Fig. 4-20 and Eqs. (4-64) and (4-66). The resulting circuit is shown in Fig. 4-22. The design equations are those given for the more general SAB circuit, namely Eqs. (4-66).

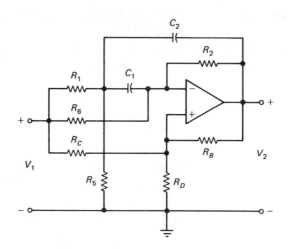

Fig. 4-22   High-pass notch network.

It is noted that the high-pass and all-pass configurations can also be obtained from the general configuration of Fig. 4-20. The high-pass configuration is obtained by setting $b_1 = b_0 = 0$ in Eqs. (4-64) and (4-66). The construction of this circuit and the derivation of its design equations are left as an exercise for the reader. Similarly, the all-pass configuration is obtained by setting $b_2 = 1$, $b_0 = a_0$, and $b_1 = -a_1$. Again, the circuit construction and the derivation of the design equations are reserved as an exercise. An alternative to the SAB all-pass circuit is given in Fig. 4-23.

The only response that cannot be realized directly with the SAB is the low-pass response. Another circuit, which is similar to the SAB, is shown in Fig. 4-24. The transfer function for this circuit is given by the relation (where $GB = \infty$)

$$H(s) = \frac{-\left(1 + \dfrac{R_D}{R_B}\right) \Big/ R_3 R_9 C_1 C_2}{s^2 + \left(\dfrac{1}{R_2 C_1} + \dfrac{1}{R_3 C_1} + \dfrac{1}{R_9 C_1} - \dfrac{R_D}{R_B}\dfrac{1}{R_9 C_2}\right)s + \dfrac{1}{R_9 R_3 C_1 C_2}\left(\dfrac{R_3}{R_2} + \dfrac{R_D}{R_B}\right)}$$

$$(4\text{-}69)$$

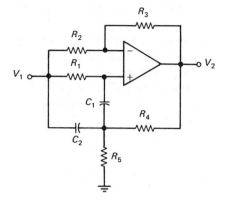

Fig. 4-23 All-pass network.

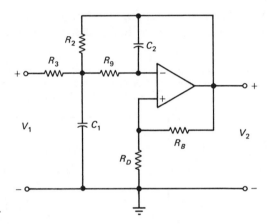

Fig. 4-24 Low-pass network.

The ideal design equations for this circuit, with $C_1 = C$, $C_2 = \gamma C$ and $R_D/R_B = \lambda$, are

$$R_9 = \frac{2(1 - \lambda/\gamma)}{b_1 C \pm C \sqrt{b_1^2 - 4(a_0 + b_0)\gamma\left(1 - \dfrac{\lambda}{\gamma}\right)}} \qquad (4\text{-}70a)$$

$$R_2 = \frac{1 + \lambda}{\gamma C^2 R_9[b_0(1 + \lambda) + a_0\lambda]} \qquad (4\text{-}70b)$$

$$R_3 = \frac{1 + \lambda}{b_0 \gamma C^2 R_9} \qquad (4\text{-}70c)$$

The design procedure is to choose λ, γ, and C, then to calculate R_9 followed by R_2 and R_3. The determination of the sensitivities and the compensation for the nominal op amp GB are left as exercises for the reader.

It is interesting to compare the overall sensitivity of a STAR high-pass notch design using the circuit in Fig. 4-22 with that of an equivalent Boctor high-pass notch design using the circuit in Fig. 4-19. This exercise follows directly from the developments cited in Example 3-6. In addition to conditions cited in this example, we shall assume that $\sigma_{GB} = 0.5$. The STAR circuit was designed to minimize its sensitivity $(M_T)_N$ to passive component variations and the Boctor circuit was designed using the minimal $S_{a(s)}^{\omega_0}$ condition Eq. (4-57). To view the effect of high-frequency operation on sensitivity, these designs were scaled in frequency such that the notch frequency varied from 10 to 10^7 rad/sec, which maintained constant ω_z/ω_0. The resulting values for $(M_T)_N$ are plotted versus the log of the notch frequency ω_z in Fig. 4-25. The following general observations are made:

1. At the lower frequencies, where the passive component sensitivities predominate, the STAR high-pass notch design is less sensitive in the overall sense.
2. At higher frequencies ($\omega_z > 10^5$ rad/sec), where the op amp GB sensitivities predominate, the Boctor design is seen to be less sensitive.

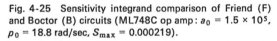

Fig. 4-25 Sensitivity integrand comparison of Friend (F) and Boctor (B) circuits (ML748C op amp: $a_0 = 1.5 \times 10^5$, $p_0 = 18.8$ rad/sec, $S_{max} = 0.000219$).

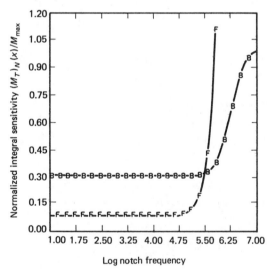

All the circuits shown in this section and others not shown can be realized with the tiny thin-film STAR hybrid integrated circuit (HIC) shown in Fig. 4-26a. This 260×760 mil HIC, comprised of two 5.1-nF α-tantalum capacitors [P28], nine larger trimmable TaN [P28] resistors, and a silicon integrated-circuit operational amplifier (which is bonded onto the surface of the ceramic substrate), is conveniently packaged in a standard 16-pin dual-in-line package (DIP). The schematic for the circuitry contained on this HIC is shown in Fig.

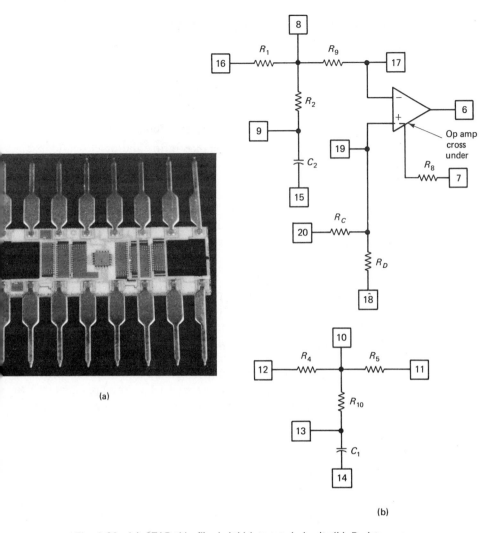

Fig. 4-26 (a) STAR thin-film hybrid integrated circuit. (b) Resistor–capacitor T's for the STAR (boxed numbers are pin numbers).

4-26b [P6(b), P28], where each square denotes one of 16 pins. Note that the terminals of each resistor and capacitor are accessible to a pair of pins. This external accessibility is important for measuring and trimming [P29] the components and for defining the circuit topology (i.e., Figs. 4-17, 4-20–4-24). The topology for each DIP is determined by a printed wiring board schematic which connects the pins appropriately to define the desired topology. This simple circuit then provides an economical means for implementing a universal type of second-order active filter building block for realizing filters of virtually arbitrary order and function. The design of higher-order filters with second-order blocks is reserved for Chapter 5.

4.4 LOW-SENSITIVITY MULTIPLE-OP AMP BIQUAD REALIZATIONS

By introducing additional op amps to the active-RC circuit, one can derive several benefits, often simultaneously. Such benefits include:

1. Lower sensitivities to both active and passive components.
2. Reduced dependence of critical resistance ratios and/or closed-loop op-amp gain(s) on the filter Q.
3. Ease in tuning, perhaps, allowing ω_0 and Q to be independently adjusted.
4. Fewer passive components.
5. A more universal filter structure, which permits general biquad realization with the minimum number of topological changes.

The most significant price paid for these features is the increased power dissipated by the filter and noise. For some applications the added power dissipation may outweigh the advantages gained from a multiple-amplifier topology. It is then good practice to carefully evaluate the filter performance and power requirements before departing from a practical SAB design. However, since low-cost monolithic integrated-circuit op amps are available with either one, two, three, or four op amps on a single chip, single-op amp biquads offer little size and cost advantage. This is particularly true when the area required to realize the passive components is reduced in the multiple-op amp implementation.

For very high Q, design benefits 1 through 3 are particularly important. Q's of 50 and higher, which are impractical for SAB realization, can be realized with one of several multiple-op amp topologies. As we show in the remaining sections of this chapter, the use of multiple op amps can lead to biquad active-filter sections which are capable of realizing very large Q values with lower sensitivities, easier-to-fabricate components, and higher pole frequencies than the single-op amp circuits discussed in previous sections. As we will see in Chapter 6, the use of multiple op amps in active switched capacitor second-order filters is virtually a requirement for practical MOS implementation.

We begin our discussion of multiple-op amp circuits by considering a pair of two-op amp filter sections. We then present in table form design information on a family of practical two-op amp biquads. This is followed by a detailed consideration of the low-sensitivity, three-op amp considerations and the general four-op amp biquad realizations.

4.4.1 Two-Op Amp Biquads

One of the simplest two-op amp circuits is shown in Fig. 4-27a. With feedback used around the two op amps resulting with amplifiers gains K_1 and

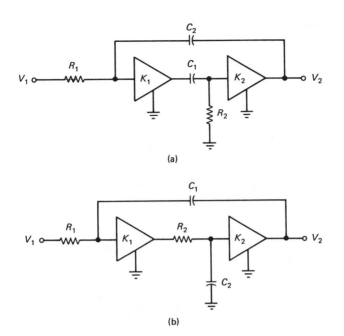

(a)

(b)

Fig. 4-27 (a) Simple two-amplifier circuit of Geffe (*Proc IEEE,* August 1969) and Soderstrand and Mitra (*Proc IEEE,* December 1969). (b) Simple two-amplifier circuit for low-pass section.

K_2, the transfer function, assuming ideal amplifiers, is given by the following relation:

$$H(s) = \frac{K_1 K_2 G_1 S_2 s}{(1 - K_1 K_2)s^2 + (G_1 S_2 + G_2 S_1)s + G_1 S_1 G_2 S_2} \qquad (4\text{-}71)$$

The center frequency ω_0 and Q for this band-pass filter are

$$\omega_0 = \sqrt{\frac{1}{R_1 R_2 C_1 C_2 (1 - K_1 K_2)}} \qquad (4\text{-}72\text{a})$$

$$Q = \frac{\sqrt{R_1 R_2 C_1 C_2 (1 - K_1 K_2)}}{R_1 C_2 + R_2 C_1} \qquad (4\text{-}72\text{b})$$

With prespecified $K_1, K_2, C_1 = C$, and $C_2 = \gamma C$, the ideal design equations are readily obtained from Eqs. (4-72):

$$R_2 = \frac{1}{2\omega_0 QC}\left[1 + \sqrt{1 - \frac{4Q^2}{1 - K_1 K_2}}\right] \qquad (4\text{-}73\text{a})$$

$$R_1 = \frac{1}{\omega_0^2 \gamma C^2 R_2 (1 - K_1 K_2)} \qquad (4\text{-}73\text{b})$$

For realizability the gain product $K_1 K_2$ is limited to

$$K_1 K_2 \leq -4Q^2 + 1 \qquad (4\text{-}74)$$

The passive component sensitivities for this network are

$$S_{R_1}^{\omega_0} = S_{R_2}^{\omega_0} = S_{C_1}^{\omega_0} = S_{C_2}^{\omega_0} = -\frac{1}{2} \tag{4-75a}$$

$$S_{K_1}^{\omega_0} = S_{K_2}^{\omega_0} = \frac{1}{2}\frac{K_1 K_2}{1 - K_1 K_2} \tag{4-75b}$$

$$S_{R_1}^{Q} = S_{C_2}^{Q} = \frac{1}{1 + \gamma(R_1/R_2)} - \frac{1}{2} \tag{4-75c}$$

$$S_{R_2}^{Q} = S_{C_1}^{Q} = \frac{\gamma(R_1/R_2)}{1 - \gamma(R_1/R_2)} - \frac{1}{2} \tag{4-75d}$$

$$S_{K_1}^{Q} = S_{K_2}^{Q} = \frac{K_1 K_2}{1 - K_1 K_2} \tag{4-75e}$$

Also, the ω_0 and Q sensitivities to the op amp GB_i, for GB_1 and GB_2, can be determined to be

$$S_{GB_1}^{\omega_0} \simeq \frac{1}{2}\frac{R_1 C_2 + R_2 C_1}{R_1 R_2 C_1 C_2 (1 - K_1 K_2)}\frac{K_1}{GB_1} = \frac{1}{2}\frac{\omega_0}{Q}\frac{K_1}{GB_1} \tag{4-75f}$$

$$S_{GB_2}^{\omega_0} \simeq \frac{1}{2}\frac{R_1 C_2 + R_2 C_1}{R_1 R_2 C_1 C_2 (1 - K_1 K_2)}\frac{K_2 + 1}{GB_2} = \frac{1}{2}\frac{\omega_0}{Q}\frac{K_2 + 1}{GB_2} \tag{4-75g}$$

$$S_{GB_1}^{Q} \simeq \frac{1}{R_1 C_2 + R_2 C_1}\frac{K_1}{GB_1} = Q\frac{K_1 \omega_0}{GB_1} \tag{4-75h}$$

$$S_{GB_2}^{Q} \simeq \frac{1}{R_1 C_2 + R_2 C_1}\frac{K_2 + 1}{GB_2} = Q\frac{(K_2 + 1)\omega_0}{GB_2} \tag{4-75i}$$

For ω_0 normalized to unity, it is noted that by setting $C_1 = C_2 = 1/2Q$ and $K_1 = -K_2 = \sqrt{4Q^2 - 1}$, we obtain from Eqs. (4-73), $R_1 = R_2 = 1$. Substituting these values into the sensitivities yields

$$S_{K_1}^{\omega_0} = S_{K_2}^{\omega_0} \simeq \tfrac{1}{2} \tag{4-76a}$$

$$S_{R_1}^{Q} = S_{R_2}^{Q} = S_{C_1}^{Q} = S_{C_2}^{Q} = 0 \tag{4-76b}$$

$$S_{K_1}^{Q} = S_{K_2}^{Q} \simeq 1 \tag{4-76c}$$

$$S_{GB_1}^{\omega_0} \simeq \frac{\omega_0}{GB_1}, \qquad S_{GB_2}^{\omega_0} \simeq \frac{\omega_0}{GB_2} \tag{4-76d}$$

$$S_{GB_1}^{Q} \simeq 2Q^2\frac{\omega_0}{GB_1}, \qquad S_{GB_2}^{Q} \simeq 2Q^2\frac{\omega_0}{GB_2} \tag{4-76e}$$

In single-op amp circuits, such as in Fig. 4-5, the gain K is proportional to Q^2. This high value of gain limits the single-op amp circuit to low-frequency, low-Q applications. However, by using two op amps, the gain requirement has been rationed between two amplifiers and causes a reduction in the gain burden carried by any one amplifier. Thus, if the single-op amp circuit is restricted to $Q \leq 5$, the two-op amp circuit should be capable of realizing Q values of $Q \leq$

10. It is noted that in Eq. (4-76), many of the $S_{x_i}^Q$ sensitivities have been forced to zero. For high-Q design, where variations in the center frequency dominate the overall filter sensitivity, such design measures are of only academic interest. In practice, the Q sensitivities are deemed sufficiently low when $|S_{x_i}^Q| \leq 1$. Attempts to force them lower usually result in overdesign, with the ensuing unnecessary sacrifice in other performance criteria and/or cost.

A second simple two-op amp circuit is shown in Fig. 4-27b, where the gains K_1 and K_2 are op amp-derived voltage-gain amplifiers. This circuit, assuming an ideal op amp, realizes the following low-pass transfer function:

$$\frac{V_2}{V_1} = \frac{K_1 K_2 G_1 G_2 S_1 S_2}{s^2 + [(G_1 S_1 + G_2 S_2 (1 - K_1 K_2)]s + G_1 G_2 S_1 S_2} \qquad (4\text{-}77)$$

where

$$G_i = \frac{1}{R_i} \quad \text{and} \quad S_i = \frac{1}{C_i} \quad \text{for } i = 1, 2$$

This circuit is readily derived from the positive-feedback, single-op amp network of Fig. 4-5 by inserting a second op amp to isolate RC networks R_1, C_1 and R_2, C_2. The design equations, with $\text{GB}_1 = \text{GB}_2 = \infty$, are

$$R_2 = \frac{1}{2} \frac{a_1}{a_0} \frac{1}{\gamma C} \left[1 + \sqrt{1 + \frac{4a_0}{a_1^2} (K_1 K_2 - 1)} \right] \qquad (4\text{-}78a)$$

$$R_1 = \frac{1}{a_0 \gamma C^2 R_2} \qquad (4\text{-}78b)$$

where $C_1 = C$, $C_2 = \gamma C$, and $K_1 K_2$ are assumed given. Also, for realizability,

$$K_1 K_2 > 1 - \frac{a_1^2}{4a_0} = 1 - \frac{1}{4Q^2} \qquad (4\text{-}79)$$

Note that when $K_1 K_2 < 1$, K_1 and K_2 are realized as inverting amplifiers. The sensitivities for this circuit are

$$S_{R_1}^{\omega_0} = S_{R_2}^{\omega_0} = S_{C_1}^{\omega_0} = S_{C_2}^{\omega_0} = -\tfrac{1}{2} \qquad (4\text{-}80a)$$

$$S_{C_1}^Q = S_{R_1}^Q = -\frac{1}{2} - \frac{1}{1 + \dfrac{R_1}{R_2} \dfrac{1}{\gamma} (1 - K_1 K_2)} \qquad (4\text{-}80b)$$

$$S_{C_2}^Q = S_{R_2}^Q = -\frac{1}{2} - \frac{1}{1 - K_1 K_2 + (R_2/R_1)\gamma} \qquad (4\text{-}80c)$$

$$S_{K_1}^Q = S_{K_2}^Q = \frac{-K_1 K_2}{1 - K_1 K_2 + (R_2/R_1)\gamma} \qquad (4\text{-}80d)$$

$$S_{\text{GB}_1}^{\omega_0} \simeq S_{\text{GB}_1}^{\tilde{Q}} \simeq \frac{1}{2} \frac{K_1^2 K_2}{R_2 C_2} \frac{1}{\text{GB}_1} \qquad (4\text{-}80e)$$

$$S_{\text{GB}_2}^{\omega_0} \simeq S_{\text{GB}_2}^{\tilde{Q}} \simeq \frac{1}{2} \frac{K_2^2 K_1}{R_2 C_2} \frac{1}{\text{GB}_2} \qquad (4\text{-}80f)$$

Comparing the component spreads and sensitivities of this two-op amp network with those of its single-op amp counterpart in Fig. 4-5, we find (Prob. 4.16) that little or no improvement in these areas has been achieved. This is simply due to the manner in which high Q is achieved in positive-feedback circuits. Since high Q is achieved via cancellation rather than large closed-loop amplifier gain, portioning the gain among multiple amplifiers is of little consequence. In contrast, for negative-feedback circuits when high Q is realized with large closed-loop amplifier gain, portioning this gain among multiple op amps can improve component spread and high-frequency performance. This was demonstrated in the discussion regarding the two-op amp circuit in Fig. 4-27. It is then not surprising that all the low-sensitivity multiple-op amp active filters discussed in the next section achieve high Q with negative feedback.

We conclude this section by presenting in Table 4-1 six circuits for the realization of all special second-order transfer functions in Fig. 4-28a to f. These circuits are based on the GIC inductance simulation circuit of Fig. 2-33. All the biquads in Table 4-1 have low passive Q sensitivities, and their dependence on the amplifier GB is minimized when the two op amps are matched. For further details and discussion of these circuits, the reader is referred to [B7].

TABLE 4-1 Second-order two-op amp circuits[a]
$$H(s) = N(s)/D(s)$$

Fig. 4-28	Type	Numerator of the Transfer Functions $N(s)$	
a	Band-pass	$2\left(\dfrac{\omega_0}{Q}\right)s$	
b	High-pass	$2s^2$	
c	All-pass	$s^2 - s\left(\dfrac{\omega_0}{Q}\right) + \omega_0^2$	
d	Low-pass notch	$\left(\dfrac{\omega_0}{\omega_n}\right)^2(s^2 + \omega_n^2)$	$\alpha = \dfrac{\omega_n}{\omega_0} \geq 1$
e	High-pass notch	$2\left(\dfrac{2 - \alpha^2}{3 - \alpha^2}\right)(s^2 + \omega_n^2)$	$\alpha = \dfrac{\omega_n}{\omega_0} \leq 1$
f	Low-pass	$2\omega_0^2$	

[a] In all cases, $\omega_0 = 1/RC$ and $D(s) = s^2 + s(\omega_0/Q) + \omega_0^2$.

ω_0 and Q enhancement of the two-op amp circuits can be obtained by straightforward analysis. If we assume matched amplifiers (i.e., $GB_1 = GB_2 = GB$) and the amplifier gains given by $a(s) \simeq GB/s$, the actual expressions for Q and ω_0 in Fig. 4-28a denoted by $\tilde{Q}$ and $\tilde{\omega}_0$ are given by

$$\tilde{\omega}_0 \simeq \omega_0\left[1 - \frac{2\omega_0}{GB}\right] \tag{4-81}$$

$$\tilde{Q} \simeq Q\left[1 - \frac{2\omega_0}{GB} + 2Q\left(\frac{2\omega_0}{GB}\right)^2\right] \tag{4-82}$$

Note also that in Fig. 4-28a the gain at the center frequency is fixed at $H_0 = 2$, as can readily be seen from Table 4-1. This fixed gain provides a disadvantage, in that it cannot be adjusted to optimize the filter's dynamic range (Section 5.3).

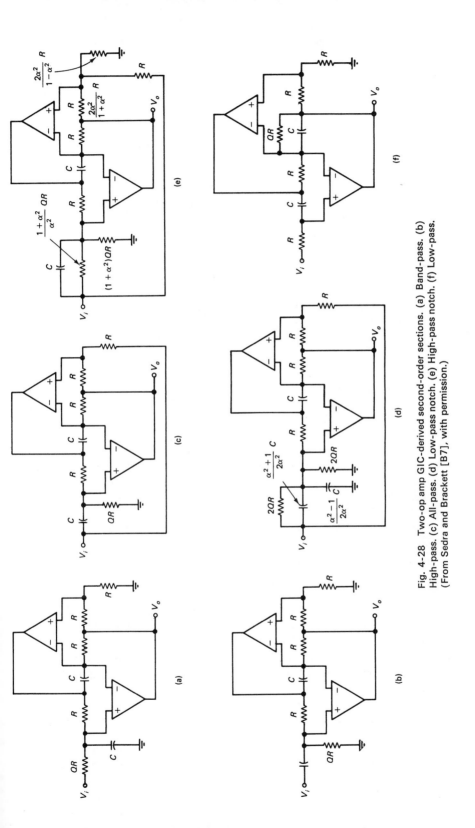

Fig. 4-28 Two-op amp GIC-derived second-order sections. (a) Band-pass. (b) High-pass. (c) All-pass. (d) Low-pass notch. (e) High-pass notch. (f) Low-pass. (From Sedra and Brackett [B7], with permission.)

The gain can be made adjustable by setting the two resistors, having a common node with V_0 in Fig. 4-28a to $(H_0 - 1)R$. This choice, however, results in a drastic Q enhancement, which renders it unacceptably large for $H_0 \neq 2$ even for applications with moderate values of gain and pole Q. Figure 4-29 demon-

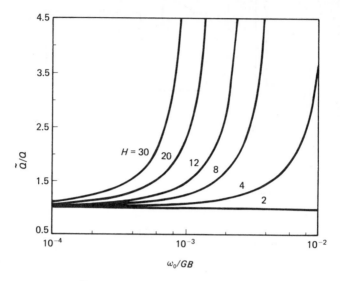

Fig. 4-29 $\tilde{Q}/Q$ as a function of $\omega_0/$GB and H for $Q = 30$. (From C.F. Choi and R. Schaumann, *Proc. 1980 IEEE Int. Symp. Circuits Syst.,* April 1980, p. 574.)

strates the extremely large sensitivity of $\tilde{Q}$ to $\omega_0/$GB for $H_0 \neq 2$. A simple passive compensation method which makes $\tilde{Q}/Q \simeq 1$ for all H_0 and all $\omega_0/$GB over a wide frequency range is available (see Prob. 4.17) The compensation, however, does not alleviate the problem of ω_0 deviation, which could be taken care of by predistortion.

Finally, a comparison of two-op amp circuits with a three-op amp circuit is made in Fig. 4-30a and b. These figures show the effect of GB on Q and ω_0 for a nominal value of $Q = 10$ corresponding to Figs. 4-44a and 4-45a. Figure 4-30c–e shows the gain, phase, and transfer function sensitivity of these two-op amp circuits as compared to a circuit with three op amps. This figure corresponds to Fig. 4-43 for various three-op amp circuits assuming similar statistics. Notice that the performance of the two-op amp circuit of Fig. 4-27a is very poor, much poorer then those of the GIC-derived circuit of Fig. 4-28a. Also note that the performance of the GIC-derived circuit is poorer than those of the three-op amp circuit shown in Fig. 4-42. More is said about three-op amp biquads in the next section.

4.4.2 Three-Op Amp Biquads

Lower sensitivities, higher Q's, more flexibility, and easier tuning can be achieved with any one of several three-op amp circuits than with those using

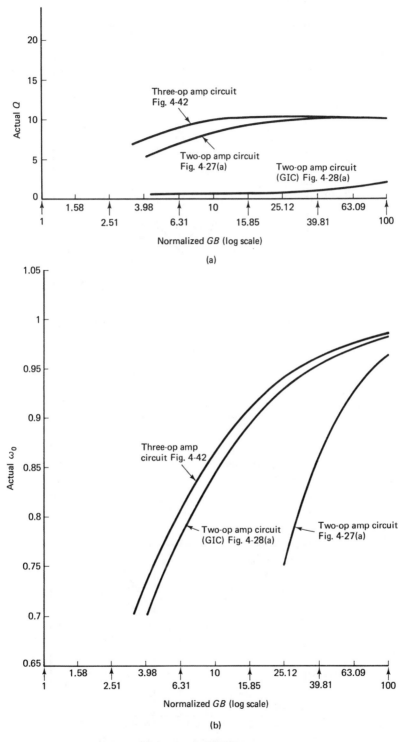

Fig. 4-30 (a) Effect of GB on Q for nominal $Q = 10$ for two-op amp and three-op amp circuits. (b) Effect of GB on ω_0 for nominal $Q = 10$. (c) Gain sensitivity versus center frequency. (d) Phase sensitivity versus center. (e) Transfer function sensitivity versus center frequency.

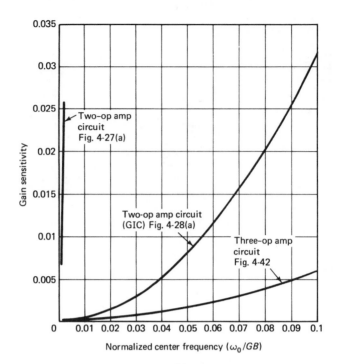

(c)

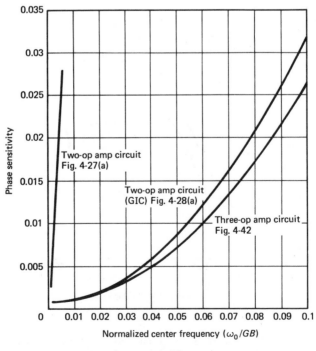

(d)

Fig. 4-30   (cont.)

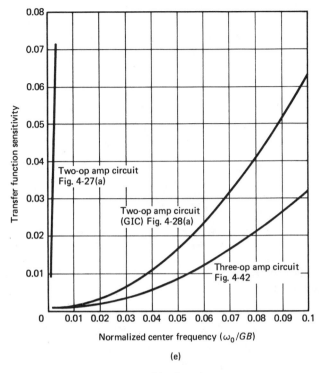

Fig. 4-30 (cont.)

fewer op amps. Here we shall consider in detail only some of the more popular circuits: the state-variable realization of Kerwin, Huelsman, and Newcomb (KHN) [P7]; the three-op amp biquad of Tow and Thomas (TT) [P8], [P9]; the Akerberg and Mossberg [P14]; and, the differential input/differential output op amp circuit of Tarmy and Ghausi (TG) [P10]. Circuits introduced by Mikhael and Bhattacharyya (MB) [P11]; Wilson, Bedri, Bowron [P13]; modified Tarmy-Ghausi [P12]; and Padukone, Mulawka, and Ghausi [P15] are reserved for the discussion on general biquad synthesis in the next section. These multiple-op amp circuits are also compared with respect to certain performance parameters in Section 4.4. Let us now consider each of these circuits individually, starting with the KHN state-variable circuit shown in Fig. 4-31a.

The **Kerwin Huelsman and Newcomb (KHN)** network is directly derived from the state-variable solution of a second-order linear differential equation with constant coefficients, following the concepts described in Section 2.4. One useful feature of this circuit is that low-pass, band-pass, and high-pass transfer functions are simultaneously available from output terminals V_1, V_2, and V_3 respectively. Assuming ideal op amps, we have

$$H_{\text{LP}}(s) = \frac{V_1}{V_S} = \frac{K_1/K_2 C_1 C_2 R_4 R_5}{D(s)} = \frac{b_0}{s^2 + a_1 s + a_0} \qquad (4\text{-}83a)$$

$$H_{\text{BP}}(s) = \frac{V_2}{V_S} = \frac{-(K_1/K_2 C_1 R_4)s}{D(s)} = \frac{b_1 s}{s^2 + a_1 s + a_0} \tag{4-83b}$$

$$H_{\text{HP}}(s) = \frac{V_3}{V_S} = \frac{(K_1/K_2)s^2}{D(s)} = \frac{b_2 s^2}{s^2 + a_1 s + a_0} \tag{4-83c}$$

where

$$K_1 = 1 + \frac{R}{R_3}, \qquad K_2 = 1 + \frac{R_1}{R_2} \tag{4-84}$$

$$D(s) = s^2 + \frac{K_1(K_2 - 1)}{K_2 C_1 R_4} s + \frac{K_1 - 1}{C_1 C_2 R_4 R_5} \tag{4-85}$$

For Eq. (4-85) we identify

$$\omega_0 = \sqrt{\frac{K_1 - 1}{C_1 C_2 R_4 R_5}} \tag{4-86a}$$

$$Q = \frac{K_2}{K_1} \frac{1}{K_2 - 1} \sqrt{(K_1 - 1) \frac{C_1 R_4}{C_2 R_5}} \tag{4-86b}$$

Specifying values for K_1, K_2, C_1, $= C$, and $C_2 = \gamma C$, the design equations for R_4 and R_5 are written

$$R_5 = \frac{a_1}{a_0 \gamma C} \frac{K_2(K_1 - 1)}{K_1(K_2 - 1)} = \frac{1}{\omega_0 \gamma CQ} \frac{K_2(K_1 - 1)}{K_1(K_2 - 1)} \tag{4-87a}$$

$$R_4 = \frac{K_1 - 1}{a_0 \gamma C^2 R_9} = \frac{Q}{\omega_0} \frac{K_1(K_2 - 1)}{K_2 C} \tag{4-87b}$$

Also, the gain-level coefficients b_0, b_1, and b_2 can be expressed in terms of ω_0, Q, and the circuit components as follows:

$$b_0 = \frac{K_1}{K_2} \frac{\omega_0^2}{K_1 - 1} \Longrightarrow |H_{\text{LP}}(j\omega)| = \frac{K_1}{K_2} \left| \frac{1}{K_1 - 1} \right| \quad \text{at } \omega = 0 \tag{4-88a}$$

$$b_1 = -\frac{\omega_0}{Q} \frac{1}{K_2 - 1} \Longrightarrow |H_{\text{BP}}(j\omega_0)| = \left| \frac{1}{K_2 - 1} \right| \tag{4-88b}$$

$$b_2 = \frac{K_1}{K_2} \Longrightarrow |H_{\text{HP}}(j\omega)| = \frac{K_1}{K_2} \quad \text{at } \omega = \infty \tag{4-88c}$$

Note that K_1/K_2 can be chosen to achieve the desired gain level according to Eqs. (4-88), or to optimize component spreads or sensitivity.

The sensitivities of ω_0 and Q to the passive components can be written (Prob. 4.22)

$$S_{R_3}^{\omega_0} = S_{R_4}^{\omega_0} = S_{R_5}^{\omega_0} = S_{C_1}^{\omega_0} = S_{C_2}^{\omega_0} = -S_R^{\omega_0} = -\tfrac{1}{2} \tag{4-89a}$$

$$S_{R_4}^{Q} = S_{C_1}^{Q} = \tfrac{1}{2} \tag{4-89b}$$

$$S_{R_3}^{Q} = -S_R^{Q} = \frac{1}{1 + R_3/R} \tag{4-89c}$$

$$S_{R_2}^{Q} = -S_{R_1}^{Q} = \frac{1}{1 + R_1/R_2} \tag{4-89d}$$

The following observations are made:

1. ω_0 depends on six components; thus, ω_0 is not minimally sensitive to the passive components. When the resistor variations track, the variation in ω_0 approaches a minimum. In an integrated-circuit realization, resistors track with temperature and can be highly correlated in manufacture. However, laser trimming the resistors tends to render them statistically independent in their initial values.

2. The Q sensitivities are low, with $|S_{x_i}^Q| \leq 1$. This was seen earlier to be a property of negative-feedback circuits.

To determine the effect of the finite op amp GB_i's on the performance of this filter, let us make the simplifying assumption that all the op amps are identical (i.e., $GB_1 = GB_2 = GB_3 = GB$). Furthermore, let us use the component values $C_1 R_4 = C_2 R_5 = T$, $R = R_1 = R_3 = 1$, and $R_2 = 2Q - 1$. After much algebraic manipulation, the following relations, using Eq. (2-29) for each integrator and

$$V_3 = \frac{K_1/K_2}{1 + sK_1/GB} V_s + \frac{K_1(K_2 - 1)/K_2}{1 + sK_1/GB} V_2 - \frac{K_1 - 1}{1 + sK_1/GB} V_1 \quad (4\text{-}90)$$

for the output of the input summing amplifier, we obtain the following approximate expressions for the ω_0 and Q sensitivities to the $GB(s)$:

$$S_{GB}^{\omega_0} \simeq \frac{\omega_0}{GB} \quad (4\text{-}91a)$$

$$S_{GB}^{\tilde{Q}} \simeq \frac{-4\tilde{Q}\omega_0}{GB} \quad (4\text{-}91b)$$

Note that Eq. (4-91b) implies that $1/\tilde{Q} \simeq 1/Q - 4\omega_0/GB$, where Q denotes the ideal value with $GB = \infty$ and $\tilde{Q}$ the actual value. From this relation the severe Q-enhancement effect, caused by cumulative phase lag contributed by the op amps, is all too apparent. When $4\omega_0/GB = 1/Q$ or $4Q\omega_0/GB = 1$, the circuit oscillates. Therefore, the useful range of operation for this circuit is dictated by the inequality $4Q\omega_0/GB < 0.1$. For example, if $f_0 = 1$ kHz, $GB = 2\pi(1 \text{ MHz})$ (i.e., a $\mu A741$ type of op amp), Q is limited to $Q < 25$. One can in principle improve this situation by passive lead compensation to cancel the phase lag contributed by the op amps, or can design for a sufficiently lower ideal Q that the actual Q is enhanced to the desired value. Unfortunately, these methods are not very effective, because of large variability of GB in manufacture, and with temperature. GB can vary by as much as 50% in manufacture and by a comparable amount over temperature range of 0 to 70°C. It is difficult to maintain compensation over such large variations in the parameter being compensated. Alternatively, one can use a wider-bandwidth op amp or, preferably, the double-pole single-zero compensation given in Eq. (2-31). A factor-of-10 increase is op amp gain at f_0 will increase the limit to $Q < 250$. Yet another alternative is to find a circuit structure in which the Q dependence on the op amp performance is minimized. Such a network structure will be described shortly.

Unfortunately, the overall sensitivity performance, after Eqs. (4-22), due to all the passive and active component variations, is no better, and perhaps poorer, than that achieved with the Delyiannis–Friend single-op amp circuits. The advantages realized with the present circuit is a partial decoupling of the synthesis equations which leads to simpler tuning and a more universal structure in which low-pass, band-pass, and high-pass functions are available simultaneously.

An additional requirement, which the designer must pay particular attention to in multiple-amplifier filters, is dynamic range. To be more specific, it is most desirable, when the input signal levels are expected to span a wide range (say a few millivolts peak to peak to a value slightly less than the supply voltage), that each op amp overloads at the same signal level. To see the problem more clearly, let us refer to the desired filter output as the primary output and to the other op amp outputs as the secondary outputs. The gain level of the primary output is, of course, determined by design [i.e., Eqs. (4-88)]. If the voltage at either (or both) of the secondary outputs is too high, these op amps will overload prematurely and undesirable nonlinear distortion will result. "Too high" usually means that the voltage exceeds the supply voltage or the slew rate. Overloads can result due to large signals which are either in-band or out-of-band. At the other end of the scale, in-band voltage levels that are too low at the secondary outputs can result in unnecessary noise penalties. A remedy to this situation is to scale the gain levels of the secondary outputs such that the voltage maxima of the secondary outputs are made equal to the voltage maximum of the primary output.

The scaling can be performed analytically by setting

$$\max_{\omega} | H_{\mathrm{LP}}(j\omega) | = \max_{\omega} | H_{\mathrm{BP}}(j\omega) | = \max_{\omega} | H_{\mathrm{HP}}(j\omega) | \qquad (4\text{-}92)$$

where the frequency of peak gain ω_{mi} is found by

$$\frac{\partial | H_i(j\omega) |}{\partial \omega} \bigg|_{\omega = \omega_{mi}} = 0 \qquad \text{where } i = \mathrm{LP, HP, BP}$$

For those possessing a computer-aided circuit analysis capability, the simplest procedure is to simulate the unscaled circuit and print the voltages at the output of each op amp. This step also serves as a confirmation of the correctness of the design. From the printed computer output, one then identifies the values of the voltage maxima. With sufficient degrees of freedom in the network, the scaling is then done in straightforward fashion.

After some study of the KHN network in Fig. 4-31a, it becomes apparent that there are insufficient degrees of freedom to perform this scaling while maintaining ω_0, Q, and the primary output level constant. Adding a resistor, as indicated in the alternative state-variable realization shown in Fig. 4-31b, yields

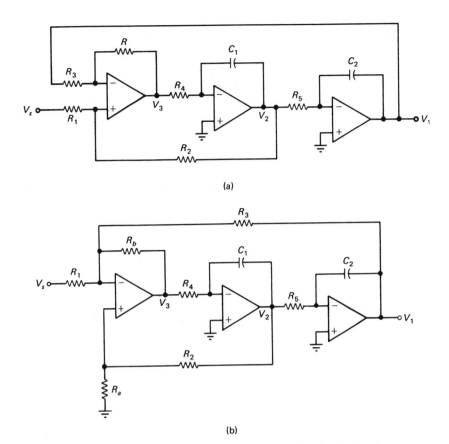

(a)

(b)

Fig. 4-31 (a) The Kerwin, Huelsman, and Newcomb (KHN) circuit. (b) KHN circuit with an additional resistor added to provide a degree of freedom for scaling.

the necessary design freedom to perform this scaling. The transfer functions for this circuit are

$$H_{\mathrm{LP}}(s) = \frac{V_1}{V_s} = \frac{-(R_b/R_1)(1/C_1C_2R_4R_5)}{D(s)} \qquad (4\text{-}93a)$$

$$H_{\mathrm{BP}}(s) = \frac{V_2}{V_s} = \frac{(R_b/R_1)(1/C_1R_4)s}{D(s)} \qquad (4\text{-}93b)$$

$$H_{\mathrm{HP}}(s) = \frac{V_3}{V_s} = \frac{-(R_b/R_1)s^2}{D(s)} \qquad (4\text{-}93c)$$

where

$$D(s) = s^2 + \frac{1 + R_b/R_1 + R_b/R_3}{1 + R_2/R_a}\frac{1}{C_1R_4}s + \frac{R_b}{R_3}\frac{1}{C_1C_2R_4R_5} \qquad (4\text{-}93d)$$

To demonstrate the scaling of this network, let us assume that V_1 is the primary output. It is our task then to scale the levels of V_2 and V_3 (i.e., the peak gain levels of H_{BP} and H_{HP}) without affecting ω_0, Q, and H_{LP}. Furthermore, since it is usually desirable to keep the capacitors fixed, the scaling is to be performed by resistor adjustments only. To adjust H_{BP} and H_{HP} by a constant multipliers μ and ν, respectively;

$$H_{BP} \longrightarrow \mu H_{BP}$$
$$H_{HP} \longrightarrow \nu H_{HP} \tag{4-94}$$

we adjust resistors R_5, R_4, R_b, and R_2 in the following way:

$$(R_5, R_4, R_b, R_2) \Longrightarrow \left(\mu R_5, \frac{\nu}{\mu} R_4, \nu R_b, \delta R_2 \right) \tag{4-95a}$$

where

$$\delta = \frac{\mu}{\nu} \left(\frac{1 + \nu \dfrac{R_b}{R_1} + \nu \dfrac{R_b}{R_3}}{1 + \dfrac{R_b}{R_1} + \dfrac{R_b}{R_3}} \right) \left(1 + \frac{R_a}{R_2} \right) - \frac{R_a}{R_2} \tag{4-95b}$$

Substituting Eqs. (4-95) into Eqs. (4-93), one can verify the validity of Eqs. (4-94) and the invariability of ω_0, Q, and H_{LP}. It is noted that typically, from (4-92) and (4-94),

$$\mu = \frac{\max\limits_{\omega} |H_{LP}(j\omega)|}{\max\limits_{\omega} |H_{BP}(j\omega)|} \quad \text{and} \quad \nu = \frac{\max\limits_{\omega} |H_{LP}(j\omega)|}{\max\limits_{\omega} |H_{HP}(j\omega)|}$$

A second circuit, with similar performances to that of the KHN circuit, is the three-op amp biquad of **Tow and Thomas (TT)**. This circuit, shown in Fig. 4-32, has both low-pass and band-pass transfer functions simultaneously available at terminals V_1 and V_2, respectively:

$$H_{LP}(s) = \frac{V_1}{V_s} = \frac{-\left(\dfrac{r_2}{r_1}\right)\dfrac{1}{R_4 C_1 R_2 C_2}}{s^2 + \dfrac{1}{R_1 C_1}s + \dfrac{r_2}{r_1}\dfrac{1}{R_2 R_3 C_1 C_2}} \tag{4-96a}$$

$$H_{BP}(s) = \frac{V_2}{V_s} = \frac{-\dfrac{1}{R_4 C_1}s}{s^2 + \dfrac{1}{R_1 C_1}s + \dfrac{r_2}{r_1}\dfrac{1}{R_2 R_3 C_1 C_2}} \tag{4-96b}$$

where the op amps are assumed ideal. The ideal design equations for this network are

$$R_2 = \frac{r_2}{r_1}\frac{1}{\omega_0^2 R_3 C_1 C_2} \tag{4-97a}$$

$$R_1 = \frac{Q}{\omega_0 C_1} \tag{4-97b}$$

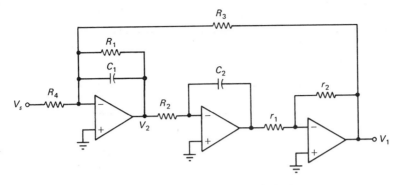

Fig. 4-32 The Tow–Thomas (TT) circuit.

$$R_4 = \frac{r_1}{r_2} \frac{\omega_0^2 R_3}{b_0} \qquad \text{for } H_{LP}$$

or

$$R_4 = \frac{1}{b_1 C_1} \qquad \text{for } H_{BP}$$

(4-97c)

As a brief example, let ω_0 be normalized to unity and assume that $r_2 = r_1 = 1$, $C_1 = C_2 = 1$, and $R_3 = 1$. Computing R_2 and R_1 yields $R_2 = 1$ and $R_1 = Q$. Thus, the largest resistor is only Q, as compared to $4Q^2$ for the Sallen and Key circuits and $\simeq Q^2/2$ for the Delyiannis–Friend circuits.

Observe that ω_0, Q, and b_0 (or b_1) are independently determined by R_2, R_1, and R_4, respectively. This decoupling of design relations is particularly useful in high-Q applications where functional tuning is needed to set those parameters accurately.

Here the tuning may be performed as follows:

1. Adjust R_2 to obtain the maximum gain, measured at node V_2, at the desired center frequency ω_0.
2. Adjust R_1 so that the gain measured at node V_2 is 3 dB down from the maximum at the band-edge frequencies, $\omega \simeq \omega_0 \pm \omega_0/2Q$.
3. Adjust R_4 so that the peak gain measured at node V_1 at frequency ω_0 is at the desired value.

The scaling of this network to ensure a wide dynamic range is left as an exercise for the reader.

An active-RC building block with inherent compensation for the finite gain–bandwidth product of the amplifier is shown in Fig. 4-33 [P14]. This circuit is referred to as the **Akerberg and Mossberg** (AM) circuit. The voltage transfer function of the circuit is written in the following convenient form:

$$\frac{V_0}{V_i} = \frac{N}{s_n^2 + s_n/Q + 1}$$

(4-98)

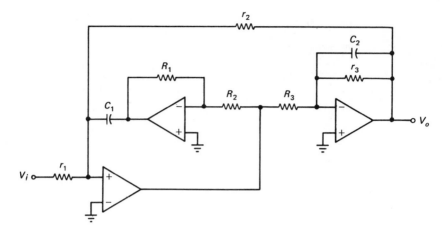

Fig. 4-33   The Akerberg–Mossberg (AM) circuit.

where $s_n = s/\omega_0$, $N = -r_2/r_1$, and

$$\omega_0 = \left(\frac{R_2}{R_1 r_2 R_3 C_1 C_2}\right)^{1/2} \tag{4-99a}$$

$$Q = r_3 \left(\frac{C_2}{C_1}\right)^{1/2} \left(\frac{R_2}{R_1 r_2 R_3}\right)^{1/2} \tag{4-99b}$$

This circuit, as we shall see later, possesses the useful characteristic that by proper design the Q can be made to be approximately independent of the gain–bandwidth product of the amplifiers. Recall that in the KHN circuit, the circuit Q was limited by the finite gain–bandwidth products of the amplifiers. For a high-Q design, the KHN circuit may oscillate, and this fact can be predicted by the inclusion of the pole of the amplifier as done before. The inclusion of the effects of finite GB, assuming that the amplifiers are identical and characterized by single-pole roll-off, namely $a(s) \simeq GB/s$, yields the following expressions for the AM circuit [P14]:

$$\frac{V_o}{V_i} = \frac{P(s_n, GB_n)}{s_n^2 + s_n/Q + 1 + (2s_n/GB_n)\sum\limits_{i=0}^{4} K_i s_n^i} \tag{4-100}$$

where $s_n = s/\omega_0$, $GB_n = GB/\omega_0$, and ω_0 is given by (4-99a). For the conditions

$$R_1 = R_2 = R_3 = r_1 = r_2 = R, \qquad r_3 = QR, \qquad C_1 = C_2$$

we have

$$P(s_n, \mathrm{GB}_n) = -\left(1 + \frac{2s_n}{\mathrm{GB}_n}\right)$$

$$K_0 = \frac{1}{Q} + 1$$

$$K_1 = \frac{3}{2} + \frac{1}{Q} + \frac{1 + 3/Q}{\mathrm{GB}_n}$$

$$K_2 = 1 + \frac{7 + 3/Q}{2\mathrm{GB}_n} + \frac{2 + (2/Q)}{(\mathrm{GB}_n)^2} \qquad (4\text{-}101)$$

$$K_3 = \frac{3}{2\mathrm{GB}_n} + \frac{3 + (1/Q)}{(\mathrm{GB}_n)^2}$$

$$K_4 = \frac{1}{(\mathrm{GB}_n)^2}$$

The Q sensitivity with respect to the gain–bandwidth product and the center frequency are given by (Prob. 4.23)

$$S_{\mathrm{GB}}^{\tilde{Q}} \simeq \frac{\omega_0}{2\mathrm{GB}} \qquad (4\text{-}102a)$$

$$S_{\mathrm{GB}}^{\tilde{\omega}_0} \simeq \frac{3}{2}\frac{\omega_0}{\mathrm{GB}} \qquad (4\text{-}102b)$$

A comparison of the $\tilde{Q}$ and $\tilde{\omega}_0$ sensitivities when the op amps are not matched (i.e., $\mathrm{GB}_1 \neq \mathrm{GB}_2 \neq \mathrm{GB}_3$) is given in Table 4-4.

The next circuit is somewhat unique among the multiple-op amp class of circuits. This circuit, shown in Fig. 4-34, was introduced by **Tarmy and Ghausi** (**TG**) [P10]. Its primary advantage is its low sensitivities to the op amp GBs, and thus its superior performance at high ω_0 and Q. Its transfer function,

Fig. 4-34   The Tarmy–Ghausi (TG) circuit.

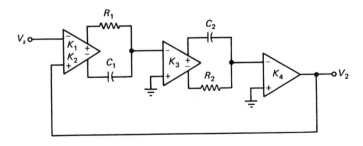

assuming ideal op amps, is given by

$$H(s) = \frac{\dfrac{K_1 K_3 K_4}{1 + K_2 K_3 K_4}(1 - T_1 s)(1 - T_2 s)}{s^2 + \left(\dfrac{1}{T_1} + \dfrac{1}{T_2}\right)\dfrac{1 - K_2 K_3 K_4}{1 + K_2 K_3 K_4}s + \dfrac{1}{T_1 T_2}} \qquad (4\text{-}103)$$

where $T_1 = R_1 C_1$ and $T_2 = R_2 C_2$. The ideal design equations for this network are as follows:

$$R_1 = R_2 = \frac{1}{\omega_0 C} \qquad \text{where } C_1 = C_2 = C \text{ is assumed given} \qquad (4\text{-}104a)$$

and with $K_3 = K_4 = 1$,

$$K_2 = \frac{2 - 1/Q}{2 + 1/Q} \simeq 1 - \frac{1}{Q} \qquad (4\text{-}104b)$$

Using $K_2 = 1 - 1/Q$ will result in an error of less than 0.5 % in the definition of the network Q for $Q > 10$.

From the relation $Q = (1 + K_2)/2(1 - K_2)$, it follows that the Q sensitivities are large, (i.e., $S_{K_2}^Q \simeq K_2 Q = Q$) but $K_2 \simeq 1$ is very stable and the effect is not as serious as those found in positive-feedback circuits.

The closed-loop amplifier gains K_1 and K_2 are realized using the differential input/differential output op amp shown in Fig. 2-8. The definitions for K_1 and K_2 are given in Eqs. (2-12). The values for the resistor ratios R_b/R_a and R_d/R_c are not unique and can be determined to optimize sensitivity. As we now show, the lower the values for these resistor ratios, the lower the amplifier gains and the sensitivities. Let us, for convenience, define $\alpha_1 = R_d/R_c$ and $\alpha_2 = R_b/R_a$ so that, from Eqs. (2-12),

$$K_1 = \frac{-\alpha_1(1 + \alpha_2)}{1 + 2\alpha_2 + \alpha_1 \alpha_2} \qquad (4\text{-}105a)$$

$$K_2 = \frac{1 + \alpha_1}{1 + 2\alpha_2 + \alpha_1 \alpha_2} \qquad (4\text{-}105b)$$

Then from the relation $K_2 = 1 - 1/Q$, we write

$$Q = \frac{1}{1 - K_2} = \frac{1 + 2\alpha_2 + \alpha_1 \alpha_2}{2\alpha_2 - \alpha_1 + \alpha_1 \alpha_2} \qquad (4\text{-}106)$$

The Q sensitivities to resistors R_a, R_b, R_c, and R_d (for α_1 and $\alpha_2 \ll 1$) are then given by

$$S_{R_b}^Q = -S_{R_a}^Q \simeq -2\alpha_2 Q \qquad (4\text{-}107a)$$

$$S_{R_d}^Q = -S_{R_c}^Q \simeq \alpha_1 Q \qquad (4\text{-}107b)$$

Thus, the smaller we make α_1 and α_2, the lower the Q sensitivities become. For example, if we set $\alpha_1 = \alpha_2 = 1/Q$, the sensitivities become

$$S_{R_b}^Q = -S_{R_a}^Q \simeq -2 \qquad (4\text{-}108a)$$

$$S_{R_d}^Q = -S_{R_a}^Q \simeq 1 \qquad (4\text{-}108b)$$

The $1/Q$ component spread is no greater than that required for the TT three-op amp biquad and it is significantly smaller than that required in the single-op amp realizations.

For completeness, the remaining passive sensitivities are

$$S_{R_1}^{\omega_0} = S_{R_2}^{\omega_0} = S_{C_1}^{\omega_0} = S_{C_2}^{\omega_0} = -\tfrac{1}{2} \qquad (4\text{-}108c)$$

$$S_{R_1}^Q = S_{C_1}^Q = \frac{1}{1 + R_1 C_1 / R_2 C_2} - \frac{1}{2} \qquad (4\text{-}108d)$$

$$S_{R_2}^Q = S_{C_2}^Q = \frac{1}{1 + R_2 C_2 / R_1 C_1} - \frac{1}{2} \qquad (4\text{-}108e)$$

for $R_1 C_1 = R_2 C_2$ and $S_{R_1}^Q = S_{C_1}^Q = S_{R_2}^Q = S_{C_2}^Q = \tfrac{1}{2}$.

It is interesting to note that:

1. The center frequency ω_0 is minimally sensitive to the passive components.
2. The Q sensitivities are low with $|S_x^Q| \leq 2$.

Thus, as far as the passive components are concerned, the TG biquad is truly a low-sensitivity network.

Let us now consider the effect of the finite op amp bandwidths on the performance of this biquad. Since all the closed-loop amplifier gains K_1, K_2, K_3, and K_4 are near unity, it is expected that the effect will be small. To establish this property, let us write the sensitivities of ω_0 and Q to the op amp GBs. The derivation of these sensitivities is very laborious and much too lengthy to be included in this book. These sensitivities, with all op amps assumed identical (i.e., $GB_1 = GB_2 = GB_3 = GB$) can be written as

$$S_{GB}^{\omega_0} \simeq \frac{3\omega_0}{2GB} \qquad (4\text{-}108f)$$

$$S_{GB}^Q \simeq \frac{-3\omega_0}{2GB} \qquad (4\text{-}108g)$$

Note that S_{GB}^Q is independent of Q and $S_{GB}^{\omega_0}$ is the same as in (4-102b). It is also noted that GB_1 determines the bandwidth of closed-loop gains K_1 and K_2, while GB_2 and GB_3 determine the bandwidths for K_3 and K_4, respectively.

The following observations are of interest:

1. $S_{GB}^{\omega_0}$ is slightly greater than that realized with the KHN and TT state-variable biquads.

2. The severe Q-enhancement effect suffered by the KHN and TT biquads has been for all practical purposes eliminated. To be more precise,

$$\frac{1}{\tilde{Q}} \simeq Q^{-1}\left(1 - \frac{3}{2}\frac{\omega_0}{\text{GB}} + \frac{3Q}{a_0}\right)$$

where a_0 denotes the op amp dc open-loop gain. For GB = 1 MHz and $a_0 = 10^5$, the change in Q is negligible for $f_0 < 20$ kHz and $Q < 3000$. Furthermore, when Q is very large, the effects of GB and a_0 tend to cancel. This is not the case for the other two circuits.

For a comparison of the $\tilde{Q}$ and $\tilde{\omega}_0$ sensitivities when the op amps are not matched, see Table 4-4.

It is noted that tuning the TG circuit, although not completely noniterative, is relatively straightforward. That is, ω_0 can be tuned by adjusting either R_1 or R_2, which has negligible effect on Q, and Q can be varied by adjusting $K_2K_3K_4$, which has no effect on ω_0 since according to (4-103), ω_0 is independent of K_i.

Despite its many advantages, the TG circuit has a main disadvantage, its use of differential output op amps. Note also that the circuit is not a true BP, LP, or HP, but for high Q applications this is not a problem. One difficulty in using differential output op amps is that many of the popular op amps, such as Fairchild's μA741, are not configured for differential output operation. However, there are differential output monolithic op amps, such as Motorola's MC1520, available at relatively modest prices. Another disadvantage is that the two differential outputs are not exactly in antiphase. A complete analysis might include this effect, which in turn depends upon the ratio of two resistors internal to the amplifier. For applications when tight sensitivity control is required, the added cost of two differential output op amps appears to be a small price to pay. To bypass this problem, attempts have been made to modify the Tarmy–Ghausi circuit in order to use only single-ended op amps. These modifications [P12], such as the one shown in Fig. 4-35, have in some manner compromised some of the low-sensitivity properties of the TG circuit. For example, the center frequency for the circuit in Fig. 4-35 depends on all 10 passive circuit elements,

Fig. 4-35 Moschytz's (M) modified TG circuit.

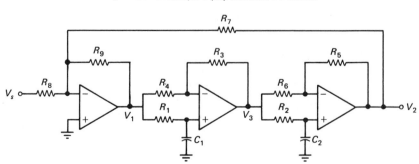

Continuous-Time Active Filters—Biquadratic Realizations

whereas the ω_0 for the TG circuit was shown to be dependent upon the minimum number of passive elements: four. A comparison of the sensitivity performance of the circuit in Fig. 4-35 with those previously considered is given in Section 4.5.1.

4.5 GENERAL BIQUADRATIC REALIZATIONS WITH MULTIPLE-OP AMP BIQUADS

A single-op amp realization of a general biquadratic transfer function [Eq. (4-63)] using the Friend circuit was shown in Section 4.3. In this section we show how easily a general biquad can be realized by simply using a summing amplifier and any one of the three-op amp biquad circuits in previous sections.

As an example, let us consider the KHN network shown in Fig. 4-31, which we noted earlier has low-pass, band-pass, and high-pass outputs simultaneously available at terminals V_1, V_2, and V_3, respectively. Thus, if we simply sum these outputs, modified by the appropriate gain and sign, we obtain the desired general biquadratic transfer function:

$$V_o = \alpha_1 V_1 + \alpha_2 V_2 + \alpha_3 V_3 = \frac{b_2 s^2 + b_1 s + b_0}{s^2 + a_1 s + a_0} V_S = \frac{K\left(s^2 + \dfrac{\omega_z}{Q_z}s + \omega_z^2\right)}{s^2 + \dfrac{\omega_p}{Q_p}s + \omega_p^2} V_S$$

$$(4\text{-}109)$$

where V_1, V_2, and V_3 are given in Eqs. (4-81) to (4-83) and α_1, α_2, and α_3 are summer gains with appropriate sign. The schematic for general KHN realization is given in Fig. 4-36. Its transfer function is given by

$$H(s) = \frac{V_o}{V_S} = \frac{K_1 K_3}{K_2 K_4} \frac{N(s)}{D(s)} = \frac{[1 + (R/R_3)][1 + (R_6/R_7)]}{[1 + (R_1/R_2)][1 + (R_8/R_9)]} \frac{N(s)}{D(s)} \qquad (4\text{-}110a)$$

Fig. 4-36 General biquad with feedforward paths.

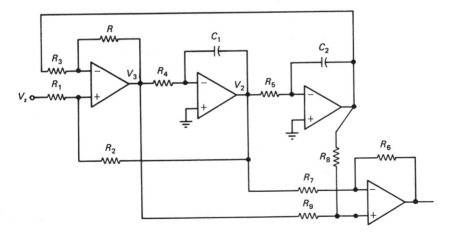

where $K_1 = 1 + R/R_3$, $K_2 = 1 + R_1/R_2$, $K_3 = 1 + R_6/R_7$, and $K_4 = 1 + R_8/R_9$; $D(s)$ is given in Eq. (4-85); and

$$N(s) = s^2 + \frac{K_4(K_3 - 1)}{K_3} \frac{1}{C_1 R_4} s + \frac{K_4 - 1}{C_1 C_2 R_4 R_5} \qquad (4\text{-}110b)$$

Recall from Eqs. (4-87) that

$$R_5 = \frac{1}{\omega_p Q_p \gamma C} \frac{K_2(K_1 - 1)}{K_1(K_2 - 1)} \qquad (4\text{-}111a)$$

$$R_4 = \frac{Q_p}{\omega_p} \frac{K_1(K_2 - 1)}{K_2 C} \qquad (4\text{-}111b)$$

where $K_1, K_2, C_1 = C$, and $C_2 = \gamma C$ are set to convenient values. Equations (4-111) completely define the coefficients of $D(s)$. To complete the design, we give in the following ideal design equations to define $N(s)$:

$$R_8 = R_9(K_1 - 1)\frac{\omega_z^2}{\omega_p^2} \qquad (4\text{-}111c)$$

$$R_7 = R_6\left\{\frac{K_1(K_2 - 1)}{K_2} \frac{Q_p \omega_z}{Q_z \omega_p}\left[1 + (K_1 - 1)\frac{\omega_z^2}{\omega_p^2}\right] - 1\right\} \qquad (4\text{-}111d)$$

Here R_9 and R_6 are conveniently used to impedance-scale the resistors in the output summing amplifier. Finally, the gains K_1 or K_2 may be used to scale the flat gain $K = K_1 K_3/K_2 K_4$.

Note that by summing all three outputs (V_1, V_2, and V_3) with a non-inverting summer, we can make $(-b_1 = a_1)$, which is required in the realization of an all-pass response. Further note that if we make $b_1 = 0$ in Eq. (4-109), by open-circuiting R_7, we realize notch and band-reject functions.

In Section 4.4 we stated the expressions for the ω_p and Q_p sensitivities. These sensitivities are obviously unaffected by the zero-formation process. Let us now determine the ω_z and $1/Q_z$ sensitivities, which can be substituted directly into Eq. (3-168c) to compute $(M_T)_N$. As noted in Section 3.7, $(M_T)_N$ is particularly suited to evaluating sensitivity in the stop band at or near a notch frequency. In the passband, the overall sensitivity is dominated by variation in ω_p and Q_p, and the sensitivity performance determined in Section 4.4 for the all-pole KHN network is duplicated in the more general circuit. The ω_z and $1/Q_z$ sensitivities are as follows:

$$S_{R_8}^{\omega_z} = S_{R_9}^{\omega_z} = S_{C_1}^{\omega_z} = S_{C_2}^{\omega_z} = S_{R_5}^{\omega_z} = -S_{R_4}^{\omega_z} = -\tfrac{1}{2} \qquad (4\text{-}112a)$$

$$S_{R_9}^{1/Q_z} = S_{C_2}^{1/Q_z} = -S_{R_8}^{1/Q_z} = -S_{C_1}^{1/Q_z} = \frac{1}{2Q_z} \qquad (4\text{-}112b)$$

$$S_{R_7}^{1/Q_z} = -S_{R_6}^{1/Q_z} = \frac{R_6}{R_7} \frac{1}{Q_z} \qquad (4\text{-}112c)$$

$$S_{R_4}^{1/Q_z} = -S_{R_5}^{1/Q_z} = \left(\frac{1}{1 + R_4/R_5}\right)\frac{1}{2Q_z} \qquad (4\text{-}112d)$$

The following is to be noted:

1. ω_z is not minimally sensitive to the passive components.
2. For a notch design $S_x^{1/Q_z} = 0$, these zero values result from the direct setting of $1/Q_z$ to zero by open-circuiting R_7. Once R_7 is opened, $1/Q_z$ remains zero no matter how large the variations in the other elements may be.

As noted in the previous discussion, one means for realizing transmission zeros in an all-pole multiple-op amp biquad is to take the weighted sum of the three-op amp outputs. This method of course requires an additional amplifier to realize the output weighted summation, as shown in Fig. 4-36. An alternative method for realizing zeros is to feed fractions of the input forward into the input of each op amp. A modified Tow–Thomas biquad which forms zeros in this manner is shown in Fig. 4-37. Because of the multiplicity of inputs, one into the negative input terminal of each op amp, this circuit is referred to as the multiple-input biquad (MIB).

Fig. 4-37 Multiple-input biquad circuit.

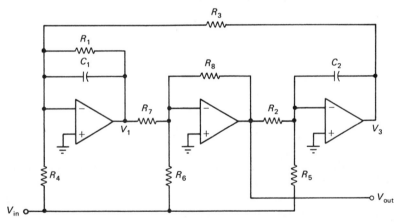

The transfer function for this circuit is

$$H(s) = \frac{V_{out}}{V_{in}} = -\frac{\dfrac{R_8}{R_6}s^2 + \dfrac{1}{R_1 C_1}\left(\dfrac{R_8}{R_6} - \dfrac{R_1 R_8}{R_4 R_7}\right)s + \dfrac{R_8}{R_3 R_5 R_7 C_1 C_2}}{s^2 + \dfrac{1}{R_1 C_1}s + \dfrac{1}{R_2 R_3 C_1 C_2}\dfrac{R_8}{R_7}} \quad (4\text{-}113a)$$

$$= -\frac{b_2 s^2 + b_1 s + b_0}{s^2 + a_1 s + a_0} \quad (4\text{-}113b)$$

The design equations for the idealized case (i.e., GB $= \infty$) from (4-113) and with

$k_1 = \sqrt{\dfrac{R_2 R_8 C_2}{R_3 R_7 C_1}}$ and $k_2 = R_7/R_8$ are as follows [P8b]:

$$R_1 = \frac{1}{a_1 C_1} \tag{4-114a}$$

$$R_2 = \frac{k_1}{C_2 \sqrt{a_0}} \tag{4-114b}$$

$$R_3 = \frac{1}{k_1 k_2} \frac{1}{C_1 \sqrt{a_0}} \tag{4-114c}$$

$$R_4 = \frac{1}{k_2 (b_2 a_1 - b_1) C_1} \tag{4-114d}$$

$$R_5 = \frac{k_1 \sqrt{a_0}}{b_0 C_2} \tag{4-114e}$$

$$R_6 = \frac{1}{b_2} R_8 \tag{4-114f}$$

where C_1, C_2, R_8, k_1, and k_2 are free parameters. The values of C_1, C_2, and R_8 control the impedance levels and are chosen to yield convenient values.

Using the element values given in Eqs. (4-114), the transfer function at the three-op amp outputs can be expressed as

$$\frac{V_{\text{out}}}{V_{\text{in}}} = -\frac{b_2 s^2 + b_1 s + b_0}{s^2 + a_1 s + a_0} \tag{4-115a}$$

$$\frac{V_1}{V_{\text{in}}} = -k_2 \frac{(b_2 a_1 - b_1)s + (b_2 a_0 - b_0)}{s^2 + a_1 s + a_0} \tag{4-115b}$$

$$\frac{V_3}{V_{\text{in}}} = -\frac{1}{k_1} \frac{\dfrac{b_0 - b_2 a_0}{\sqrt{a_0}} s + \dfrac{a_1 b_0 - a_0 b_1}{\sqrt{a_0}}}{s^2 + a_1 s + a_0} \tag{4-115c}$$

We see from Eqs. (4-115b) and (4-115c) that we can use k_1 and k_2 to equalize the maxima of secondary voltages V_1 and V_3 to the maximum of the primary voltage V_{out}. As discussed earlier, this voltage scaling operation serves to maximize the signal swing capability of the filter.

For this circuit the ω_z and $1/Q_z$ sensitivities may be stated as follows:

$$S_{R_3}^{\omega_z} = S_{R_5}^{\omega_z} = S_{R_7}^{\omega_z} = S_{C_1}^{\omega_z} = C_{C_2}^{\omega_z} = -S_{R_6}^{\omega_z} = -\tfrac{1}{2} \tag{4-116a}$$

$$S_{R_3}^{1/Q_z} = S_{R_5}^{1/Q_z} = S_{C_2}^{1/Q_z} = -S_{C_1}^{1/Q_z} = \frac{1}{2Q_z} \tag{4-116b}$$

$$S_{R_1}^{1/Q_z} = \frac{-1}{R_1} \sqrt{\frac{R_3 R_5 R_7 C_2}{R_6 C_1}} \tag{4-116c}$$

$$S_{R_4}^{1/Q_z} = \frac{R_6}{R_4 R_7} \sqrt{\frac{R_3 R_5 R_7 C_2}{R_6 C_1}} \tag{4-116d}$$

$$S_{R_6}^{1/Q_z} = -S_{R_7}^{1/Q_z} = -\frac{1}{2Q_z} - S_{R_4}^{1/Q_z} \tag{4-116e}$$

Note that for a notch response where $1/Q_z = 0$, some of the S_x^{1/Q_z} are not zero. These nonzero sensitivities are a direct result of the cancellation employed to force b_1 to zero.

It can be shown that the transmission zeros can be placed inherently on the $j\omega$-axis by feeding forward through a capacitor. However, since capacitors cannot be practically trimmed, this technique is not very practical for active-RC realizations. However, as we shall see in Chapter 6, this technique can be effectively applied to switched-capacitor filters.

It is again noted that in the passband the overall filter sensitivity is dominated by the variations in the denominator coefficients. That is, the sensitivities for KHN and MIB circuits are equivalent, in the passband, to those derived for the all-pole circuits in Section 4.4. For those applications where stop-band sensitivities are important, particularly at and about a notch, the reader is referred to Section 3.7.

Let us now consider a general biquadratic realization using the low-sensitivity TG circuit. Unlike the KHN and the TT circuits, none of the conventional responses (i.e., strictly low-pass, band-pass, or high-pass responses) are available directly from the TG network, as seen in Eq. (4-103). However, by using an extra summing amplifier, any biquadratic transfer function can be synthesized. The schematic for general biquad realization using the TG circuit is given in Fig. 4-38. The resulting transfer function, by a straightforward analysis, is given by

$$H(s) = \frac{V_0}{V_S} = \frac{N(s)}{(1 + K_2K_3K_4)T_1T_2s^2 + (T_1 + T_2)(1 - K_2K_3K_4)s + 1 + K_2K_3K_4}$$

(4-117a)

where

$$N(s) = \frac{1}{Q}\left\{(-K_5 + K_6 + K_7)s^2 - 2\left[\left(1 + \frac{1}{Q}\right)K_5 + K_7\right]s - K_5 - K_6 + K_7\right\}$$

(4-117b)

Fig. 4-38 Tarmy–Ghausi circuit with summer to realize a general biquad.

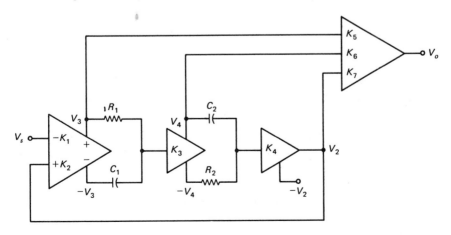

Thus, the proper choice of K_5, K_6, and K_7, realized according to Fig. 4-39, will produce the desired gain and zeros. One advantage of using differential output op amps is that negative values of K_5, K_6, and K_7 are readily achieved by making the connections to the appropriate negative output terminals $-V_3$, $-V_4$, and $-V_2$ of the K_2, K_3, and K_4, amplifiers respectively.

Fig. 4-39   Details of the summer in Fig. 4-38.

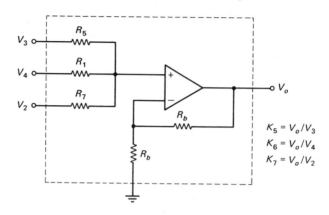

$$K_5 = V_o/V_3$$
$$K_6 = V_o/V_4$$
$$K_7 = V_o/V_2$$

For example, if $K_5 = K_7 = K$ and $K_6 = 0$, the following band-pass transfer function is realized:

$$H(s) = \frac{K}{Q} \frac{s}{s^2 + (1/Q)s + 1} \qquad (4\text{-}118a)$$

As another example, consider the realization of

$$H(s) = \frac{K}{Q} \frac{s^2 + b^2}{s^2 + (1/Q)s + 1} \qquad (4\text{-}118b)$$

Comparing Eqs. (4-117b) and (4-118b), we determine that

$$K_5 = -K_7 = -\tfrac{1}{2}K(1 + b) \qquad (4\text{-}119a)$$

$$K_6 = 2K_7\frac{1 - b^2}{1 + b^2} \qquad (4\text{-}119b)$$

yield the desired realization.

Another general biquad circuit that is proposed by **Mikhael and Bhattacharyya** [P11] is shown in Fig. 4-40 and will be referred to as the **MB** circuit. This circuit requires a relatively small resistor spread to realize a given pole Q. Like the multiple-input biquad in Fig. 4-37, the zeros are formed with a resistive feedforward network and the circuit requires only three op amps to realize a general biquadratic transfer function. The transfer functions for the two most

Continuous-Time Active Filters—Biquadratic Realizations

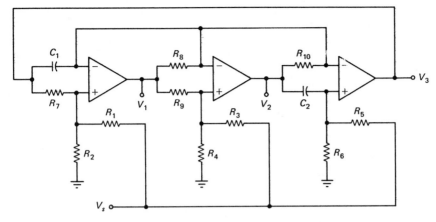

Fig. 4-40 The Mikhael–Bhattacharyya (MB) circuit.

useful output terminals, assuming ideal op amps, are:

$$\frac{V_1}{V_s} = \left\{ s^2\left[G_1\left(1 + \frac{G_4}{G_9}\right) - \frac{G_2 G_3}{G_9}\right] + s\frac{G_3 G_7 G_8}{G_9 C_1} \right.$$
$$\left. + \frac{G_7 G_{10}}{C_1 C_2}\left[\left(1 + \frac{G_4}{G_9}\right)G_5 - \frac{G_3 G_6}{G_9}\right] \right\} \Big/ D(s) \qquad (4\text{-}120\text{a})$$

$$\frac{V_3}{V_s} = \left\{ s^2 G_1 + s\frac{G_7 G_8}{G_9 C_1}\left[\left(1 + \frac{G_2}{G_7}\right)G_3 - \frac{G_1 G_4}{G_7}\right] \right.$$
$$\left. + \frac{G_7 G_{10}}{C_1 C_2}\left[\left(1 + \frac{G_2}{G_7}\right)G_5 - \frac{G_4 G_6}{G_7}\right] \right\} \Big/ D(s) \qquad (4\text{-}120\text{b})$$

where

$$D(s) = s^2(G_1 + G_2) + s\frac{G_7 G_8}{G_9 C_1}(G_3 + G_4) + \frac{G_7 G_{10}}{C_1 C_2}(G_5 + G_6) \qquad (4\text{-}120\text{c})$$

and $G_i = 1/R_i$.

One can obtain many other types of biquadratic responses from Fig. 4-40, for example: (1) by setting $G_3 = 0$ in Eq. (4-120a) or $G_3 = G_1 = 0$ in Eq. (4-120b), one can realize a notch response; or (2) by setting $G_2 = G_3 = G_6 = 0$ and $G_7 = G_1$, one can realize an all-pass transfer function; or (3) we can realize a band-pass response at V_3 by merely setting $G_1 = G_5 = G_4 = 0$ in Eq. (4-120b).

It is noted from (4-120c) that the center frequency and Q are given by

$$\omega_0 = \sqrt{\frac{G_7 G_{10}}{C_1 C_2}\frac{G_5 + G_6}{G_1 + G_2}} \qquad (4\text{-}121\text{a})$$

$$Q = \frac{R_8 R_3 R_4}{R_9(R_3 + R_4)}\sqrt{\frac{R_7 C_1(R_1 + R_2)(R_5 + R_6)}{R_6 R_{10} C_2 R_1 R_2 R_5}} \qquad (4\text{-}121\text{b})$$

A comparison of the sensitivity parameter of MB circuits is given in Section 4.5.

The universal building block for the general biquad circuit proposed by **Akerberg and Mossberg** is shown in Fig. 4-41, and will be referred to as **AM** circuit. The transfer function for the ideal infinite gain op amps, is given by

$$\frac{V_2}{V_1} = -\frac{C_3}{C_2} \frac{s^2 + s\left(\frac{1}{r_4} - \frac{R_2}{r_5 R_3}\right)\frac{1}{C_3} + \frac{R_2}{r_1 R_1 R_3 C_1 C_3}}{s^2 + s\frac{1}{r_3 C_2} + \frac{R_2}{r_2 R_1 R_3 C_1 C_2}} \tag{4-122a}$$

Fig. 4-41 The Akerberg–Mossberg general biquad network.

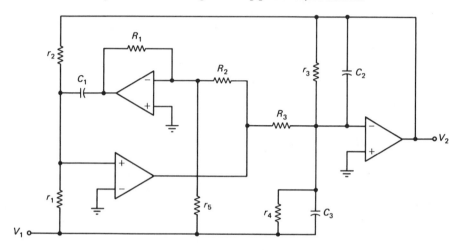

Typically, one chooses

$$R_1 = R_2 = R_3 = r_1 = r_2$$
$$r_3 \quad \text{and} \quad r_4 \gg R_1 \quad \text{and} \quad C_1 = C_2 \tag{4-122b}$$

Under these conditions and assuming identical amplifiers with one-pole roll-off, the sensitivity expressions for the pole Q and the pole frequency are given by (4-102a) and (4-102b) (see also Table 4-4).

A circuit that possesses low sensitivity to variations in both the passive components and the op amp GB was proposed by **Padukone, Mulawka, and Ghausi** [P15]. This circuit is shown in Fig. 4-42 and will be referred to as the **PMG** circuit. This circuit has the advantage of being insensitive to GB variations even if the op amps are nonidentical.

Analysis of the circuit, assuming an ideal op amp for the time being, yields the following transfer functions:

$$\frac{V_1}{V_{in}} = T_1 = \{s^2[C_2 G_6(C_3 G_2 + C_3 G_3 - C_1 G_8)] + sC_2 G_1 G_4 G_8$$
$$+ [G_4 G_5(G_2 G_9 + G_3 G_9 - G_7 G_8)]\}/D(s) \tag{4-123a}$$

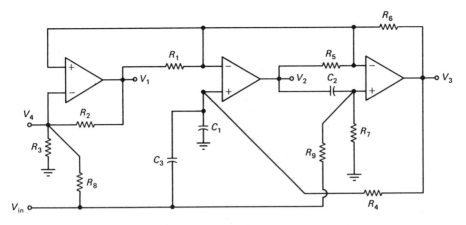

Fig. 4-42 The Padukone, Mulawka, and Ghausi (PMG) circuit.

$$\frac{V_2}{V_{in}} = T_2 = \{s^2 C_2 C_3 G_2 G_6 + s[G_2 G_6 (C_3 G_7 - C_1 G_9) + C_2 G_1 G_4 G_8]$$

$$+ [G_1 G_4 G_7 G_8 + G_4 G_9 (G_2 G_5 - G_1 G_3)]\}/D(s) \qquad (4\text{-}123\text{b})$$

$$\frac{V_3}{V_{in}} = T_3 = \{s^2 [C_2 C_3 (G_2 G_6 - G_1 G_3) + C_1 C_2 G_1 G_8]$$

$$+ s[G_2 G_5 (C_1 G_9 - C_3 G_7) + C_2 G_1 G_4 G_8] + G_2 G_4 G_5 G_9\}/D(s) \qquad (4\text{-}123\text{c})$$

$$\frac{V_4}{V_{in}} = T_4 = \{s^2 C_2 C_3 G_2 G_6 + s C_2 G_1 G_4 G_8 + G_2 G_4 G_5 G_9\}/D(s) \qquad (4\text{-}123\text{d})$$

where

$$D(s) = s^2 (C_1 + C_3) C_2 G_2 G_6 + s C_2 G_1 G_4 (G_3 + G_8) + G_2 G_4 G_5 (G_7 + G_9) \qquad (4\text{-}123\text{e})$$

From Eqs. (4-123) it is clear that most of the commonly used second-order transfer functions can be realized by choosing the element values appropriately. From (4-123e), we have

$$\omega_0 = \left[\frac{G_4 G_5 (G_7 + G_9)}{(C_1 + C_3) C_2 G_6} \right]^{1/2} \qquad (4\text{-}124)$$

$$Q = \left[\frac{(C_1 + C_3) G_5 G_6 (G_7 + G_9)}{C_2 G_4} \right]^{1/2} \frac{G_2}{G_1 (G_3 + G_8)} \qquad (4\text{-}125)$$

It can readily be shown that $S_x^{\omega_0} \leq \frac{1}{2}$ and $S_x^Q \leq 1$ for all passive components. Furthermore, for low Q sensitivity to GB despite mismatches in amplifier GBs, a good design choice is $G_5 = G_6 = G$. In other words, low-sensitivity performance is obtained by matching only two resistors: R_5 and R_6. This property is not shared by many other circuits discussed previously. The AM[P14] circuit requires that all three op amps to be identical, whereas the MB [P11] and Reddy circuits [P16] require at least two matched op amps.

4.5.1 Comparison of Practical Multiple-Op Amp Biquads

Thus for we have covered many different active-RC biquad circuits, possessing a variety of contrasting properties. It is useful to summarize this material by comparing some of the circuits discussed in this book. This comparison is given in Table 4-2. For a more detailed comparison of the sensitivities, M_α, M_β, and M_T have been computed for MB, KHN, TG, MIB, and DF band-pass realizations, with $f_p = 1$ kHz and $Q_p = 10$ and 50. Here $Q_p\omega_p/\text{GB} \ll 1$, so only the passive component variations are accounted for in this analysis. The R's and C's were assumed to be independent random variables with standard deviation $\sigma_R = \sigma_C = \sigma_{RC} = 0.01$. The computed M_α, M_β, and M_T, in which the limits of integration were $\omega_p(1 - 1/Q_p) \leq \omega \leq \omega_p(1 + 1/Q_p)$, are tabulated in Table 4-3. In the computation of the DF circuit sensitivities, $\beta = 21.16$ was used. This β value is consistent with $\sigma_{\text{GB}} = 0.2$. [P2].

The active and passive sensitivities for high-Q ($Q \geq 10$) three-op amp biquads are further compared in Table 4-4. In the table the sensitivities for identical op amps as well as unmatched op amps are also shown. In KHN and TT circuits, notice the excessive dependence of Q on the amplifier GB. In the AM circuit, exact matching of the op amps is critical; only if all three op amps are identical can the excessive dependence of Q on GB be eliminated. The MB circuit has the same disadvantage as does the AM circuit. However, unlike the AM circuit, which requires three matched op amps in the MB circuit, $S_{\text{GB}_t}^Q$ can be reduced by a factor of $\sqrt{Q}$ by matching only the first and the third op amps. The TG circuit does not require the op amps to be identical; however, since the circuit uses differential output op amps, Q can be sensitive to the differences in the antiphase outputs of the op amp. A number of circuits that exhibit low Q-sensitivity to GB without requiring the op amps to be identical have been presented by Wilson, Bedri, and Bowron (WBB) in [P13]. Among these, the band-pass positive feedback, the notch negative feedback, and the combined positive and negative feedback circuits employ null adjusted twin-T sections and therefore require the use of high-stability, low-tolerance RC components. Also, the circuits have high Q sensitivity to passive component variations of the order of $Q/2$. The all-pass configuration [P13], which does not require a twin-T section, is included in Table 4-4 for comparison. The modified Tarmy–Ghausi circuit introduced by Moschytz [P12a], which uses single-ended op amps, is also included in Table 4-4 and listed as the M circuit. The M circuit is similar to the WBB configuration and attains low $\tilde{Q}$ sensitivity to GB without requiring identical op amps. However, its passive Q sensitivities are even higher (of the order of Q) than those of the WBB circuit. The sensitivity of $\tilde{Q}$ to GB in the PMG circuit is independent of Q even in the presence of op amp mismatches. Its passive and active sensitivities are also low.

TABLE 4-2 Biquad comparison

Realization	$\omega_p \ll GB$				Effect of Finite Op Amp Bandwidth		$\dfrac{C_{max}}{C_{min}}$	$\dfrac{R_{max}}{R_{min}}$	R_{max}	No. of Op Amps					Passive Elements Define ω_0	
	$|S^{Q_p}_{R,C}|$	$|S^{\omega_p}_{R,C}|$	$|S^{\omega_z}_{R,C}|$	$|S^{1/Q_z}_{R,C}|$	$S^{Q_p}_{GB} \simeq$	$S^{\omega_p}_{GB} \simeq$				LP	HP	BP	H	AP	R's	C's
MB (Fig. 4-39)	≤ 1	≤ 1	≤ 1	$\leq 1^a$	$-\dfrac{1}{2}\dfrac{\omega_p}{GB}$	$\dfrac{1}{2}\dfrac{\omega_p}{GB}$	1	$2Q_p^{1/2}$	$RQ_p^{1/2}$	3	3	3	3	3	4	2
KHN (Fig. 4-35)	≤ 1	≤ 1	≤ 1	$\leq 1^a$	$-4\dfrac{Q_p\omega_p}{GB}$	$\dfrac{\omega_p}{GB}$	1	$2Q_p$	RQ_p	3	4	3	4	4	4	2
AM (Fig. 4-41)	≤ 1	≤ 1	≤ 1	≤ 1	$\dfrac{\omega_p}{2GB}$	$\dfrac{3}{2}\dfrac{\omega_p}{GB}$	1	Q_p	RQ_p	3	4	3	4	4	2	2
TG (Fig. 4-37)	≤ 2	$\geq 1^b$	≥ 1	≥ 1	$3\dfrac{\omega_p}{2GB}$	$\dfrac{3}{2}\dfrac{\omega_p}{GB}$	1	Q_p	RQ_p	4	4	4	4	4	2	2
MIB (Fig. 4-36)	≤ 1	≥ 1	≥ 1	≥ 1	$-4\dfrac{Q_p\omega_p}{GB}$	$\dfrac{\omega_p}{GB}$	1	Q_p	RQ_p	3	3	3	3	—	4	2
DF (Fig. 4-21)	≥ 1	≥ 1	≥ 1	≥ 1	$-\sqrt{\dfrac{GB}{\omega_p}\dfrac{\sigma_{R,C}}{\sigma_{GB}}}^c$	$\sqrt{\dfrac{GB}{\omega_p}\dfrac{\sigma_{R,C}}{\sigma_{GB}}}$	1	$\tfrac{1}{2}Q_p^2$	$\tfrac{1}{2}RQ_p^2$	1^d	1	1	1	—	2	2
Two op amps (Fig. 4-28)	≤ 1	≤ 1	≤ 1	≤ 1	$2Q_p\dfrac{\omega_p}{GB}$	$\dfrac{\omega_p}{GB}$	1	Q_p	RQ_p	2	2	2	2	2	4	2

a For the notch realizations, $|S^{1/Q_z}_{R,C}| = 0$.
b All cases given by ≥ 1 are not bounded.
c Computed with

$$\beta = \beta_{opt} = \frac{4GB}{\omega_p}\frac{8\sigma_R^2}{\sigma_{GB}^2} + \sigma_C^2\left(\frac{4GB}{\omega_p}\right)\frac{\sigma_{R,C}}{\sigma_{GB}}$$

where $\sigma_{R,C} = \sigma_R = \sigma_C$.
d The low-pass function cannot be realized with the DF circuit.

TABLE 4-3 M_α, M_β, and M_T for $\omega_p(1 - 1/Q_p) \leq \omega \leq \omega_p(1 + 1/Q_p)$

Sensitivity Measures:	Uncorrelated Case					
	M_α		M_β		M_T	
Q_p:	10	50	10	50	10	50
TG	0.0014	0.0071	0.0030	0.0151	0.0044	0.0222
DF	0.0019	0.0100	0.0032	0.0165	0.0051	0.0265
KHN, MIB (TT), MB	0.0021	0.0106	0.0045	0.0226	0.0066	0.0332

Table 4-4 provides a comparison of active and passive sensitivity of each circuit but does not show which circuit has the lowest overall sensitivity. For this purpose, the sensitivity measures M_α, M_β, and M_T were calculated for the circuits in Table 4-5. The calculations were for $Q = 50$, $f_0 = 1$ kHz, and $f_0 = 20$ kHz, assuming op amps with a nominal GB of 1 MHz and 2% mismatch in GBs (i.e., $GB_1 = 1.02$ MHz, $GB_2 = 1$ MHz, $GB_3 = 0.98$ MHz), and the element variation statistics were taken to be $\sigma_C = \sigma_R = 0.2\%$, $\sigma_{GB} = 10\%$, with $\rho_{i_j} = 0$. The frequency range of interest is taken as $\omega_1 = \omega_0(1 - 1/2Q)$ and $\omega_2 = \omega_0(1 + 1/2Q)$, with ω_0 normalized to unity.[2]

For classification purposes, all the three-op amp biquads can be classified into one of the following three categories:

1. The two-integrator-loop-based circuits, such as the KHN, TT, and AM circuits.
2. The all-pass-based circuits, such as the TG, M, and WBB circuits.
3. The GIC-based circuits, such as the MB and PMG circuits.

(The reader is encouraged to redraw the MB and PMG circuits in the form of Fig. 4-28 to appreciate this; see Prob. 4.32).

From Table 4-5 the following observations are made:

1. At low frequencies (i.e., $f_0 = 1$ kHz) all the circuits except the M and TG circuits have the same M_T (i.e., $M_T = 9.4 \times 10^{-4}$). The M circuit has the highest M_α sensitivity, since its passive Q sensitivity is the highest. The TG circuit has the lowest sensitivity, since it uses the minimal number of components.
2. At high frequencies the active sensitivities $S_{GB_i}^{\omega_0}$ and $S_{GB_i}^{Q}$ become significant and therefore the circuits exhibit higher M_α, M_β, and M_T sensitivities. The TG circuit still has the lowest sensitivity.

[2]For the design equations, the following were assumed:

For KHN network: $R = 1$ $R_2 = 2Q - 1$
For MB network: $G_4 = 0$
For PMG network: $G_3 = G_9 = 0$ and
  $C_3 = 0$

For WBB network: $G_1 = G_0$, $G_2 = G_3$,
  $R_1C_1 = R_2C_2 = 1$
For M network: $G_8 \ll G_7$, G_9
For TG network: $G_1S_1 = G_2S_2$,
  $G_a/G_b = G_c/G_d = 1/Q$

TABLE 4-4 Active and passive ω_0 and Q sensitivities of the various networks[a]

| | Active Sensitivities for $Q \gg 1$ | | | | | | | | | | | | | | | | Passive Sensitivities | |
| | Unmatched Op Amps | | | | | | | | | | | | Identical Op Amps | | | | | |
Circuit	$S_{a_{01}}^{\omega_0}$	$S_{a_{02}}^{\omega_0}$	$S_{a_{03}}^{\omega_0}$	$S_{a_{01}}^{Q}$	$S_{a_{02}}^{Q}$	$S_{a_{03}}^{Q}$	$S_{GB_1}^{\omega_0}$	$S_{GB_2}^{\omega_0}$	$S_{GB_3}^{\omega_0}$	$S_{GB_1}^{Q}$	$S_{GB_2}^{Q}$	$S_{GB_3}^{Q}$	$S_{a_0}^{\omega_0}$	$S_{a_0}^{Q}$	$S_{GB}^{\omega_0}$	S_{GB}^{Q}	$S_{x_p}^{\omega_0}$	$S_{x_p}^{Q}$
Kerwin, Huelsman, Newcomb (KHN)	$\dfrac{0.5}{a_{01}}$	$\dfrac{0.5}{a_{02}}$	$\dfrac{1}{a_{03}}$	$\dfrac{Q}{a_{01}}$	$\dfrac{Q}{a_{02}}$	$\dfrac{-1}{a_{03}}$	$\dfrac{\omega_0}{2GB_1}$	$\dfrac{\omega_0}{2GB_2}$	$\simeq 0$	$\dfrac{-\omega_0 Q}{GB_1}$	$\dfrac{-\omega_0 Q}{GB_2}$	$\dfrac{-2\omega_0 Q}{GB_3}$	$\dfrac{2}{a_0}$	$\dfrac{2Q}{a_0}$	$\dfrac{\omega_0}{GB}$	$\dfrac{-4Q\omega_0}{GB}$	≤ 1	≤ 1
Tow–Thomas (TT)	$\dfrac{0.5}{a_{01}}$	$\dfrac{1}{a_{02}}$	$\dfrac{0.5}{a_{03}}$	$\dfrac{Q}{a_{01}}$	$\dfrac{1}{a_{02}}$	$\dfrac{Q}{a_{03}}$	$\dfrac{\omega_0}{2GB_1}$	$\simeq 0$	$\dfrac{\omega_0}{2GB_3}$	$\dfrac{-\omega_0 Q}{GB_1}$	$\dfrac{-2\omega_0 Q}{GB_2}$	$\dfrac{-\omega_0 Q}{GB_3}$	$\dfrac{2}{a_0}$	$\dfrac{2Q}{a_0}$	$\dfrac{\omega_0}{GB}$	$\dfrac{-4Q\omega_0}{GB}$	≤ 1	≤ 1
Akerberg–Mossberg (AM)	$\dfrac{-1}{a_{01}}$	$\dfrac{0.5}{a_{02}}$	$\dfrac{0.5}{a_{03}}$	$\dfrac{-1}{a_{01}}$	$\dfrac{2Q}{a_{02}}$	$\dfrac{Q}{a_{03}}$	$\simeq 0$	$\dfrac{\omega_0}{GB_2}$	$\dfrac{\omega_0}{2GB_3}$	$\dfrac{2\omega_0 Q}{GB_1}$	$\dfrac{-\omega_0 Q}{GB_2}$	$\dfrac{-\omega_0 Q}{GB_3}$	0	$\dfrac{3Q}{a_0}$	$\dfrac{3\omega_0}{2GB}$	$\dfrac{\omega_0}{2GB}$	≤ 1	≤ 1
Mikhael–Bhattacharyya (MB)	$\dfrac{-1}{a_{01}}$	$\simeq 0$	$\dfrac{1}{a_{03}}$	$\dfrac{Q}{a_{01}}$	$\dfrac{\sqrt{Q}}{a_{02}}$	$\dfrac{-1}{a_{03}}$	$\dfrac{\omega_0}{2GB_1}$	$\simeq 0$	$\simeq 0$	$\dfrac{2\omega_0 Q}{GB_1}$	$\dfrac{2\omega_0\sqrt{Q}}{GB_2}$	$\dfrac{-2\omega_0 Q}{GB_3}$	0	$\dfrac{Q}{a_0}$	$\dfrac{\omega_0}{2GB}$	$\dfrac{-\omega_0}{2GB}$	≤ 1	≤ 1
Tarmy–Ghausi (TG)	0	0	0	$\dfrac{Q}{a_{01}}$	$\dfrac{Q}{a_{02}}$	$\dfrac{Q}{a_{03}}$	$\dfrac{\omega_0}{2GB_1}$	$\dfrac{\omega_0}{2GB_2}$	$\dfrac{\omega_0}{2GB_3}$	$\dfrac{-\omega_0}{GB_1}$	$\dfrac{-\omega_0}{GB_2}$	$\dfrac{-\omega_0}{GB_3}$	0	$\dfrac{3Q}{a_0}$	$\dfrac{3\omega_0}{2GB}$	$\dfrac{-3\omega_0}{2GB}$	≤ 1	≤ 2
Wilson, Bedri, Bowron (all-pass) (WBB)	0	0	0	$\dfrac{Q}{a_{01}}$	$\dfrac{2Q}{a_{02}}$	$\dfrac{2Q}{a_{03}}$	$\dfrac{\omega_0}{2GB_1}$	$\dfrac{\omega_0}{GB_2}$	$\dfrac{\omega_0}{GB_3}$	$\dfrac{-\omega_0}{2GB_1}$	$\dfrac{-\omega_0}{GB_2}$	$\dfrac{-\omega_0}{GB_3}$	0	$\dfrac{5Q}{a_0}$	$\dfrac{5\omega_0}{2GB}$	$\dfrac{5\omega_0}{2GB}$	≤ 1	$\leq \dfrac{Q}{2}$
Moschytz (M) (modified TG)	0	0	0	$\dfrac{2Q}{a_{01}}$	$\dfrac{2Q}{a_{02}}$	$\dfrac{2Q}{a_{03}}$	$\dfrac{\omega_0}{GB_1}$	$\dfrac{\omega_0}{GB_2}$	$\dfrac{\omega_0}{GB_3}$	$\dfrac{-\omega_0}{GB_1}$	$\dfrac{-\omega_0}{GB_2}$	$\dfrac{-\omega_0}{GB_3}$	0	$\dfrac{6Q}{a_0}$	$\dfrac{3\omega_0}{GB}$	$\dfrac{-3\omega_0}{GB}$	≤ 1	$\leq Q$
Padukone, Mulawka, Ghausi (PMG)	0	0	0	$\dfrac{-\sqrt{Q}}{a_{01}}$	$\dfrac{2Q}{a_{02}}$	$\dfrac{2Q}{a_{03}}$	$\simeq 0$	$\dfrac{\omega_0}{GB_2}$	$\dfrac{\omega_0}{GB_3}$	$\simeq 0$	$\dfrac{-\omega_0}{GB_2}$	$\dfrac{-\omega_0}{GB_3}$	0	$\dfrac{4Q}{a_0}$	$\dfrac{2\omega_0}{GB}$	$\dfrac{-2\omega_0}{GB}$	≤ 1	≤ 1

[a] a_{0i} and GB_i denote the open-loop dc gain and the gain-bandwidth product of the ith op amp. x_p denotes a resistance or a capacitance. For the case of identical op amps, $a_{01} = a_{02} = a_{03} = a_0$ and $GB_1 = GB_2 = GB_3 = GB$. $\bar{\omega}_0$ and $\bar{Q}$ are the actually realized values of ω_0 and Q, taking a_{0i} and GB_i into consideration.

245

TABLE 4-5   Gain, phase, and transfer function sensitivities at f_0 = 1 kHz and 20 kHz

Sensitivity Measures:	M_α		M_β		M_T	
f_0 (kHz):	1	20	1	20	1	20
KHN, TT	0.00017	0.0010	0.00077	0.0012	0.00094	0.0022
AM	0.00017	0.0011	0.00077	0.0016	0.00094	0.0027
MB	0.00017	0.0012	0.00077	0.0011	0.00094	0.0023
TG	0.00011	0.0002	0.00052	0.0009	0.00063	0.0011
WBB	0.00027	0.0005	0.00067	0.0018	0.00094	0.0023
M	0.00052	0.0008	0.00073	0.0023	0.00125	0.0031
PMG	0.00017	0.0004	0.00077	0.0018	0.00094	0.0022

Figure 4-43a–c shows the gain, phase, and transfer function sensitivities for the circuits listed in Table 4-4. The calculations are, as in Table 4-5, for $Q = 50$, $\sigma_R = \sigma_C = 0.2\%$, $\sigma_{GB} = 10\%$, $\omega_1 = 1 - 1/2Q$, and $\omega_2 = 1 + 1/2Q$. Note that at low frequencies, such that $\sigma_x^2(\omega_0/GB)^2 \ll \sigma_{GB}^2$ (where $x = R, C$), the sensitivities are about the same and do not vary by much. However, as the center

Fig. 4-43   Sensitivity versus center frequency normalized with respect to gain–bandwidth product for various three-op amp circuits (Q = 50). (a) Gain sensitivity. (b) Phase sensitivity. (c) Transfer function sensitivity.

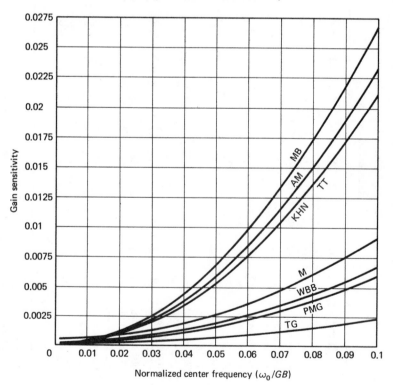

(a)

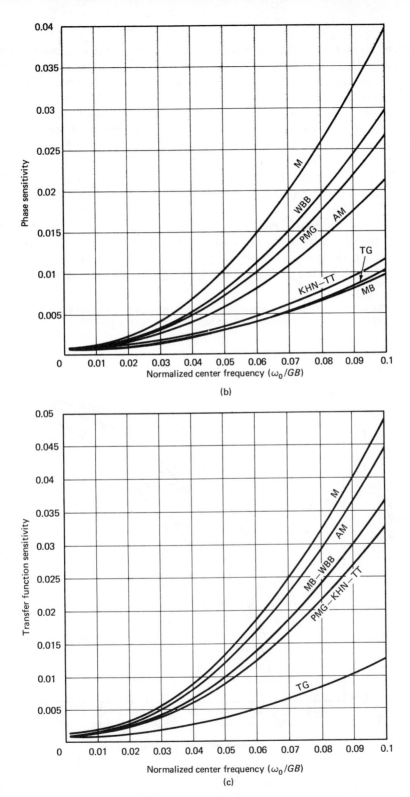

Fig. 4-43 (cont.)

frequency is increased, the active sensitivities become significant, leading to an increase in M_α, M_β, and M_T. Since the sensitivity increases considerably as the center frequency increases, this also puts a limit on the high center frequency operation of the active-RC filters. At very high frequencies, active-R filters can become attractive, as discussed in Section 4.6. The Q enhancement and the center frequency deviations are shown in Figs. 4-44 and 4-45. In these figures the amplifier GBs are assumed to be $GB_i = GB/\eta_i$, $i = 1, 2, 3$, with $GB = 1$ MHz, $\eta_1 = 0.98$, $\eta_2 = 1$, and $\eta_3 = 1.02$.

Figure 4-44a and b shows the effect of GB on the pole frequency ω_0 for pole Q values of 10 and 100. Figures 4-45a and b show the effect of GB on pole Q for nominal Q values of 10 and 100, respectively. Note that even for $GB/\omega_0 = 100$, the KHN and TT circuits are unstable and cannot be designed for a Q of 100, whereas the MB and AM circuits are becoming unstable for $GB/\omega_0 \le 10$. For nominal values of $Q = 10$, $\omega_0 = 1$, and $GB/\omega_0 = 5.0$, the actual values of ω_0 and Q denoted by $\tilde{\omega}_0$ and $\tilde{Q}$ are listed in Table 4-6.

Fig. 4-44 (a) Effect of finite gain–bandwidth product on ω_0 for nominal $Q = 10$. (b) Effect of finite gain–bandwidth product on ω_0 for nominal $Q = 100$.

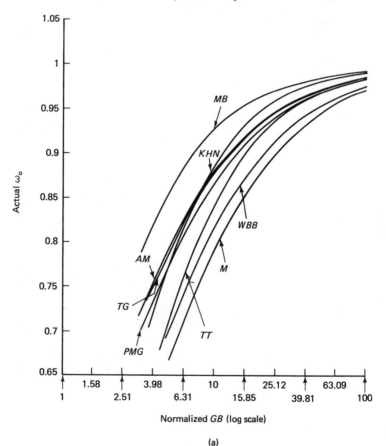

(a)

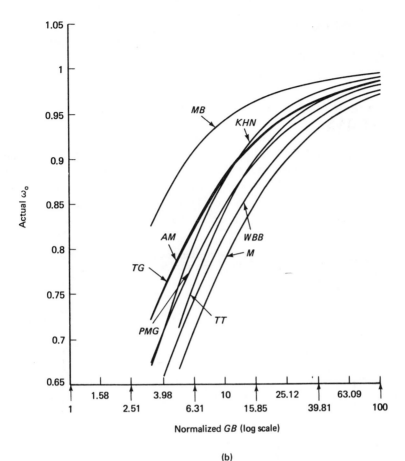

(b)

Fig. 4-44 (cont.)

TABLE 4-6

	KHN	TT	AM	MB	TG	WBB	M	PMG
$\tilde{\omega}_0$			0.793	0.857	0.788	0.701	0.667	0.772
$\tilde{Q}$	Unstable	Unstable	16.373	23.532	11.177	9.77	9.678	10.656

Throughout the past four sections of this chapter, the predominant performance criterion, beyond the filter shape, has been sensitivity. Sensitivity, we recognize, is very important because it determines the degree to which a circuit is able to perform in a practical environment and its ability to withstand degradations in its components with time. Also, to a large extent sensitivity determines the ultimate production cost of a filter through the circuit yield. Yield is a means of measuring the success of the production process and is simply the ratio of those circuits produced whose responses meet specifications (i.e., the desired response $\pm$ a specified tolerance) to the total number of circuits produced.

Obviously, if we can maintain a uniform fabrication difficulty and increase the number of acceptable circuits produced (i.e., increase yield), the unit cost of the circuit is reduced. Since yield and lifetime are directly related to sensitivity, low-sensitivity networks are also expected to be low-cost filters.

4.6 ACTIVE-R AND ACTIVE-C BIQUAD REALIZATIONS

Besides sensitivity, other means by which we can reduce the cost of the circuit is to reduce its size or the area on the chip (real estate) and simplify the fabrication process. By eliminating the capacitors, at least two for each second-order section, we have the potential for doing both. These capacitors, because of their high values (typically $C \gg 100$ pF), are the most impractical values to fabricate in

Fig. 4-45 (a) Effect of finite gain–bandwidth product on Q for nominal $Q = 10$. (b) Effect of finite gain–bandwidth product on Q for nominal $Q = 100$.

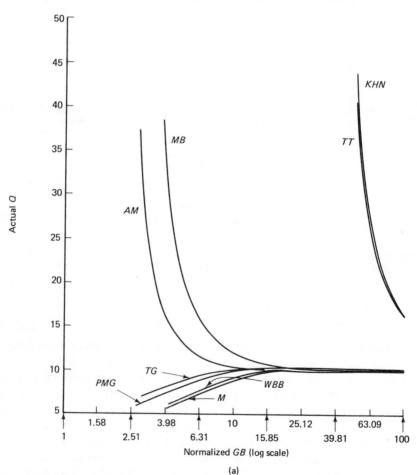

(a)

Continuous-Time Active Filters—Biquadratic Realizations

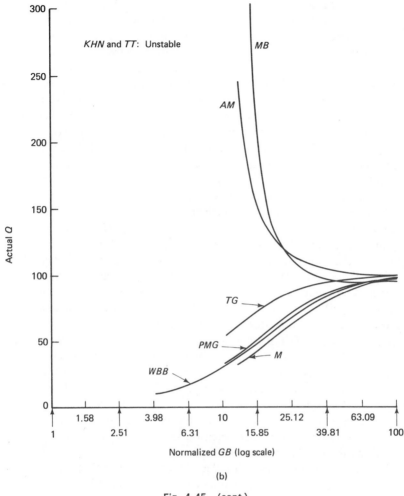

KHN and TT: Unstable

MB

AM

TG

PMG

M

WBB

Actual Q

Normalized GB (log scale)

1.58   2.51   3.98   6.31   10   15.85   25.12   39.81   63.09   100

1

(b)

Fig. 4-45   (cont.)

monolithic IC form and consume a substantial portion of the filter's chip area or real estate. Among the various methods that have been proposed for the implementation of monolithic filters, active-R filters [P22–P24, P26] received early attention because they employ no capacitors other than the small capacitors used to internally compensate the op amps [see Eqs. (2-26)–(2-28)]. In addition, they are more suitable for operation at higher frequencies ($f_0 >$ 20 kHz) than are comparable active-RC filters. This may be their prime function as seen later in the section. In active-R circuits, the amplifiers are used as integrators and filter parameters, such as pole frequency ω_0, pole quality factor Q, and flat gain, are established by the GBs of the amplifiers and ratios of resistors. Since resistor ratios can be maintained to high precision with hybrid or monolithic technology, one is left with the task of finding a means to stabilize the amplifier GBs, both in manufacture and with temperature.

As we were made aware of in previous sections of this chapter, the parasitic op amp pole begins to be a nonignorable source of error for active-RC filters when the desired pass band exceeds 10 kHz. In fact, for some topologies, one needs $Q\omega_p \ll GB$ to avoid enhancement and, perhaps, instability. Active-R filters can be made to operate satisfactorily at frequencies up to approximately $GB/4$. For example, using $\mu A741$-type op amps with GB $\simeq$ 1 MHz, active-R realizations can be used to realize passbands centered at frequencies up to 250 kHz. The high value of Q will also set a limit on these values. The dynamic range for these filters will depend upon the slew rate and noise specifications of the op amp. The primary disadvantage is the inability to hold GB accurately in manufacture or with temperature. For example, as noted in earlier examples, $\sigma_{R,C}$ can be held in manufacture to a range of 0.1 to 5%. With laser trimming, one can manufacture filters with precision comparable to $\sigma_{R,C} = 0.1\%$. More important, by proper choice of materials and preparation, thin-film R's and C's can be made to track with temperature so that the RC product can be held to within $\pm$20ppm ($\pm$0.002%) over a temperature range of 0 to 70°C. The op amp GB, on the other hand, is not inherently a precise quantity. In manufacture, σ_{GB} ranges from 10 to 50%. Manufacturing tolerances can be compensated by tuning; however, potentially more damaging is the wide variation in GB with temperature: $\pm$10% over a range of 0 to 70°C. One effective means for remedying this almost fatal temperature variation is to use phase-lock-loop methods to electronically lock the GB of a "programmable" op amp to some fixed reference. Perhaps more effective is to develop a monolithic integrator which is designed for use in active-R filters [P26]. Currently, active-R filters using standard op amps are limited in their usefulness at lower frequencies due to the fact that a closed-loop pole f_p must be pulled from the frequency GB of the amplifier to the desired lower frequency. This method of operation results in large resistor ratios and poor signal-to-noise performance when the desired pole frequency $f_p \ll GB$, as will be demonstrated later in this section.

For realization in MOS technology, it is well known that ratios of capacitors can be implemented more economically and accurately than can ratioed resistors. In addition, MOS capacitors can be realized which are highly linear and temperature-stable. Perhaps more important is that MOS filters can be manufactured using processes that are, by and large, identical to those used to manufacture high-volume digital components such as memories and microprocessors. One means for implementing MOS filters is to modify the active-R filters by replacing every resistor with a capacitor. In MOS technology, capacitor ratios can be held to tight tolerances with excellent temperature stability. In addition, dc paths, implemented via large, but imprecise, resistors, must be established in order to provide proper bias currents to the amplifier and dc feedback. This type of filter is referred to in the literature as an active-C filter [P25]. Since active-C filter topologies are obtained directly from active-R topologies, the design techniques, limitations, and performance for these two types of filters will be very similar.

As we discuss in Chapter 6, one means for exploiting the precision and economy offered by MOS implementation lies in the use of active switched capacitor filters. In contrast to active-RC and active-R filters which are continuous analog networks, active switched capacitor filters are analog sampled-data networks. As noted in Section 1.9, sampled-data filters require high-frequency continuous analog prefiltering and postfiltering for anti-aliasing protection (i.e., to band-limit the input spectrum to make reconstruction possible) and, when a smooth output is desired, for reconstructing the desired analog signal. By operating the sampled-data filter at high sampling rates (> 100 kHz) we can employ low-order anti-aliasing and reconstruction filters which are much lower in precision than that required for the sampled-data filter. Active-C filters appear to be suited to providing these filters economically and technology-compatible on the same chip with the sampled-data switched capacitor filter. In addition, for high-frequency operation, strong arguments can be made for implementing MOS filters using active-C methodology and the phase-locking methods employed successfully in precision active-R implementations.

It is the objective of this section to present some of the topologies, design techniques, performance characteristics, and limitations of the active-R and active-C classes of filters. We confine the discussion here to second-order transfer function realization. High-order filters can be implemented using the methods discussed in Chapter 5.

4.6.1 Minimum-Element Active-R Biquads

Consider first the circuit shown in Fig. 4-46, which has been proposed by **Rao–Srinivasan**. Hereafter, this circuit is referred to as the **RS** circuit [P22]. This circuit is seen to be impressively simple, requiring only two op amps and two resistors, when compared to the previously discussed active-RC filters. The op amp gains $a_i(s)$ will be represented by the one-pole model Eq. (2-24):

$$a_i(s) = \frac{a_{0i} p_{0i}}{s + p_{0i}} = \frac{GB_i}{s + p_{0i}} \tag{4-126}$$

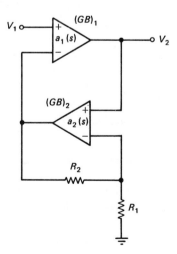

Fig. 4-46 Minimum-element
active-R circuit ($\omega_0 \simeq$ GB).

Most commercially available units are internally compensated to have a 6-dB/octave roll-off for stability reasons, as discussed in Chapter 2. Typically, the break frequency p_{0i} is much less than the desired filter operating frequency. For example, an internally compensated μA741 op amp has a $p_{0i} \simeq 20$ rad/sec, whereas active-R realizations operate at $\omega > 2\pi(10 \text{ kHz})$. Thus, if we assume $\omega \gg p_{0i}$, then, with negligible error, we can approximate $a_i(s)$ by Eq. (2-25), repeated here for convenience:

$$a_i(s) \simeq \frac{GB_i}{s} \qquad (4\text{-}127)$$

This approximate model for $a_i(s)$ together with $R_i = \infty$ and $R_0 = 0$ (see Fig. 2-14) will be used to characterize the op amp throughout the remainder of this section.

Using Eq. (4-127), we can write the transfer function for the circuit in Fig. 4-46 as follows:

$$H(s) = \frac{V_2}{V_1} = \frac{GB_1(s + GB_2/K)}{s^2 + (GB_2/K)s + GB_1 GB_2} \qquad (4\text{-}128)$$

where $K = 1 + R_2/R_1$. From Eq. (4-128), we can readily identify

$$\omega_0 = \sqrt{GB_1 GB_2} \qquad (4\text{-}129a)$$

$$Q = K\sqrt{\frac{GB_1}{GB_2}} \qquad (4\text{-}129b)$$

Note that $\omega_0 \simeq GB$ in Eq. (4-129a) is only good in theory; in practice this is much too high. The higher order op amp poles will cause response error and possibly instability.

For identical op amps (i.e., $GB_1 = GB_2 = GB$), Eq. (4-129b) implies that the resistor ratio $R_2/R_1 = Q - 1$ is a resistor spread no worse than required for most active-RC realizations. The ω_0 and Q sensitivities can be written by inspection from Eqs. (4-129):

$$S^{\omega_0}_{GB_1} = S^{\omega_0}_{GB_2} = \tfrac{1}{2} \qquad (4\text{-}130a)$$

$$S^{\omega_0}_{R_1} = S^{\omega_0}_{R_2} = 0 \qquad (4\text{-}130b)$$

$$S^{Q}_{GB_1} = -S^{Q}_{GB_2} = \tfrac{1}{2} \qquad (4\text{-}130c)$$

$$S^{Q}_{R_2} = -S^{Q}_{R_1} = \frac{K - 1}{K} \qquad (4\text{-}130d)$$

There are several general observations that we can make from the analysis of this simple circuit:

1. A minimum of two op amps (or integrators) are required to realize a second-order function in an active-R implementation. Thus, a single-amplifier active-R biquad is not possible.

2. The sensitivities $S^{\omega_0}_{x_i}$ and $S^{Q}_{x_i}$ are very low: on just the merits of Eqs.

(4-130), the reader might conclude that this active-R circuit is much less sensitive than the best active-RC circuits.

3. However, if we note that, according to Eq. (2-27), GB $= g_m/C$, then $\omega_0 = \sqrt{g_{m1}g_{m2}/C_1C_2}$ is dependent on four (the minimum number) passive and active parameters. Although the $\sum_{gm,c}(S_{x_i}^{\omega_0})^2$ is minimum, as in the DF and TG active-RC circuits, the normalized variation $\Delta\omega_0/\omega_0$ will be intrinsically larger than in the active-RC equivalents, due to the imprecision of parameters g_m and C. Testimony to the variability of g_m and C is the typical GB standard deviation of $\sigma_{GB} \simeq 20\%$. In an integrated circuit in which several op amps are fabricated on a single chip, the GBs tend to be highly correlated in manufacture and track with temperature. This tracking tends to make ratioed GBs less variable than a stand-alone GB, yet it tends to increase the liability of GB product dependence.

4. Q depends upon the ratio of two GBs and the ratio of two resistors. Since ratio parameters can be manufactured with greater precision and held more constant with temperature than are single R's, GBs, and GB products, the Q instability is a second-order effect in comparison to variations in ω_0. In addition, we know that variations in ω_0 are Q times more important in determining overall sensitivity than is Q. Thus, in active-R implementations, the overall performance of the filter is almost entirely dependent on our ability to stabilize ω_0.

5. The circuit in Fig. 4-46 does not represent a practical implementation when the amplifiers $a(s)$ are conventional fixed-GB op amps. In the absence of a variable GB, there is no means for adjusting GB to obtain a desired ω_0. Of course, in the design of a monolithic filter, the compensation capacitors of the amplifiers can be altered directly on the photo mask to realize a desired ω_0.

6. The high midband gain $[H(j\omega_0) = K(GB_1/GB_2) = Q$ when $GB_1 = GB_2 = GB]$, coupled with the slew rate of the op amps, severely limits the dynamic range of this filter.

Another circuit arrangement is given in Fig. 4-47. Using Eq. (4-127) and assuming that $R_0 \simeq 0$ and $R_i \simeq \infty$ for each op amp, the transfer function for this circuit is determined as follows:

$$H(s) = \frac{V_2}{V_1} = \frac{s + GB_2/K}{s^2 + GB_2 s/K + GB_1 GB_2/K} \tag{4-131}$$

where $K = 1 + R_2/R_1$. We note the similarity between the transfer functions in Eqs. (4-132) and (4-128). The ω_0 and Q for the circuit are given by

$$\omega_0 = \sqrt{\frac{GB_1 GB_2}{K}} \tag{4-132}$$

$$Q = \sqrt{\frac{KGB_1}{GB_2}} \tag{4-133}$$

Observe that ω_0 varies as $1/\sqrt{K}$ and Q with $\sqrt{K}$. Thus, we are able to tune ω_0 by adjusting K; however, we must also accept the corresponding change in Q. Thus, the tuning of ω_0 and Q is not independent. This circuit has one other disadvantage, in that for $GB_1 = GB_2$,

$$K = Q^2 \quad \text{or} \quad \frac{R_2}{R_1} = Q^2 - 1 \qquad (4\text{-}134)$$

That is, the resistor spread has increased by a factor Q over that of the previous circuit. Furthermore, in the previous circuit, $S_{R_1}^{\omega_0} = S_{R_2}^{\omega_0} = 0$; however, in this circuit, $S_{R_1}^{\omega_0} = -S_{R_2}^{\omega_0} = -\frac{1}{2}[(K-1)/K]$. Thus, the overall sensitivity of this circuit will be slightly higher than that of circuit in Fig. 4-46.

It is instructive to view the mechanism that enables us to reduce the center frequency ω_0 from $\sqrt{GB_1 GB_2}$ to some lower value $\sqrt{GB_1 GB_2/K}$. To make this observation, let us write the expression for the output V_3 of amplifier a_2 in Fig. 4-47:

$$V_3 = \frac{a_2(s)}{K} V_2 - \frac{a_2(s)}{K} V_3 \qquad (4\text{-}135)$$

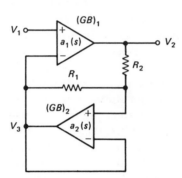

Fig. 4-47 Minimum-element active-R circuit $(\omega_0 \simeq GB/\sqrt{K})$.

From Eq. (4-135), we see that V_2 is attenuated by K at the noninverting input to a_2 and an equally attenuated V_3 is fed back through the inverting input. For $\omega_0^2 \ll \sqrt{GB_1 GB_2}$, K must be large; thus, the signal attenuation at the inputs to a_2 is correspondingly large. Unfortunately, amplifier noise is processed with gain $a_2(s)$. Therefore, to reduce the center frequency from $\sqrt{GB_1 GB_2}$ by a factor K, the price paid is that the signal-to-noise performance is also reduced by a factor K. Of course, there is also the area penalty for the resistor spread K. It is then important for the designer to carefully consider the liability of this signal-to-noise reduction when considering an active-R design where $\omega_0/\sqrt{GB_1 GB_2} \ll 1$.

Although the circuits examined here are extremely simple, the properties observed are generally found in all active-R circuits. Let us now examine some more general active-R biquadratic implementations.

4.6.2 General State-Variable Active-R Biquads

This class of active-R circuits, like the RS circuits, uses two op amps; however, its derivation stems from state-variable theory [P23, P24]. Here the op amps, with gain function GB/s, serve as the integrators in the state-space realization. For example, consider the state-space active-R realization shown in Fig. 4-48. Much like the KHN active-RC circuit, the active-R state-variable network provides all the primary transfer functions—low-pass, band-pass, and

Fig. 4-48   State-variable active-R biquad.

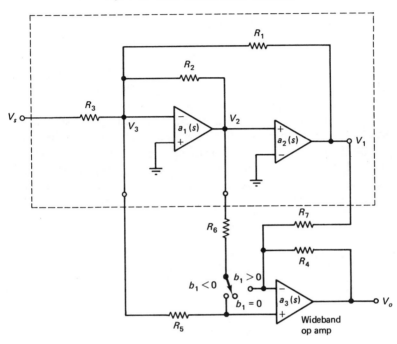

high-pass—simultaneously at terminals V_1, V_2, and V_3, respectively. Analyzing this circuit, we find

$$H_{LP}(s) = \frac{V_1}{V_s} = \frac{-GB_1GB_2/K_3}{s^2 + (GB_1/K_2)s + GB_1GB_2/K_1} \qquad (4\text{-}136a)$$

$$H_{BP}(s) = \frac{V_2}{V_s} = \frac{-sGB_1/K_3}{s^2 + (GB_1/K_2)s + GB_1GB_2/K_1} \qquad (4\text{-}136b)$$

$$H_{HP}(s) = \frac{V_3}{V_s} = \frac{s^2/K_3}{s^2 + (GB_1/K_2)s + GB_1GB_2/K_1} \qquad (4\text{-}136c)$$

where

$$K_1 = 1 + \frac{R_1}{R_2} + \frac{R_1}{R_3} \qquad (4\text{-}136d)$$

$$K_2 = 1 + \frac{R_2}{R_1} + \frac{R_2}{R_3} \qquad (4\text{-}136e)$$

$$K_3 = 1 + \frac{R_3}{R_1} + \frac{R_3}{R_2} \qquad (4\text{-}136f)$$

The equations for ω_0 and Q are then

$$\omega_0 = \sqrt{\frac{GB_1 GB_2}{K_1}} \qquad (4\text{-}137)$$

$$Q = K_2 \sqrt{\frac{GB_2}{K_1 GB_1}} \qquad (4\text{-}138)$$

Note that, in contrast to the RS active-R circuits, the state-variable circuits can be tuned by adjusting R_1 to tune ω_0 and R_2 to tune Q. $H_{BP}(j\omega_0)$ is fixed for chosen ω_0 and Q. The tuning, however, requires an iterative adjustment of R_1, R_2, and R_3 since each of the filter parameters ω_0, Q, and $H(j\omega_0)$ depend upon all three resistor values. With $GB_i = g_{mi}/C_i$, the sensitivities $S_{x_i}^{\omega_0}$ and $S_{x_i}^{Q}$ are comparable to those given for KHN and TT active-RC circuits.

A general biquad realization is readily obtained by simply combining a wide-band summing amplifier with the state-variable active-R biquad as shown in Fig. 4-48. In order for the summing circuit to behave properly,[3] $GB_3 \gg GB_1$, GB_2. Thus, for frequencies well above the high-frequency cutoff of the filter, amplifier $a_3(s)$ is approximately an ideal op amp. For example, if $a_1(s)$ and $a_2(s)$ are μA741 op amps with $GB = 1$ MHz, $a_3(s)$ should have $GB_3 \geq 10$ MHz. If $GB_3 \gg GB_1$, GB_2, the transfer function at the output of the general biquad is simply

$$H(s) = \frac{V_0}{V_s} \simeq \frac{K_4 V_3}{V_s} + \frac{K_5 V_2}{V_s} + \frac{K_6 V_1}{V_s} = \frac{b_2 s^2 + b_1 s + b_0}{s^2 + a_1 s + a_0} \qquad (4\text{-}139a)$$

where

$$K_4 = \begin{cases} 1 + \dfrac{R_4}{R_6 + R_7} & \text{for } b_1 > 0 \\[2mm] 1 + \dfrac{R_4}{R_7} & \text{for } b_1 = 0 \\[2mm] \dfrac{R_6}{R_6 + R_7}\left(1 + \dfrac{R_4}{R_7}\right) & \text{for } b_1 < 0 \end{cases} \qquad (4\text{-}139b)$$

$$K_5 = \begin{cases} \dfrac{-R_4}{R_6} & \text{for } b_1 > 0 \\[2mm] 0 & \text{for } b_1 = 0 \\[2mm] \dfrac{R_5}{R_5 + R_6}\left(1 + \dfrac{R_4}{R_6}\right) & \text{for } b_1 < 0 \end{cases} \qquad (4\text{-}139c)$$

$$K_6 = \frac{-R_4}{R_7} \qquad \text{for all } b_1 \qquad (4\text{-}139d)$$

In general, it is desirable to have a topology in which the sensitivities $S_{x_i}^{\omega_0}$ and $S_{x_i}^{GB}$ are low; the ω_0, Q, and flat gain can be tuned noniteratively; and sufficient freedom to scale the gains of the op amp outputs to maximize dynamic

---

[3]Note that K is the gain of the closed-loop summing amplifier and also that the closed-loop bandwidth $\omega_c \gg \omega_0$, where $\omega_c \simeq GB/K$ and ω_0 is the filter center frequency.

range. The scaling procedure outlined in Section 4.5 for multiple-op amp active-*RC* biquads is applicable here. An active-*R* structure [P23] that provides this freedom is shown in Fig. 4-49a and b.

Fig. 4-49 (a) and (b) Active-*R* sections with sufficient degrees of freedom for the design of ω_0, Q, and H_B.

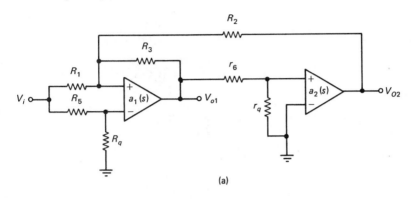

(a)

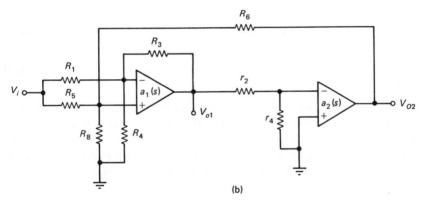

(b)

The transfer functions for op amp outputs V_{01} and V_{02} in Fig. 4-49a are

$$H_1(s) = \frac{V_{01}}{V_i} = \frac{sGB_1(a_5 - a_1)}{s^2 + sGB_1a_3 + GB_1GB_2a_2\alpha_6} \qquad (4\text{-}140a)$$

$$H_2(s) = \frac{V_{02}}{V_i} = \frac{GB_1GB_2(a_5 - a_1)\alpha_6}{s^2 + sGB_1a_3 + GB_1GB_2a_2\alpha_6} \qquad (4\text{-}140b)$$

where

$$a_1 = \left(1 + \frac{R_1}{R_2} + \frac{R_1}{R_3}\right)^{-1} \qquad (4\text{-}140c)$$

$$a_2 = \frac{R_1}{R_2}a_1 \qquad (4\text{-}140d)$$

$$a_3 = \frac{R_1}{R_3}a_1 \qquad (4\text{-}140e)$$

$$a_5 = \left(1 + \frac{R_5}{R_8}\right)^{-1} \qquad (4\text{-}140\text{f})$$

$$\alpha_6 = \left(1 + \frac{r_6}{r_8}\right)^{-1} \qquad (4\text{-}140\text{g})$$

Similarly, for the circuit in Fig. 4-49b, the transfer functions are

$$H_1(s) = \frac{sGB_1(\hat{a}_5 - \hat{a}_1)}{s^2 + sGB_1\hat{a}_3 + GB_1GB_2\hat{a}_6\hat{\alpha}_2} \qquad (4\text{-}141\text{a})$$

$$H_2(s) = \frac{-GB_1GB_2(\hat{a}_5 - \hat{a}_1)\hat{\alpha}_2}{s^2 + sGB_1\hat{a}_3 + GB_1GB_2\hat{a}_6\hat{\alpha}_2} \qquad (4\text{-}141\text{b})$$

where

$$\hat{a}_1 = \left(1 + \frac{R_1}{R_3} + \frac{R_1}{R_4}\right)^{-1} \qquad (4\text{-}141\text{c})$$

$$\hat{a}_3 = \frac{R_1}{R_3}\hat{a}_1 \qquad (4\text{-}141\text{d})$$

$$\hat{a}_5 = \left(1 + \frac{R_5}{R_6} + \frac{R_5}{R_8}\right)^{-1} \qquad (4\text{-}141\text{e})$$

$$\hat{a}_6 = \frac{R_5}{R_6}\hat{a}_5 \qquad (4\text{-}141\text{f})$$

$$\hat{\alpha}_2 = \left(1 + \frac{r_2}{r_4}\right)^{-1} \qquad (4\text{-}141\text{g})$$

The design equations for these circuits are summarized in Tables 4-7 and 4-8, respectively. In these tables it is noted that H_B denotes the midband gain for the bandpass function and H_L the dc gain of the low-pass function. Further, by proper choice of components, H_B and H_L can either be made inverting or noninverting, as desired. It is left as an exercise for the reader to show that $|S_{x_i}^{\omega_0}| \leq \frac{1}{2}$ and $|S_{x_i}^{Q}| \leq 1$.

TABLE 4-7   Design equations for circuit in Fig. 4-49a

Parameters	Circuit Fig. 4-49a
$H_B > 0$ $(G_1 = 0)$	$\left(1 + \frac{R_3}{R_2}\right)\left(1 + \frac{R_5}{R_8}\right)^{-1}$
H_L	$\left(1 + \frac{R_2}{R_3}\right)\left(1 + \frac{R_5}{R_8}\right)^{-1}$
$H_B < 0$ $(G_5 = 0$ and $R_8 = 0)$	$-\frac{R_3}{R_1}$
H_L	$-\frac{R_2}{R_1}$
ω_0^2	$GB_1GB_2\frac{R_1}{R_2}\left(1 + \frac{R_1}{R_2} + \frac{R_1}{R_3}\right)^{-1}\left(1 + \frac{r_6}{r_8}\right)^{-1}$
Q	$\frac{\omega_0}{GB_1}\left(1 + \frac{R_3}{R_1} + \frac{R_3}{R_2}\right)$

TABLE 4-8 Design equations for circuit in Fig. 4-49b

Parameters	Circuit Fig. 4-49b
$H_B > 0 \ (G_1 = G_8 = 0)$	$\left(1 + \frac{R_3}{R_4}\right)\left(1 + \frac{R_5}{R_6}\right)^{-1}$
H_L	$-\frac{R_6}{R_5}$
$H_B < 0 \ (G_5 = G_8 = 0 \text{ and } R_6 = 0)$	$-\frac{R_3}{R_1}$
H_L	$\left(1 + \frac{R_1}{R_3} + \frac{R_1}{R_4}\right)^{-1}$
ω_0^2	$GB_1 GB_2 \frac{R_5}{R_6}\left(1 + \frac{R_5}{R_6} + \frac{R_5}{R_8}\right)^{-1}\left(1 + \frac{r_2}{r_4}\right)^{-1}$
Q	$\frac{\omega_0}{GB_1}\left(1 + \frac{R_3}{R_1} + \frac{R_3}{R_4}\right)$

Two other useful active-R circuits where the optimization of the dynamic range and the design for noniterative tuning can be achieved simultaneously are shown in Fig. 4-50a and b.

The voltage transfer functions for the circuit in Fig. 4-50a can be written in the following forms:

$$H_1(s) = \frac{K_1 s}{s^2 + (\omega_0/Q)s + \omega_0^2} \tag{4-142}$$

$$H_2(s) = \frac{K_2}{s^2 + (\omega_0/Q)s + \omega_0^2} \tag{4-143}$$

where

$$K_1 = (GB_2)\frac{G_{10}}{G_{10} + G_{11} + G_{12}} \tag{4-144a}$$

$$K_2 = (GB_1)(GB_2)\frac{G_5}{G_5 + G_6}\frac{G_{10}}{G_{10} + G_{11} + G_{12}} \tag{4-144b}$$

$$\omega_0^2 = (GB_1)(GB_2)\frac{G_5}{G_5 + G_6}\frac{G_8}{G_8 + G_9} \tag{4-144c}$$

$$\frac{\omega_0}{Q} = (GB_2)\frac{G_{11}}{G_{10} + G_{11} + G_{12}} \tag{4-144d}$$

For the circuit in Fig. 4-50b, we have

$$H_1(s) = \frac{K_3 s + K_0}{s^2 + (\omega_0/Q)s + \omega_0^2} \tag{4-145}$$

$$H_2(s) = \frac{K_4}{s^2 + (\omega_0/Q)s + \omega_0^2} \tag{4-146}$$

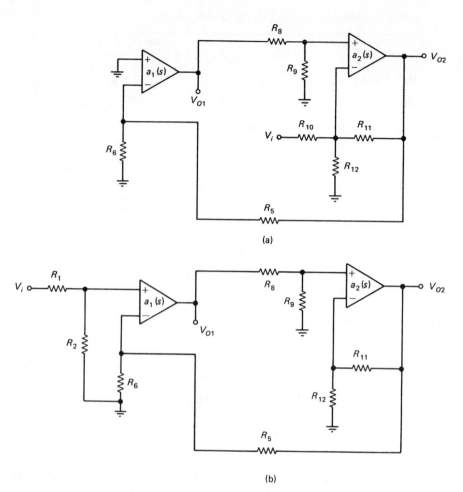

Fig. 4-50   (a) and (b) Active-R circuits with good dynamic range and noniterative tuning.

where

$$K_3 = (\text{GB}_1) \frac{G_1}{G_1 + G_2} \qquad (4\text{-}147a)$$

$$K_0 = (\text{GB}_1)(\text{GB}_2) \frac{G_{11}}{G_{11} + G_{12}} \frac{G_1}{G_1 + G_2} \qquad (4\text{-}147b)$$

$$K_4 = (\text{GB}_1)(\text{GB}_2) \frac{G_8}{G_8 + G_9} \frac{G_1}{G_1 + G_2} \qquad (4\text{-}147c)$$

$$\omega_0^2 = (\text{GB}_1)(\text{GB}_2) \frac{G_8}{G_8 + G_9} \frac{G_5}{G_5 + G_6} \qquad (4\text{-}147d)$$

$$\frac{\omega_0}{Q} = (\text{GB}_2) \frac{G_{11}}{G_{11} + G_{12}} \qquad (4\text{-}147e)$$

The design equations for maximum dynamic range and noniterative tuning are given as follows. Let us define for convenience $\beta = \omega_0/\text{GB} < 1$ and assume that $\text{GB}_1 = \text{GB}_2$. For the circuit in Fig. 4-50a, we have

$$\frac{G_9}{G_8} = \frac{G_6}{G_5} = \frac{1}{\beta} - 1 \qquad (4\text{-}148\text{a})$$

$$\frac{G_{11}}{G_{10}} = \frac{1}{H_0} \qquad (4\text{-}148\text{b})$$

$$\frac{G_{12}}{G_{10}} = \frac{1 + (\beta/Q)(H_0 + 1)}{H_0 \beta/Q} \qquad (4\text{-}148\text{c})$$

where H_0 in the center frequency gain and is limited by $H_0 < Q/\beta - 1$. The tuning sequence are (1) set ω_0 by R_6, (2) set H_0 by R_{11}, and (3) set Q by R_{12}. For the circuit in Fig. 4-50b, the design equations are

$$\frac{G_9}{G_8} = \frac{G_6}{G_5} = \frac{1}{\beta} - 1 \qquad (4\text{-}149\text{a})$$

$$\frac{G_2}{G_1} = \frac{Q}{\beta H_0} - 1 \qquad (4\text{-}149\text{b})$$

$$\frac{G_{12}}{G_{11}} = \frac{Q}{\beta} - 1 \qquad (4\text{-}149\text{c})$$

The conditions on H_0 and Q are

$$H_0 < \frac{Q}{\beta} \qquad \text{and} \qquad Q > \beta \qquad (4\text{-}149\text{d})$$

The tuning sequence in this case is as follows: (1) set ω_0 by R_6, (2) set Q by R_{11} and (3) set H_0 by R_{12}.

●

EXAMPLE 4-5

Consider the design of a second-order band-pass filter for the following specifications: $f_0 = 50$ kHz, $Q = 10$, and $H_0 = 5$ using op amps with $\text{GB} = 1$ MHz. For the circuit in Fig. 4-50a, from Eq. (4-148) we obtain

$$\frac{G_6}{G_5} = 19, \qquad \frac{G_9}{G_8} = 19, \qquad \frac{G_{11}}{G_{10}} = \frac{1}{5}, \qquad \frac{G_{12}}{G_{10}} = 38.8$$

For the circuit in Fig. 4-50b, from Eq. (4-149) we obtain

$$\frac{G_2}{G_1} = 39, \qquad \frac{G_6}{G_5} = 19, \qquad \frac{G_9}{G_8} = 19, \qquad \frac{G_{12}}{G_{11}} = 199$$

The inclusion of the Q-enhancement effects will change the design values as discussed later.

●

Much has been said in previous sections about the effect of finite op amp bandwidth on the performance of active-RC filters. In fact, the op amp GB was treated as a parasitic, in active-RC, which we compensate for by predistorting the nominal design. In active-R, the op amp is modeled as an integrator (i.e., GB is a parameter in the design). In active-R the parasitic effects arising from the higher-order, nondominant op amp poles beyond the unity gain–bandwidth frequency cannot be ignored. The accuracy of the analysis could be improved by employing a two-pole op amp model. As discussed in Chapter 2, the gain error that we incur by ignoring the nondominant poles is relatively harmless. However, ignoring the added phase shift, particularly when arising within a feedback loop, can result in unpredicted instability. As given in Eq. (2-32), a simple model for accurately accounting for the phase contributions of nondominant poles is to multiply the dominant pole model by an excess phase term $e^{-s/mGB}$ which leaves the gain uneffected. The constant parameter m, which is larger than 1, takes into account the excess phase shift (e.g., for a National Semiconductor LM 348N, $m \simeq 2$). Let us now reexamine our active-R filter using the model of Eq. (2-32):

$$a(s) \simeq \frac{a_0 e^{-s/mGB}}{1 + s/p_0} \simeq \frac{GB}{s} e^{-s/mGB} \qquad (4\text{-}150)$$

To facilitate this analysis, let us divide the numerator and denominator of Eq. (4-141a) by $GB_1 GB_2 = GB^2$ and rewrite $H_1(s)$ in the form

$$H_1(s) = \frac{[H_B \omega_0/(GBQ)]a^{-1}}{a^{-2} + [\omega_0/GBQ]a^{-1} + (\omega_0/GB)^2} \qquad (4\text{-}151)$$

where we have set $s/GB = a^{-1}$ in (4-141a), since this approximation was used earlier. Now substituting the more accurate approximation equation (4-150) for the gain $a(s)$ in Eq. (4-151), we obtain

$$H_1(s) = \frac{[H_B \omega_0/(GBQ)](s/GB)e^{\tau s}}{\dfrac{s^2}{(GB)^2} e^{2\tau s} + s\dfrac{\omega_0}{GBQ}\dfrac{e^{\tau s}}{GB} + \dfrac{\omega_0^2}{(GB)^2}}$$

$$= \frac{H_B(\omega_0/Q)se^{-\tau s}}{s^2 + s(\omega_0/Q)e^{-\tau s} + \omega_0^2 e^{-2\tau s}} \qquad (4\text{-}152)$$

where $\tau = 1/mGB$. For $\omega\tau \ll 1$, $e^{-\tau s} \simeq 1 - \tau s$, and we can approximate $H_1(s)$ by

$$H_1(s) \simeq \frac{H_B(\omega_0/Q)s(1 - \tau s)}{s^2 + s(\omega_0/Q)(1 - \tau s) + \omega_0^2(1 - 2\tau s)}$$

$$= \frac{H_B(\omega_0/Q)s(1 - \tau s)}{s^2(1 - \tau\omega_0/Q) + s[(\omega_0/Q) - 2\omega_0^2\tau] + \omega_0^2} \qquad (4\text{-}153)$$

The approximation $e^{-\tau s} = 1 - \tau s$ can be made at the outset in (4-150):

$$a(s) = \frac{GB}{s(1 + s/mGB)} \qquad (4\text{-}154)$$

where the excess phase shift contribution is approximated by a second pole. Denoting the actual center frequency and quality factor by $\tilde{\omega}_0$ and $\tilde{Q}$, respectively, and the ideal quantities (i.e., $\tau = 0$) by ω_0 and Q, was can write

$$\tilde{\omega}_0 \simeq \omega_0 \left(1 + \frac{\tau\omega_0}{2Q}\right) = \omega_0 \left(1 + \frac{\omega_0}{2QmGB}\right) \tag{4-155a}$$

$$\tilde{Q} \simeq \frac{Q}{1 - 2Q\omega_0\tau} = \frac{Q}{1 - 2Q\omega_0/mGB} \tag{4-155b}$$

From Eqs. (4-155) we see that ω_0 is only negligibly affected, whereas Q is seen to be severely enhanced. This Q enhancement can be nominally compensated for using techniques described in previous sections for similar compensation in active-RC biquads. Nevertheless, exact compensation is difficult to achieve as op amp parameters vary in manufacture and with temperature. For example, from Eq. (4-155b), we can solve for Q in terms of $\tilde{Q}$:

$$Q = \frac{\tilde{Q}}{1 + 2\tilde{Q}\omega_0/mGB} \tag{4-156}$$

Equations (4-155) are also valid for the circuits in Figs. 4-50a and b. Owing to Q enhancement, the value of H_0 will also be changed to $\hat{H}_0$:

$$H_0 = \hat{H}_0 \frac{\hat{Q}}{Q} \tag{4-157}$$

Thus, in the design example given earlier for a desired $\tilde{Q} = 10$ at $f_0 = 50$ kHz with GB $= 1$ MHz and $m = 2$, we must design the ideal circuit for $Q = 6.667$. Using the new value of Q (i.e., $Q = 6.667$), the design values for the example circuits are as follows:

For Fig. 4-50a: $G_2/G_{10} = 25.67$ and $G_{12}/G_{11} = 25.47$ and the other element values remain the same.

For Fig. 4-50b: $G_2/G_1 = 25.67$ and $G_{12}/G_{11} = 132.33$ and the other element values remain the same.

It is interesting to determine the statistical sensitivity of these two active-R networks. For $\rho_{R_iR_j} = 0.7$, $\rho_{GB_1}\rho_{GB_2} = 0.7$ and $\rho_{R_iGB_j} = 0$, $\sigma_{R_i}^2 = \sigma_{GB_i}^2 = 10^{-4}$ and the frequency band $0.9 \leq \omega \leq 1.1$ [i.e., $\omega_0(1 \pm 1/Q)$], we obtain the following results:

For Fig. 4-50a:

$$M_\alpha = 1.594 \times 10^{-3}, \qquad M_\beta = 3.381 \times 10^{-3}, \qquad M_T = 4.976 \times 10^{-3}$$

For Fig. 4-50b:

$$M_\alpha = 1.594 \times 10^{-3}, \qquad M_\beta = 3.381 \times 10^{-3}, \qquad M_T = 4.975 \times 10^{-3}$$

It should be noted that with GB tolerance similar to those in passive components, the sensitivity of active-R networks is better than that in active-RC

networks. However, with the GB tolerance poorer than the passive components (e.g., $\sigma^2_{GB_i} = 10^{-3}$), the resulting sensitivities for the active-R circuits are:
For Fig. 4-50a:

$$M_\alpha = 1.336 \times 10^{-2}, \qquad M_\beta = 2.837 \times 10^{-2}, \qquad M_T = 4.172 \times 10^{-2}$$

For Fig. 4-50b:

$$M_\alpha = 1.246 \times 10^{-2}, \qquad M_\beta = 2.645 \times 10^{-2}, \qquad M_T = 3.892 \times 10^{-2}$$

These sensitivities are considerably poorer than the active-RC networks.

It is interesting to note a statistical sensitivity comparison of the active-R and active-RC biquads for high frequencies, as shown in Fig. 4-51. In this figure

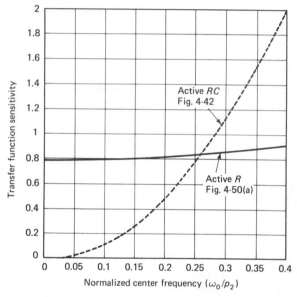

Fig. 4-51 Transfer function sensitivity of active RC and active R versus center frequency normalized with respect to the second pole of an op amp, mGB ($m = 2$).

the active-R circuit of Fig. 4-50a and the active-RC circuit of Fig. 4-42 are compared assuming the following parameter variations: $\sigma_R = \sigma_C = 0.2\%$, $\sigma_{GB} = \sigma_{p_2} = 10\%$, where $p_2 = m$GB (with $m = 2$) and $Q = 50$.[4] The frequency band of interest is $\omega_1 = \omega_0 (1 - 1/2Q)$ and $\omega_2 = \omega_0(1 + 1/2Q)$ with ω_0 normalized to unity. We note in Fig. 4-51 that for $\omega_0 \gtrsim 0.25p_2 = 0.5$GB, the active-$R$ sensitivity performance is better than that of active-RC biquad circuit.

---

[4]The reader is reminded that a high value of Q for active-R circuit necessitates predistortion and the use of enhancement condition Equation (4-156).

Continuous-Time Active Filters—Biquadratic Realizations

4.6.3 Active-C Biquads

As noted in the introduction to this section, active-R filters can be readily transformed to active-C filters [P25] by substituting C's for R's in any active-R topology. To maintain proper dc bias for the op amps and dc stability, large simulated resistors arbitrary in value are connected to the op amp input to provide stable dc operation for the op amps. In Fig. 4-51, the active-R biquad in Fig. 4-49a is converted to an active-C filter. The stabilization resistors, shown in Fig. 4-52 with dashed connections, can be implemented by suitably biased

Fig. 4-52 Active-C biquad.

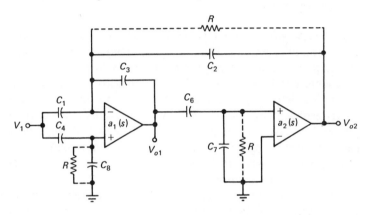

JFETs. The transfer functions for this filter are as given in Eqs. (4-132a) and (4-132b), where

$$a_5 = \frac{C_4}{C_4 + C_5} \tag{4-158a}$$

$$a_1 = \frac{C_1}{C_1 + C_2 + C_3} \tag{4-158b}$$

$$a_3 = \frac{C_3}{C_1 + C_2 + C_3} \tag{4-158c}$$

$$a_2 = \frac{C_2}{C_1 + C_2 + C_3} \tag{4-158d}$$

$$\alpha_6 = \frac{C_6}{C_6 + C_7} \tag{4-158e}$$

Active-C filters clearly possess the same precision limitations and Q-enhancement effects as their active-R equivalents. Also, the phase-locking techniques mentioned in Section 4.6.3 are in principle equally applicable here. As mentioned earlier, one filter application that appears particularly suitable for implementation in active-C methodology is anti-aliasing and reconstruction filters for MOS switched capacitor filters. When the sampling rate for the switched capacitor filter is sufficiently high (100 kHz for a 3 kHz switched capacitor filter), anti-aliasing and reconstruction can be achieved with second-order continuous filters. Moreover, these filters are low-Q and have precision requirements compatible with non-phase-locked active-C filters. The degree to which active-C filters will be employed for this application will depend upon how active-C realizations compete in terms of substrate area and performance with active-RC alternative (such as the Sallen and Key circuit in Fig. 4-5). It appears that, although as the sampling rates increases toward 1 MHz, either first- or second-order active-C implementation becomes an attractive option. Switched capacitor filters are discussed in detail in Chapter 6.

4.7 SUMMARY

In this chapter we considered the design and performance of single-amplifier and multiple-amplifier active-RC and active-$R(C)$ biquad circuits. Negative-feedback single-amplifier circuits were shown to have low-Q sensitivities to passive components but large ω_0 and Q sensitivities to the op amp GB. The latter property restricted negative feedback circuits to low Q. On the other hand, positive-feedback circuits were shown to offer the ability of trade-off between active and passive sensitivities and can be designed for higher Q than the negative-feedback circuits. It was further shown that by employing both negative and positive feedback, an optimum compromise between these conflicting requirements can be achieved.

By isolating the passive-RC networks with additional amplifiers, the design equations were seen to be decoupled, thus rendering a noniterative tuning sequence. Furthermore, multiple types of transfer functions could be made available simultaneously. A comprehensive comparison of various three op amp biquads available in literature has been presented. The comparisons include sensitivity as well as enhancement considerations. The TG and PMG circuits were shown to be particularly suited, because of their low sensitivities, to the passive components and op amp GBs (for unmatched op amps) and to high-Q applications. However, because of the lower power consumption, the DF single-amplifier biquad has proven to be a most suitable workhorse circuit for $Q \leq 30$. For larger Q's, low sensitivities and the ability to tune ω_0, Q, and flat gain noniteratively are particularly important: Both features are found in the TG, MB, and PMG circuits.

For frequencies above the audio range, active filters can be realized using the op amp as an integrator. The transfer function coefficients can then be adjusted by either ratioed resistors in an active-R implementation or by ratioed capacitors in an active-C implementation. In either case the filter precision is intrinsically poor because of the unavoidable dependence of ω_0 on the product of two op amp GBs. Furthermore, since the GBs vary, the filter transfer function are subject to wide variations with temperature. These variations can, however, be reduced significantly with phase-locking methods. One application that appears suited to an active-C implementation are low-Q, second-order anti-aliasing and reconstruction filters to provide input spectrum band limiting and output smoothing for sampled data-switched capacitor filters. For high sampling rates, the precision demanded of such filters is compatible with the intrinsic precision capabilities of active-C (active-R) filters.

REFERENCES

Books

B1. GHAUSI, M., *Electronic Circuits*. New York: Van Nostrand Reinhold, 1971, Chap. 8.

B2. HUELSMAN, L., ed., *Active Filters*. New York: McGraw-Hill, 1970.

B3. HEINLEIN, W., AND W. HOLMES, *Active Filters*. Englewood Cliffs, N.J.: Prentice-Hall, 1974.

B4. MITRA, S., *Analysis and Synthesis of Linear Active Networks*. New York: Wiley, 1969.

B5. MOSCHYTZ, G., *Linear Integrated Networks Design*. New York: Van Nostrand Reinhold, 1975.

B6. DARYANANI, G., *Principles of Active Network Synthesis and Design*. New York: Wiley, 1976.

B7. SEDRA, A., AND P. BRACKETT, *Filter Theory and Design: Active and Passive*. Portland, Oreg.: Matrix, 1978.

B8. SCHAUMANN, R., SODERSTRAND, M. A., AND LAKER, K. R., eds., *Modern Active Filter Design*, IEEE Press, 1981.

Papers

P1. SALLEN, P. R., AND E. L. KEY, "A Practical Method of Designing RC Active Filters," *IRE Trans. Circuit Theory*, CT-2 (1955), 74–85.

P2. FLEISCHER, P., "Sensitivity Minimization in a Single Amplifier Biquad Circuit," *IEEE Trans. Circuits Syst.*, CAS-23 (January 1976), 45–55.

P3. (a) SEDRA, A., "Generation and Classification of Single Amplifier Filters," *Int. J. Circuit Theory Appl.*, 2 (March 1974), 51–67. (b) SEDRA, A., AND L. BROWN, "A Refined Classification of Single Amplifier Filters," *Int. J. Circuit Theory Appl.*, 7 (March 1979), 127–137.

P4. DELYIANNIS, T., "High Q-Factor Circuit with Reduced Sensitivity," *Electron. Lett.*, 4 (December 1968), 577.

P5. Tow, J., "Design Formulas for Active-RC Filters Using Op Amp Biquad," *Electron. Lett.*, 5 (July 1969), 339–341.

P6. (a) FRIEND, J., "SAB: Single Amplifier Biquad," *1970 IEEE Int. Symp. Circuit Syst. Dig. Pap.*, p. 179. (b) FRIEND, J., "STAR: An Active Biquadratic Filter Section," *IEEE Trans. Circuits Syst.*, CAS-22 (February 1975), 115–121.

P7. KERWIN, W. T., L. P. HUELSMAN, AND R. W. NEWCOMB, "State Variable Synthesis for Insensitive Integrated Circuit Transfer Functions," *IEEE J. Solid-State Circuits*, SC-2 (September 1967), 87–92.

P8. (a) Tow, J., "A Step-by-Step Active Filter Design," *IEEE Spectrum*, 6 (December 1969), 64–68. (b) FLEISCHER, P. E., AND J. Tow, "Design Formulas for Biquad Active Filters Using Three Operational Amplifiers," *Proc. IEEE*, 61, no. 5 (May 1973), 662–663.

P9. (a) THOMAS, L. C., "The Biquad: Pt. I. Some Practical Design Considerations," *IEEE Trans. Circuit Theory*, CT-18 (May 1971), 350–357; (b) THOMAS, L. C., "The Biquad: Pt. II. 1 Multipurpose Active Filtering System," *IEEE Trans. Circuit Theory*, CT-18 (May 1971), 358–361.

P10. TARMY, R., AND M. S. GHAUSI, "Very High-Q Insensitive Active RC Networks," *IEEE Trans. Circuit Theory*, CT-17 (August 1970), 358–366.

P11. MIKHAEL, W. B., AND B. B. BHATTACHARYYA, "A Practical Design for Insensitive RC-Active Filters," *IEEE Trans. Circuits Syst.*, CAS-22 (May 1975), 407–415.

P12. (a) MOSCHYTZ, G., "High-Q Factors Insensitive Active RC Network, Similar to the Tarmy–Ghausi Circuit but Using Single-ended Operational Amplifiers," *Electron. Lett.*, 8 (1972), 458–459. (b) REVANKAR, G., K. SHANKAR, AND R. DATAR, "A Modified Moschytz High-Q Active Filter," *Int. J. Electron.*, 1977, 117–120.

P13. WILSON, G., Y. BEDRI, AND P. BOWRON, "RC-Active Methods with Reduced Sensitivity to Amplifier Gain Bandwith Product," *IEEE Trans. Circuits Syst.*, CAS-21 (September 1974), 618–626.

P14. AKERBERG, D., AND K. MOSSBERG, "A Versatile Active RC Building Block with Inherent Compensation for the Finite Bandwidth of the Amplifier," *IEEE Trans. Circuits Syst.*, CAS-21 (January 1974), 75–78.

P15. PADUKONE, P., J. MULAWKA, AND M. S. GHAUSI, "An Active Biquadratic Section with Reduced Sensitivity to Operational Amplifier Imperfections," *J. Franklin Inst.*, Vol. 30, No. 1 (1980), 27–40.

P16. REDDY, M., "An Insensitive Active-RC Filter for High Q and High Frequencies," *IEEE Trans. Circuits Syst.*, CAS-23 (July 1976), 429–433.

P17. BRUTON, L. T., "Multiple-Amplifier RC-Active Filter Design with Emphasis on GIC Realizations," *IEEE Trans. Circuits Syst.*, CAS-25 (October 1978), 830–845.

P18. BUDAK, A., AND D. PETRELA, "Frequency Limitations of Active Filters Using Operational Amplifiers," *IEEE Trans. Circuit Theory*, CT-19 (July 1972), 322–328.

P19. HESZBERGER, A., AND E. SIMONYI, "Comments on Practical Design for Insensitive RC-Active Filters," *IEEE Trans. Circuits Syst.*, CAS-23 (May 1976), 326–328.

P20. ARONHIME, P. B., "Effects of Finite Gain-Bandwidth Product on Three Recently Proposed Quadratic Networks," *IEEE Trans. Circuits Syst.*, CAS-24 (November 1977), 657–660.

P21. BOCTOR, S. A., "A Novel Second Order Canonical RC-Active Realization of High-Pass Notch Filter," *IEEE Trans. Circuits Syst.*, CAS-22, No. 5 (May 1975), 397–404.

P22. (a) R. RAO, K., AND S. SRINIVASAN, "Low-Sensitivity Active Filters Using the Operational Amplifier Pole," *Proc. IEEE*, December 1974, pp. 1713–1714. (b) RAO, K., AND S. SRINIVASAN, "A Band-Pass Filter Using Operational Amplifier Pole," *IEEE Trans. Solid-State Circuits*, SC-8 (June 1973), 245–246.

P23. BRAND, J. R., AND R. SCHAUMANN, "Active-R Filters: Review of Theory and Practice," *IEEE J. Electron. Circuits Syst.*, 2, no. 4 (July 1978), 89–101.

P24. SODERSTRAND, M. A., "Active-R Ladders: High Order Low Sensitivity Active-R Filters without External Capacitors," *IEEE Trans. Circuits Syst.*, CAS-25 (December 1978), 1032–1038.

P25. SCHAUMANN, R., AND J. R. BRAND, "Design Method for Monolithic Analog Filters," *Electron. Lett.*, 14, no. 22 (October 1978), 710–711.

P26. TAN, K. S., AND P. R. GRAY, "Fully Integrated Analog Filters Using Bipolar-JFET Technology," *IEEE J. Solid-State Circuits*, SC-13, no. 6 (December 1978), 814–821.

P27. RAO, K. R., et al, "A Novel 'Follow the Master' Filter," *Proc. IEEE*, 65 (1977), 1725–1726.

P28. WOROBEY, W., AND J. RUTKIEWICZ, "Tantalum Thin-Film *RC* Circuit Technology for a Universal Active Filter," *IEEE Trans. Parts, Hybrids Packag.*, PHP-12, no. 4 (December 1976), 276–282.

P29. LOPRESTI, P. V., "Optimum Design of Linear Tuning Algorithms," *IEEE Trans. Circuits Syst.*, CAS-24, no. 3 (March 1977), 144–151.

PROBLEMS

4.1 For the circuit shown in Fig. 4-1:
   (a) Derive Eq. (4-1).
   (b) If $K_2 = 2$ and $K_1 = 1$, $Y_1 = Y_A = 1 + s$, and

$$Y_2 = s + 1 + \frac{4s}{s+1}, \qquad Y_B = \frac{(2 + \sqrt{2})s}{s+1}$$

   determine the transfer function $H(s)$ and sketch the magnitude response.
   (c) Compare part (b) with the case where K_1 changes by $+10\%$ and K_2 does not change.
   (d) Repeat part (c) if K_2 changes by $+10\%$ and K_1 does not vary.

4.2 For the circuit in Fig. 4-5:
   (a) Derive Eq. (4-5), where $K = 1 + R_a/R_b$, for a noninverting op amp.
   (b) Design the circuit for a Butterworth response with $\omega_{3dB} = 2\pi(10^3)$ rad/sec. Assume that $C_1 = C_2 = 0.1$ μF and $R_a = R_b = 1$ kΩ.
   (c) Determine the passive sensitivities of ω_0 and Q, that is, $S_{x_i}^{\omega_0}$ and $S_{x_i}^{Q}$, where x_i are the R's and C's.
   (d) Determine the active sensitivities $S_K^{\omega_0}$ and S_K^{Q}.

4.3 Determine the pole locations corresponding to the design in Prob. 4.2 if the op amp actual gain is given by Eq. (4-15) with GB $= 2\pi \times 10^5$ rad/sec.

4.4 (a) Repeat the design in Prob. 4.2(b) but instead of equal capacitor values, assume equal resistor values (i.e., $R_1 = R_2 = 1$ kΩ and $R_a = R_b = 1$ kΩ).
   (b) Determine the pole locations if the actual gain is given by Eq. (4-15) with GB $= 2\pi \times 10^5$ rad/sec.

4.5 Verify the statistical sensitivity measures M_α, M_β, and M_T obtained in Example 4-2. Show the details of your work.

4.6 (a) Design the circuit in Fig. 4-8 for a Chebyshev response with $\frac{1}{2}$-dB ripple. The desired passband in $\omega_C = 2\pi \times 10^3$ rad/sec. The amplifier gain is realized by an inverting op amp. Let $R_1 = R_3$ and $C_1 = C_2$.
   (b) Determine the pole locations of the designed circuit if the op amp is not ideal and is characterized by $a(s) = GB/s$ with GB $= 2\pi \times 10^5$ rad/sec.

4.7 For the circuit in Fig. 4-10, derive the expression for the transfer function given in Eq. (4-33). Design the circuit for a center frequency $f_0 = 10^3$ Hz and $Q = 5$.

If the op amp is not ideal, and with a GB = 10^5 Hz, determine the actual center frequency and the actual Q of the designed circuit.

4.8 Design the circuit of Fig. 4-11 for a center frequency $\omega_0 = 10^4$ rad/sec and $Q = 10$. The gain K is realized with a noninverting op amp (i.e., $K = 1 + R_a/R_b$). The capacitor values are to be equal. The resistor ratios are to be minimized in this design. If the op amp GB = 10^6 rad/sec, determine the actual values of ω_0 and Q.

4.9 (a) For the circuit shown in Fig. P4-9, show that the transfer function is given by [B7].

$$\frac{V_o}{V_i} = \frac{-sb\omega_0/2.2Q}{s^2 + s(\omega_0/Q) + \omega_0^2}$$

where $b = 4.41Q^2 - 1$.

Fig. P4-9

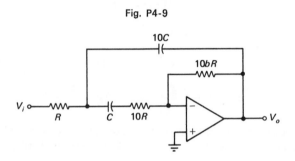

(b) Design the circuit for $\omega_0 = 10^4$ rad/sec and $Q = 10$ and determine the mid-band gain of the circuit.

4.10 Derive Eq. (4-52).

4.11 Use the circuit in Fig. P4-11 to realize the following transfer function:

$$T(s) = \frac{H_B s}{s^2 + (\omega_0/Q)s + \omega_0^2}$$

Fig. P4-11

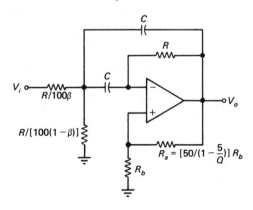

(a) For $\omega_0 = 1$, $Q = 20$, and $H_B = 25$, show the design values.
(b) Show that in this realization the value of H_B is constrained by $H_B \geq \omega_0 Q^2$.

4.12 Derive the voltage transfer function for the circuit shown in Fig. P4-12 for the following cases:
(a) $Y = \gamma G = \gamma/R$; α, β, and γ are constants. Show that $\omega_p > \omega_z$.
(b) $Y = \gamma Cs$, and show that $\omega_p < \omega_z$.
(c) Verify Eq. (4-41) as a special case of the circuit above.

Fig. P4-12

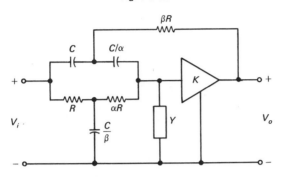

4.13 For the circuit shown in Fig. P4-11, the circuit elements are chosen as shown [B7].
(a) Show that the circuit provides a complete design information for a band-pass design:

$$\frac{V_o}{V_i} = \frac{s\beta\omega_0(10.2 - 1/Q)}{s^2 + s(\omega_0/Q) + \omega_0^2}$$

with $\omega_0 = 10/RC$.
(b) Design the circuit for $\omega_0 = 1$, $Q = 30$, and the mid-band gain $|H(j\omega_0)| = 2.0$.

4.14 (a) For the circuit shown in Fig. 4-27a, derive Eq. (4-71).
(b) For $C_1 = C_2 = 1/2Q$, $R_1 = R_2 = 1$, and

$$K_1 K_2 = -(4Q^2 - 1)$$

design the circuit for $Q = 10$, $\omega_0 = 1$.
(c) Determine the actual values of Q and ω_0 if inverting and noninverting op amps are used to implement the voltage gains K_1 and K_2. Let the normalized GB $= 50\omega_0$.

4.15 Derive the active sensitivities of the network given in Fig. 4-27a. In other words, words, derive Eqs. (4-76d) and (4-76e).

4.16 Determine the statistical sensitivity measures M_α, M_β and M_T for the circuit in Fig. 4-27a designed for $Q = 50$ and $f_0 = 1$ kHz; assume that GB $= 1$ MHz. The element variations are independent and the statistics are assumed to be a normal distribution with $\sigma_C = \sigma_R = 0.2\%$ and $\sigma_{GB} = 10\%$. The frequency range of interest is from $f_1 = f_0(1 - 1/2Q)$ to $f_2 = f_0(1 + 1/2Q)$.

4.17 For the GIC-derived circuit shown in Fig. P4-17, show that the transfer function is given by

$$\frac{V_o}{V_i} = \frac{s(1 + G_5/G_4)G_7/C_6}{s^2 + sG_7/C_6 + G_1G_3G_5/C_2C_6G_4}$$

Fig. P4-17

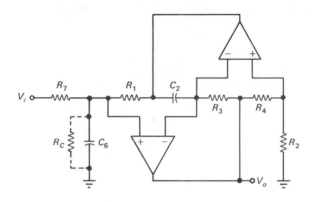

If $G_3 = G_4 = G_1 = G_5 = G$ and $G_7 = G/Q$, design the circuit for $f_0 = 5$ kHz and $Q = 30$. What is the midband gain $|H(j\omega_0)| = H$? What are the actual values of Q and ω_0 if the op amp GB $= 2\pi \times 10^6$ rad/sec?

4.18 Determine the statistical sensitivity measures M_α, M_β, and M_T for the circuit shown in Fig. 4-28a designed for $Q = 50$, $f_0 = 1$ kHz, and GB $= 1$ MHz. The element variation statistics are assumed to be normally distributed with $\sigma_R = \sigma_C = 0.2\%$ and $\sigma_{GB} = 10\%$ and uncorrelated. The frequency range of interest is from $f_1 = f_0(1 - 1/2Q)$ to $f_2 = f_0(1 + 1/2Q)$.

4.19 Consider the circuit shown in Fig. P4-17 and let

$$C_2 = C_6 = C, \qquad G_3 = G_4, \qquad G_1 = G_5 = G, \qquad G_7 = \frac{G}{Q}$$

(a) Design the circuit for $f_0 = 5$ kHz, $Q = 30$, and $H = 6$.
(b) Determine the actual values of Q and ω_0 if the op amp gain–bandwidths are the same with GB $= 2\pi \times 10^6$ rad/sec.
(c) Repeat part (b) if a compensating resistor $R_c = 50$ kΩ is connected across the capacitor C_6.

4.20 Derive Eqs. (4-81) and (4-82).

4.21 Design the circuit of Fig. 4-28d to realize the transfer function

$$H(s) = \frac{H(s^2 + a_0)}{s^2 + b_1 s + b_0}$$

with b_1, b_0, and H given in Table B-6:

$$b_1 = 0.8291 \qquad a_0 = 2.7698$$
$$b_0 = 1.6232 \qquad H = 0.5567$$

4.22 Derive Eqs. (4-89).

4.23 Derive Eqs. (4-102).

4.24 For the circuit [P8(b)] shown in Fig. P4-24, show that for the realization of a biquad given as follows:

$$T(s) = -K\left(\frac{s^2 + cs + d}{s^2 + as + b}\right)$$

Fig. P4-24

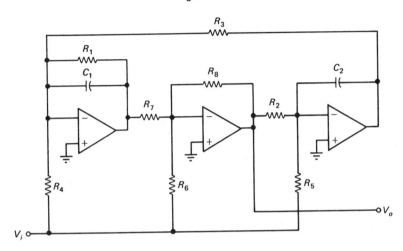

the design equations are

$$C_1 = C_2 = 1$$

$$R_1 = \frac{1}{a}$$

$$R_2 = R_3 = R_7 = R_8 = \frac{1}{\sqrt{b}} \qquad R_4 = \frac{1}{K(a-c)}$$

$$R_5 = \frac{\sqrt{b}}{Kd} \qquad\qquad R_6 = \frac{1}{K\sqrt{b}}$$

4.25 For the KHN circuit shown in Fig. 4-31a, assuming

$$G_1 = G_3 = G_4 = G_5 = G, \qquad C_1 = C_2 = C = 1,$$

$$\frac{G}{C} = \omega_0, \qquad G_2 = \frac{\omega_0}{2Q-1}, \qquad a_i(s) = \frac{GB}{s}$$

(a) Show that the polynomial that describes the pole locations is given by

$$D(s) = \frac{2s_n^5}{GB_n^3} + s_n^4\left[\frac{5}{GB_n^2} + \frac{4}{GB_n^3}\right] + s_n^3\left[\frac{4}{GB_n} + \frac{6}{GB_n^2} + \frac{2}{GB_n^3}\right]$$

$$+ s_n^2\left[1 + \frac{(2+1/Q)}{GB_n} + \frac{1}{GB_n^2}\right] + s_n\frac{1}{Q}\left[1 + \frac{1}{GB_n}\right] + 1$$

where $s_n = s/\omega_0$ and $GB_n = GB/\omega_0$.

(b) Plot the root-locus of the dominant poles of $D(s)$ as GB_n varies from ∞ to 1; assume that $Q = 10$.

(c) Verify that the circuit becomes unstable for low values of GB_n.

4.26 Plot the actual Q and the actual ω_0 of the dominant pole pair given by $D(s)$ (in Prob. 4.25) against ω_0/GB, taking nominal values of $Q = 10$ and $\omega_0 = 1$, and confirm that they agree with curves shown in Figs. 4-44a and 4-45a.

4.27 For the PMG circuit shown in Fig. 4-42, let

$$G_2 = G_4 = G_5 = G_6 = G_7 = G_9 = G$$

$$G_1 = G_3 + G_8 = G/\sqrt{Q}, \qquad C_1 + C_3 = C_2 = \frac{G}{\omega_0}, \quad a_i(s) = \frac{GB}{s}$$

to show that the natural frequencies of the circuit are given by

$$D(s) = \frac{s_n^5}{GB_n^3}\left(1 + \frac{1}{Q} + \frac{1}{\sqrt{Q}}\right) + s_n^4 \left[\frac{4 + 2/\sqrt{Q} - 1/Q}{GB_n^2} + \frac{2(1 + 1/Q)(2 + 1/\sqrt{Q})}{GB_n^3}\right]$$

$$+ s_n^3 \left[\frac{3 + 1/\sqrt{Q}}{GB_n} + \frac{2(4 + 2/\sqrt{Q} - 1/Q)}{GB_n^2} + \frac{(1 + 1/\sqrt{Q})(2 + 1/\sqrt{Q})}{GB_n^3}\right]$$

$$+ s_n^2 \left[1 + \frac{(4 - 1/\sqrt{Q} - 1/Q)}{GB_n} + \frac{4 + 2/\sqrt{Q} - 1/Q)}{GB_n^2}\right]$$

$$+ s_n\left[1 + \frac{3 + 1/\sqrt{Q}}{GB_n}\right] + 1$$

4.28 Repeat Prob. 4.25(b) for the circuit in Fig. 4-42. Verify that the circuit is stable for all values of GB_n in the range $1 \le GB_n \le \infty$ for $Q = 10$.

4.29 Plot the actual Q and the actual ω_0 of the dominant pole pair of PMG circuit given by $D(s)$ (in Prob. 4.27) against ω_0/GB, taking the nominal values of $Q = 10$ and $\omega_0 = 1$ and confirm that they agree with the curves shown in Figs. 4-44a and 4-45a.

4.30 Calculate M_α, M_β, and M_T for the AM circuit shown in Fig. 4-33 for $Q = 50$ and $f_0 = 20$ kHz, assuming that $GB = 1$ MHz. Assume the element variations to be statistically independent with $\sigma_R = \sigma_C = 0.2\%$ and $\sigma_{GB} = 10\%$. The frequency range of interest is given by $\omega_1 = \omega_0(1 - 1/2Q)$ and $\omega_2 = \omega_0(1 + 1/2Q)$. Check your results with those given in Fig. 4-43a to c.

4.31 Repeat Prob. 4.30 for the TT circuit given in Fig. 4-32.

4.32 Show that the MB and PMG circuits are GIC-based circuits. Redraw the circuits and label the two GICs in each case.

4.33 Design the active–R circuit shown in Fig. 4-50a for the following specifications:

$$f_0 = 20\,\text{kHz}, \; Q = 20, \;\; \text{and} \;\; H_0 = 4$$

Assume the op amps to be identical with the following parameters: $GB = 2\pi(10^6)$ rad/sec. Choose convenient element values to minimize the spread where you have choices.

What are the actual values of f_0, Q, and H_0 if excess phase is taken into account (i.e., if $m = 3$)?

chapter five

High-Order Filter Realization

5.1 INTRODUCTION

In Chapter 4 we considered the realization of second-order or biquadratic filter sections. These filter sections are often used as the building blocks for high-order filter realization. By "high-order" filter we mean a filter whose order is greater than or equal to 4. Since a third-order filter can be realized by simply adding a passive RC network to the active biquadratic section, such a network serves as an additional building block for odd-order filters.

As mentioned in Chapter 4, there are several ways one can realize a high-order filter; however, they all fall into one of the following five realization classes.

1. In the *direct synthesis* approach a high-order filter $H(s)$ is realized using a particular configuration comprised of one or more active elements and passive-RC networks.

2. In the *cascade realization* approach, a high-order filter is realized by cascading two or more active biquadratic sections (e.g., those described in Chapter 4). This method is widely used in industry because cascade filters are relatively straightforward to design and tune.

3. In the *inductance simulation* approach, an inductance is simulated by an active-RC network such as a capacitor-loaded gyrator; thus, classical lossless passive-ladder filter realizations can be transformed directly into active-filter realizations. The transformed active filter retains many of the desirable properties of the passive realization.

4. In the *generalized immittance* approach, which closely parallels the inductance simulation method, classical RLC synthesis is followed with a

frequency-dependent impedance (or admittance) transformation. This transformation provides a transfer function invariant scaling. The scaling introduces a new active element, the frequency-dependent negative resistor (FDNR). In Section 2.9 the FDNR was shown to be readily realizable with active networks. The primary advantage derived from this approach is to reduce the number of op amps otherwise required in certain realizations derived by inductance simulation.

5. In the *multiple-loop feedback (and feedforward)* approach, the high-order filter is realized by interconnecting first or second-order filter sections into a given multiple-loop feedback topology. This method retains the modularity of cascade designs while achieving the desirable properties of filters realized by the inductance simulation and generalized immittance approaches.

In this chapter these methods will be discussed in some detail. As in Chapter 4, the op amp is the active building block for all filter realizations. Examples of some of the mentioned realization approaches listed can be found in industrial products.

As mentioned previously, the most widely used method is the cascade method. The cascade approach is then followed by the multiple-loop feedback, inductance simulation, and generalized immittance methods. The cascade and multiple-loop feedback methods are treated in greater detail, because these two compatible methods use essentially the same building blocks (e.g., an active biquad network) and provide the designer with sufficient freedom to realize efficient, manufacturable active filters. In the spirit of efficiency, by using single-amplifier biquads, these two approaches allow the implementation of high-order filters with a minimum number of op amps. Although much is said in the literature regarding the small size and cheapness of op amps, they still consume power and are the predominate sources of noise. With the increasing costs of energy, the energy consumed over the lifetime of a filter will cost more than the production of the filter. Thus, those approaches that require the fewest op amps while not unduly sacrificing performance should receive our greatest attention.

5.2 DIRECT SYNTHESIS APPROACH

In this approach a specified high-order transfer function is realized directly by a simple circuit configuration and appropriately placed RC networks, determined by classical driving-point immittance synthesis. Several configurations are possible; however, in this book we consider only one sample configuration, as this approach is not practical and leads to poor sensitivity. Once the desired driving-point immittances have been determined, the design is straightforward. Let us now consider the two-op amp configuration shown in Fig. 5-1. The voltage transfer function for this circuit is given by

$$\frac{V_o}{V_i} = \frac{Y_1 Y_5 - Y_2 Y_3}{Y_3 Y_6 - Y_4 Y_5} \tag{5-1}$$

where Y_i are RC driving-point admittance functions. For convenience and ease

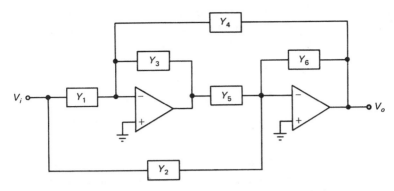

Fig. 5-1 Two op amp high-order filter configuration. The Y_i blocks represent the admittances for RC one-port networks.

of design, we choose

$$Y_3 = Y_5 = 1 \tag{5-2a}$$

In other words, Y_3 and Y_5 are conductances normalized to unity. The use of Eqs. (5-2) in (5-1) yields the simple expression

$$H(s) = \frac{V_o}{V_i} = \frac{Y_1 - Y_2}{Y_6 - Y_4} \tag{5-2b}$$

Now suppose that a high-order voltage transfer function is specified in the following form:

$$H(s) = \frac{N(s)}{D(s)} = \frac{b_m s^m + b_{m-1} s^{m-1} + \ldots + b_1 s + b_0}{s^n + a_{n-1} s^{n-1} + \ldots + a_1 s + a_0} \tag{5-3}$$

where $m \leq n$. Both $N(s)$ and $D(s)$ can be divided by a polynomial $Q(s)$ without a change in $H(s)$:

$$H(s) = \frac{N(s)/Q(s)}{D(s)/Q(s)} \tag{5-4}$$

where the polynomial $Q(s)$ in Eq. (5-4) possesses distinct negative real roots. The reason for the introduction of $Q(s)$ and its form will become apparent shortly. Specifically, we introduce

$$Q(s) = \prod_{i=1}^{l} (s + \sigma_i) \tag{5-5}$$

with degree $Q(s) \geq \deg D(s) - 1$ (i.e., $l \geq n - 1$). From Eqs. (5-4) and (5-3) we can identify

$$\frac{N(s)}{Q(s)} = Y_1 - Y_2 \tag{5-6}$$

$$\frac{D(s)}{Q(s)} = Y_6 - Y_4 \tag{5-7}$$

Let us now briefly discuss the reasons for the choice of $Q(s)$ in Eq. (5-5) and the resulting relations in Eqs. (5-6) and (5-7). Since the roots of $Q(s)$ are all negative real and distinct by choice in Eq. (5-5), we have, from a partial fraction expansion,

$$\frac{N(s)}{Q(s)} = \underbrace{\left(k_\infty^+ + \frac{k_0^+}{s} + \sum_i \frac{k_i^+}{s + \sigma_i}\right)}_{Z_{RC}^a} - \underbrace{\left(k_\infty^- + \frac{k_0^-}{s} + \sum_j \frac{k_j^-}{s + \sigma_j}\right)}_{Z_{RC}^b} \qquad (5\text{-}8)$$

where k_i^+ and k_j^- are all real and positive numbers. Note that for a minimum element (canonic) network, Z_{RC}^a and Z_{RC}^b contain no common poles (e.g., $k_0 = s\frac{N}{Q}\Big|_{s=0}$ may be either positive or negative, hence belonging to either Z_{RC}^a or Z_{RC}^b). The expressions inside each set of parentheses are readily recognized to be the driving-point impedance of an RC network, as shown in Fig. 5-2a. Now consider the partial fraction expansion of $N(s)/sQ(s)$:

$$\frac{N(s)}{sQ(s)} = Z_{RC}^c(s) - Z_{RC}^d(s) \qquad (5\text{-}9)$$

or

$$\frac{N(s)}{Q(s)} = sZ_{RC}^c(s) - sZ_{RC}^d(s) \qquad (5\text{-}10a)$$

$$= \underbrace{\left(k_0^+ + k_\infty^+ s + \sum_i \frac{k_i^+ s}{s + \sigma_i}\right)}_{Y_{RC}^A} - \underbrace{\left(k_0^- + k_\infty^- s + \sum_j \frac{k_j^- s}{s + \sigma_j}\right)}_{Y_{RC}^B} \qquad (5\text{-}10b)$$

Note that k_i in Eqs. (5-8) and (5-10b) are not the same and the immittances are therefore labeled by different superscripts. The expression inside each set of parentheses is recognized to be the driving-point admittance of an RC network, as shown in Fig. 5-2b (see also Appendix A).

Thus, for a given $N(s)/D(s)$, we obtain circuit element values from Eqs. (5-6), (5-7), and (5-10c). It should be noted that since the choice of σ_i and the degree of $Q(s)$ in Eq. (5-5b) are arbitrary, the decompositions given in Eqs. (5-9) and (5-10) are not unique. This nonuniqueness can be utilized in minimizing circuit elements and/or sensitivity.

To illustrate the method, let us consider the following simple example.

●

EXAMPLE 5-1

Consider the realization of a three-pole Butterworth transfer function using the circuit in Fig. 5-1:

$$H(s) = \frac{1}{s^3 + 2s^2 + 2s + 1} \qquad (5\text{-}11)$$

High-Order Filter Realization

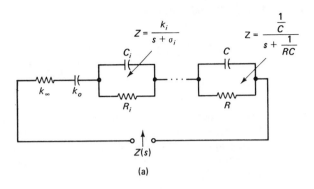

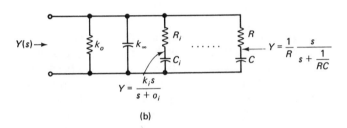

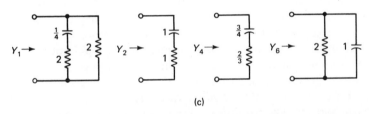

(c)

Fig. 5-2 (a) RC network with driving-point impedance of the form Z_{RC}^a, Z_{RC}^b in Eq. (5-8). (b) RC network with driving-point admittance of the form Y_{RC}^A, Y_{RC}^B in Eq. (5-10b). (c) RC networks corresponding to the driving-point admittances in Eqs. (5-14) and (5-15). (Element values in ohms and farads.)

In this case the degree of $D(s)$ is 3. Hence, one possible choice for $Q(s)$ is a polynomial of degree 2:

$$Q(s) = (s+1)(s+2) \tag{5-12}$$

Note that we have arbitrarily chosen $\sigma_1 = 1$, $\sigma_2 = 2$. The identifications are as follows:

$$\frac{N(s)}{Q(s)} = \frac{1}{(s+1)(s+2)} = \frac{1}{2} + \frac{s/2}{s+2} - \frac{s}{s+1} \tag{5-13a}$$

$$\frac{D(s)}{Q(s)} = \frac{s^3 + 2s^2 + 2s + 1}{(s+1)(s+2)} = \frac{s^2 + s + 1}{(s+2)}$$
$$= s + \frac{1}{2} - \frac{3s/2}{s+2} \tag{5-13b}$$

From Eqs. (5-6), (5-7), and (5-13), we have

$$Y_1 = \frac{1}{2} + \frac{s/2}{s+2} \qquad Y_2 = \frac{s}{s+1} \qquad (5\text{-}14)$$

$$Y_6 = s + \frac{1}{2} \qquad Y_4 = \frac{3s/2}{s+2} \qquad (5\text{-}15)$$

The circuit element values corresponding to Eqs. (5-14) and (5-15) are as shown in Fig. 5-2c. The reader could equally well use the circuit configuration of Fig. 4-1 to realize a general high-order transfer function. Many other configurations of this type can be found in the early literature on active filters. Problems (5.3) and (5.4) illustrate two such examples.

•

Comments are in order regarding the performance and practicality of direct-form realizations. First, it is evident that high-order transfer functions can be implemented with very few op amps: one or two (such as the configuration in Fig. 5-1). However, this efficiency in op amps does not come without sacrifice. First, both the numerator and denominator in Eq. (5-3) are derived by a cancellation of terms. As we observed in Chapter 4 for positive-feedback biquads, such cancellations result in large passive-component sensitivities, particularly for narrowband filters. The poor sensitivity performance alone will render this approach impractical in many applications. Recall from Chapter 4 that the use of multiple op amps in biquad circuits introduced isolation which tended to reduce the labor of tuning. In direct implementations of high-order filters, the transfer function coefficient to component-value relationships is very complex and interrelated. The tuning for such a network has to be determined by digital computer. Also damaging is that in many cases, as the order increases, so do the component spreads. The requirement for very large component spreads will all but eliminate integration in hybrid form. Nevertheless, because of the small number of op amps required, direct implementation may be an available candidate for some moderate order (≤ 5) designs with unchallenging requirements. However, in most applications, the reader will find the direct implementations unsatisfactory for one or more of the reasons stated above.

5.3 CASCADE REALIZATIONS

The simplest, and perhaps still the most widely used synthesis procedure for realizing high-order filters is to cascade two or more essentially isolated biquadratic sections. The number of biquadratic sections for an nth-order filter is simply $n/2$, for n even. For n odd the realization requires $(n-1)/2$ second-order sections and one first-order section or $(n-3)/2$ second-order sections and one third-order section. The realization is shown in Fig. 5-3, where T_i denotes the transfer function for the ith biquadratic filter section, which can be implemented by any one of the biquadratic circuits discussed in Chapter 4. An analysis of

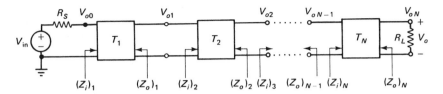

Fig. 5-3   Schematic for a cascade of N noninteracting filter sections with transfer functions T_i.

Fig. 5-3 quickly reveals the importance of using noninteracting sections (T_i) in cascade synthesis. Let us first consider the output of the first section V_{o1}:

$$V_{o1} = \left[\frac{(Z_i)_1}{(Z_i)_1 + R_s}\right]\left[\frac{(Z_i)_2}{(Z_i)_2 + (Z_o)_1}\right]T_1 V_{in} \tag{5-16}$$

where Z_i and Z_o represent the section input and output impedances, respectively. Then the output voltage of the second block V_{o2} is similarly written

$$V_{o2} = \left[\frac{(Z_i)_1}{(Z_i)_1 + R_s}\right]\left[\frac{(Z_i)_2}{(Z_i)_2 + (Z_o)_1}\right]\left[\frac{(Z_i)_3}{(Z_i)_3 + (Z_o)_2}\right]T_1 T_2 V_{in} \tag{5-17}$$

Extending this argument we can arrive at the expression for the output voltage for a cascade of N sections:

$$V_{oN} = V_o = \left[\frac{(Z_i)_1}{(Z_i)_1 + R_s}\right]\left[\frac{R_L}{R_L + (Z_o)_N}\right]\prod_{k=1}^{N-1}\left[\frac{(Z_i)_{k+1}}{(Z_i)_{k+1} + (Z_o)_k}\right]$$
$$\times \left(\prod_{j=1}^{N} T_j\right)V_{in} \tag{5-18}$$

If the section output impedances $(Z_o)_k$ are very low and their input impedances $(Z_i)_k$ are very large, more specifically if $(Z_o)_k \ll (Z_i)_{k+1}, (Z_o)_N \ll R_L$ and $(Z_i)_1 \gg R_s$, then Eq. (5-18) reduces to

$$V_{oN} = V_o \simeq \prod_{j=1}^{N} T_j V_{in} \tag{5-19}$$

from which we obtain the desired transfer function,

$$H = \frac{V_o}{V_{in}} = \prod_{j=1}^{N} T_j \tag{5-20}$$

Thus, the transfer function for a cascade of sections T_j is simply the product of the individual section transfer functions, when the input impedance of each section is always much larger then the output impedance of the preceding section or, in other words, the sections are noninteracting. Fortunately, these conditions are satisfied by most biquad filter sections, as they utilize op amps, which ensures noninteraction. Note that throughout this chapter H is used to denote the overall transfer function for a high-order filter (e.g., a cascade filter) and T_i to denote the ith-section transfer function (i.e., a biquadratic transfer function). Later it will become convenient to define H_k to be the transfer function from the input to the kth-section output of the cascade:

$$H_k = \prod_{j=1}^{k} T_j \quad \text{where } k = 1, 2, \ldots, N \tag{5-21}$$

We note that $H_1 = T_1$ and $H_N = H$. The general form of the overall transfer function can be expressed as

$$H(s) = \frac{\prod_{i=1}^{m/2} (b_{i2}s^2 + b_{i1}s + b_{i0})}{\prod_{j=1}^{n/2} (a_{j2}s^2 + a_{j1}s + a_{j0})} \quad \text{for } m, n \text{ even} \tag{5-22}$$

$$= \frac{\prod_{i=1}^{(m-1)/2} (b_{i2}s^2 + b_{i1}s + b_{i0})(b_1s + b_0)}{\prod_{j=1}^{(n-1)/2} (a_{j2}s^2 + a_{j1}s + a_{j0})(a_1s + a_0)} \quad \text{for } m, n \text{ odd} \tag{5-23}$$

The transfer functions for m even, n odd and m odd, n even are of similar form and are not cited here, to avoid undue repetition. To synthesize the cascade filter one must appropriately group together complex pole–zero pairs, which correspond to the section transfer functions T_j in Eq. (5-20), to form

$$H(s) = \prod_{j=1}^{N} T_j = \prod_{j=1}^{n/2} \left(\frac{b_{j2}s^2 + b_{j1}s + b_{j0}}{a_{j2}s^2 + a_{j1}s + a_{j0}} \right) \tag{5-24a}$$

for n even, and

$$H(s) = \prod_{j=1}^{N} T_j = \left[\prod_{j=1}^{(n-1)/2} \left(\frac{b_{j2}s^2 + b_{j1}s + b_{j0}}{a_{j2}s^2 + a_{j1}s + a_{j0}} \right) \right] \left(\frac{b_1s + b_0}{a_1s + a_0} \right) \tag{5-24b}$$

for n odd. Note that by simply multiplying $T_N \times T_{N-1}$ of Eq. (5-24b), we have a third-order section. As seen in Eqs. (5-23) and (5-24), each biquad section is characterized by T_i of the following form:

$$T_i = \frac{b_{i2}s^2 + b_{i1}s + b_{i0}}{a_{i2}s^2 + a_{i1}s + a_{i0}} = \frac{k_i[s^2 + (\omega_{zi}/Q_{zi})s + \omega_{zi}^2]}{s^2 + \omega_{pi}/Q_{pi}s + \omega_{pi}^2} \tag{5-25}$$

and is realized according to the techniques given in Chapter 4. The first-order section in Eq. (5-24b), used when n is odd, is simply realized with a passive-RC network. It is noted that in deriving Eqs. (5-24), the roots of $N(s)$ and $D(s)$ of Eqs. (5-23) were appropriately pairwise grouped together to form the section transfer functions T_i. This pairwise grouping, often referred to as *pole–zero pairing*, is an important phase in the design of a cascade filter, since it ultimately determines the dynamic range of the filter and often affects the filter's sensitivity. Both of these problems are discussed in more detail in Sections 5.3.1 and 5.3.2. Once this pairing has been determined, the design reduces to the repeated synthesis of Eq. (5-25), as discussed in Chapter 4.

High-Order Filter Realization

5.3.1 Sensitivity Considerations

Recall that in Section 3.8 we derived the sensitivity measures for a cascade filter. Let us then begin our discussion here with Eq. (3-176), which we rewrite for convenience:

$$M_T = \int_{\omega_1}^{\omega_2} \sum_{k=1}^{N} (S_{\mathbf{x}_k}^{T_k})^{t*} \mathbf{P}_k (S_{\mathbf{x}_k}^{T_k}) \, d\omega \tag{5-26}$$

where in Eq. (5-26),

$$S_{\mathbf{x}_k}^{T_k} = \mathbf{d}_{T_k} = [S_{x_{k1}}^{T_k} \ldots S_{x_{kq}}^{T_k}]^t$$

with q denoting the total number of components in section k and

$$S_{x_{ki}}^{T_k} = \frac{x_{ki}}{T_k} \frac{\partial T_k}{\partial x_{ki}}$$

Also, $\mathbf{P}_k$ denotes the kth-section convariance matrix:

$$\mathbf{P}_k = E\left[\left(\frac{\Delta \mathbf{x}_k}{\mathbf{x}_k}\right) \left(\frac{\Delta \mathbf{x}_k}{\mathbf{x}_k}\right)^t \right]$$

We note that it is often desirable to compute and display the integrand S_T of M_T in Eq. (5-26), where

$$S_T = \sum_{k=1}^{N} (S_{\mathbf{x}_k}^{T_k})^{t*} \mathbf{P}_k (S_{\mathbf{x}_k}^{T_k}) \tag{5-27}$$

Let us express the kth-section sensitivity vector $S_{\mathbf{x}_k}^{T_k}$ in terms of the respective contributions from the numerator N_k and denominator D_k; that is, from Eq. (3-9),

$$S_{\mathbf{x}_k}^{T_k} = S_{\mathbf{x}_k}^{N_k} - S_{\mathbf{x}_k}^{D_k} \tag{5-28}$$

where $T_k = N_k/D_k$. Substituting Eq. (5-28) into Eq. (5-26), we obtain M_T in terms of the section numerator and denominator sensitivities:

$$M_T = \int_{\omega_1}^{\omega_2} \sum_{k=1}^{N} \{ (S_{\mathbf{x}_k}^{N_k})^{t*} \mathbf{P}_k (S_{\mathbf{x}_k}^{N_k}) + (S_{\mathbf{x}_k}^{D_k})^{t*} \mathbf{P}_k (S_{\mathbf{x}_k}^{D_k})$$
$$- 2 \, \mathrm{Re} \, [(S_{\mathbf{x}_k}^{N_k})^{t*} \mathbf{P}_k (S_{\mathbf{x}_k}^{D_k})] \} \, d\omega \tag{5-29}$$

with Eqs. (5-28) and (5-29), we can appreciate some of the reasoning behind some of the pole–zero pairing rules that have appeared in the literature and also begin to comprehend the difficulty in applying them in any general, consistent manner. Two contradicting rules have been proposed for reducing sensitivity in cascade active filters. They may be stated as follows:

1. Pair the high-Q poles with those zeros farthest away such that $|p_i - z_i|$ is maximum.
2. Pair the high-Q poles with the closest zeros such that $|p_i - z_i|$ is minimum.

In some special cases, sensitivity may be independent of the pole–zero pairing. Much of the reasoning for proposing rule 1 is obtained from Eq. (5-29) and the knowledge that the more sensitive (i.e., most critical) sections are those that possess the highest pole Q's. The relationship between Q and sensitivity is discussed in detail in various sections of Chapter 3.

For example, if the N_Q high-Q pole pairs (D_k's) are paired with those zeros (N_k's) farthest away then, in the pass band

$$|S_{x_k}^{N_k}| \ll |S_{x_k}^{D_k}| \qquad \text{for } k = 1, 2, \ldots, N_Q \qquad (5\text{-}30)$$

which reduces M_T to

$$M_T^{(1)} \simeq \int_{\omega_1}^{\omega_2} \left\{ \sum_{k=1}^{N_Q} (S_{x_k}^{D_k})^{t*} P_k (S_{x_k}^{D_k}) + \sum_{k=N_Q+1}^{N} (S_{x_k}^{T_k})^{t*} P_k (S_{x_k}^{T_k}) \right\} d\omega \qquad (5\text{-}31)$$

where the superscript (1) refers to the application of pairing rule 1. Thus, by applying rule 1, the effect of N_Q section numerators on the individual section sensitivities has been reduced. Referring again to the overall sensitivity expressed in Eq. (5-29), it appears that by pairing the high-Q poles with the closest zeros (i.e., rule 2), an advantage can be gained from the natural cancellation in Eq. (5-29) such that

$$M_T^{(2)} < M_T^{(1)} \qquad (5\text{-}32)$$

However, to achieve the cancellation required to obtain the result in Eq. (5-32), the coefficients of N_i and D_i must be essentially controlled by common elements or elements that are statistically correlated. In other words, we require that the variations in N_k track the variations in D_k. When N_k and D_k share no common elements and the elements are statistically independent [i.e., $P_{k(ij)} = \rho_{kij} \sigma_{xki} \sigma_{xkj} = 0$], Eq. (5-29) becomes

$$M_T = \int_{\omega_1}^{\omega_2} \left[\sum_{k=1}^{N} (S_{x_k}^{N_k})^{t*} P_k (S_{x_k}^{N_k})^{t} + (S_{x_k}^{D_k})^{t*} P_k (S_{x_k}^{D_k}) \right] d\omega \qquad (5\text{-}33)$$

The benefit can only be derived from the application of rule 2 when N_i and D_i are statistically or otherwise dependent. In the special case where all D_i are high-Q and all zeros are far removed, then $|S_{x_k}^{N_k}| \ll |S_{x_k}^{D_k}|$ for all N, and M_T in Eq. (5-29) reduces to

$$M_T^{(3)} = \int_{\omega_1}^{\omega_2} \sum_{k=1}^{N} (S_{x_k}^{D_k})^{t*} P_k (S_{x_k}^{D_k}) \, d\omega \qquad (5\text{-}34)$$

Thus, for this case, $M_T^{(3)}$ is virtually independent of the pole–zero pairing.

Clearly, as far as sensitivity is concerned, the optimum pole–zero pairings and their effect on the overall sensitivity depend on both the transfer function to be realized and the individual section implementations. Thus, an accepted but unfortunate fact of life is that no *general rule* for pole–zero pairing in cascade realizations can be stated. The procedure to minimize M_T is to simply insert all possible combinations of pole–zero pairs and select the ordering that

High-Order Filter Realization

results in the smallest M_T. For moderate order, order ≤ 10, this is not an unmanageable task. In general, computer optimization is required. To summarize, this sensitivity optimization depends on several factors:

1. The choice of biquad section topology.
2. The component variation statistics (i.e., variances σ_{xkl} and correlation coefficients ρ_{klj}).
3. The frequency range of interest, $\omega_1 \leq \omega \leq \omega_2$, which are the limits of integration in Eq. (5-29).

A significant change in any of these factors can alter the optimum pole–zero pairing for a particular filter realization.

To appreciate the problem, consider the effects of pole–zero pairing on the sensitivity of the following filters. The first filter is the fifth-order D4-channel bank filter described in detail in Chapter 1. The transfer function for this filter, per Eq. (1-57), is

$$H(s) = \frac{K(s^2 + \omega_{z_1}^2)(s^2 + \omega_{z_2}^2)}{(s - p_0)(s - p_1)(s - p_1^*)(s - p_2)(s - p_2^*)} \tag{5-35}$$

where

$$\omega_{z1} = 4.32 \times 10^4 \text{ rad/sec}$$
$$\omega_{z2} = 2.92 \times 10^4 \text{ rad/sec}$$
$$p_0 = -1.68 \times 10^4 \text{ rad/sec} \tag{5-36}$$
$$p_1, p_1^* = (-0.97 \pm j1.75) \times 10^4 \text{ rad/sec}$$
$$p_2, p_2^* = (-0.236 \pm j2.224) \times 10^4 \text{ rad/sec}$$

The real pole p_0 can be synthesized by means of a passive RC network and hence will not be considered any further. There are two possible pole–zero pairings or decompositions. One choice, distinguished by the superscript A, is as follows:

$$H(s) = T_1^A T_2^A \tag{5-37a}$$

where

$$T_1^A = \frac{K_1[s^2 + (43.2)^2]}{s^2 + 19.4s + 400.34} \tag{5-37b}$$

$$T_2^A = \frac{K_2[s^2 + (29.2)^2]}{s^2 + 4.72s + 507.33} \tag{5-37c}$$

which corresponds to pairing according to rule 2. It is noted that the gain constants K_1 and K_2 are chosen such that $K_1 K_2 = K$ and such that the maximum signal levels are equalized at the output of each section of the filter. This choice ensures that no one section will be overdriven or underdriven ahead of the other. Therefore, selecting K_1 and K_2 in this manner is an important step toward maximizing the dynamic range for the filter. A more complete discussion of the dynamic range consideration in cascade filters will be given shortly. Note that the frequency is normalized by 10^3 in Eqs. (5-37).

The second choice, designated by the superscript B, is as follows:

$$H = T_1^B T_2^B \tag{5-38a}$$

where

$$T_1^B = \frac{K_1[s^2 + (29.2)^2]}{s^2 + 19.4s + 400.34} \tag{5-38b}$$

$$T_2^B = \frac{K_2[s^2 + (43.2)^2]}{s^2 + 4.72s + 507.33} \tag{5-38c}$$

which corresponds to pairing according to rule 1.

Once the decompositions (either A or B) and the gain constants K_1 and K_2 are determined, the design reduces to the individual synthesis of each biquad (T_1 and T_2), as discussed in Chapter 4. The overall frequency response and the individual frequency responses of each section are shown in Fig. 5-4a and b, respectively. The biquad transfer functions were realized using the multiple op-amp feedforward topology given in Fig. P4-24. The normalized element values for these realizations, obtained from the design equations (see Prob. 4.24), are listed in Table 5-1.

TABLE 5-1[a]

	T_1^A	T_2^A	T_1^B	T_2^B
C_1	1	1	1	1
C_2	1	1	1	1
R_1	0.0515	0.2119	0.0515	0.2119
R_2	0.05	0.0444	0.05	0.0444
R_3	0.05	0.0444	0.05	0.0444
R_4	0.0515	0.2119	0.0515	0.0515
R_5	0.0107	0.0264	0.0235	0.0121
R_6	0.05	0.0444	0.05	0.0444
R_7	0.05	0.0444	0.05	0.0444
R_8	0.05	0.0444	0.05	0.0444

[a]The gains of the individual blocks are assumed to be unity (i.e., $K_1 = K_2 = 1$).

To determine the best pairing, from a sensitivity standpoint, the sensitivity measure M_T in Eq. (5-29) was evaluated for both the A and B designs. In these computations the desired band of interest (i.e., the limits of integration for M_T) was assumed to $0 \leq \omega \leq 18.85$ and the component variations were assumed to be independent, identically distributed random variables with standard deviations $\sigma = 1\%$ for the passive elements and the op amps are assumed to be ideal.[1]

[1]If the tolerances of the op amp gain–bandwidth products are included (e.g., σ_{GB} = 20%), the sensitivity results corresponding to Eq. (5-39) remain essentially the same. This is because the center frequency is far less than GB of the op amp. The sensitivity becomes large as ω_0 is close to GB (see Fig. 4-43).

High-Order Filter Realization

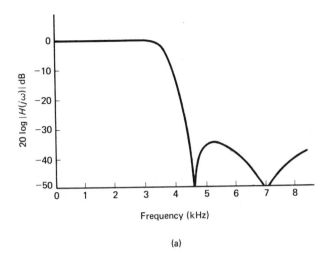

(a)

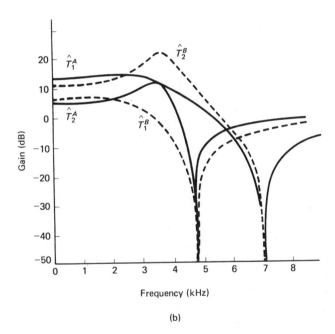

Frequency (kHz)

(b)

Fig. 5-4 (a) Gain response for $H(s)$ in Eq. (5-35). (b) Gain responses for $\hat{T}_1^A$, $\hat{T}_2^A$ in Eq. (5-38) from decomposition A and for $\hat{T}_1^B$, $\hat{T}_2^B$ in Eq. (5-39) from decomposition B. Note that $\hat{T}_i = T_i/K_i$.

To assist the reader in computing M_T, the element sensitivities for the A and B biquads are given in Table 5-2 (see Prob. 5.6). In Table 5-2, $N_1 = b_1 - \omega^2 = 1866.24 - \omega^2$, $N_2 = b_2 - \omega^2 = 852.64 - \omega^2$, and

$$D_1 = (d_1 - \omega^2) + j\omega c_1 = (400.34 - \omega^2) + j\omega 19.4$$
$$D_2 = (d_2 - \omega^2) + j\omega c_2 = (507.33 - \omega^2) + j\omega 4.72$$

TABLE 5-2

	T_1^A	T_2^A	T_1^B	T_2^B
$S_{C_1}^{T_i}$	$\dfrac{-b_1}{N_1} + \dfrac{d_1 + j\omega c_1}{D_1}$	$\dfrac{-b_2}{N_2} + \dfrac{d_2 + j\omega c_2}{D_2}$	$\dfrac{-b_2}{N_2} + \dfrac{d_1 + j\omega c_1}{D_1}$	$\dfrac{-b_1}{N_1} + \dfrac{d_2 + j\omega c_2}{D_2}$
$S_{C_2}^{T_i}$	$\dfrac{d_1}{D_1} - \dfrac{b_1}{N_1}$	$\dfrac{d_2}{D_2} - \dfrac{b_2}{N_2}$	$\dfrac{d_1}{D_1} - \dfrac{b_2}{N_2}$	$\dfrac{d_2}{D_2} - \dfrac{b_1}{N_1}$
$S_{R_1}^{T_i}$	$j\omega c_1 \left(\dfrac{1}{D_1} - \dfrac{1}{N_1}\right)$	$j\omega c_2 \left(\dfrac{1}{D_2} - \dfrac{1}{N_2}\right)$	$j\omega c_1 \left(\dfrac{1}{D_1} - \dfrac{1}{N_2}\right)$	$j\omega c_2 \left(\dfrac{1}{D_2} - \dfrac{1}{N_1}\right)$
$S_{R_2}^{T_i}$	$\dfrac{d_1}{D_1}$	$\dfrac{d_2}{D_2}$	$\dfrac{d_1}{D_1}$	$\dfrac{d_2}{D_2}$
$S_{R_3}^{T_i}$	$S_{C_2}^{T_i}$	$S_{C_1}^{T_i}$	$S_{C_2}^{T_i}$	$S_{C_1}^{T_i}$
$S_{R_4}^{T_i}$	$\dfrac{j\omega c_1}{N_1}$	$\dfrac{j\omega c_2}{N_2}$	$\dfrac{j\omega c_1}{N_2}$	$\dfrac{j\omega c_2}{N_1}$
$S_{R_5}^{T_i}$	$\dfrac{-b_1}{N_1}$	$\dfrac{-b_2}{N_2}$	$\dfrac{-b_2}{N_2}$	$\dfrac{-b_1}{N_1}$
$S_{R_6}^{T_i}$	$\dfrac{\omega^2 - j\omega c_1}{N_1}$	$\dfrac{\omega^2 - j\omega c_2}{N_2}$	$\dfrac{\omega^2 - j\omega c_1}{N_2}$	$\dfrac{\omega^2 - j\omega c_2}{N_1}$
$S_{R_7}^{T_i}$	$\dfrac{d_1}{D_1} - \dfrac{b_1 - j\omega c_1}{N_1}$	$\dfrac{d_2}{D_2} - \dfrac{b_2 - j\omega c_2}{N_2}$	$\dfrac{d_1}{D_1} - \dfrac{b_2 - j\omega c_1}{N_2}$	$\dfrac{d_2}{D_2} - \dfrac{b_1 - j\omega c_2}{N_1}$
$S_{R_8}^{T_i}$	$\dfrac{b_1 - \omega^2}{N_1} - \dfrac{d_1}{D_1}$	$\dfrac{b_2 - \omega^2}{N_2} - \dfrac{d_2}{D_2}$	$\dfrac{b_2 - \omega^2}{N_2} - \dfrac{d_1}{D_1}$	$\dfrac{b_1 - \omega^2}{N_1} - \dfrac{d_2}{D_2}$

Performing the integration for M_T numerically over a field of 51 points using Simpson's rule[2] yields the following results:

$$M_T^A = 0.0193 < 0.0213 = M_T^B \tag{5-39}$$

From the inequality Eq. (5-39), pairing the highest-Q poles with the closest zeros yields the best sensitivity.

As another example illustrating the effects of pole–zero pairing on the sensitivity of a cascade design, consider the various possible designs for the following band-pass transfer function:

$$H(s) = K\frac{s(s^2 + 0.25)(s^2 + 2.25)}{(s^2 + 0.2s + 1.01)(s^2 + 0.09s + 0.83)(s^2 + 0.1s + 1.18)} \tag{5-40}$$

The pole–zero pattern for this transfer function is shown in Fig. 5-5. For convenience, the transfer function is normalized so that $\omega_0 = 1$ rad/sec and the

---

[2]The Simpson's rule for integration is given by the following formula:

$$M_T = \frac{h}{3}\left[S_T(1) + \sum_{\substack{i=2 \\ i=\text{even}}}^{N_{\text{points}}-1} 4S_T(i) + \sum_{\substack{i=3 \\ i=\text{odd}}}^{N_{\text{points}}-2} 2S_T(i) + S_T(N_{\text{points}}) \right]$$

where N_{points} is odd and h is the distance between successive points. In the equation above, $N_{\text{points}} = 51$ and $h = (18.85 - 0)/50 = 0.377$.

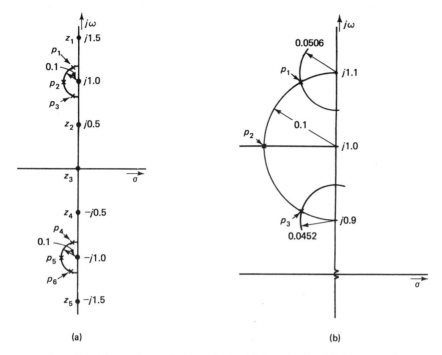

Fig. 5-5 s-Plane pole–zero locations for $H(s)$ in Eq. (5-40). (a) Pole–zero positions. (b) Location of poles p_1, p_2 and p_3.

bandwidth is 0.2 rad/sec. The overall frequency response of the filter is sketched in Fig. 5-6.

The six possible pole–zero pairings are as follows:

$$H^A(s) = \left(\frac{s^2 + 0.25}{s^2 + 0.09s + 0.83}\right)\left(\frac{s^2 + 2.25}{s^2 + 0.1s + 1.18}\right)\left(\frac{s}{s^2 + 0.2s + 1.01}\right) \quad (5\text{-}41\text{a})$$

$$H^B(s) = \left(\frac{s^2 + 2.25}{s^2 + 0.09s + 0.83}\right)\left(\frac{s^2 + 0.25}{s^2 + 0.1s + 1.18}\right)\left(\frac{s}{s^2 + 0.2s + 1.01}\right) \quad (5\text{-}41\text{b})$$

$$H^C(s) = \left(\frac{s}{s^2 + 0.09s + 0.83}\right)\left(\frac{s^2 + 2.25}{s^2 + 0.1s + 1.18}\right)\left(\frac{s^2 + 0.25}{s^2 + 0.2s + 1.01}\right) \quad (5\text{-}41\text{c})$$

$$H^D(s) = \left(\frac{s^2 + 0.25}{s^2 + 0.09s + 0.83}\right)\left(\frac{s}{s^2 + 0.1s + 1.18}\right)\left(\frac{s^2 + 2.25}{s^2 + 0.2s + 1.01}\right) \quad (5\text{-}41\text{d})$$

$$H^E(s) = \left(\frac{s^2 + 2.25}{s^2 + 0.09s + 0.83}\right)\left(\frac{s}{s^2 + 0.1s + 1.18}\right)\left(\frac{s^2 + 0.25}{s^2 + 0.2s + 1.01}\right) \quad (5\text{-}41\text{e})$$

$$H^F(s) = \left(\frac{s}{s^2 + 0.09s + 0.83}\right)\left(\frac{s^2 + 0.25}{s^2 + 0.1s + 1.18}\right)\left(\frac{s^2 + 2.25}{s^2 + 0.2s + 1.01}\right) \quad (5\text{-}41\text{f})$$

Note that H^A corresponds to pairing the highest-Q poles with the closest zeros (i.e., rule 2) and that H^B corresponds to pairing the highest-Q poles with the

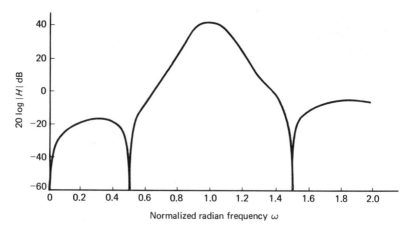

Fig. 5-6   Gain response for $H(s)$ in Eq. (5-40).

zeros farthest away (i.e., rule 1). The other pairings do not correspond to either rule.

The designs for the biquad transfer functions in Eqs. (5-41) were carried out assuming realization with the multiple op amp biquad in Fig. 4-37. The determination of the element values for these designs is left to the exercises. The element values for the best design will be given later.

To determine the best pole–zero pairing, from a sensitivity point of view, M_T was evaluated for all six cases, assuming the same component variations as in the preceding example. The integrands S_T of measure M_T for the six pole–zero pairings are displayed in Fig. 5-7 over the frequency range $0.75 \le \omega \le 1.25$. Clearly, in the vicinity of the center frequency, H^A yields the lowest M_T. More specifically, integrating the S_T's over the frequency range $0.80 \le \omega \le 1.20$ yields the values for M_T listed in Table 5-3.

TABLE 5-3

Cases	A	B	C	D	E	F
Sensitivity (M_T)	0.02027	0.02170	0.02060	0.02065	0.02145	0.02124

The normalized element values, corresponding to Fig. 4-37, which implement transfer function H^A are listed in Table 5-4.

It should be pointed out that when sufficient computer time is available, one can alternatively compute the S_T and M_T via Monte Carlo analysis. The use of Monte Carlo analysis to compute exact sensitivities is discussed in Section 3.10.

The only drawback of the cascade design method is that the resulting designs are inherently more sensitive to component variations than are the ladder-derived analogs and multiple-loop feedback topologies discussed later in this chapter. More specifically, it can be shown that on a statistical basis,

High-Order Filter Realization

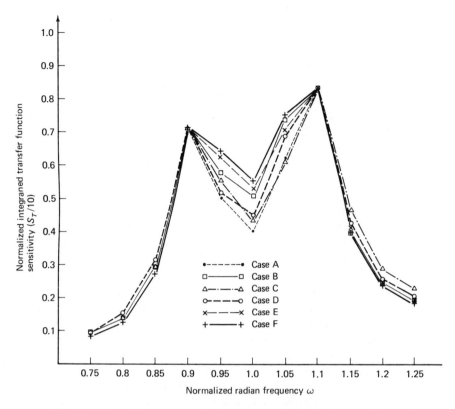

Fig. 5-7 Plots of S_T versus radian frequency, ω, for the six decompositions of $H(s)$ in Eq. (5-41).

TABLE 5-4

	T_1^A	T_2^A	T_3^A
C_1	1	1	1
C_2	1	1	1
R_1	11.1	10	5
R_2	1.0976	0.9206	0.995
R_3	1.0976	0.9206	0.995
R_4	11.1	10	1
R_5	3.644	0.4828	∞
R_6	1.0976	0.9206	∞
R_7	1.0976	0.9206	0.995
R_8	1.0976	0.9206	0.995

the variations in the section transfer functions T_i tend to be additive and produce a larger variation in H. Nevertheless, because of the simplicity of the design and tuning, the cascade method has been by far the most widely used approach and still retains considerable popularity.

5.3.2 Dynamic-Range Considerations

An equally important, and at times a more crucial test to the viability of a filter realization is its dynamic range. By dynamic range we mean that range of input signal levels, usually expressed in dB, the filter can accommodate without incurring a distorted output. That is, a filter which can handle input signals varying in level from 1 mV to 1 V without distortion is said to have a dynamic range of 60 dB. Active filters typically have a dynamic range between 70 and 100 dB. The maximum signal levels are limited by the slew rate and saturation of the operational amplifier and the minimum signal levels are limited by the system noise. Thus, to achieve good dynamic range is a twofold problem: (1) to ensure that signal levels within the cascade do not become excessively large to overdrive op amps in one or more sections, and (2) to ensure that in-band signals do not become so small that they become depressed into the noise. The pole–zero pairings, ordering of sections, and choice of section gain constants usually have a significant effect on the overall dynamic range.

Here we are concerned with maximum overall and minimum in-band signal levels observed at the output of each biquad section. Actually, when using multiple-amplifier biquads these levels should be examined at the output of every op amp. Moreover, in Chapter 4, we discussed design techniques for maximizing the dynamic range within a biquad. In this chapter we assume that the internal levels can be appropriately treated once the desired pole–zero pairings, ordering, and gain constants have been chosen. The dynamic range for outputs internal to the biquad can also be affected by the pole–zero pairing and ordering. For those cases where unsatisfactory dynamic range conditions occur internal to one or more biquads, the pole–zero pairings and/or ordering can be altered accordingly.

To mathematically characterize and maximize the dynamic range for a cascade of biquads, consider the following definitions. First we define

$$S_{k\,\min} = \min_{\omega_{\mathrm{CL}} \le \omega \le \omega_{\mathrm{CH}}} |H_k| = \min_{\omega_{\mathrm{CL}} \le \omega \le \omega_{\mathrm{CH}}} \left| \prod_{i=1}^{k} T_i \right| \qquad (5\text{-}42)$$

(ω_{CL} and ω_{CH} are the low- and the high-frequency band edges) as the minimum in-band signal at the ith-section output, and

$$S_{k\,\max} = \max_{0 \le \omega < \infty} |H_k| = \max_{0 \le \omega < \infty} \left| \prod_{i=1}^{k} T_i \right| \qquad (5\text{-}43)$$

at the maximum signal level at the ith-section output. Then a useful figure of merit for the dynamic range at the output of each stage of the cascade is

$$r_j = \frac{S_{j\,\max}}{S_{j\,\min}} \qquad \text{for } j = 1, 2, \ldots, N \qquad (5\text{-}44)$$

The object is to pair the poles and zeros and order the section such that the maximum r_j is minimized:

$$\min_{\substack{\text{pole–zero pairings and ordering}}} (r_{\text{max}}) = \min (\max_{1 \leq j \leq N} r_j) \qquad (5\text{-}45)$$

By minimizing r_{max} we tend to equalize the N-section dynamic ranges, thus minimize the opportunity for the extremes of the input levels to overdrive or underdrive one or more sections ahead of the others. We note that large signal levels can cause op amps to overdrive at any frequency; thus, we must be concerned with the maximum signal level regardless of its frequency of occurrence. On the other hand, since signal amplification, if any, is only applied in-band, we need only be concerned with low signal levels and signal-to-noise ratios in this region. By "in-band" we mean within the pass band of the overall cascade.

As a simple example to demonstrate the identification of $S_{j\text{min}}$ and $S_{j\text{max}}$, let the first stage for a cascade design with passband $\omega_{\text{CL}} \leq \omega \leq \omega_{\text{CH}}$ be a high-pass notch biquad with transfer function

$$H_1 = T_1 = \frac{K_1(s^2 + \omega_z^2)}{s^2 + (\omega_p/Q_p)s + \omega_p^2} \qquad (5\text{-}46)$$

where $\omega_p > \omega_z$. If we sketch H_1 as a function of frequency, we can readily identify the desired maximum and minimum signal levels relative to the overall passband $\omega_{\text{CL}} \leq \omega \leq \omega_{\text{CH}}$. These levels are illustrated in Fig. 5-8. We note that,

Fig. 5-8 Second-order low-pass notch gain response illustrating $S_{l\text{max}}$ and $S_{l\text{min}}$.

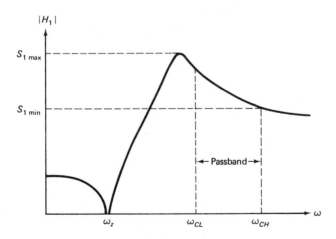

although the peak occurs outside the overall passband, it is identified as $S_{1\text{max}}$. Also, since the minimum level occurs out-of-band, $S_{1\text{min}}$ occurs at the band edge: ω_{CH}.

From the previous discussion, it should be clear that three operations must be performed in order to optimize the dynamic range[3] for a cascade active filter:

1. Determine the optimum pole–zero pairings for the N-section T_k's.
2. Determine the optimum ordering for the interconnected sequence of N sections.
3. Determine the gain constants K_i such that the transmission levels at the output of each section are consistent with the performance objectives. More will be said shortly about choosing the K_i.

It is noted that operations 1 and 2 involve the minimization of r_{max}. Recall that by minimizing r_{max}, we tend to equalize the N-section dynamic ranges. After completion of 1 and 2 we can then adjust the gain constants K_i, subject to the constraint that $\prod_{i=1}^{N} K_i = K$, where K defines the overall gain level. In general, operations 1 and 2 are interrelated; however, usually a good suboptimum design can be achieved by treating these operations independently in the order given. That is, pair the section poles and zeros such that the largest difference between the maximum signal level and the maximum in-band signal level is minimized on an individual-section basis. In other words, make the individual $|T_k|$ as flat as possible. This objective suggests pairing poles and zeros according to rule 2 (i.e., pair the highest-Q poles with the closest zeros). Once the pole–zero pairings have been determined, the sections are ordered such that r_{max} is minimized. In addition, when possible, it is often desirable to place a low-pass or band-pass section at the beginning of the cascade. This placement will prevent strong high-frequency noise from entering the succeeding stages of the cascade. When the input is expected to have a sizable dc or low frequency, out of band (eg. 60 Hz power line coupling) components, it will be desirable to use either a high-pass or a band-pass as an initial stage. This placement will prevent these components from possibly causing an overload in succeeding low-pass sections. Finally, it is usually desirable to place either a high pass or a band-pass at the end of the cascade. This placement will tend to eliminate dc offsets that have accumulated through the preceding sections and to reduce internally generated low-frequency noise. Of course, there is little we can do about the dc offset and low-frequency noise from the last section. It is noted that dc offsets and low-frequency $(1/f)$ noise are generated by the op amps.

Before proceeding with a detailed discussion regarding the selection of the gain constants K_i, let us consider an example illustrating the pole–zero pairing and section-ordering operations. Returning to the fourth-order transfer function in Eq. (5-35), let us reexamine the pole–zero pairings in Eqs. (5-37) and (5-38) from the point of view of dynamic range. For this example we assume that the desired overall gain constant is

$$K = K_1 K_2 = 0.128 \tag{5-47}$$

---

[3]It is noted here that in some cases the optimization of dynamic range includes a sensitivity trade-off.

which corresponds to a low-frequency gain $|H(j0)| = 1$. To determine the best pole–zero pairing for dynamic range, consider the following gain-normalized section transfer functions:

$$\hat{T}_1^A = \frac{T_1^A}{K_1} = \frac{s^2 + (43.2)^2}{s^2 + (19.4)s + 400.34} \tag{5-48a}$$

$$\hat{T}_2^A = \frac{T_2^A}{K_2} = \frac{s^2 + (29.2)^2}{s^2 + (4.72)s + 507.33} \tag{5-48b}$$

$$\hat{T}_1^B = \frac{T_1^B}{K_1} = \frac{s^2 + (29.2)^2}{s^2 + (19.4)s + 400.34} \tag{5-48c}$$

$$\hat{T}_2^B = \frac{T_2^B}{K_2} = \frac{s^2 + (43.2)^2}{s^2 + (4.72)s + 507.33} \tag{5-48d}$$

At this point the gain constants K_i are left arbitrary and are determined during the final step of the design. The magnitude responses for each of the transfer functions given in Eq. (5-48) are plotted in Fig. 5-4b. The solid lines denote $|\hat{T}_1^A|$, $|\hat{T}_2^A|$, and the dashed lines, $|\hat{T}_1^B|$, $|\hat{T}_2^B|$. Note that pairing the high-Q poles with the closest zeros (T_1^A, T_2^A) results in the flatter section gain responses. Table 5-5 contains the $|\hat{T}_k|$ maximum and the $|\hat{T}_k|$ in-band ($0 \le f \le 3.3$ kHz) mini-

TABLE 5-5

Section	Max $\lvert\hat{T}_k\rvert$ ($0 < f < \infty$)	Min $\lvert\hat{T}_k\rvert$ ($0 < f < 3.3$ kHz)	Max/Min
$\hat{T}_1^A$	4.974	4.662	1.067
$\hat{T}_2^A$	3.516	1.681	2.092
$\hat{T}_1^B$	2.135	1.048	2.037
$\hat{T}_2^B$	12.997	3.679	3.533

mum. The maximum-to-minimum ratios are also given. Recall that the frequencies are normalized (i.e., $f/10^3$). As expected from our previous discussion, pole–zero pairings T_1^A and T_2^A are optimum. To determine the ordering, the $S_{j\min}$, $S_{j\max}$, and r_j are listed in Table 5-6. For completeness, the cascade sequences $H = T_1^B T_2^B$ and $H = T_2^B T_1^B$ are also included. From the $r_{\max}$ column in Table 5-6 we conclude that the sequence $H = \hat{T}_1^A \hat{T}_2^A$ minimizes $r_{\max}$.

TABLE 5-6

Cascade Sequence	$S_{1\min}$	$S_{1\max}$	$S_{2\min}$[a]	$S_{2\max}$[a]	r_1	r_2	$r_{\max}$
$\hat{T}_1^A \hat{T}_2^A$	4.662	4.974	1.0	1.54	1.067	1.540	1.540
$\hat{T}_2^A \hat{T}_1^A$	1.681	3.516	1.0	1.54	2.092	1.540	2.092
$\hat{T}_1^B \hat{T}_2^B$	1.048	2.135	1.0	1.54	2.037	1.540	2.037
$\hat{T}_2^B \hat{T}_1^B$	3.679	12.997	1.0	1.54	3.533	1.540	3.533

[a]$S_{2\min}$ ($0 \le f \le 3.3$ kHz) occurs at $f = 0$ and $S_{2\max}$ occurs at $f \simeq 3.26$ kHz.

As mentioned previously, the final step in the process is to assign the gain constants K_i to adjust the gain levels at the output of each section. The criteria for allocating these gain levels will depend upon the application; however, one typical criterion is to minimize the maximum signal level within the cascade, that is select the K_i such that

$$\min_{1 \leq k \leq N} \left(\max_{0 \leq \omega < \infty} |H_k| \right) \leq \frac{(V_o)_{\max}}{(V_{\text{in}})_{\max}} \tag{5-49}$$

Clearly, to determine the viability of the design, the maximum input signal level $(V_{\text{in}})_{\max}$ to be processed must be determined from a consideration of the intended application. The maximum output level $(V_o)_{\max}$ is either determined by op amp slew-rate limitations or saturation. Limitations in op amp dynamic range, dictated by slew rate and saturation, are discussed in Appendix C. To satisfy Eq. (5-49) implies that

$$\max_{0 \leq \omega < \infty} |H_1| = \max_{0 \leq \omega < \infty} |H_2| = \ldots = \max_{0 \leq \omega < \infty} |H_{N-1}| = \max_{0 \leq \omega < \infty} |H| \tag{5-50}$$

In other words, the gain constants K_i are chosen such that H_k maxima are equal:

$$K_1 = \frac{\max\limits_{0 \leq \omega < \infty} |H|}{\max\limits_{0 \leq \omega < \infty} |\hat{T}_1|} \tag{5-51a}$$

$$K_2 = \frac{1}{K_1} \frac{\max\limits_{0 \leq \omega < \infty} |H|}{\max\limits_{0 \leq \omega < \infty} |\hat{T}_1 \hat{T}_2|} \tag{5-51b}$$

or, in general,

$$K_i = \frac{1}{\prod\limits_{k=1}^{i-1} K_k} \left(\frac{\max\limits_{0 \leq \omega < \infty} |H|}{\max\limits_{0 \leq \omega < \infty} \left| \prod\limits_{k=1}^{i} \hat{T}_k \right|} \right) \quad \text{for } i = 1, \ldots, N-1 \tag{5-51c}$$

$$K_N = \frac{K}{\prod\limits_{k=1}^{N-1} K_k} \tag{5-51d}$$

It is noted that the $\max\limits_{0 \leq \omega < \infty} |H_k|$ and $\max\limits_{0 \leq \omega < \infty} |H|$ are most readily obtained by scanning the printed output from a computer-aided computation of the gain scaled amplitude responses $|H/K|$ and the $\left| H_k \Big/ \prod\limits_{i=1}^{k} K_i \right|$.

Another criterion, which is useful when the intended application demands the processing of very small signals (e.g., systems used to monitor biological functions), is to select the K_i's to maximize the minimum signal level in the cascades:

$$\max_{1 \leq k \leq N} \left(\min_{\omega_{\text{CL}} \leq \omega \leq \omega_{\text{CH}}} |H_k| \right) \geq \frac{(V_o)_{\min}}{(V_{\text{in}})_{\min}} \tag{5-52}$$

For the applications where Eq. (5-52) is useful, the minimum ouput level $(V_o)_{\min}$ is determined by the system noise, which in turn limits the minimum input signal amplitudes which can be reliably processed. As mentioned previously, our concern is only with small in-band signals. To satisfy Eq. (5-52) implies that

$$
\min_{\omega_{\mathrm{CL}} \leq \omega \leq \omega_{\mathrm{CH}}} |H_1| = \min_{\omega_{\mathrm{CL}} \leq \omega \leq \omega_{\mathrm{CH}}} |H_2| = \cdots = \min_{\omega_{\mathrm{CL}} \leq \omega \leq \omega_{\mathrm{CH}}} |H_{N-1}|
$$
$$
= \min_{\omega_{\mathrm{CL}} \leq \omega \leq \omega_{\mathrm{CH}}} |H| \tag{5-53}
$$

That is, we equalize the minimum signal levels. Satisfying Eq. (5-53) ensures that no section output will be underdriven ahead of the other sections. The K_i values that satisfy Eq. (5-53) are given by the following expressions:

$$
K_1 = \frac{\displaystyle\min_{\omega_{\mathrm{CL}} \leq \omega \leq \omega_{\mathrm{CH}}} |H|}{\displaystyle\min_{\omega_{\mathrm{CL}} \leq \omega \leq \omega_{\mathrm{CH}}} |\hat{T}_1|} \tag{5-54a}
$$

$$
K_2 = \frac{1}{K_1} \frac{\displaystyle\min_{\omega_{\mathrm{CL}} \leq \omega \leq \omega_{\mathrm{CH}}} |H|}{\displaystyle\min_{\omega_{\mathrm{CL}} \leq \omega \leq \omega_{\mathrm{CH}}} |\hat{T}_1 \hat{T}_2|} \tag{5-54b}
$$

or, in general,

$$
K_i = \frac{1}{\displaystyle\prod_{k=1}^{i-1} K_k} \frac{\displaystyle\min_{\omega_{\mathrm{CL}} \leq \omega \leq \omega_{\mathrm{CH}}} |H|}{\displaystyle\min_{\omega_{\mathrm{CL}} \leq \omega \leq \omega_{\mathrm{CH}}} \left| \prod_{k=1}^{i} \hat{T}_k \right|} \tag{5-54c}
$$

$$
K_N = \frac{K}{\displaystyle\prod_{k=1}^{N-1} K_k} \tag{5-55}
$$

Again the minima of H_k and H are readily determined with the aid of a digital computer or programmable calculator.

As an example, let us computed the gain constants for the fourth-order filter considered previously for pole–zero pairing and section ordering. Recall that after pairing and ordering, we obtained

$$
H = T_1^A T_2^A \tag{5-56}
$$

where H, T_1^A, and T_2^A are given by Eqs. (5-35) and (5-37), respectively. For this example we assume that the desired criterion is to minimize the maximum section output levels. Computing K_1 and K_2 according to Eqs. (5-51) yields

$$
K_1 = \frac{\displaystyle\max_{0 < \omega < \infty} |H|}{\displaystyle\max_{0 \leq \omega < \infty} |\hat{T}_1^A|} = \frac{1}{4.974} = 0.201 \tag{5-57a}
$$

$$
K_2 = \frac{K}{K_1} = \frac{0.128}{0.201} = 0.635 \tag{5-57b}
$$

where $\displaystyle\max_{0 \leq \omega < \infty} |\hat{T}_1^A|$ is given in Table 5-5.

Once the pole–zero pairing, ordering, and gain constants have been selected, the individual biquad sections can be designed according to the procedures discussed in Chapter 4. Unlike sensitivity, dynamic range optimization is essentially independent of the biquad realization. However, as discussed in Chapter 4, gain levels at the outputs of internal op amps of multiple-op amp biquad sections must be similarly optimized.

5.4 ACTIVE-LADDER REALIZATIONS

Active-ladder synthesis is simply the implementation of passive ladder designs with active networks. Synthesis procedures and explicit design formulas for realizing passive-ladder networks are found in abundance throughout the literature (see the references in Appendix A). Designs in normalized component values for many generic filter shapes such as Chebyshev and elliptic can be found tabulated in filter handbooks (Appendix B). Without doubt there is more literature available for designing passive-ladder filters than any other type of filter. Thus, because of this large body of knowledge and experience with passive ladders, it seems quite natural that active filters be designed to emulate their behavior.

Let us now briefly examine the passive-ladder filter, especially the doubly terminated LC ladder structure shown in Fig. 5-9. To realize an all pole high-pass transfer function $H(s)$, where $H(s) = V_o/V_{in}$, we simply assign inductors to the shunt arms and capacitors to the series arms; that is, referring to Fig. 5-9,

$$Y_i = \frac{1}{sL_i} \quad \text{for } i = 1, 3, 5, \ldots, n-1 \tag{5-58a}$$

$$Z_i = \frac{1}{sC_i} \quad \text{for } i = 2, 4, \ldots, n \tag{5-58b}$$

To realize an all-pole low-pass $H(s)$, we assign capacitors to the shunt arms and inductors to the series arms. That is, referring to Fig. 5-9,

$$Y_i = sC_i \quad \text{for } i = 1, 3, 5, \ldots, n-1 \tag{5-59a}$$

$$Z_i = sL_i \quad \text{for } i = 2, 4, \ldots, n \tag{5-59b}$$

Fig. 5-9 Doubly terminated passive-ladder configuration.

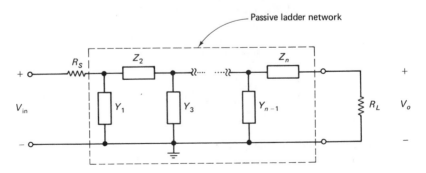

High-Order Filter Realization

For band-pass realization we apply the low-pass to band-pass transformation to the low-pass design (see Section 1.5), resulting in

$$Y_i = sC + \frac{1}{sL} = \frac{C(s^2 + \omega_0^2)}{s} \qquad (5\text{-}60a)$$

$$Z_i = sL + \frac{1}{sC} = \frac{L(s^2 + \omega_0^2)}{s} \qquad (5\text{-}60b)$$

where $\omega_0 = \sqrt{1/LC}$ and Y_i, Z_i refer to the ladder immittances in Fig. 5-9. Note that passive-ladder filters can also use symmetric lattice sections. Such techniques are quite convenient for designing quartz crystal filters. Since this subject matter is beyond the scope of this book, the interested reader is encouraged to examine the sampling of literature on the subject cited at the end of the chapter [B5].

In addition to the large reservoir of experience in passive-ladder synthesis, the prime motivator for active-ladder realization is the low sensitivity to passive parameter variations exhibited by passive-ladder filters. As we would expect, active-ladders inherit by-and-large the low-sensitivity properties of the passive filters which they simulate. To establish the low-sensitivity property of doubly terminated ladder networks, consider the simple differential sensitivity definition given in Eq. (3-6) and the general doubly terminated ladder filter in Fig. 5-10.

Fig. 5-10 Schematic representation for a general doubly terminated ladder filter.

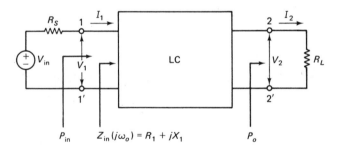

The input impedance of the circuit, as seen from the input terminals 1-1, may be expressed as a function of $s = j\omega$ as follows:

$$Z_{in}(j\omega) = R_1(\omega) + jX_1(\omega) \qquad (5\text{-}61)$$

From the input impedance and the input current, the input power can be determined to be

$$P_{in} = |I_1|^2 \, \text{Re}\,(Z_{in}) \qquad (5\text{-}62a)$$

$$= \frac{V_1^2 R_1}{(R_s + R_1)^2 + X_1^2} = P_o \qquad (5\text{-}62b)$$

Since the LC network is lossless, P_{in} in Eq. (5-62) is equal to the corresponding output power, P_o. An estimate of the sensitivity can be obtained by calculating the differential change ∂P_o, which results from differential variations in the network components x. Differentiating P_o with respect to x yields

$$\frac{\partial P_o}{\partial x} = \frac{\partial P_o}{\partial R_1}\frac{\partial R_1}{\partial x} + \frac{\partial P_o}{\partial X_1}\frac{\partial X_1}{\partial x} \tag{5-63}$$

Using Eq. (5-62b), the partial derivatives in Eq. (5-63) are readily determined as follows:

$$\frac{\partial P_o}{\partial R_1} = \frac{R_s^2 - R_1^2 + X_1^2}{[(R_1 + R_s)^2 + X_1^2]^2} V_1^2 \tag{5-64a}$$

$$\frac{\partial P_o}{\partial X_1} = \frac{-2R_1 X_1}{(R_1 + R_s)^2 + X_1^2} V_1^2 \tag{5-64b}$$

Normally, the filter is designed such that the power transferred to the load is maximum. This condition, referred to as the *maximum power transfer condition*, is obtained when the terminating impedances Z_S and Z_L are conjugate-matched to the input and output impedances, respectively, of the LC two-port. From Fig. 5-10, we have

$$Z_{\text{in}} = Z_s^* \longrightarrow R_1 = R_s \quad \text{and} \quad X_1 = 0 \tag{5-65}$$

These conditions are satisfied at the frequencies at which the filter transmission is maximum. Substituting the maximum power transfer condition, Eq. (5-65), into Eqs. (5-64) yields

$$\frac{\partial P_o}{\partial R_1} = 0 \quad \text{and} \quad \frac{\partial P_o}{\partial X_1} = 0 \tag{5-66}$$

Equation (5-66) implies that the differential sensitivity in Eq. (5-64) is zero:

$$S_X^{P_o} = \frac{X}{P_o}\frac{\partial P_o}{\partial X} = 0 \tag{5-67}$$

The frequencies at which the maximum power transfer condition is satisfied will, of course, lie in the filter passband. Since the sensitivity is zero at these frequencies, it is expected that the sensitivity will be very small in the vicinity of these frequencies.

A few comments are in order regarding the result expressed in Eq. (5-66):

1. It is noted that the power gain $G_P = P_o/P_{\text{in}}$ is directly related to the gain response $|V_2/V_1|$:

$$G_P = \frac{P_o}{P_{\text{in}}} = \frac{|V_2|^2/R_L}{|I_1|^2 R_1} = \left|\frac{V_2}{V_{\text{in}}}\right|^2 \frac{4R_s}{R_L} \tag{5-68}$$

Thus, the P_o sensitivity nulls at the frequencies of maximum power transfer imply that the gain response sensitivities to R_1 and X_1 are also nulled. However,

Eq. (5-66) does not imply that the corresponding transfer function sensitivities are similarly nulled. More will be said about this later.

2. Filters designed for flat magnitude response (e.g., Butterworth or Chebyshev) achieve maximum power transfer over most of the passband frequencies. Thus, very low gain response sensitivity is typically achieved over much of the passband. However, there are a host of other filter types that achieve near-maximum power transfer over a small portion of the passband (e.g., Bessel or Gaussian). For these filters the in-band gain sensitivity will be larger than that for a flat passband filter. For this class of filters it has been shown that optimized lossy (RLC) ladder networks can achieve lower in-band sensitivities. Moreover, for these applications where response variations outside the passband (e.g., in the transition band) are of concern, optimized lossy ladder networks can be designed to provide lower sensitivity than their lossless LC counterparts. More will be said about lossy ladder filters in Section 5.7.

In general, the doubly terminated lossless ladder is less sensitive than the singly terminated lossless ladder. Figure 5-11 displays Monte Carlo analyses

Fig. 5-11 Standard deviation of the transfer function versus frequency for the singly terminated and doubly terminated LC ladder realizations of a third-order Butterworth low-pass filter.

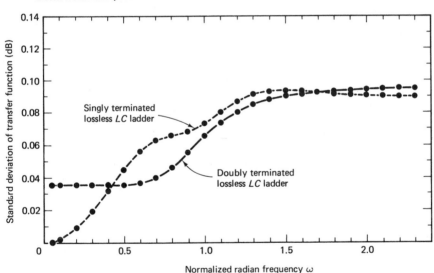

(500 samples) for third-order Butterworth lowpass doubly terminated and singly terminated lossless LC ladder filters. The component variations are assumed uniformly distributed with tolerances of $\pm 1.0\%$ [P25]. Note that the zero sensitivity at dc, for the singly terminated case, is simply due to the fact that at dc (where all capacitances look like open circuits and inductances look like short circuits), the source is directly across the load R_L.

5.4.1 Active-Ladder Realizations
with Inductance Simulation
and Generalized Impedance Converters

In this section we consider the synthesis of active filters derived from passive *RLC* prototype filters by directly replacing inductors with equivalent active *RC* networks. The building block for these implementations is the generalized-impedance converter (GIC) shown in Fig. 2-33 and repeated in Fig. 5-12. The utility of this element is evident from the impedance function Z_{in}.

Fig. 5-12 Schematic for a generalized impedance converter (GIC).

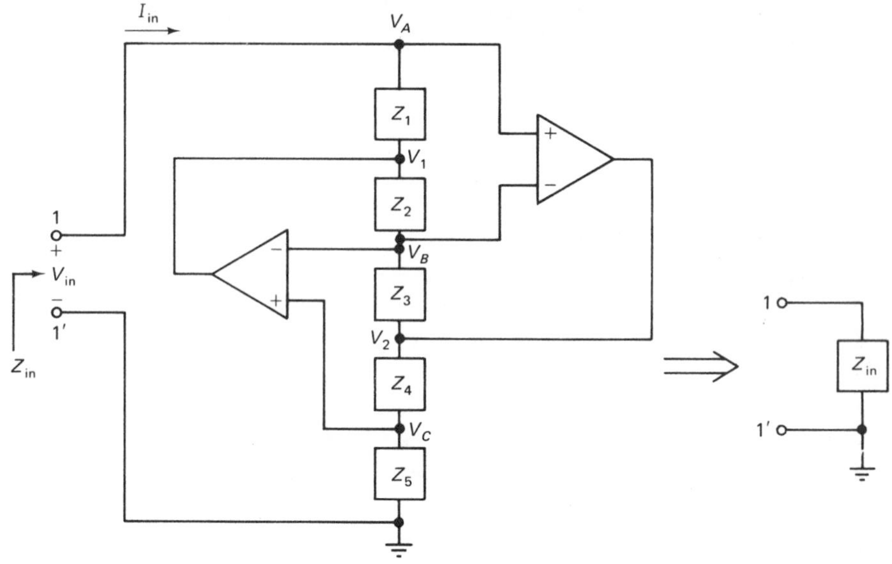

This impedance function can be derived from the nodal equations written at nodes V_A, V_B, and V_C:

$$I_{in} = V_A Y_1 - V_1 Y_1 \qquad (5\text{-}69\text{a})$$

$$(Y_2 + Y_3)V_B = Y_2 V_1 + Y_3 V_2 \qquad (5\text{-}69\text{b})$$

$$(Y_4 + Y_5)V_C = Y_4 V_2 \qquad (5\text{-}69\text{c})$$

and for ideal op amps

$$V_A = V_B = V_C = V_{in} \qquad (5\text{-}69\text{d})$$

Solving for $Z_{in} = V_{in}/I_{in}$, we obtain

$$Z_{in} = \frac{Y_2 Y_4}{Y_1 Y_3 Y_5} = \frac{Z_1 Z_3 Z_5}{Z_2 Z_4} \qquad (5\text{-}70)$$

From Eq. (5-70), we note:

1. When $Z_1 = Z_3 = Z_5 = Z_2 = R$ and $Z_4 = 1/sC$, then

$$Z_{in} = sR^2C \qquad (5\text{-}71)$$

Equation (5-71) is the impedance for an inductor of value

$$L = R^2C \qquad (5\text{-}72)$$

Clearly, we can interchange the roles of Z_2 and Z_4, with $Z_2 = 1/sC$ and $Z_4 = R$, to obtain Eq. (5-71). However, the choice of $Z_4 = 1/sC$ yields an inductance simulation which is most tolerant to the finite op amp GB [B1].

2. When $Z_2 = Z_4 = Z_5 = R$ and $Z_1 = Z_3 = 1/sC$, then

$$Z_{in} = \frac{1}{s^2RC^2} = \frac{1}{s^2D} \qquad (5\text{-}73)$$

In Eq. (5-73), with $s = j\omega$, $Z_{in} = -1/D\omega^2$; hence, (5-73) is the impedance for a frequency-dependent negative resistance (FDNR) of value $-1/D\omega^2$ with

$$D = RC^2 \qquad (5\text{-}74)$$

We can, of course, obtain Eq. (5-73) by setting any two of the impedances Z_1, Z_3, Z_5 to $1/sC$. However, it has been shown that the choice $Z_1 = Z_3 = 1/sC$ results in an FDNR with smaller sensitivities to the finite op amp gain bandwidths. This topic is covered in Section 2.9.

5.4.2 Inductance Simulation

As we have shown, the GIC is capable of realizing grounded inductors and grounded FDNRs. Floating inductors are realized by cascading two gyrators, with a capacitor connected across the common port, as shown in Fig. 5-13a. Underneath each gyrator we give its short-circuit admittance, matrix $[y_{ij}]$. If the gyrators are not identical, parasitic elements will appear, as indicated by the dashed elements in Fig. 5-13b. These parasitic elements can be expressed in terms of the gyrator conductances, i.e.

$$L_1 = \frac{C}{G_1^a(G_2^a - G_1^b)} \qquad (5\text{-}75a)$$

$$L_2 = \frac{C}{G_1^b(G_2^b - G_1^a)} \qquad (5\text{-}75b)$$

$$\alpha = \frac{G_1^aG_1^b - G_2^aG_2^b}{Cs} \qquad (5\text{-}75c)$$

For identical gyrators (i.e., $G_1^a = G_1^b = G_2^a = G_2^b$) the parasitic elements vanish; hence, we obtain a floating inductor as shown in Fig. 5-13c. Other solutions

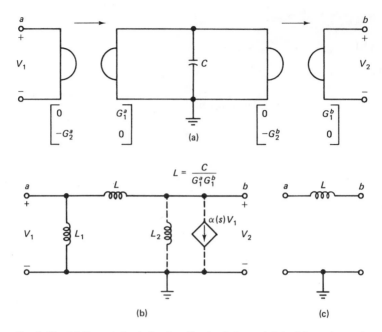

$$L = \frac{C}{G_1^a G_1^b}$$

Fig. 5-13 (a) Gyrator simulation for a floating inductor (c). In (b) are shown the parasitic elements that arise when the two gyrators in part (a) are not identical.

for the realization of floating inductors have been proposed. For a discussion of these solutions the interested reader is referred to [P5, P6]. Let us now see how these components can be used to realize active filters. The design begins with a selection of an appropriate passive prototype network. Typically, the prototype selected is a doubly terminated LC ladder network, such as the circuits shown in Fig. 5-14. Methods and tables for designing passive filters of these types can be found in several widely used textbooks and handbooks (see the references in Appendix B). To synthesize the high-pass filter in Fig. 5-14a,

Fig. 5-14 (a) Doubly terminated fifth-order high-pass LC ladder filter. (b) Doubly terminated fifth-order low-pass LC ladder filter.

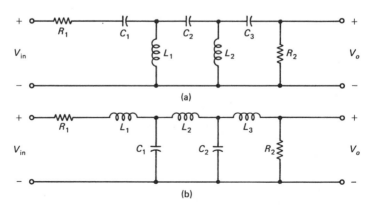

High-Order Filter Realization

we simply replace the inductors L_1 and L_2 with appropriately terminated GICs, as shown in Fig. 5-15. The GIC-simulated inductance values are then

$$L_1 \longrightarrow RR_{L1}C_{L1} \quad \text{and} \quad L_2 \longrightarrow RR_{L2}C_{L2} \tag{5-75d}$$

Fig. 5-15 Active-RC GIC simulation of the high-pass LC ladder filter in Fig. 5-14a.

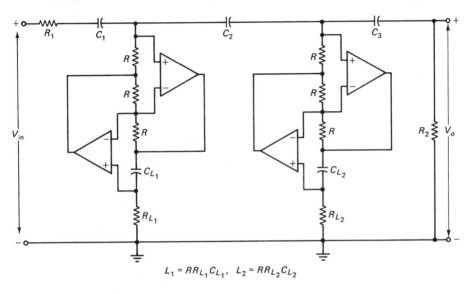

$$L_1 = RR_{L_1}C_{L_1}, \quad L_2 = RR_{L_2}C_{L_2}$$

For the low-pass filter, the floating inductors are replaced by the GIC circuit in Fig. 5-13.

It is noted that GIC-derived circuits typically require more op amps than either cascade or multiple-loop feedback designs using single-op amp biquads. As mentioned previously, there is increasing motivation to reduce the number of op amps, to reduce unit cost, power dissipation, and filter degradation due to second-order effects. There are several methods for reducing the number of op amps required to implement GIC-derived active-ladder networks.

For example, one could use a single op amp realization of a gyrator [P6] and/or use frequency transformations or use dual networks that require fewer floating inductors. Unfortunately, single-op amp inductors and FDNRs are either nonideal or derive their ideal behavior from cancellation.

5.4.3 Transformation of Elements Using Frequency-Dependent Negative Resistors

Consider the doubly terminated passive ladder of Fig. 5-10, or for that matter, any passive RLC network. The voltage ratio function of the network is dimensionless:

$$\frac{V_o}{V_i} = \frac{f_1(Z_i)}{f_2(Z_i)} \tag{5-76}$$

where Z_i are impedance functions with the dimension of ohms. If *every* impedance is scaled or transformed by some function or transformation $W(s)$, one
can rewrite Eq. (5-76) as follows:

$$\frac{V_o}{V_i} = \frac{Wf_1(Z_i)}{Wf_2(Z_i)} = \frac{f_1(Z_i)}{f_2(Z_i)} \qquad (5\text{-}77)$$

Thus, the voltage ratio function is unaltered by this transformation. One
transformation of particular interest is

$$W = \frac{K}{s} \qquad (5\text{-}78)$$

Multiplying each impedance in an *RLC* circuit by $W = K/s$ (i.e., $Z_i' = WZ_i$)
transforms resistors into capacitors, inductors into resistors, and capacitors into
frequency-dependent negative resistors (FDNRs), as shown in Fig. 5-16. The

Fig. 5-16   Impedance transformations $Z_i' = WZ_i$ where
$W = K/s$.

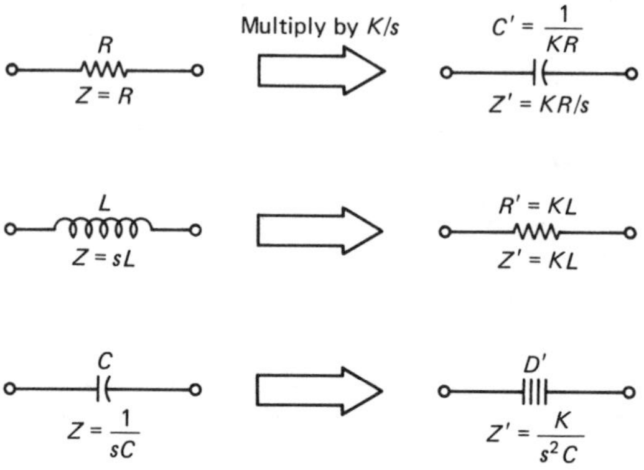

FDNR element is readily realizable as discussed earlier, by Fig. 5-12 (see also
Fig. 2-34), with the input impedance given by Eq. (5-73). From the discussion
above it is clear that a passive-*RLC* filter design can be readily transformed to
an active ladder using FDNRs. For example, the circuit of Fig. 5-14b is transformed as in Fig. 5-17a. The choice of *K* is arbitrary and can be chosen to render
convenient or practical element values. Comparing the fifth-order low-pass *RLC*
circuit in Fig. 5-14b with the transformed RCD circuit in Fig. 5-17a, we observe
that an active-ladder implementation of the *RLC* circuit requires three *floating*
inductors, whereas the equivalent RCD circuit requires two *grounded* FDNRs.
Since a floating inductor requires either two gyrators or two GICs (see Fig. 5-13),
the RCD fifth-order low-pass filter implementation results in a significant
savings in op amps. In particular the active *RLC* implementation would require
as many as 12 op amps, four op amps for each floating inductor, whereas the
active RCD implementation requires only four op amps, two for each grounded

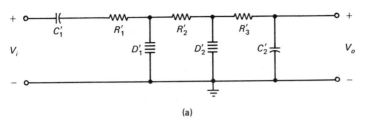

(a)

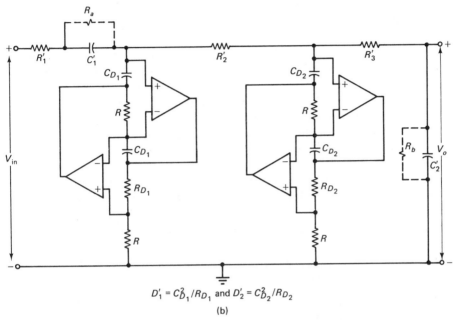

$$D'_1 = C^2_{D_1}/R_{D_1} \text{ and } D'_2 = C^2_{D_2}/R_{D_2}$$

(b)

Fig. 5-17 (a) Fifth-order low-pass ladder filter using grounded FDNRs. (b) Active-*RC* GIC simulation of the fifth-order low-pass filter in part (a).

FDNR, as shown in Fig. 5-17b. On the other hand, multiplying the impedances of the fifth-order high-pass network in Fig. 5-14a by K/s results in an RCD network comprised of three floating FDNRs. A floating FDNR is implemented with two GICs, as shown in Fig. 5-18. We may conclude from this exercise that

Fig. 5-18 GIC simulation of a floating FDNR.

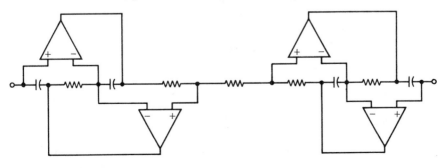

passive low-pass networks are most efficiently implemented as RCD active ladders and that passive high-pass networks are most efficiently implemented as *RLC* active ladders. Since band-pass networks will require some floating inductors or FDNRs, the advantage obtained with active *RLC* versus active RCD must be examined on a case-by-case basis. It is noted that in any case, GIC inductors and FDNRs should be designed, according to methods given in Section 2.9, to minimize the errors in simulation due to the finite op amp GBs.

Let us return to the RCD network in Fig. 5-17b, since a careful examination of this circuit reveals two serious problems:

1. The noninverting terminals of some op amps do not have a dc path to ground. Such a path is needed to supply these op amps with the required bias currents.

2. Typically, an active filter is embedded in other electronic circuitry which has been designed to drive resistive loads. Thus, the capacitive source and load required for the RCD implementation may be pathological for the interfacing electronics.

Problem 2 can be overcome by simply buffering the input and output with unity-gain amplifiers (see Fig. 2-7). However, this does not eliminate the biasing problem. One solution that has been suggested is to place large resistors (R_a, R_b) across C_1' and C_2' in Fig. 5-17b. The insertion of these resistors will provide the

Fig. 5-19 GIC implementation for a floating inductor using Gorski–Popiel's embedding technique.

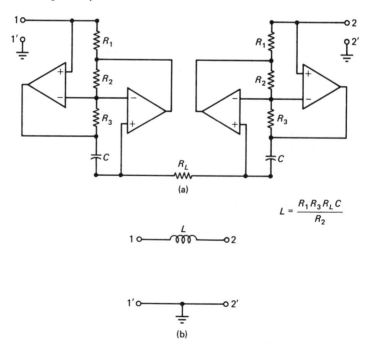

$$L = \frac{R_1 R_3 R_L C}{R_2}$$

(a)

(b)

High-Order Filter Realization

necessary dc paths to ground, at the expense, however, of some error in the gain response.

It should be noted that there is a technique, referred to as *embedding*, whereby *LC* networks are embedded between two GICs in order to arrive at an op amp efficient implementation [P8]. This technique is demonstrated in Figs. 5-19, 5-20, and 5-21 for simulating a floating inductor, a T-network of inductors, and a π-network of inductors, respectively. The embedded network is not restricted to being all resistive; in fact, any complex *LC* network can be embedded. For example, let us use the embedding technique to implement the RLC, fifth-order low-pass network in Fig. 5-14b. It is convenient to embed only the *LC* portion of this network, as shown in Fig. 5-22. This embedding leaves the resistive terminations unchanged, thus overcomes the serious difficulties mentioned previously. This implementation requires four GICs (i.e., eight op amps). In contrast, a cascade implementation, involving single-amplifier biquads, would require no more than three op amps (at the expense of increased sensitivity). However, there are indirect methods of simulating doubly terminated *LC* networks via signal flow graphs (e.g., [P10]), which use first-order and biquad sections. The resulting active-ladder networks are commonly referred to

Fig. 5-20 GIC implementation for a *T*-network of inductors using Gorski–Popiel's embedding technique.

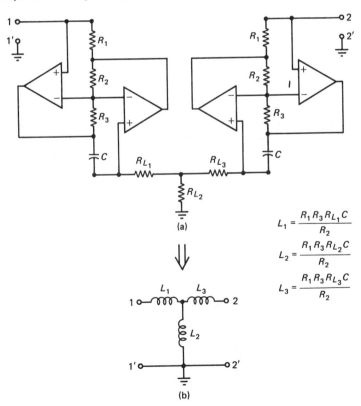

$$L_1 = \frac{R_1 R_3 R_{L_1} C}{R_2}$$

$$L_2 = \frac{R_1 R_3 R_{L_2} C}{R_2}$$

$$L_3 = \frac{R_1 R_3 R_{L_3} C}{R_2}$$

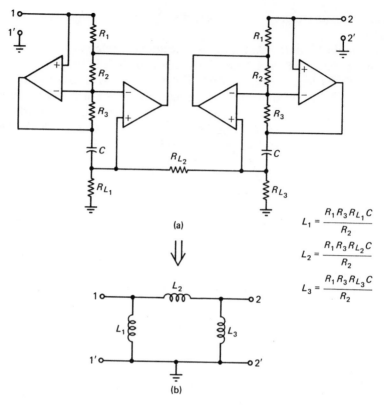

$$L_1 = \frac{R_1 R_3 R_{L_1} C}{R_2}$$

$$L_2 = \frac{R_1 R_3 R_{L_2} C}{R_2}$$

$$L_3 = \frac{R_1 R_3 R_{L_3} C}{R_2}$$

Fig. 5-21 GIC implementation for a π-network of inductors using Gorski–Popiel's embedding technique.

as leapfrog or multiple-loop feedback active filters. When the individual sections are biquads, the overall structure is often referred to as a coupled biquad. These implementations typically require fewer op amps then GIC implementations and the low sensitivity of the passive ladder network is maintained. These implementations are considered in more detail in Section 5.6.

Before concluding this section, we remark further on the sensitivity of passive-ladder filters. It should be pointed out that doubly terminated lossless filter does not necessarily possess the best sensitivity performance over the entire bandwidth. They do have lowest sensitivity at the frequencies of maximum power transfer and in the close neighborhood of these frequencies, but not necessarily at other frequencies. Figure 5-23 shows a Monte Carlo analysis of a sixth-order Butterworth band-pass filter with a normalized center frequency at 1 Hz and 3-dB bandwidth of 0.04 Hz (i.e., an overall $Q = 25$). First we observe that the doubly terminated lossless active ladder is less sensitive at all frequencies than is the singly terminated active-ladder implementation. However, at the band edges superior sensitivity performance is achieved with the (sensitivity optimized) doubly terminated active lossy ladder implementation. In fact, if

High-Order Filter Realization

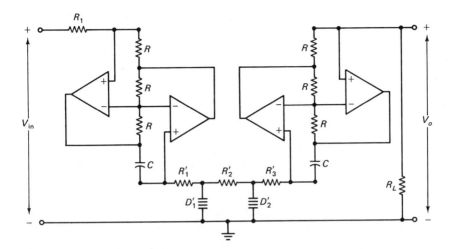

Fig. 5-22 Application of Gorski–Popiel's embedding technique to the implementation of the low-pass LC ladder filter in Fig. 5-14b.

Fig. 5-23 Standard deviations of the transfer function versus frequency for various realizations of a sixth-order Butterworth band-pass filter.

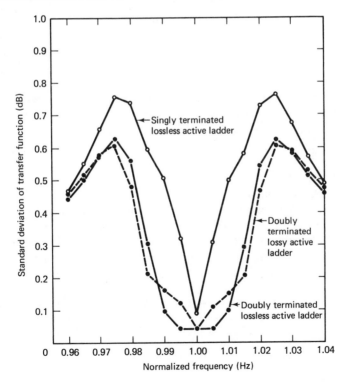

the sensitivity is averaged over the band $0.96 \leq \omega < 1.04$, the lossy ladder achieves a lower average sensitivity [P14]. So for those applications in which only the in-band response variations are of concern, the lossless ladder is clearly superior.

5.5 MULTIPLE-LOOP FEEDBACK (AND FEEDFORWARD) METHODS

Multiple-loop feedback (MF) topologies arose from a search for realizations of transfer functions of order $n \geq 4$ which retain the convenient and practical modular implementation of the cascade approach but provide superior sensitivity performance. When the desired transfer function is a high-order symmetrical band-pass (BP) or band-reject (BR) function, MF realizations result in low-sensitivity, op amp-efficient designs. By symmetrical BP and BR functions, we mean BP and BR functions derived from a low-pass (LP) prototype function via LP-to-BP and LP-to-BR transformations, respectively [i.e., Eqs. (1-61) and (1-64)].

To realize an MF filter according to one of the flowgraphs in Fig. 5-24, cascaded biquads are coupled together through an external resistive feedback (and perhaps feedforward network). Figure 5-24 displays all the feedback topologies and design philosophies which have appeared in the literature: follow-the-leader-feedback (FLF) [P16], generalized follow-the-leader feedback (GFLF), primary-resonator-block (PRB), shifted-companion-form (SCF) [P17], modified leapfrog (MLF) [P14], leapfrog (LF) [P10–12, P13], coupled biquad (CB), inverse FLF (IFLF), and the minimum-sensitivity-feedback (MSF) topologies. The biquadratic T_i's are of the form of Eq. (5-25). Although only the configurations for sixth-order filter design are shown, all but MSF can be generalized to arbitrary order by simply adding the appropriate number of biquad sections and repeating the feedback patterns shown in Fig. 5-24. Observe that for fourth-order design, with two biquad sections, all topologies but FLF[4] reduce to the same graph.

The improved sensitivity performance of MF circuits over that of cascade design (not surprising to those familiar with feedback systems) is demonstrated by the Monte Carlo simulations shown [P17] in Fig. 5-25 for several realizations of a sixth-order Butterworth band-pass response. These curves represent the standard deviation (in decibels) of the gain variation $|\Delta|H||$ as a function of frequency. The active devices are considered ideal, and all passive elements (R's and C's) are assumed to be random variables, uniformly distributed about their nominal value with a tolerance of $\pm 1.732\%$ (equivalent to a standard

---

[4]The first feedback loop, F_{11}, has historically been part of the FLF topology. This loop has little effect on filter sensitivity; as a result, it has been omitted in other topologies. Nevertheless, this loop does provide added design and tuning flexibility. When appropriate this loop can be used as a means to scale the section quality factors. It is noted that a self-loop, F_{ii}, can be placed around any section within the MF topologies in Fig. 5-24.

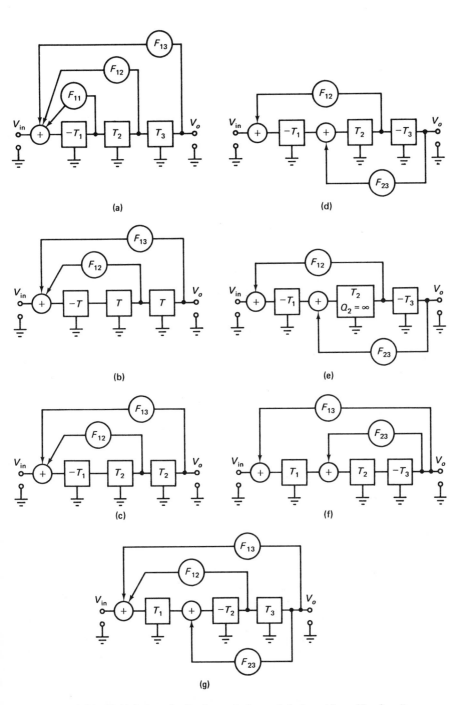

Fig. 5-24 Multiple-loop feedback topologies and design philosophies for all-pole band-pass realization (shown for sixth order). (a) FLF. (b) PRB. (c) SCF. (d) MLF. (e) LF, CB. (f) IFLF. (g) MSF.

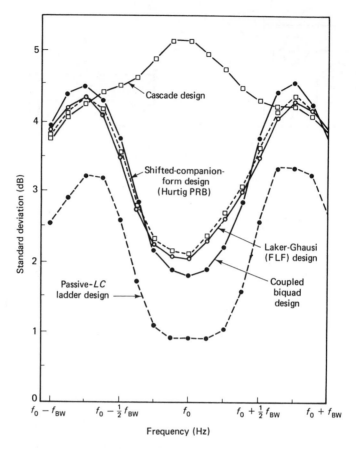

Fig. 5-25 Simulated variations of the sixth-order Butterworth band-pass filter (1.732% passive component tolerances).

deivation $\sigma = 1\%$). Note the comparison of the MF realizations CB, PRB, FLF, and SCF with the cascade design at the upper limit and the resistor terminated passive-LC ladder design at the lower limit. The much reduced passband sensitivities, as compared to cascade designs, are typical of MF band-pass filters. Note that, at the band edge, the CB realization has poor sensitivity. The factor-of-2 (or so) difference in sensitivity between the active MF designs and the passive LC ladder is primarily due to the fact that active RC resonance requires at least four elements, and LC resonance requires only two. Thus, more components contribute to the variance in active filters. For practical active filter realizations, this component-count disadvantage exists regardless of the synthesis approach.

In the following, the synthesis methods for MF active filters with mini-mized sensitivity and maximized dynamic range are considered. The solution of the design equations, together with the optimization of sensitivity and dynamic range, requires, in general, a digital computer program. However, for

High-Order Filter Realization

simple cases, say $n \leq 6$, a programmable calculator is adequate. Simple synthesis methods for realizing suboptimal filters and a number of design tables will be presented. The performance of the different MF design techniques will be compared and their salient features will be pointed out.

5.5.1 General Remarks About Multiple-Loop Feedback (MF) Filter Synthesis

In succeeding pages of this chapter, the design of specific MF filters will be discussed in detail. To make this treatment efficient and to avoid unnecessary repetition, this subsection will be used to introduce and define the concepts that will be commonly applied in all cases.

Figure 5-24 shows the various MF topologies discussed in the literature. Note that among the six configurations in Fig. 5-24a–f, there are really only three topologies; that is, PRB and SCF are special cases of FLF, LF and LF(CB) are special cases of MLF. In addition, all the networks in Fig. 5-24 are seen to be special cases of MSF. Since MSF filters are always derived from a full graph, containing all possible feedback loops (including the self-loops F_{ii} if desired), the generality of MSF shown for $n = 6$ in Fig. 5-24 extends to arbitrary n. The total of seven networks reflects differences only in the design philosophy associated with some topologies. Accordingly, only the FLF, LF, MLF, IFLF, and MSF topologies will be discussed in detail, and the special procedure leading to the remaining configurations will be examined briefly.

Most of the reported work in MF filters has been devoted to realizing narrow-band, symmetrical band-pass (BP) responses which are derivable from a known low-pass (LP) prototype via an LP-to-BP transformation. This interest in narrow-band BP responses is primarily due to the need to reduce the intrinsically large sensitivities of cascade realizations which increase with the filter quality factor $\hat{Q}$ (see Section 3.6). For symmetrical BP responses, the mathematics associated with the MF filter synthesis simplify significantly, but in most cases computer or calculator aids are still required. It is emphasized, however, that MF filters are not restricted to BP requirements. For example, MF techniques have been shown to be effective in reducing the sensitivities of band-reject (BR) filters [P22].

If the desired BP transfer function $H(s)$ of the form of Eq. (5-3) is symmetrical (n even), then by means of the LP-to-BP transformation, the filter can be designed directly from the LP prototype transfer function. The LP prototype transfer function takes the form

$$H_{\mathrm{LP}}(p) = \frac{N_{\mathrm{LP}}(p)}{D_{\mathrm{LP}}(p)} = \frac{c_M p^M + c_{M-1} p^{M-1} + \ldots + c_1 p + c_0}{p^N + d_{N-1} p^{N-1} + \ldots + d_1 p + d_0} \qquad (5\text{-}79)$$

where $M \leq N = n/2$ and p is the LP frequency variable normalized such that the 3-dB bandwidth of H_{LP} equals a given value γ. Note that in general the tabulated LP prototypes have nonunity normalized 3-dB bandwidths, except for the Butterworth case, where $\gamma = 1$. If $H(s)$ is an all-pole BP function, as is the case for our later examples and comparisons, the LP prototype becomes

$$H_{LP}(p) = \frac{N_{LP}(p)}{D_{LP}(p)} = \frac{H_{MN}d_0}{p^N + d_{N-1}p^{N-1} + \ldots + d_1p + d_0} \quad (5\text{-}80)$$

Note that the normalized BP transfer function, Eq. (5-3), is obtained from Eq. (5-80) via the LP-to-BP transformation (see Section 1.6) $p = \gamma\hat{Q}(s^2 + 1)/s$, where $\hat{Q} = \omega_0/\Delta\omega_{3dB}$ denotes the overall BP filter quality factor, s is normalized such that the BP center frequency (ω_0) equals unity, $\Delta\omega_{3dB}$ is the 3-dB bandwidth, and γ is the normalized bandwidth of the LP prototype.

When the MF networks in Fig. 5-24 are used to realize all-pole BP functions, the biquad T_i's are also BP functions, as they would be for a cascade design, i.e.

$$T_i(s) = \frac{N_i(s)}{D_i(s)} = \frac{H_{Bi}s/Q_i}{s^2 + s/Q_i + 1} = H_{Bi}\hat{T}_i(s) \quad (5\text{-}81)$$

In contrast to cascade designs, each T_i is synchronously tuned to the same center frequency ω_0 $(\omega_0 = 1)$. One can show that stagger-tuning the sections, even slightly, will degrade at least the passband sensitivities [P30].

To realize the finite transmission zeros implied in Eqs. (5-3) and (5-79), there are several methods, three of which are shown in Fig. 5-26.[5] In principle, any one of the zero-forming methods in Fig. 5-26 can be used with any one of the MF networks in Fig. 5-24 to realize a general symmetrical BP response. With the first method the finite transmission zeros are obtained as in cascade fashion, with general biquad T_i's of the form of Eq. (5-25). Forming transmission zeros in this manner, however, does severely complicate the synthesis equations, necessitating a very careful design without which the sensitivity performance can be quite poor. Therefore, to get good sensitivity performance, a complex synthesis procedure involving sensitivity minimization is required. This method is, however, particularly useful when the desired response is nonsymmetric (i.e., it is not derivable from a LP prototype). The two remaining methods use feedforward to realize the desired transmission zeros. The technique in Fig. 5-26b is to tap, scale, and sum the input voltage and each of the section output voltages. In Fig. 5-26c the technique is to feed scaled replicas of the input signal forward to each of the section inputs. When the desired BP response is symmetrical, all-pole synchronously tuned BP biquad sections are typically used. It is noted that when $m < n$ in Eq. (5-3), use of the method of Fig. 5-26c can result in the elimination of the output summing amplifier. One can combine method (a) with (b) or (c) to obtain a more general synthesis technique with several new design freedoms (see also Section 5.6). These freedoms can perhaps be exploited to improve the stop-band sensitivity performance. At the time of this writing, no attempt has been made to optimize the sensitivity behavior of such a general multiple-loop feedback–feedforward structure.

The feedforward techniques have been found to work well for moderate-

---

[5]For another approach where the transmission zeros are derived from the general biquads, see Section 5.6.3. This approach is general and the desired BP response need not be symmetrical.

High-Order Filter Realization

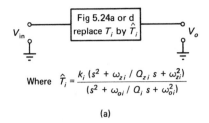

Where $\hat{T}_i = \dfrac{k_i \, (s^2 + \omega_{zi} \, / \, Q_{zi} \, s + \omega_{zi}^2)}{(s^2 + \omega_{oi} \, / \, Q_i \, s + \omega_{oi}^2)}$

(a)

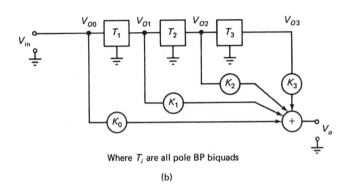

Where T_i are all pole BP biquads

(b)

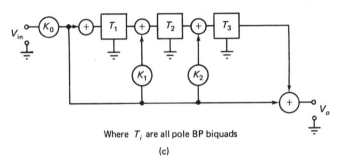

Where T_i are all pole BP biquads

(c)

Fig. 5-26 Methods for realizing finite transmission zeros in multiple-feedback active filters [To preserve generality, the feedback paths are not shown in parts (b) and (c).] (a) GFLF, GLF—finite transmission zero biquad sections. (b) FLF, SCF, CB—feedforward summation. (c) CB, SCF—feedforward.

order filters where $n \leq 8$. However, when the techniques are applied to filters of higher order, the reliance on subtraction to achieve transmission nulls tends to limit the filter stop-band performance [P22], and also the synthesis algorithm can result in impractical element values. Nevertheless, the feedforward techniques are quite adequate for a wide variety of requirements, particularly where stop-band sensitivities are not critical. An example, demonstrating experimentally the performance of this design, is given later.

The summations in Fig. 5-26 are implemented by means of resistive or, if need be, op amp summers as in Fig. 5-27. Depending on the particular topology of the sections (T_i), the interstage summations in Fig. 5-26c can fre-

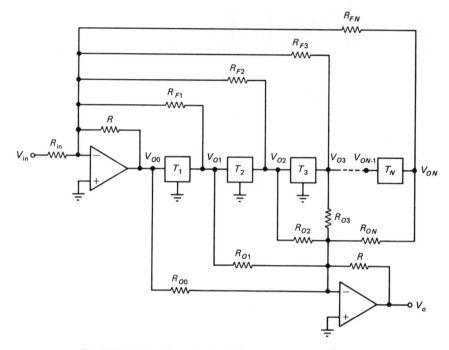

Fig. 5-27   FLF configuration for $2N$th-order band-pass realization.

quently be incorporated into the input stages of the biquad sections. In any event, the weighting factors K_i, as well as the feedback factors F_{ij} in Fig. 5-24, are always expressed as resistor ratios.

5.5.2 Sensitivity and Dynamic-Range Considerations

The statistical multiparameter sensitivity measures, to be used for the comparison and optimization, were discussed in Chapter 3. These measures, which have been proven to agree quite favorably with the Monte Carlo analysis but require much less computation time, have been shown to be valuable analytical tools for both the evaluation and design of MF filters. One has the option of considering either magnitude response (gain, α) sensitivity, or phase (β) sensitivity, or transfer function (T) sensitivity. The measures can readily be expressed for any MF network comprised of BP sections in the following general form (see Section 3.4):

$$M_\alpha = E \int_{\omega_1}^{\omega_2} \left| \frac{\Delta \alpha}{\alpha} \right|^2 d\omega \simeq \int_{\omega_1}^{\omega_2} \sum_{i=1}^{N} S_{\alpha i} \, d\omega = \int_{\omega_1}^{\omega_2} S_\alpha \, d\omega \qquad (5\text{-}82a)$$

$$M_\beta = E \int_{\omega_1}^{\omega_2} |\Delta \beta|^2 \, d\omega \simeq \int_{\omega_1}^{\omega_2} \sum_{i=1}^{N} S_{\beta i} \, d\omega = \int_{\omega_1}^{\omega_2} S_\beta \, d\omega \qquad (5\text{-}82b)$$

$$M_T = E \int_{\omega_1}^{\omega_2} \left| \frac{\Delta H}{H} \right|^2 d\omega \simeq \int_{\omega_1}^{\omega_2} \sum_{i=1}^{N} S_{T i} \, d\omega = \int_{\omega_1}^{\omega_2} S_T \, d\omega \qquad (5\text{-}82c)$$

High-Order Filter Realization

where $M_T = M_\alpha + M_\beta$ and $N(j\omega)/D(j\omega) = \alpha(\omega) \exp [j\beta(\omega)]$ from Eq. (5-3). In Eq. (5-82), E denotes the expected value, ω_1 and ω_2 define a designer-specified frequency band, $\Delta H = H' - H$, $H' = H(j\omega, \mathbf{x} + \Delta\mathbf{x})$, $\Delta\alpha = \alpha' - \alpha$, $\Delta\beta = \beta' - \beta$, $\mathbf{x}$ is a vector containing the nominal values of the network elements, and the vector $\Delta\mathbf{x}$ denotes the multiparameter random element variations. The quantities S_{Ti}, $S_{\alpha i}$, and $S_{\beta i}$ reflect the contribution of the ith section to the overall sensitivity measures. Recall that the sensitivity measures M_T, M_α, M_β were derived for a three-section FLF topology in Section 3.9. Let us briefly review the physical meaning attached to these measures. The integrands $S_{\alpha,\beta,T} \simeq E | \cdot |^2$ approximate the statistical second moments, as a function of frequency, for the complex random functions $|\Delta\alpha/\alpha|$, $|\Delta\beta|$, and $|\Delta T/T|$, respectively. When the component nominal and mean values are identical, it can be shown that $S_{\alpha,\beta,T}$ reduces to the variances (i.e., standard deviation squared) of these random functions. Plotting $S_{\alpha,\beta,T}$ versus frequency produces curves similar to those shown in Fig. 5-25. The functional or integral sensitivity measures $M_{\alpha,\beta,T}$ then denote "average sensitivity" values for $S_{\alpha,\beta,T}$ over the specified frequency band $\omega_1 \le \omega \le \omega_2$. The functional sensitivity measures are particularly applicable to minimizing sensitivity with nonlinear programming techniques. Various sensitivity measures are considered in detail in Chapter 3.

Much of the practical value attached to these sensitivity measures is due to their ability to serve as figures of merit for quickly screening or comparing active filter designs or objective functions for sensitivity optimization. Consequently, one is encouraged to apply judicious approximations to simplify Eqs. (5-82). Usually, the following five simplifying assumptions are valid in narrow-band ($\hat{Q} \ge 10$) BP design: (1) only sensitivities within the passband and transition bands are of interest; (2) statistical correlation between circuit elements, imposed either by the manufacturing process or by tuning exists only within each section; (3) the sensitivities of the section denominator dominate over those of the section numerator; (4) variations in the section center frequencies are the dominant effects that determine the section denominator sensitivities ($S_x^{D_i}$); and (5) in active-RC designs, variations in the passive components dominate, over the active element variations for $\omega_0 \ll$ GB. These assumptions are usually valid, at voice frequencies, for many of the low-sensitivity active-RC biquads in use today (see Chapter 4). Applying assumptions (1) through (3) for the computation of $S_{\alpha i}$, $S_{\beta i}$ and S_{Ti} to Eq. (5-82) results in the following approximate expressions:

$$S_{Ti} = \left[\frac{\mathcal{L}_i}{D}S_x^{D_i}\right]^{t*} (\mathbf{P})_i \left[\frac{\mathcal{L}_i}{D}S_x^{D_i}\right] \tag{5-83a}$$

$$S_{\alpha i} = \left[\mathrm{Re} \left(\frac{\mathcal{L}_i}{D}S_x^{D_i}\right) \right]^t (\mathbf{P})_i \left[\mathrm{Re} \left(\frac{\mathcal{L}_i}{D}S_x^{D_i}\right) \right] \tag{5-83b}$$

$$S_{\beta i} = \left[\mathrm{Im} \left(\frac{\mathcal{L}_i}{D}S_x^{D_i}\right) \right]^t (\mathbf{P})_i \left[\mathrm{Im} \left(\frac{\mathcal{L}_i}{D}S_x^{D_i}\right) \right] \tag{5-83c}$$

where superscripts t and $*$ denote, respectively, the matrix transpose and complex-conjugate operations. The quantities

$$\mathcal{L}_i = D_i \frac{\partial D}{\partial D_i} \tag{5-84}$$

are complex, frequency-dependent scale factors which depend on the feedback topology, and are readily determined from the denominator $D(s)$ of the transfer function $H(s)$. The matrix $(\mathbf{P})_i$ denotes the covariance matrix which contains the component statistics of section i.

Further, applying assumptions (4) and (5) to Eq. (5-83) yields the following expression for S_{Ti}, $S_{\alpha i}$, and $S_{\beta i}$ in terms of the passive-element statistics (correlation coefficients ρ_{rs} and standard deviations σ_{x_r}):

$$
\begin{aligned}
S_{Ti} &\simeq 4\left|\frac{\mathcal{L}_i}{DD_i}\right|^2\left[\sum_{r=1}^{l_i}(S_{x_r}^{\omega_{0i}})^2\sigma_{x_r}^2 + \sum_{r=1}^{l_i-1}\sum_{s=r+1}^{l_i}(S_{x_r}^{\omega_{0i}})(S_{x_s}^{\omega_{0i}})\rho_{rs}\sigma_{x_r}\sigma_{x_s}\right] \\
&= \left|\frac{\mathcal{L}_i}{DD_i}\right|^2\kappa_i
\end{aligned}
\tag{5-85a}
$$

$$
\begin{aligned}
S_{\alpha i} &\simeq 4\left(\operatorname{Re}\frac{\mathcal{L}_i}{DD_i}\right)^2\left[\sum_{r=1}^{l_i}(S_{x_r}^{\omega_{0i}})^2\sigma_{x_r}^2 + \sum_{r=1}^{l_i-1}\sum_{s=r+1}^{l_i}(S_{x_s}^{\omega_{0i}})(S_{x_r}^{\omega_{0i}})\rho_{rs}\sigma_{x_r}\sigma_{x_s}\right] \\
&= \left(\operatorname{Re}\frac{\mathcal{L}_i}{DD_i}\right)^2\kappa_i
\end{aligned}
\tag{5-85b}
$$

$$
\begin{aligned}
S_{\beta i} &\simeq 4\left(\operatorname{Im}\frac{\mathcal{L}_i}{DD_i}\right)^2\left[\sum_{r=1}^{l_i}(S_{x_r}^{\omega_{0i}})^2\sigma_{x_r}^2 + \sum_{r=1}^{l_i-1}\sum_{s=r+1}^{l_i}(S_{x_r}^{\omega_{0i}})(S_{x_s}^{\omega_{0i}})\rho_{rs}\sigma_{x_r}\sigma_{x_s}\right] \\
&= \left(\operatorname{Im}\frac{\mathcal{L}_i}{DD_i}\right)^2\kappa_i
\end{aligned}
\tag{5-85c}
$$

The total number of circuit elements l_i, determining the ith-section center frequency ω_{0i} $(\omega_{0i} = 1)$ in Eqs. (5-85) is typically 4 or 6 for most useful active RC biquads]. The reader should not be discouraged by the apparent complexity of Eqs. (5-85), since in most practical cases $|S_{x_r}^{\omega_{0i}}|$ are independent of x_r (as per Chapter 4, typically $|S_{x_r}^{\omega_{0i}}| = \frac{1}{2}$). If we further assume that all passive elements possess the same standard deviation σ (i.e., $\sigma_{x_i} = \sigma$ for all R's and C's), then Eqs. (5-85) reduce to the following simple forms:

$$
S_{Ti} \simeq \kappa\left|\frac{\mathcal{L}_i}{DD_i}\right|^2
\tag{5-86a}
$$

$$
S_{\alpha i} \simeq \kappa\left[\operatorname{Re}\left(\frac{\mathcal{L}_i}{DD_i}\right)\right]^2
\tag{5-86b}
$$

$$
S_{\beta i} \simeq \kappa\left[\operatorname{Im}\left(\frac{\mathcal{L}_i}{DD_i}\right)\right]^2
\tag{5-86c}
$$

where

$$
\kappa = 4\sigma^2\left[\sum_{r=1}^{l_i}(S_{x_r}^{\omega_0})^2 + \sum_{r=1}^{l_i-1}\sum_{s=r+1}^{l_i}\rho_{rs}(S_{x_r}^{\omega_0})(S_{x_s}^{\omega_0})\right]
\tag{5-86d}
$$

with the $\omega_{0i} = \omega_0$ normalized to unity. Note that κ depends on the biquad topology and the correlation coefficients ρ_{rs}. For example, when the elements are uncorrelated ($\rho_{rs} = 0$ for all r and s), $\kappa = 4\sigma^2$ for Deliyannis-Friend and Tarmy–Ghausi biquads, and $\kappa = 6\sigma^2$ for the state-variable-type biquads.

The measures Eqs. (5-82) may also be used to evaluate response variations due to temperature drifts, where element changes are highly correlated not only within, but also between all sections T_i in the MF filter.

In order to obtain reliable results from the expression above, consistent with those obtained by Monte Carlo methods, attention must be paid to the choice of functions on which the sensitivity measures are based—$\Delta T/T'$ and $\Delta\alpha/\alpha'$—and, as discussed in Section 3.7, on the convergence properties of the Taylor series of the integrands in Eqs. (5-82).

As suggested earlier, the measures can be used in either integral $(M_{\alpha,\beta,T})$ or in integrand $S_{\alpha,\beta,T} = \sum_{i=1}^{N} S_{\alpha i,\beta i,Ti}$ form. In integrand form, one may plot sensitivity as a function of frequency, to obtain a detailed characterization of a filter's variability. When used in integral form, $M_{\alpha,\beta,T}$ provide convenient functionals for sensitivity minimization and design screening. It is interesting to note that minimizing M_α, M_β, and M_T individually usually results in the same optimum design.

As shown later, MF design equations are obtained by equating the transfer function V_o/V_{in} of one of the circuits in Fig. 5-24, where T_i is given in Eq. (5-81), to a prescribed BP characteristic and comparing the coefficients. The process yields equations for the unknown section quality factors Q_i, the gain constant H_{Bi}, and the feedback factors F_{ij}. Not all of these parameters are determined by the design equations, and thus the free parameters can be used in optimization.

It is customary in MF design to use $F_{ij} = F_{ij}(Q_i, H_{Bi})$ for satisfying the synthesis equations so that Q_i and H_{Bi} remain, within practical limits, free variables. Typically, sensitivity is minimized by adjusting in an iterative manner the section Q_i values subject to constraints, such as topological realizability conditions and practical fabrication requirements. For example, $F_{ij}(Q_i, K_i)$ must clearly be real and finite, and $Q_i > 0$; also, one may specify that $Q_i \leq Q_M$, where Q_M is some maximum realizable Q value, obtained, for example, by considerations of element value spread. Mathematically, sensitivity is optimized by minimizing any one of the sensitivity measures in Eqs. (5-82) as a function of the N variables Q_i, $i = 1, \ldots, N$; the constraints are handled in terms of an appropriately defined penalty function [P16]. The degree to which this minimization can improve sensitivity performance over a good educated "guess" is very much topology-dependent. However, it should be noted that a poor MF design can be more sensitive than the cascade design we are seeking to improve [P19]. Thus, it is usually good practice to perform a complete sensitivity analysis or Monte Carlo simulation before a design is finalized.

Once the Q_i values have been determined, the remaining free parameters [i.e., the gain constant K_i in Eq. (5-81)] are adjusted to maximize signal swing. Just as in cascade designs, the objective of this step is to maximize the undistorted peak-to-peak signal handling capability of the filter, which of course determines the upper limit of the dynamic range. This procedure is particularly important for operation at higher frequencies (e.g., for $f_0 > 10$ kHz), where signal-level limits are determined by op amp slewing. To ensure the full signal-handling capability of the filter, it is necessary to control the voltage maxima at the output of all op amps such that for increasing signal levels, no op amp behaves nonlinearly sooner than any other one. Clearly, this implies that the gains H_{Bi} should be chosen such that the voltage maxima at all critical nodes within the filter are equal [i.e., Eq. (5-50)]. Typically, these critical nodes may be assumed to be the outputs of the sections T_i and of the summers.

The assumption that the output voltage of a section T_i is the critical voltage is certainly true for single-amplifier biquads (SABs), where the section output corresponds to the amplifier output. In multiamplifier biquads it is assumed that the largest signal occurs at the section output. Once the desired section gain levels have been determined, the dynamic range for the individual sections can be optimized independently, as discussed in Chapter 4.

In this case the requirement can be put into the form

$$\max_{0 \leq \omega < \infty} |H_i(j\omega)| = \max_{0 \leq \omega < \infty} |H(j\omega)| \quad \text{for all } i \leq N - 1 \quad (5\text{-}87)$$

where $H_i(j\omega)$, $i = 1, \dots, N$, are the voltage transfer functions (V_{oi}/V_{in}) from the input to the output of the ith section and where $H(j\omega)$ is the specified transfer function for the total filter.

We note that computational effort will be reduced if Eq. (5-87) is evaluated for the equivalent prototype low-pass functions (i.e., before using the BP-to-LP transformation). This is, of course, possible only for symmetrical functions $H(s)$. With the definitions

$$H(j\omega) = H_N(j\omega) = H_{MN}\hat{H}_N(j\omega) \quad (5\text{-}88\text{a})$$

$$M_N = \max_{0 \leq \omega \leq \infty} |\hat{H}_N(j\omega)| \quad (5\text{-}88\text{b})$$

where $H_{MN} = |H_N(j\omega_0)|$ is the desired midband gain of the total filter, Equation (5-87) can be rewritten as

$$\max_{0 \leq \omega \leq \infty} |H_i(j\omega)| = H_{MN}M_N \quad \text{for } i \leq N - 1 \quad (5\text{-}89)$$

For all-pole BP filters, the functions $H_i(s)$ for any of the MF topologies depend on the section gains H_{Bi} defined in Eq. (5-81), so that via Eq. (5-87) those values of H_{Bi} can be determined which result in a uniform signal level throughout the MF topology.

Satisfaction of Eqs. (5-87) or (5-89) maximizes signal swing (i.e., the upper limit imposed on dynamic range). The lower limit, the noise floor, can always be improved by reducing the noise sources within the MF filter, notably by decreasing the impedance level and the noise contributions of the amplifiers. The degree to which MF design parameters, such as Q_i, can be used to reduce noise (with a corresponding trade-off in sensitivity performance) has not yet been investigated.

Observe that this procedure is independent of and does not affect the already completed sensitivity minimization. The specific equations and design tables for a number of MF all-pole BP filters are given in the subsequent sections.

Our comments thus far regarding dynamic range have been directed primarily toward all-pole BP MF filter design. When finite transmission zeros are realized via feedforward summation (Fig. 5-26b and c), nothing in the design process changes in principle, except that the output of the feedforward summer must also be included in satisfying Eqs. (5-87) or (5-89). When transmission zeros are realized in cascade fashion (Fig. 5-26), using general biquad sections of the form of Eq. (5-25), the process of sensitivity and dynamic range optimization becomes considerably more complex. For this class of MF filters there is the additional problem of pole–zero pairing, which affects to some extent both sensitivity and dynamic range. A further complication is that the feedback factors and pole–zero pairing are interdependent; thus, a change in pole–zero pairing requires a complete redesign of the filter. Pole–zero pairing in cascade active filters has been discussed in the literature (see Section 5.3), but little is known about its effects in MF filters.

The concepts introduced here will be applied to specific MF topologies in the subsequent sections. Synthesis equations will be derived which allow the designer to determine, for a given transfer function $H(s)$ or $H_{LP}(p)$, the feedback and feedforward factors F_{ij} and K_i in Figs. 5-24 and 5-26 and the design parameters (usually just Q_i and H_{Bi}) for the sections $T_i(s)$. Once the biquad design parameters are determined, the reader can obtain the appropriate realizations as discussed in Chapter 4. It is noted again that any cascadable biquad can be used to realize $T_i(s)$.

For clarity of presentation, a unified approach is used in the sequel for all topologies. The approach may not be the most efficient or the most elegant one for every case; however, its execution will result in a filter that performs well. When appropriate, alternative synthesis methods are also recommended to the reader. Some illustrative examples are provided in sufficient detail for the reader to verify synthesis equations, and if desired, to build or simulate the resulting filters.

5.6 FOLLOW-THE-LEADER-FEEDBACK (FLF)-TYPE FILTERS

The follow-the-leader-feedback (FLF)-type filters derive their name from the configuration shown in Fig. 5-24a to c and their similarity to the popular game that children play. The same is true with the leapfrog (LF)-type filters shown in Fig. 5-24d and e. These names are given in the literature for convenience to distinguish the various feedback topologies among the multiple-loop feedback configurations. In the following subsections the design methods with examples for the various FLF-type filters will be given.

5.6.1 FLF Synthesis

The FLF-type circuits are generally represented by the circuit schematic shown in Fig. 5-27. To remove some interaction and simplify tuning, the feedback and feedforward coefficients F_{1j} and K_i are implemented by single-input op amp summers. Straightforward analysis of Fig. 5-27 yields

$$H_N(s) = \frac{V_{oN}}{V_{in}} = \frac{-\alpha \prod_{j=1}^{N} T_j(s)}{1 + \sum_{k=1}^{N} F_{1k} \prod_{j=1}^{k} T_j(s)} \tag{5-90}$$

$$H_i(s) = \frac{V_{oi}}{V_{in}} = H_N(s) \left[\prod_{j=i+1}^{N} T_j^{-1}(s) \right] \quad \text{for } i = 0, 1, \ldots, N-1 \tag{5-91}$$

$$H(s) = \frac{V_o}{V_{in}} = -\sum_{j=0}^{N} K_j H_j(s) = \alpha \frac{K_0 + \sum_{j=1}^{N} K_j \prod_{k=1}^{j} T_k(s)}{1 + \sum_{j=1}^{N} F_{1j} \prod_{k=1}^{j} T_k(s)} \tag{5-92a}$$

where

$$\alpha = \frac{R}{R_{in}}, \qquad F_{1j} = \frac{R}{R_{fj}}, \qquad K_j = \frac{R}{R_{oj}} \tag{5-92b}$$

In some situations, where T_i with both positive and negative gains are not available, the single-input summers can be replaced with differential-input summers. For all-pole BP designs, the output summing amplifier is not needed and $H(s) = -H_N(s)$ in Eq. (5-90).

In particular for $N = 3$, the expression in (5-92a) is given by

$$H(s) = \frac{V_o}{V_i} = \frac{K_o + K_1 T_1 + K_2 T_1 T_2 + K_3 T_1 T_2 T_3}{1 + F_{11} T_1 + F_{12} T_1 T_2 + F_{13} T_1 T_2 T_3}$$

where, for $R = R_{in}$, $\alpha = 1$.

The FLF synthesis equations are obtained by inserting T_i from Eq. (5-81) into Eq. (5-92), performing a BP-to-LP transformation and comparing coefficients between Eqs. (5-92a) and (5-79). (For asymmetrical BP functions, where BP-to-LP transformation cannot be performed, one compares coefficients between Eqs. (5-92) and (5-24). In principle, the procedure does not change. Performing the BP-to-LP transformation whenever possible reduces the degree of the functions and thereby simplifies the algebra significantly.)

This procedure can be readily written in matrix form as follows:

$$\mathbf{Lt} = \hat{\mathbf{d}} \tag{5-93a}$$

$$\mathbf{Lk} = \hat{\mathbf{c}} \tag{5-93b}$$

$$k_0 = c_N \tag{5-93c}$$

where $\mathbf{L}$ is an $N \times N$ lower triangular matrix whose nonzero elements contain the unknown parameters $q_i = \gamma \hat{Q} / Q_i$:

$$L_{ik} = \frac{1}{(i-k)!} \left[\sum_{j_1=k+1}^{N} \sum_{j_2=k+1}^{N} \cdots \sum_{j_{i-k}=k+1}^{N} \left(\prod_{r=1}^{i-k} q_r \right) \right] \tag{5-94}$$

High-Order Filter Realization

for $i > k$ and $j_1 \neq j_2 \neq \ldots \neq j_{i-k}$. We note that the parameter q_i is introduced to simplify the notation in the synthesis equations. The vector $\mathbf{t}$ is a $N \times 1$ vector of "pseudo loop gains," with elements

$$t_{1j} = F_{1j} \prod_{k=1}^{j} H_{Bk} q_k, \qquad j = 1, \ldots, N \tag{5-95a}$$

and the $N \times 1$ vector $\mathbf{k}$ contains the terms

$$k_j = K_j \prod_{k=1}^{j} H_{Bk} q_k, \qquad j = 1, \ldots, N \tag{5-95b}$$

and $k_o = K_o$. The $N \times 1$ vectors $\hat{\mathbf{d}}$ and $\hat{\mathbf{c}}$ depend, respectively, on the denominator and numerator coefficients of Eq. (5-79) and on q_i. The vectors $\hat{\mathbf{d}}$ and $\hat{\mathbf{c}}$ are of the following form:

$$\hat{\mathbf{d}}^t = [\hat{d}_{N-1} \quad \hat{d}_{N-2} \quad \ldots \quad \hat{d}_1 \quad \hat{d}_0] \quad \text{and} \quad \hat{\mathbf{c}}^t = [\hat{c}_{N-1} \quad \hat{c}_{N-2} \quad \ldots \quad \hat{c}_1 \quad \hat{c}_0]$$

where

$$\hat{d}_{N-i} = d_{N-i} - \frac{1}{i!}\left[\sum_{j_1=1}^{N} \sum_{j_2=1}^{N} \ldots \sum_{j_i=1}^{N} \left(\prod_{r=1}^{i} q_r \right) \right] \tag{5-96a}$$

$$\hat{c}_{N-i} = c_{N-i} - \frac{c_N}{i!}\left[\sum_{j_1=1}^{N} \sum_{j_2=1}^{N} \ldots \sum_{j_i=1}^{N} \left(\prod_{r=1}^{i} q_r \right) \right] \tag{5-96b}$$

To illustrate the structure of Eqs. (5-93), they are written below for $N = 3$ [i.e., for a sixth-order BP $H(s)$]:

$$\begin{bmatrix} 1 & 0 & 0 \\ q_2 + q_3 & 1 & 0 \\ q_2 q_3 & q_3 & 1 \end{bmatrix} \begin{bmatrix} t_{11} \\ t_{12} \\ t_{13} \end{bmatrix} = \begin{bmatrix} d_2 - (q_1 + q_2 + q_3) \\ d_1 - (q_1 q_2 + q_1 q_3 + q_2 q_3) \\ d_0 - q_1 q_2 q_3 \end{bmatrix} \tag{5-97a}$$

$$\begin{bmatrix} 1 & 0 & 0 \\ q_2 + q_3 & 1 & 0 \\ q_2 q_3 & q_3 & 1 \end{bmatrix} \begin{bmatrix} k_1 \\ k_2 \\ k_3 \end{bmatrix} = \begin{bmatrix} c_2 - (q_1 + q_2 + q_3)c_3 \\ c_1 - (q_1 q_2 + q_1 q_3 + q_2 q_3)c_3 \\ c_0 - q_1 q_2 q_3 c_3 \end{bmatrix} \tag{5-97b}$$

and

$$k_o = K_o = c_3$$

The lower triangularity of matrix $\mathbf{L}$ in Eqs. (5-93) assures that for any LP prototype and any selection of Q_i values, there exist finite t_{1j}'s and k_j's. This property is particularly important when iterative methods are used to optimize the filter's sensitivity.

The arbitrariness of the sections' Q_i values is used to select the parameters q_i such that the sensitivity measures in Eqs. (5-82) or (5-85) are minimized. To evaluate the quantities in Eqs. (5-83), (5-85), or (5-86), one must choose a section topology (to calculate $S_x^{p_i}$) and determine the quantities $\mathcal{L}_i$ of Eq. (5-84), which for FLF design can readily be written in the following recursive form:

$$\mathcal{L}_1 = \prod_{j=1}^{N} D_j \tag{5-98a}$$

$$\mathcal{L}_i = \mathcal{L}_{i-1} + t_{1,i-1}\left(\frac{j\omega}{\gamma Q}\right)^{i-1} \prod_{j=1}^{N} D_j \qquad \text{for } 2 \leq i \leq N \tag{5-98b}$$

Once the q_i values are determined, the parameters t_{1j} and k_j can be calculated from Eqs. (5-93). It is clear from Eq. (5-95) that the resistor ratios F_{1j} and K_j in Eq. (5-92b) can still be obtained for an arbitrary set of gains H_{Bk}, which are chosen to satisfy Eq. (5-87) for a maximized dynamic range.

Using Eqs. (5-81) and (5-91) and defining

$$M_i = \max_{0 \leq \omega < \infty} \left| \hat{H}_N(j\omega) \prod_{j=i+1}^{N} \hat{T}_j^{-1}(j\omega) \right| \qquad \text{for } i = 0, \ldots, N-1 \tag{5-99}$$

which can readily be evaluated by computer, Eq. (5-89) results in

$$M_i H_{MN} \prod_{j=i+1}^{N} H_{Bj}^{-1} = M_N H_{MN} \tag{5-100}$$

or

$$H_{Bi} = \frac{M_{i-1}}{M_i} \qquad \text{for } i = 1, \ldots, N \tag{5-101a}$$

Approximate but relatively accurate values, which permit the computer evaluation of Eq. (5-99) to be bypassed, are

$$H_{Bi} \simeq (1 + q_i^{-2})^{1/2} \tag{5-101b}$$

H_{B1} is determined from Eq. (5-101) only if the input op amp summer in Fig. 5-27 is used; if a resistive summer is employed instead, H_{B1} is arbitrary. Finally, the parameter α in Eq. (5-92b) sets the overall gain H_{MN} of the function in Eq. (5-90) as

$$-H_{MN} = -\frac{\alpha}{d_N} \prod_{j=1}^{N} H_{Bj} q_j \tag{5-102}$$

When Eq. (5-92a) is used, $\alpha = 1$ since all the coefficients c_i in Eq. (5-79) are already realized via Eq. (5-93b). If this gain overdrives the output op amp, α must be reduced appropriately.

In summary, for a given LP prototype, element statistics (σ_{xi}, ρ_{rs}), and frequency band $\omega_1 \leq \omega \leq \omega_2$ for the sensitivity minimization, the design procedure for minimum-sensitivity FLF BP filters can be stated as follows:

1. Adjust the Q_i values and compute t_{1j} to minimize either M_α or M_β or M_T over the frequency band $\omega_1 \leq \omega \leq \omega_2$. Note that t_{1j}'s must be computed from Eq. (5-93a) with each iteration of the Q_i's to compute sensitivity [i.e., $\mathcal{L}_i$ in Eq. (5-98b)].

2. From the optimum Q_i's, compute the feedforward k_i's according to Eqs. (5-93b) and (5-93c).

3. Adjust gain constants H_{Bi} to equalize the voltage maxima according to Eqs. (5-99) to (5-101).

4. From the Q_i's and H_{Bi}'s, compute the feedback (F_{1j}'s) and feedforward (K_j's) factors according to Eqs. (5-95a) and (5-95b), respectively, and α according to Eq. (5-102).

5. Determine the resistor values R_{in}, R_{Fj}, and R_{oj} (for all j) from the computed α, F_{1j}'s, and K_j's according to $R_{in} = -R/\alpha$, $R_{Fj} = R/F_{1j}$, and $R_{oj} = R/K_j$, where R determines the impedance level and is usually set to some convenient value such as 10 kΩ.

6. Compute the biquad section element values to realize the BP T_i's specified by Q_i, H_{Bi} and ω_0, according to design procedures given in Chapter 4. Note how the biquad sections are designed independently, as they would be in cascade synthesis.

Illustrative examples

●

EXAMPLE 5-2

Consider the FLF sensitivity minimization when the desired response is a fourth-order bandpass function, obtained from a second-order low-pass prototype (i.e., $N = 2$). For $N = 2$, the minimization of M_T can be performed with pencil and paper. The FLF sensitivity measure, when $N = 2$, is written

$$M_T = \int_{\omega_1}^{\omega_2} \frac{1}{|D|^2}\left(|D_2|^2 \kappa_1 + \left|D_1 + \frac{j\omega_n}{\gamma\hat{Q}}t_{11}\right|^2 \kappa_2\right) d\omega \qquad (5\text{-}103)$$

with $D_i = (1 - \omega_n^2) + j\omega_n(1/Q_i)$ and $t_{11} = d_1 - \gamma\hat{Q}(1/Q_1 + 1/Q_2)$. Substitution of D_i and t_{11} into the sensitivity expression yields

$$M_T = \int_{\omega_1}^{\omega_2} \frac{1}{D^2}\left\{\left[(1 - \omega_n^2)^2 + \omega_n^2 \frac{1}{Q_2^2}\right]\kappa_1\right.$$
$$\left. + \left[(1 - \omega_n^2)^2 + \omega_n^2\left(\frac{d_1}{\gamma\hat{Q}} - \frac{1}{Q_2}\right)^2\right]\kappa_2\right\} d\omega \qquad (5\text{-}104)$$

For simplicity, let $\kappa_1 = \kappa_2 = 4\sigma^2$; this is equivalent to realizing the FLF response using active biquads identical in type (i.e., Fig. 4-17 or Fig. 4-34) and with identical element σ's. With this simplification the sensitivity reduces to

$$M_T = \int_{\omega_1}^{\omega_2} \frac{4\sigma^2}{|D|^2}\left\{2(1 - \omega_n^2)^2 + \omega_n^2\left[\frac{d_1^2}{(\gamma\hat{Q})^2} - \frac{2d_1}{\gamma\hat{Q}Q_2} + \frac{2}{Q_2^2}\right]\right\} d\omega \qquad (5\text{-}105)$$

Evaluating the gradient of Eq. (5-105) yields

$$\frac{\partial M_T}{\partial(1/Q_2)} = \int_{\omega_1}^{\omega_2} \frac{4\sigma^2}{|D|^2}\left[\omega_n^2\left(\frac{4}{Q_2} - \frac{2d_1}{\gamma\hat{Q}}\right)\right] d\omega \qquad (5\text{-}106)$$

For minimum sensitivity, the gradient described in Eq. (5-106) is set to zero (with Q_i's constrained such that $Q_i < \infty$ for $i = 1, 2$), yielding

$$\frac{1}{Q_2} = \frac{1}{2}\frac{d_1}{\gamma\hat{Q}} \Longrightarrow Q_2 = \frac{2}{d_1}\gamma\hat{Q} \qquad (5\text{-}107a)$$

Substituting Eq. (5-107a) into the expression for t_{11}, with t_{11} set to zero, we obtain

$$Q_1 = \frac{2}{d_1}\gamma\hat{Q} \qquad (5\text{-}107b)$$

Observe that in the expression for M_T, the measure is independent of Q_1; hence, $\partial M_T/\partial(1/Q_1) = 0$. This derivative is zero, in general, for all N. Substituting Eq. (5-107a) into Eq. (5-105) yields the following minimum M_T:

$$\min(M_T) = \sigma^2\left(\frac{d_1}{\gamma\hat{Q}}\right)^2 \int_{\omega_1}^{\omega_2} \frac{1}{|D|^2}\,d\omega \qquad (5\text{-}108)$$

EXAMPLE 5-3

As another example, let us design a sixth-order ($N = 3$) FLF BP filter ($\omega_0 = 1$, $\hat{Q} = 25$ and $H_{\text{MN}} = 1$), with LP prototype given by

$$H_{\text{LP}}(p) = \frac{-1}{p^3 + 2p^2 + 2p + 1} \qquad (5\text{-}109)$$

(i.e., $c_3 = -1$, $d_2 = 2$, $d_1 = 2$, and $d_0 = 1$); also, $H_{\text{LP}}(p)$ has been normalized such that the bandwidth $\gamma = (\omega_{3\text{dB}})_n = 1$. For a minimum gain sensitivity S_α, it has been found (see Table 5-7) that the optimum values for Q_1, Q_2, and Q_3 are

$$Q_1 = 43.100, \qquad Q_2 = 44.650, \qquad Q_3 = 29.075$$

Hence,

$$q_1 = \frac{\gamma\hat{Q}}{Q_1} = 0.580, \qquad q_2 = \frac{\gamma\hat{Q}}{Q_2} = 0.560, \qquad q_3 = \frac{\gamma\hat{Q}}{Q_3} = 0.860$$

From Eq. (5-97a) we evaluate the t_{1i}'s:

$$t_{11} = d_2 - (q_1 + q_2 + q_3) = 2 - 0.580 - 0.560 - 0.860 = 0$$

$$t_{12} = d_1 - (q_1q_2 + q_1q_3 + q_2q_3) - t_{11}(q_2 + q_3)$$

$$= 2 - (0.325 + 0.499 + 0.482) - 0 = 0.694$$

$$t_{13} = d_0 - q_1q_2q_3 - q_2q_3t_{11} - q_3t_{12}$$

$$= 1 - 0.279 - 0 - 0.597 = 0.124$$

The gain constants H_{Bi} for the BP biquad T_i's are computed according to Eq. (5-101b):

$$H_{\text{B1}} = \sqrt{1 + 2.973} = 1.993, \qquad H_{\text{B2}} = \sqrt{1 + 3.189} = 2.047,$$

$$H_{\text{B3}} = \sqrt{1 + 1.352} = 1.534$$

Given the t_{ij}'s and the H_{Bi}'s, the feedback factors F_{ij} are computed according to Eq. (5-95a):

$$F_{11} = 0, \qquad F_{12} = \frac{t_{12}}{H_{B1}H_{B2}q_1 q_2} = 0.524,$$

$$F_{13} = \frac{t_{13}}{H_{B1}H_{B2}H_{B3}q_1 q_2 q_3} = 0.071$$

Also, from Eq. (5-102),

$$\alpha = \frac{H_{MN}d_N}{H_{B1}H_{B2}H_{B3}q_1 q_2 q_3} = 0.572$$

Thus, if $R = 10 \text{ k}\Omega$, the feedback resistors R_{Fi} and R_{in} are

$$R_{F1} = \infty, \qquad R_{F2} = \frac{10 \text{ k}\Omega}{0.524} = 19.084 \text{ k}\Omega, \qquad R_{F3} = \frac{10 \text{ k}\Omega}{0.071} = 140.85 \text{ k}\Omega,$$

$$R_{in} = \frac{10 \text{ k}\Omega}{0.572} = 17.483 \text{ k}\Omega$$

•

5.6.2 Shifted-Companion-Form (SCF) and Primary-Resonator-Block (PRB) Synthesis

As noted previously, SCF and PRB are special cases of FLF. The SCF differs from FLF design essentially by eliminating the self-loop around block T_1 (i.e., $F_{11} = 0$) and the assignment of identical sections to T_2 through T_N. This implies [see Eqs. (5-93a) and (5-97a)] that $q_i = d_{N-1} - \sum_{i=2}^{N} q_i$ and $q_2 = \dots q_N = q$. From this point the SCF design follows closely the FLF design procedure outlined in Section 5.6.1.

In PRB design all sections T_i are assumed to be identical; thus, $Q_i = Q$ and $H_{Bi} = H_B$, $i = 1, \dots, N$, and $F_{11} = 0$. Using identical sections provides three manufacturing advantages: (1) increases the volume serviced by each biquad, (2) reduces the engineering time required to design the filter, and (3) minimizes the maximum section Q_i. With these simplifications the PRB synthesis equations are readily obtained from the FLF case:

$$Q = \frac{N\gamma\hat{Q}}{d_{N-1}} \qquad \text{(implying that } t_{11} = 0\text{)} \qquad (5\text{-}110)$$

$$t_{1i} = \hat{d}_{N-i} - \sum_{k=1}^{i-1} L_{ik}t_{1k} \qquad (5\text{-}111\text{a})$$

$$k_i = \hat{c}_{N-i} - \sum_{k=1}^{i-1} L_{ik}t_{1k} \qquad (5\text{-}111\text{b})$$

$$k_0 = c_N \qquad (5\text{-}111\text{c})$$

where (recall that $q = \gamma \hat{Q}/Q$)

$$\hat{d}_{N-i} = d_{N-i} - \frac{N!}{(i)! \, (N-i)!}(q)^i \qquad (5\text{-}112a)$$

$$\hat{c}_{N-i} = c_{N-i} - \frac{c_N N!}{(i)! \, (N-i)!} \qquad (5\text{-}112b)$$

$$L_{ik} = \frac{(N-k)!}{(i-k)! \, (N-i)!} \qquad (5\text{-}112c)$$

We note that PRB design, because of the simple and systematic nature of Eqs. (5-110) and (5-111), can be readily performed with even modest-capacity programmable calculators. Because of the broad minimum that typifies FLF designs [P18], sensitivity is only modestly sacrificed in PRB circuits, where $Q_i = Q$ is chosen a priori rather than through sensitivity minimization.

If equality among the sections is strictly adhered to, dynamic range cannot be optimized since the H_{Bi}'s are then identical. However, some suboptimum ought to be found because dynamic range can suffer significantly for an improper selection of H_{Bi}, because of the wide range of voltage maxima in FLF(PRB) designs. A very good choice is the approximate solution indicated in Eq. (5-101b):

$$H_{Bi} = H_B = \sqrt{1 + q^{-2}} = \sqrt{1 + \left(\frac{Q}{\gamma \hat{Q}}\right)^2} \qquad (5\text{-}113)$$

which is valid for symmetrical BP response with $T(s)$ given in Eq. (5-81). In summary, the PRB synthesis procedure can be stated as follows:

1. Compute Q from Eq. (5-110).
2. Compute t_{1j} and k_j according to Eqs. (5-111).
3. Adjust H_{Bi}, according to Eq. (5-113), to achieve approximate equalization of the voltage maxima.
4. From the Q and H_{Bi} values, compute the F_{1j} and the K_j using Eqs. (5-95) [i.e., $F_{1j} = t_{1j}(H_B q)^{-j}$ and $K_j = k_j(H_B q)^{-j}$].
5. Compute R_{fj} and R_{oj}, from the F_{1j} and K_j, according to Eq. (5-92b); α determines the overall gain, as in Eq. (5-102).
6. Compute the biquad element values for each section to realize the T_i's, according to the methods prescribed in Chapter 4.

●

EXAMPLE 5-4

Let us consider the PRB design of the sixth-order Butterworth response considered in the previous FLF example. The specifications for this filter $\omega_0 = 2\pi \times 10^3$ rad/sec, $\hat{Q} = 20$, and $H_{MN} = 1$. Using Eq. (5-110) the three section Qs are readily determined:

$$Q = \frac{N\gamma \hat{Q}}{d_{N-1}} = \frac{3(20)}{2} = 30 \quad \text{and} \quad q = \frac{\gamma \hat{Q}}{Q} = 0.667$$

From Eqs. (5-111a) and (5-112), we obtain the t_{1i}'s:

$$t_{11} = 0$$

$$t_{12} = d_1 - \frac{3!}{2!\,1!}q_2 = 2 - 3(0.667)^2 = 0.665$$

$$t_{13} = d_0 - \frac{3!}{3!\,0!}q_3 - \frac{1!}{1!\,0!}qt_{12} = 1 - (0.667)^3 - 0.667(0.665)$$

$$= 0.260$$

Then according to Eq. (5-113),

$$H_B = H_{B1} = H_{B2} = H_{B3} = \sqrt{1 + q^{-2}} = \sqrt{1 + (1.5)^2} = 1.803$$

Using Eq. (5-95a), the feedback factors are calculated to be

$$F_{11} = 0, \qquad F_{12} = \frac{t_{12}}{(H_B q)^2} = 0.460, \qquad F_{13} = \frac{t_{13}}{(H_B q)^3} = 0.149$$

Also from Eq. (5-102),

$$\alpha = \frac{H_{MN}d_N}{(H_B q)^3} = 0.575$$

Choosing $R = 10$ kΩ in Fig. 5-27, the resistors R_{F1}, R_{F2}, R_{F3}, and R_{in} are determined to be

$$R_{F1} = \infty, \qquad R_{F2} = \frac{10 \text{ k}\Omega}{0.460} = 21.739 \text{ k}\Omega, \qquad R_{F3} = \frac{10 \text{ k}\Omega}{0.149} = 67.114 \text{ k}\Omega,$$

$$R_{in} = \frac{10 \text{ k}\Omega}{0.575} = 17.391 \text{ k}\Omega$$

Let us implement this filter using the single-amplifier biquad in Fig. 4-17. All three biquads have the following identical specifications: $\omega_{0i} = 2\pi \times 10^3$ rad/sec, $Q_i = 30$, and $H_{Bi} = 1.803$ for $i = 1, 2, 3$.

Using the design equations Eqs. (4-53), (4-47b), and (4-66), with $b_0 = b_2 = 0$, $b_1 = H_{Bi}\,\omega_{0i}/Q_i$, $G_6 = G_7 = G_c = 0$, and $R_A = R_D$, the element values for the three identical bandpass SABs are determined as follows. First, the capacitors C_1 and C_2 are arbitrarily assigned values of 0.01 μF and R_D is set to 5 kΩ. Then to minimize the SAB sensitivities, $\beta_{opt} = R_2/R_1 = R_2/(R_4 \| R_5)$ is computed according to Eq. (4-53). Assuming that $\sigma_R = \sigma_C = 0.5\%$, $\sigma_{GB} = 50\%$, and GB $= 2\pi \times 10^6$ rps for both SAB sections, yields $\beta_{opt} \simeq 40$. The resistors R_4, R_5, R_2, and R_B are now computed using Eqs. (4-47b) and (4-66):

$$R_4 = 276.65 \text{ k}\Omega, \qquad R_5 = 2.5391 \text{ k}\Omega, \qquad R_2 = 100.64 \text{ k}\Omega,$$

$$R_B = 111.86 \text{ k}\Omega$$

As noted in Section 4.2.6, voltage divider R_4 and R_5 is needed to independently adjust the gain constant.

The schematic for the PRB filter is shown in Fig. 5-28. Note that the band-pass T_i's given by Eqs. (4-63) and (4-64) have an inverted gain [this is explicitly shown in Eq. (4-45a)]; hence, to derive the proper sign around the

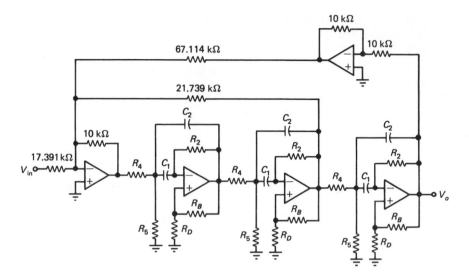

Fig. 5-28   PRB implementation of a sixth-order Butterworth band-pass filter. The sections are realized as Deliyannis–Friend single-amplifier biquads (see Fig. 4-17).

R_{F3} feedback loop, an inverting amplifier is needed. Nevertheless, only five op amps are required to consummate this design.

●

5.6.3 Generalized FLF (GFLF) Synthesis

The GFLF method uses the FLF topology in Fig. 5-24a with general biquad T_i's of the form given in Eq. (5-25). Its transfer function can be readily obtained in terms of the individual section transfer functions $T_i = N_i/D_i$. For an n-section design, from Eq. (5-90), the transfer function is given by

$$H(s) = \frac{N(s)}{D(s)} = \frac{-\alpha \prod_{j=1}^{k} N_j(s)}{\prod_{j=1}^{n} D_j(s) + \sum_{k=1}^{n-1} F_{1k} \prod_{j=1}^{k} N_j(s) \prod_{i=k+1}^{n} D_i(s) + F_{1N} \prod_{j=1}^{n} N_j(s)}$$

(5-114)

From Eq. (5-114) it is seen that the GFLF transmission zeros are derived from the zeros of T_i in cascade fashion, as shown in Fig. 5-26a. Unfortunately, however, biquad transmission zeros enter into the denominator D of transfer function H via the external feedback. As mentioned earlier, this condition complicates the nature of the feedback signals and places extra burden on the stability and ideality of the section transmission zeros. Nevertheless, GFLF designs offer the ability to realize nonsymmetrical transfer functions and an improved stop-band attenuation, relative to all-pole filters of the same degree. Also, reduced stop-band sensitivities in high-order filters may be obtained. A further by-product is that GFLF filters can often be designed with sections of

High-Order Filter Realization

lower Q than cascade circuits, a feature that can be used to reduce noise and element value spreads within the biquad sections.

One procedure for synthesizing GFLF filters is to successively reduce the desired transfer function by degree 2 with the extraction of a biquadratic section. The extraction process begins with section T_n and terminates when all feedback gains F_{1j} and sections T_i have been determined. Exercising complete freedom in selecting F_{1j}'s can result in biquad sections with right-half-plane poles. To ensure all T_i's have left-half-plane poles, certain restrictions must be placed on the values of F_{1j}.

GFLF design procedure Let the desired voltage transfer function be given by

$$H(s) = \frac{V_{out}}{V_{in}} = \frac{N(s)}{D(s)} \qquad (5\text{-}115)$$

To indicate that n biquadratic sections are still to be determined at this stage of synthesis, we use the notation

$$H(s) = H^{(n)}(s) = \frac{N^{(n)}(s)}{D^{(n)}(s)} \qquad (5\text{-}116)$$

so that $N^{(n)}(s) = N(s)$ and $D^{(n)}(s) = D(s)$. Referring then to Fig. 5-29, we have

$$H^{(n)}(s) = \frac{\tilde{H}^{(n)}(s)}{1 + F_{1n}\tilde{H}^{(n)}(s)} \qquad (5\text{-}117)$$

$$= \frac{\tilde{N}^{(n)}(s)}{\tilde{D}^{(n)}(s) + F_{1n}\tilde{N}^{(n)}(s)} \qquad (5\text{-}118)$$

where

$$\tilde{H}^{(n)}(s) = \frac{\tilde{N}^{(n)}(s)}{\tilde{D}^{(n)}(s)} \qquad (5\text{-}119)$$

Equating the numerator and denominator of (5-116) and (5-118), we have

$$\tilde{N}^{(n)}(s) = N^{(n)}(s) \qquad (5\text{-}120a)$$

$$\tilde{D}^{(n)}(s) = D^{(n)}(s) - F_{1n}N^{(n)}(s) \qquad (5\text{-}120b)$$

The right-hand side of (5-120a) and (5-120b) is completely known for a chosen value of F_{1n}, and hence $\tilde{H}^{(n)}(s)$ can be calculated. For $\tilde{H}^{(n)}(s)$ to be a stable transfer function, it can be shown that F_{1n} must satisfy the following conditions:[6]

$$\alpha_n \leq F_{1n} \leq \beta_n \qquad (5\text{-}121a)$$

where

$$-1 \leq \alpha_n \leq 0 \quad \text{and} \quad 0 \leq \beta_n \leq 1 \qquad (5\text{-}121b)$$

[6]These conditions ensure the stability of each biquadratic block used in the configuration. In principle, unstable biquads can be used in an MF filter. However, to facilitate tuning and to insure stability in the event one or more feedback loops are disconnected, we recommend the use of stable biquads.

Note that there is considerable latitude in the choice of F_{1n}, which can be utilized to minimize sensitivity.

Having determined $\tilde{H}^{(n)}(s)$, its numerator and denominator can now be factored into a product of biquadratic terms. By selecting a quadratic term from the numerator as $N_n(s)$ and one from the denominator as $D_n(s)$, the end section $T_n(s) = N_n(s)/D_n(s)$ is realized. Denoting the remaining transfer function as $H^{(n-1)}(s)$, we get

$$\tilde{H}^{(n)}(s) = H^{(n-1)}(s)T_n(s) \tag{5-122}$$

This completes one step of synthesis and we are left with a transfer function $H^{(n-1)}(s)$ whose degree is 2 less than the original. By repeating the process until all F_{ij}'s and T_j's are determined, the required transfer function is realized. The design procedure is outlined as follows:

1. Select a value of F_{1n} and compute the resulting $\tilde{D}^{(n)}$ according to Eq. (5-120b).
2. Factor $\tilde{D}^{(n)}$ into its n second-order factors, $D_i(s)$, for $i = 1, \ldots, n$.
3. Select a D_j from the $\tilde{D}^{(n)}$ factors in Eq. (5-120b).
4. Compute $D^{(n-1)} = \tilde{D}^{(n)}/D_j$ and $N^{(n-1)} = \tilde{N}^{(n)}/N_k$, form the relation $\tilde{D}^{(n-1)} = D^{(n-1)} - F_{1,n-1}N^{(n-1)}$, and return to step 1.

Note that at the end of the ith cycle one obtains two transfer functions, a biquad T_{n-i+1} and another, probably more complex, transfer function $H^{(n-i)} = N^{(n-i)}/D^{(n-i)}$ whose degree is two less than that of the preceding $H^{(n-i+1)}$. Also, $\tilde{D}^{(n-i+1)} = D^{(n-i+1)} - F_{1,n-i+1}N^{(n-i+1)}$ as i increases from 1 to n. The i design steps provide the following synthesis information: (a) the i feedback factors F_{1n-i+1} to F_{1n} and (b) the biquad section transfer functions T_{n-i+1} to T_n. The synthesis process illustrates the considerable amount of freedom one has in selecting F_{1j}, D_j, and the (N_i, D_i) pairings. These selections should be made with the following criteria in mind: (1) realizability of D_i's and hence of $T_i(s)$, (2) sensitivity, and (3) dynamic range. For good dynamic range the voltage maxima must be equalized per Eq. (5-87). The procedure should closely parallel that outlined for the FLF designs, with the final result being the values of the gains K_i in Eq. (5-25). Pole–zero pairings affect both sensitivity and dynamic range. The relationship between these interdependent factors is complicated and unknown at this time.

To gain insight into the mechanics of the GFLF synthesis process and to provide a vehicle for comparison with FLF, consider the design of a two-section (fourth-order) bandpass (BP) filter, characterized by the following low-pass (LP) prototype:

$$H(p) = \frac{K(p^2 + c_0)}{p^2 + d_1 p + d_0} = \frac{N(p)}{D(p)} \tag{5-123}$$

where p is the low-pass prototype complex frequency variable and $c_0 > d_0$. Applying the LP-to-BP transformation, $p = \gamma \hat{Q}(s^2 + 1)/s$ (ω_0 is normalized to

High-Order Filter Realization

unity), to Eq. (5-123) yields the desired BP transfer function:

$$H(s) = \frac{K\left(s^4 + \left[2 + \dfrac{c_0}{(\gamma\hat{Q})^2}\right]s^2 + 1\right)}{s^4 + \dfrac{d_1}{\gamma\hat{Q}}s^3 + \left[2 + \dfrac{d_0}{(\gamma\hat{Q})^2}\right]s^2 + \dfrac{d_1}{\gamma\hat{Q}}s + 1} \qquad (5\text{-}124)$$

For symmetrical BP designs, much of the GFLF synthesis can be carried out most simply with the LP prototype. To begin the synthesis, referring to Fig. 5-29a and Eqs. (5-120), we form

$$\tilde{N}^{(2)}(p) = N^{(2)}(p) = N(p) \qquad (5\text{-}125a)$$

$$\tilde{D}^{(2)}(p) = D^{(2)}(p) - F_{12}N^{(2)}(p) = D(p) - F_{12}N(p) \qquad (5\text{-}125b)$$

Fig. 5-29 Schematic of GFLF decomposition of $H(s)$.

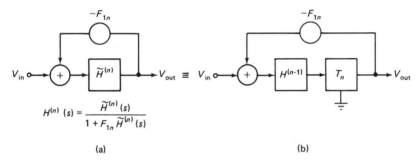

(a)

(b)

Substituting into Eq. (5-125b) the expressions for $N(p)$ and $D(p)$ in Eq. (5-123) yields

$$\begin{aligned}
\tilde{D}^{(2)}(p) &= p^2 + d_1 p + d_0 - F_{12}K(p^2 + c_0) \\
&= (1 - F_{12}K)p^2 + d_1 p + (d_0 - F_{12}Kc_0) \\
&= (1 - F_{12}K)\left[p^2 + \left(\frac{d_1}{1 - F_{12}K}\right)p + \frac{d_0 - F_{12}Kc_0}{1 - F_{12}K}\right]
\end{aligned} \qquad (5\text{-}126)$$

Thus,

$$H^{(2)}(p) = \frac{\dfrac{K}{1 - F_{12}K}(p^2 + c_0)}{p^2 + \left(\dfrac{d_1}{1 - F_{12}K}\right)p + \dfrac{d_0 - F_{12}Kc_0}{1 - F_{12}K}} = \frac{\alpha(p^2 + \tilde{c}_0)}{p^2 + \tilde{d}_1 p + \tilde{d}_0} \qquad (5\text{-}127)$$

is the open-loop LP prototype transfer function and

$$\tilde{H}^{(2)}(s) = \frac{\alpha\left(s^4 + \left[2 + \dfrac{\tilde{c}_0}{(\gamma\hat{Q})^2}\right]s^2 + 1\right)}{s^4 + \dfrac{\tilde{d}_1}{\gamma\hat{Q}}s^3 + \left[2 + \dfrac{\tilde{d}_0}{(\gamma\hat{Q})}\right]s^2 + \dfrac{\tilde{d}_1}{\gamma\hat{Q}}s + 1} \qquad (5\text{-}128a)$$

$$= T_1(s)T_2(s) \qquad (5\text{-}128b)$$

is the corresponding open-loop BP function, where

$$\alpha = \frac{K}{1 - F_{12}K} \tag{5-129a}$$

$$\tilde{c}_0 = c_0 \tag{5-129b}$$

$$\tilde{d}_1 = \frac{d_1}{1 - F_{12}K} > 0 \Longrightarrow F_{12} < \frac{1}{K} \tag{5-129c}$$

$$\tilde{d}_0 = \frac{d_0 - F_{12}Kc_0}{1 - F_{12}K} > 0 \Longrightarrow F_{12} < \frac{d_0}{Kc_0} < 1 \tag{5-129d}$$

Note that for stability $\tilde{d}_1$ and $\tilde{d}_0$ must be positive; hence, as indicated in Eq. (1-129d), $F_{12} < d_0/Kc_0$ and $< 1/K$. Thus, the open-loop pole positions are dependent upon the chosen value of F_{12}.

A practical choice for F_{12} is that which makes the poles of $\tilde{H}^{(2)}(p)$ real and equal. This choice for F_{12} will equalize the pole Q's for the two biquad sections in the BP filter. That equal Q's minimize the maximum Q was mentioned previously. Since both the component spreads and sensitivity increase proportionally with Q, the choice of equal Q's tends to render the individual biquads more manufacturable. For real and equal LP prototype poles, F_{12} is chosen so that

$$\tilde{d}_1^2 = 4\tilde{d}_0 \tag{5-130}$$

Substituting Eqs. (5-129c) and (5-129d) into Eq. (5-130) yields the following quadratic equation in F_{12}:

$$(KF_{12})^2 - \frac{d_0 + c_0}{c_0}(KF_{12}) + \frac{4d_0 - d_1^2}{4c_0} = 0 \tag{5-131a}$$

Solving for F_{12} yields

$$F_{12} = \frac{1}{2c_0K}[d_0 + c_0 - \sqrt{(c_0 - d_0)^2 + d_1^2 c_0}] < \frac{d_0}{c_0 K} \tag{5-131b}$$

Upon calculating F_{12} in Eq. (5-132), we then determine $\tilde{d}_1$, $\tilde{d}_0$, and α. Substituting these values into Eq. (5-127) and factoring the resultant into biquadratic functions T_1 and T_2 completes the GFLF synthesis process. The biquad sections are then designed according to procedures given in Chapter 4.

●

EXAMPLE 5-5

Consider the GFLF synthesis of the BP filter, with $\omega_0 = 1$ and $\hat{Q} = 24.356$, defined by the following LP prototype:

$$H(p) = \frac{0.784(p^2 + 18.9756)}{p^2 + 1.4126p + 1.0565} \tag{5-132}$$

The normalized 3-dB bandwidth for $H(p)$ from (5-132) is $\gamma = 1.0731$; hence,

$\gamma\hat{Q} = (1.073)(24.356) = 26.1342$. Designing the GFLF for equal-Q biquad sections, F_{12} is computed according to Eq. (5-131):

$$F_{12} = 0.0365 \qquad (5\text{-}133)$$

Hence,

$$\alpha = 0.8071, \qquad \tilde{c}_0 = 18.9756, \qquad \tilde{d}_1 = 1.4542, \qquad \tilde{d}_0 = 0.5287 \qquad (5\text{-}134)$$

Substituting Eq. (5-134) into Eq. (5-128a) and factoring according to Eq. (5-128b) yields

$$T_i = \frac{H_{Bi}(s^2 + \omega_{zi}^2)}{s^2 + \omega_{0i}/Q_i s + \omega_{0i}^2} \qquad (5\text{-}135)$$

where

Section 1: $H_{B1} = 0.8984$, $\omega_{z1} = 1.0868$, $Q_1 = 35.9431$, $\omega_{01} = 1$

Section 2: $H_{B2} = 0.8984$, $\omega_{z2} = 0.9201$, $Q_2 = 35.9431$, $\omega_{02} = 1$

The block diagram design, with $R_{f2} = R/F_{12}$, is shown in Fig. 5-30.

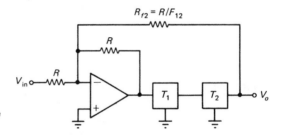

Fig. 5-30 Block diagram for the GFLF design in Example 5-5.

EXAMPLE 5-6

As a comparison, let us repeat the design in Example 5-4, as an FLF with feed-forward summation. This circuit is shown in Fig. 5-31. In the FLF the T_i's are BP biquad transfer functions of the form

$$T_i = \frac{-H_{Bi}/Q_i s}{s^2 + 1/Q_i s + 1} \qquad (5\text{-}136)$$

For this example, we choose the section Q_i's to minimize the in band transfer function sensitivity (M_T). Note that, to the first order, the effects of the transmission zeros can be ignored in the sensitivity minimization. Recall that in Example 5-2 we established the optimum Q values for a two-section all-pole FLF filter. Hence, using Eqs. (5-107) and following the FLF design procedure outlined in Section 5.6.1, we obtain

Section 1: $H_{B1} = 1.7334$ and $Q_1 = 37.0016$

Section 2: $H_{B2} = 1.7334$ and $Q_1 = 37.0016$

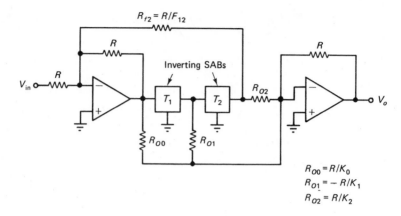

Fig. 5-31 Block diagram for the FLF design in Example 5-6.

Also, $t_{12} = 0.5576$, $k_0 = 0.784$, $k_1 = -1.1075$, and $k_2 = 14.8769$. Note that the inverting T_i in Eq. (5-136) implies that k_1 must be negative in the summing scheme employed in Fig. 5-31. Finally, using Eqs. (5-95),

$$F_{12} = 0.3720, \qquad K_0 = 0.7840, \qquad K_1 = -0.7389, \qquad K_2 = 8.1068$$

●

The sensitivity M_T was computed, over the passband $0.99 \leq \omega \leq 1.01$, for both the GFLF and FLF designs. The passive component variations were assumed to be independent and Gaussian-distributed with $\sigma = 0.1\%$. Furthermore, the biquad sections were assumed to be realized with the single-amplifier biquads in Fig. 4-18 for the FLF and Figs. 4-22 and 4-23 for the GFLF. These results are

$$M_T^{\mathrm{GFLF}} = 7.00 \times 10^{-5} \qquad (5\text{-}137a)$$

$$M_T^{\mathrm{FLF}} = 6.00 \times 10^{-5} \qquad (5\text{-}137b)$$

Since in GFLF the zeros of the T_i's enter into the definition of the poles, because of the feedback, the improved FLF sensitivity is not surprising. The expectation for this comparison to reverse for higher-order designs was mentioned previously. In addition, recall that GFLF, unlike FLF, is not restricted to symmetrical BP design.

5.6.4 Inverse Follow-the-Leader-Feedback (IFLF) Synthesis

Although FLF and IFLF circuits are topologically different, an IFLF design can be readily obtained from an FLF design through a simple transformation, which completely preserves the FLF sensitivity behavior.

An IFLF filter, using N sections and resistor summers, is shown sche-

High-Order Filter Realization

matically in Fig. 5-32. This circuit is described by

$$H(s) = H_N(s) = \frac{V_o}{V_{in}} = \frac{-\alpha \prod\limits_{j=1}^{N} T_j(s)}{1 + \sum\limits_{k=1}^{N} F_{kN} \prod\limits_{j=k}^{N} T_j(s)} \tag{5-138a}$$

$$H_i(s) = \frac{V_{oi}}{V_{in}} = -H_N(s) \cdot \frac{1 + \sum\limits_{j=i+1}^{N} F_{jN} \prod\limits_{k=j}^{N} T_k(s)}{\prod\limits_{j=i+1}^{N} (G_j/G_{Tj}) T_j(s)}, \quad i = 1, \ldots, N-1 \tag{5-138b}$$

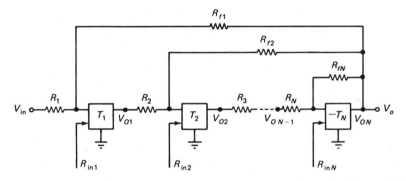

Fig. 5-32   IFLF configuration for 2Nth-order all-pole band-pass design.

where (with $G_{Ti} = G_i + G_{fi} + G_{in,i}$)

$$\alpha = \prod_{i=1}^{N} \frac{G_i}{G_{Ti}}, \qquad F_{iN} = \frac{G_{fi}}{G_i} \prod_{j=1}^{N} \frac{G_j}{G_{Tj}}, \qquad i = 1, \ldots, N \tag{5-139}$$

For all-pole filters, $T_i(s)$ is given in Eq. (5-81).

A comparison of Eqs. (5-138a) and (5-90) shows that a minimum-sensitivity IFLF filter can be obtained directly from a similarly optimized FLF filter by setting

$$T_j^{(\text{IFLF})}(s) = T_{N+1-j}^{(\text{FLF})}(s) \tag{5-140}$$

That is, the optimum values of quality factors and loop gains are

$$Q_j^{(\text{IFLF})} = Q_{N+1-j}^{(\text{FLF})}, \qquad t_{jN}^{(\text{IFLF})} = t_{1,N+1-j}^{(\text{FLF})} \tag{5-141}$$

where t_{jN} is defined as

$$t_{jN} = F_{jN} \prod_{k=j}^{N} H_{Bk} q_k, \qquad j = 1, \ldots, N \tag{5-142}$$

and $q_k = \gamma \hat{Q}/Q_k$. From Eq. (5-142) it is clear that the gains H_{Bj} remain free parameters which can be chosen to maximize dynamic range according to Eq.

(5-87). Recalling Eqs. (5-81), (5-142), and (5-138b), and defining,

$$M_i = \max_{0 \le \omega < \infty} \left| \frac{H_N(j\omega)\left[1 + \sum_{j=i+1}^{N} t_{jN} \prod_{k=j}^{N} \hat{T}_k(j\omega)\right]}{\prod_{j=i+1}^{N} \hat{T}_j(j\omega)} \right| \qquad \text{for } i < N \qquad (5\text{-}143)$$

which can be readily evaluated by computer, results with Eq. (5-89) in the gain constants

$$H_{Bi} = \left(\frac{G_{Ti}}{G_i}\right)\frac{M_{i-1}}{M_i}, \qquad \text{for } i = 2, \ldots, N \qquad (5\text{-}144a)$$

which equalize the signal levels throughout the filter. From Eqs. (5-138a), (5-124), (5-127), and (5-129a), one determines

$$H_{B1} = \frac{H_{MN}M_N}{M_1}\frac{G_{T1}}{G_1}\,|D(j\omega_0)| \qquad (5\text{-}144b)$$

which sets the desired gain H_{MN}. $D(j\omega_0)$ is, of course, the denominator of Eq. (5-138a) evaluated at $s = j\omega_0$.

One interesting feature of IFLF circuits is that the range of the internal signals varies only moderately, as compared to FLF (see Fig. 5-42). It is noted that the design of FLF filters is somewhat easier than that of equivalent IFLF circuits, even if the transformations (5-140) and (5-141) are used, because of the dependence of H_{Bi} and F_{iN} on the input admittances $G_{i,\text{in}}$ of the blocks $T_i(s)$. Notice in particular that the synthesis becomes very complicated if these admittances are frequency-dependent.

5.7 LEAPFROG (LF)-TYPE FILTERS

Leapfrog (LF)-type filters refer to a somewhat broader class of active filter topologies which simulate passive-ladder networks on a voltage basis. Furthermore, like the GIC simulations, leapfrog active filters can be derived directly from an appropriately terminated passive-ladder network. The leapfrog implementation then retains the low sensitivity of the passive ladder.

5.7.1 Leapfrog Synthesis via Signal-Flow-Graph Simulation

In a previous section we described a method for implementing active-ladder networks via the impedance simulation of the inductor or via FDNR elements. The fact that such implementations typically require an excessive number of op amps was already mentioned. In addition, the maximization of dynamic range is difficult in GIC structures. In particular, dynamic range considerations can require the GIC components to be selected in conflict with the op amp GB compensation relations given in Chapter 2. Thus, to maximize dynamic range will lead to some sacrifice in the active-element sensitivities.

An alternative method of simulation, which does not suffer from this potential dynamic range—sensitivity conflict—is to simulate voltage–current relations in the passive-ladder network via signal flow graphs. This simulation will result, for low-pass passive ladders, in active networks comprised of damped and lossless integrators and inverting amplifiers. For band-pass design, using the LP-to-BP transformation, the damped and lossless integrator sections become BP sections of finite and infinite Q, respectively. Since the biquad sections can be implemented using single-amplifier biquads (e.g., Fig. 4-21), this type of passive-ladder simulation will tend to minimize the number of op amps required in the active-RC realization.

Let us then consider the flow-graph simulation of the fifth-order all-pole low-pass resistive-terminated ladder network in Fig. 5-33. The low-pass ladder is particularly convenient to work with since the realizations for the other transfer function types (i.e., BP, HP, and BR) can be derived from the active LP ladder via the appropriate frequency transformation (see Chapter 1). To begin, let us write the voltage and current relations for the network in Fig. 5-33:

$$I_{in} = \frac{1}{R_a}(V_{in} - V_1) \tag{5-145a}$$

$$V_1 = \frac{1}{sC_1}(I_{in} - I_2) \tag{5-145b}$$

$$I_2 = \frac{1}{sL_2}(V_1 - V_2) \tag{5-145c}$$

$$V_2 = \frac{1}{sC_3}(I_2 - I_4) \tag{5-145d}$$

$$I_4 = \frac{1}{sL_4}(V_2 - V_3) \tag{5-145e}$$

$$V_3 = V_0 = \frac{1}{sC_5}(I_4 - I_6) \tag{5-145f}$$

$$I_6 = \frac{1}{R_b}V_0 \tag{5-145g}$$

To write the signal flow graph, it is convenient to have all the dependent variables as voltages. This operation is readily completed if we simply multiply

Fig. 5-33 Fifth-order all-pole low-pass doubly terminated LC ladder network.

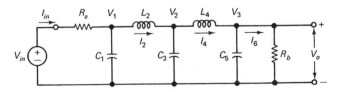

each current in Eqs. (5-145) by an arbitrary resistance R:

$$RI_{in} = \frac{R}{R_a}(V_{in} - V_1) \Longrightarrow \hat{V}_{in} = \frac{R}{R_a}(V_{in} - V_1) \tag{5-146a}$$

$$V_1 = \frac{1}{sC_1}\frac{R}{R}(I_{in} - I_2) \Longrightarrow V_1 = \frac{1}{sC_1 R}(\hat{V}_{in} - \hat{V}_2) \tag{5-146b}$$

$$RI_2 = \frac{R}{sL_2}(V_1 - V_2) \Longrightarrow \hat{V}_2 = \frac{R}{sL_2}(V_1 - V_2) \tag{5-146c}$$

$$V_2 = \frac{1}{sC_3}\frac{R}{R}(I_2 - I_4) \Longrightarrow V_2 = \frac{1}{sC_3 R}(\hat{V}_2 - \hat{V}_4) \tag{5-146d}$$

$$RI_4 = \frac{R}{sL_4}(V_2 - V_3) \Longrightarrow \hat{V}_4 = \frac{R}{sL_4}(V_2 - V_3) \tag{5-146e}$$

$$V_0 = \frac{1}{sC_5}\frac{R}{R}(I_4 - I_6) \Longrightarrow V_0 = \frac{1}{sC_5 R}(\hat{V}_4 - \hat{V}_6) \tag{5-146f}$$

$$RI_6 = \frac{R}{R_b}V_0 \Longrightarrow \hat{V}_6 = \frac{R}{R_b}V_0 \tag{5-146g}$$

Note that the redefinition of the currents I_{in}, I_2, I_4, and I_6 has resulted in the "dummy" voltage variables $\hat{V}_{in}$, $\hat{V}_2$, $\hat{V}_4$, and $\hat{V}_6$. Referring to the voltage relations Eqs. (5-146), we can draw the signal flow graph shown in Fig. 5-34. Note that the flow graph is comprised of integrator blocks, summers, inverting amplifiers (-1), and finite-gain amplifiers $(R/R_a$ and $R/R_b)$. In order for the integrators to be efficiently implemented as active-RC networks (Fig. 2-10a), it is convenient to introduce a sign inversion into each of their transfer functions, as shown in Fig. 5-35. Furthermore, the first and last integrators, with local feedback loops $-R/R_a$ and $-R/R_b$, in Fig. 5-34 are recognized to be damped integrators, as shown in Fig. 5-35. A general observation can be made at this point: that the first and last sections of an active ladder will be damped integrators due to the resistive terminations. All internal sections (i.e., sections 2 through $N - 1$) are lossless integrators, emulating the lossless character of the internal LC ladder network.

The corresponding equations for Fig. 5-35 are

$$V_1 = \frac{R_a/R}{1 + sC_1 R_a}\hat{V}_2 - \frac{V_{in}}{1 + sC_1 R_a} \tag{5-147a}$$

$$\hat{V}_2 = \frac{R}{sL_2}V_2 - \frac{R}{sL_2}V_1 \tag{5-147b}$$

$$V_2 = \frac{\hat{V}_4}{sC_3 R} - \frac{\hat{V}_2}{sC_3 R} \tag{5-147c}$$

$$\hat{V}_4 = \frac{R}{sL_4}V_0 - \frac{R}{sL_4}V_2 \tag{5-147d}$$

$$V_0 = \frac{-R_b/R}{1 + sC_5 R_b}\hat{V}_4 \tag{5-147e}$$

High-Order Filter Realization

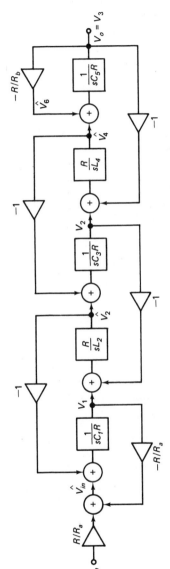

Fig. 5-34 Signal flow graph for the fifth-order low-pass ladder network in Fig. 5-33. All blocks are lossless integrators.

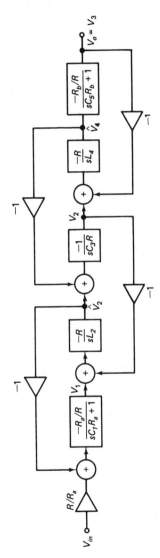

Fig. 5-35 Flow graph in Fig. 5-34 reconfigured so that the first and last blocks, which account for the resistive terminations in Fig. 5-33, are lossy integrators.

An active-RC implementation of the signal flow graph in Fig. 5-35 is shown in Fig. 5-36. Note that the (-1) gains in Fig. 5-34 are implemented with inverting amplifiers. The component values for the active-RC implementation in Fig. 5-36 can be determined in terms of R_a, R_b, R_3, C_1, L_2, C_3, L_4, and C_5

Fig. 5-36 Active-RC implementation of the flow graph in Fig. 5-35.

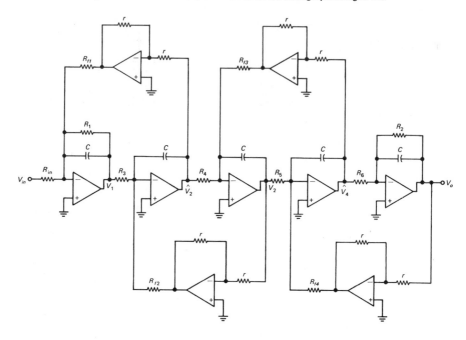

in the signal flow graph in Fig. 5-35 by equating the voltage relations obtained at the outputs of the integrator blocks. The corresponding equations for Fig. 5-36 are [for comparison with Eqs. (5-147), the same equation numbers are used but with primes]

$$V_1 = \frac{R_1/R_{f1}}{1 + sCR_1}\hat{V}_2 - \frac{R_1/R_{in}}{1 + sCR_1}V_{in} \qquad (5\text{-}147a')$$

$$\hat{V}_2 = \frac{V_2}{sCR_{f2}} - \frac{V_1}{sCR_3} \qquad (5\text{-}147b')$$

$$V_2 = \frac{\hat{V}_4}{sCR_{f3}} - \frac{\hat{V}_2}{sCR_4} \qquad (5\text{-}147c')$$

$$\hat{V}_4 = \frac{V_0}{sCR_{f4}} - \frac{V_3}{sCR_5} \qquad (5\text{-}147d')$$

$$V_0 = \frac{R_2/R_6}{1 + sCR_2}\hat{V}_4 \qquad (5\text{-}147e')$$

It is usually convenient to equalize the capacitance values in the active-RC implementation; hence, we set $C_1 = C_2 = C_3 = C_4 = C_5 = C$ as shown in

Fig. 5-36. Equating like terms in Eqs. (5-147), we obtain the following relations for the resistors:

$$V_1: \quad R_1 = \frac{C_1 R_a}{C}, \quad R_{\text{in}} = R_1, \quad \text{and} \quad R_{f1} = \frac{R_1 R}{R_a} \qquad \text{(5-148a)}$$

$$\hat{V}_2: \quad R_3 = \frac{L_2}{RC} \quad \text{and} \quad R_{f2} = R_3 \qquad \text{(5-148b)}$$

$$V_2: \quad R_4 = \frac{C_3 R}{C} \quad \text{and} \quad R_{f3} = R_4 \qquad \text{(5-148c)}$$

$$\hat{V}_4: \quad R_5 = \frac{L_4}{RC} \quad \text{and} \quad R_{f4} = R_5 \qquad \text{(5-148d)}$$

$$V_0: \quad R_2 = \frac{C_5 R_b}{C} \quad \text{and} \quad R_6 = \frac{R_2 R}{R_b} \qquad \text{(5-148e)}$$

A symmetric BP biquad is obtained from the all-pole low-pass flow graph in Fig. 5-35, when a LP to BP transformation is applied to the integrator blocks in Fig. 5-35. Performing this transformation results in the flow graph shown in Fig. 5-37. The block T_i in the transformed BP flow graph are now BP sections, with T_i of the form

$$T_i = \frac{K_i(\omega_0/Q_i)s}{s^2 + (\omega_0/Q_i)s + \omega_0^2} \qquad i = 1, 5 \qquad \text{(5-149a)}$$

$$T_j = \frac{K_j s}{s^2 + \omega_0^2} \qquad j = 2, 3, 4 \qquad \text{(5-149b)}$$

Fig. 5-37 Flow graph for a tenth-order band-pass ladder network obtained by applying the low-pass to band-pass transformation to the low-pass flow graph in Fig. 5-35. The resulting blocks T_i are band-pass biquads. Note that only T_1 and T_5 have finite Q.

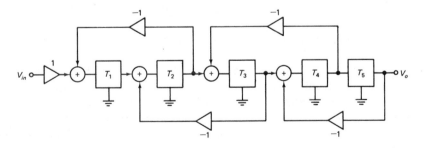

One result of the voltage analogy between the active-RC BP structure in Fig. 5-37 and the passive ladder is that all but the first and last biquad sections have infinite pole Qs, emulating the lossless internal structure of the passive ladder. A second and more significant result is that the low passband magnitude response sensitivities associated with properly matched passive ladder are preserved. In fact, when properly terminated, the magnitude response sensitivities for passive-ladder filters achieve minima at frequencies of maximum power transfer and experience indicates that this property is transferred to active ladders. Thus, for LF filters which achieve maximum power transfer at distinct frequencies, over a band about the center frequency, the passband magnitude response sensitivities are indeed very low. Butterworth and Chebyshev filters fall within this class. The low passband sensitivity for Butterworth LF designs was illustrated in Fig. 5-25.

Not all filters posses flat or equiripple passband, nor is one always solely interested in passband magnitude-response sensitivities. When requirements demand, transition and stop band magnitude response sensitivities, or phase or transfer function sensitivities, can be of prime interest. For these situations, a more general RLC ladder, or analogously a modified leapfrog (MFL) active ladder provides the freedom to vary the internal Q_i's to optimize the filter to meet such requirements. To emphasize the analogy between the passive-and active-ladder structures, sixth-order BP doubly terminated LC, RLC, active LF, and active MLF circuits are shown together in Fig. 5-38. Observe that the active ladder can be implemented with either passive or active summers. As illustrated in Fig. 5-38d, when both the inverting and noninverting BP section outputs are available, external inverting amplifiers are not needed.

An additional remark regarding the distinction between the leapfrog LF and coupled biquad (CB) design philosophies is in order. Both LF and CB use the LF topology in Fig. 5-24e for symmetrical all-pole BP design. For more general realizations, the two design philosophies are quite different. LF designs are synthesized in much the same manner as one might synthesize a passive ladder, and the T_i's are general biquads of the form of Eq. (5-25). Thus, the general LF, like the GFLF, derives its finite transmission zeros in cascade manner from the biquads T_i's. Furthermore, both symmetrical and non-symmetrical transfer functions can in principle be realized with general LF circuits. The CB, like FLF, uses BP T_i's and derives its finite zeros using either one of the feedforward techniques illustrated in Fig. 5-26b and c. As a result, only all-pole CBs are truly passive-ladder-derived.

5.7.2 Coupled Biquad (CB) Synthesis

When an appropriate prototype LC ladder is not available or when it is desired to deviate from the maximum power transfer design to optimize sensitivity outside the passband, it is convenient to derive the transfer function for the CB circuit in terms of the section transfer functions T_i and the feedback factors F_{ij}. Furthermore, analyzing the CB circuit in this way facilitates the maximization of dynamic range.

The transfer function for the general CB circuit (i.e., the configuration in Fig. 5-38c extended to N sections) is best derived using Mason's formula (see Appendix C in [B4]):

$$H(s) = \frac{V_o}{V_{in}} = \frac{\alpha \prod_{i=1}^{N} T_i(s)}{\Delta_N(s)} \tag{5-150}$$

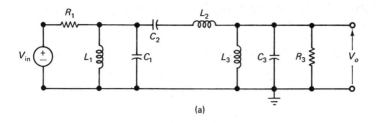

(a)

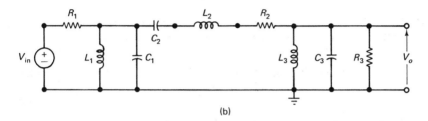

(b)

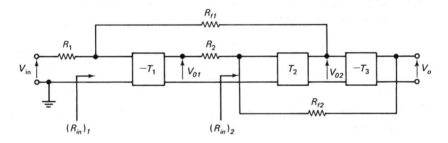

CB(LF): $Q_2 = \infty$, MLF: $Q_2 \neq \infty$

(c)

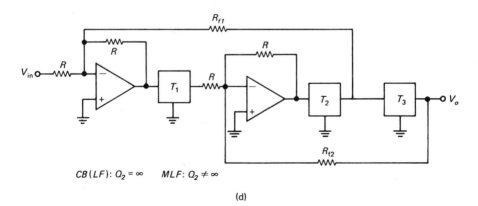

CB(LF): $Q_2 = \infty$ MLF: $Q_2 \neq \infty$

(d)

Fig. 5-38 (a) Doubly terminated sixth-order band-pass LC ladder network. (b) Doubly terminated sixth-order band-pass RLC ladder network. (c) MLF(LF) active-RC implementations of parts (a) and (b) with resistive summers. (d) MLF(LF) active-RC implementations of parts (a) and (b) with active summers.

where $\Delta_N(s)$ is the determinant of the feedback configuration and α is given by a ratio of resistors. For example, for $N = 4$,

$$\Delta_4(s) = 1 + F_{12}T_1T_2 + F_{23}T_2T_3 + F_{12}F_{34}T_1T_2T_3T_4 \qquad (5\text{-}151)$$

$$F_{12} = \frac{G_{f1}G_2}{G_{T1}G_{T2}}, \qquad F_{23} = \frac{G_{f2}G_3}{G_{T2}G_{T3}}, \qquad F_{34} = \frac{G_{f3}}{G_{T3}} \qquad (5\text{-}152a)$$

$$\alpha = \frac{G_1G_2G_3}{G_{T1}G_{T2}G_{T3}} \qquad (5\text{-}152b)$$

where $G_{Ti} = G_i + G_{Fi} + (G_{in})_i$ for $i = 1, 2, 3$. Similarly, the transfer function from input V_{in} to the output of section i can be expressed as

$$H_i(s) = \frac{V_{oi}}{V_{in}} = \frac{\Delta_{Ni}(s)H_N(s)}{\displaystyle\prod_{j=i+1}^{N} (G_j/G_{Tj})T_j(s)}, \qquad i < N \qquad (5\text{-}153)$$

where $H_N(s)$ is given in Eq. (5-150) and $\Delta_{Ni}(s)$ is the partial determinant of the feedback configuration with all feedback loops touching the forward path from V_{in} to V_{oi} removed. For example for $N = 4$,

$$\Delta_{41} = 1 + F_{23}T_2T_3 + F_{34}T_3T_4, \qquad \Delta_{42} = 1 + F_{34}T_3T_4, \qquad \Delta_{43} = 1 \qquad (5\text{-}154)$$

The case $N = 3$ is obtained from Eqs. (5-150) to (5-154) by setting $F_{34} = 0$ (i.e., $R_3 = G_{f3} = 0$ and $T_4 = 1$).

For symmetrical all-pole LF BP designs, all sections T_i are synchronously tuned to ω_0, and $Q_i = \infty$, $i \neq 1, N$. When $F_{ij} = +1$, the CB structure in Fig. 5-38c is directly equivalent to a passive LC ladder. Note that in Fig. 5-38c the required sign inversion around the two two-section feedback loops is achieved in blocks T_1 and T_3 which is different from the block diagram given in Fig. 5-37. To achieve $F_{ij} = 1$ requires replacing the resistive summers in Fig. 5-38c with op amp summers; however, in many active sections the desired summation function is naturally achieved at the input of the section, so that additional op amps are not always necessary.

Using Eq. (5-153) with (5-150) in the feedforward topology of Fig. 5-26b results in the transfer function of a $2N$th-order CB design; that is, defining $g_j = G_j/G_{Tj}$,

$$\begin{aligned} H(s) = \frac{V_o}{V_{in}} &= K_o + \sum_{i=1}^{N} K_i H_i(s) \\ &= \frac{1}{\Delta_N(s)}\left[K_o\,\Delta_N(s) + \sum_{i=1}^{N} K_i\,\Delta_N(s) \prod_{j=1}^{i} g_jT_j(s) \right] \end{aligned} \qquad (5\text{-}155)$$

which, of course, is useful only for symmetrical BP transfer functions $H(s)$.

In principle, one can derive synthesis equations for LF and CB design by equating, respectively, the coefficients of Eqs. (5-150) and (5-155) with those of Eq. (5-20). The equations derived in this manner become nonlinear for $N > 3$ and thus are difficult to solve and time-consuming in the sensitivity optimization.

High-Order Filter Realization

However, their utility in examining and maximizing dynamic range was mentioned previously. The most straightforward means for arriving at a LF or CB design is to emulate the signal flow graph for a properly terminated LC ladder. For symmetrical BP design, an all-pole CB design can be readily derived from a low-pass LC ladder prototype. This process yields the denominator in Eq. (5-155). To obtain the desired transmission zeros, the feedforward gains can be derived in a manner similar to Eq. (5-93).

The statistical multiparameter sensitivity measures in Eqs. (5-82) and (5-86) can be used to screen LF and CB designs; for example, when T_i are BP sections, the topology-dependent $\mathcal{L}_i$ in Eq. (5-84) can be written, for $N = 3$ as

$$\mathcal{L}_1 = D_1 D_2 D_3 + \left(\frac{j\omega}{\gamma\hat{Q}}\right)^2 t_{23} D_1 \qquad (5\text{-}156a)$$

$$\mathcal{L}_2 = D_1 D_2 D_3 \qquad (5\text{-}156b)$$

$$\mathcal{L}_3 = D_1 D_2 D_3 + \left(\frac{j\omega}{\gamma\hat{Q}}\right)^2 t_{12} D_3 \qquad (5\text{-}156c)$$

where

$$t_{ij} = F_{ij}(\gamma\hat{Q})^2 \prod_{k=i}^{j} a_k \qquad (5\text{-}157)$$

For finite values of Q_k, $a_k = H_{Bk}/Q_k$ and for $Q_k = \infty$, $a_k = H_{Bk}$. The N section case is quite complicated and no general expressions are available for $\mathcal{L}_i$. It is noted that CB and LF all-pole symmetrical realizations do have one degree of freedom for sensitivity minimization, either Q_1 or Q_N. Some improvements in transition band and stop band sensitivities can be obtained in this manner, as discussed in [P14].

No matter how the LF and CB circuits are derived, the T_i's must be properly scaled according to Eq. (5-87) to achieve good dynamic range. The procedure for equalizing the signal maxima is identical to the one used for MLF filters and will be discussed in the next section.

5.7.3 Modified Leapfrog (MLF) Synthesis

An MLF active ladder, as noted earlier, is the analog of a doubly terminated RLC ladder, as illustrated in Fig. 5-38. Because of the internal resistors in the passive circuit, all blocks $T_i(s)$ in an MLF filter, in contrast to LF design, have finite pole Q's. These additional degrees of freedom are used to minimize the sensitivity equations (5-82) to (5-86). MLF has an additional advantage, as compared to LF, in that parasitic losses, due to the finite gain–bandwidth (GB) of the op amps and capacitor dissipation factors, can be taken into account by predistorting the ideal, finite Q_i values. If a high-frequency active filter realization is desired, MLF is particularly advantageous, since infinite Q active -R biquads can give rise to nonlinear oscillations. The equations (5-150) to (5-157), derived for LF design, apply similarly to the MLF topology, with the understanding that the blocks $T_i(s)$ now have finite Q_i values. Unfortunately, however, the insight and systematic procedures available for LC ladder synthesis do not exist for RLC synthesis. Therefore, for MLF filters, a procedure for symmetrical BP design has been used which parallels that outlined earlier for FLF synthesis. The method, based on Eqs. (5-150) and (5-155), is to select the free Q_i values to minimize the desired sensitivity measure (one of Eqs. (5-82)). It has been found that for some designs both passband and stop band sensitivities can be improved by an appropriate choice of Q_i. A sensitivity comparison is given in Section 5.9. Once the optimal values of Q_i are known, the synthesis equations, obtained by equating Eqs. (5-150) and (5-21), are solved for the loop gains t_{ij}, defined in Eq. (5-157). As in the FLF cases, the gain constants H_{Bi} remain free parameters which are chosen to satisfy Eq. (5-88).

With Eqs. (5-81) and (5-152) in mind, we define

$$M_i = \max_{0 \le \omega < \infty} \left| \frac{\hat{H}_N(j\omega) \, \Delta_{Ni}(j\omega)}{\prod\limits_{j=i+1}^{N} \hat{T}_j(j\omega)} \right| \qquad \text{for } i < N \qquad (5\text{-}158)$$

which with Eq. (5-89) yields the gain constants

$$H_{Bi} = \left(\frac{G_{Ti}}{G_i} \right) \frac{M_{i-1}}{M_i}, \qquad \text{for } i = 2, \ldots, N \qquad (5\text{-}159\text{a})$$

which equalize the signal level in the MLF structure. From Eqs. (5-150) and (5-159a), one determines

$$H_{B1} = H_{MN} \frac{M_N}{M_1} \frac{G_{T1}}{G_1} |D(j\omega_0)| \qquad (5\text{-}159\text{b})$$

which sets the desired midband gain H_{MN}.

As pointed out before, dynamic range maximation of LF and CB filters proceeds in exactly the same fashion; care must be taken only to avoid possible difficulties due to the constraint $Q_i = \infty$, $i \ne 1, N$.

The MLF design procedure follows essentially the same sequence of steps 1 to 6 as the FLF method summarized at the end of Section 5.6.1, except that the equations cited there must be replaced by the corresponding ones in this section. The essential difference is that, in contrast to the linear synthesis equations (5-93a) for FLF, the corresponding equations for MLF become nonlinear for $N \ge 4$. For any given design, the coefficients of the nonlinear MLF synthesis equations depend on the set q_i provided by the sensitivity minimization process. This q_i set may fall into a region in q space for which the equations have no real or finite solutions, t_{ij}, (i.e., a design is impossible) [P19]. A satisfactory general solution to the MLF synthesis problem requires further investigation.

To illustrate the different interconnections required to construct an MLF (LF) filter from a cascade of biquads, consider the schematics for sixth-order MLF realization in Fig. 5-39. In Fig. 5-39a, the biquad used is the multiple-input biquad in Fig. 4-37. Note that in this biquad both inverting and noninverting

High-Order Filter Realization

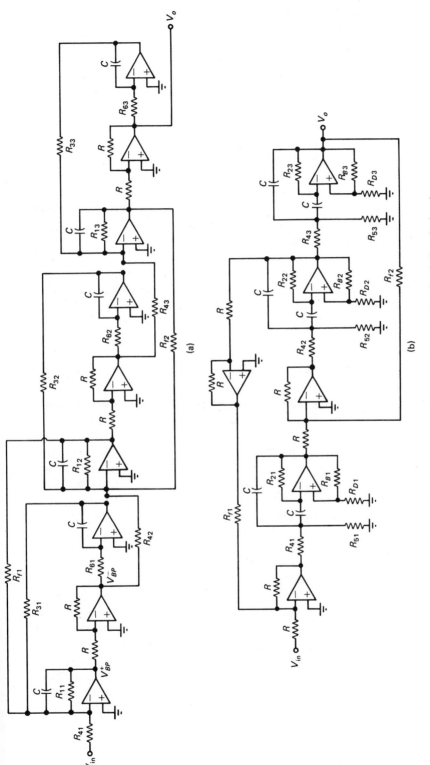

Fig. 5-39 Schematics for the active-RC implementation of a sixth-order band-pass MLF filter (a) using the multiple-input band-pass biquad in Fig. 4-37 and (b) using the single-amplifier band-pass biquad in Fig. 4-17.

BP outputs are available. Furthermore, the summing operations can be performed by the input stages of sections 1 and 2. This realization is seen to require nine op amps. In contrast, the implementation in Fig. 5-39b uses the single-amplifier biquad in Fig. 4-17 (with $G_6 = G_7 = G_c = 0$). In this implementation two active summers and one inverting amplifier are used to complete the realization. The SAB implementation is seen to require six op amps. Later we show how to eliminate the summing and inverting amplifiers. The reader is urged to verify the transfer functions for these circuits.

5.8 MINIMUM-SENSITIVITY-FEEDBACK (MSF) TOPOLOGY

MSF represents the most general form of MF design; in principle, all the aforementioned MF topologies are special cases of MSF. One can thus visualize MSF not only as a distinct MF topology but as a general formalism for MF synthesis [P26].

The purpose of the MSF development was to determine the feedback configuration resulting in the lowest possible sensitivity. In particular, it was to be investigated whether good stop band and passband sensitivities could be achieved simultaneously in the same filter. This objective was based on the following observations:

1. Optimized LF and MLF active filters yield the lowest sensitivities in the passband.

2. Optimized FLF active filters generally yield sensitivities lower than MLF designs at the band edges and in the stop band.

These observations are illustrated by the magnitude response sensitivity spectra in Fig. 5-40 for sixth-order LF, MLF, and FLF BP designs with $\hat{Q} = 25$. For these curves, the T_i's were assumed to be realized with Tarmy–Ghausi (TG) biquads with R's and C's normally distributed, correlated random variables with standard deviations of 0.2 and correlation coefficients $\rho_{R_iR_j} = \rho_{C_iC_j} = 0.8$ and $\rho_{R_iC_j} = 0$. In normalized form, however, these curves are essentially independent of the biquad realization as long as the assumptions stated in Section 5.5.1 for the evaluation of Eqs. (5-83) are valid, which is usually the case for most practical active RC biquads, at voice frequencies.

With the full freedom provided by MSF, one can achieve, to some degree, improvements in both passband and stop band sensitivities over those of other MF topologies. Using the FLF topology as a starting structure, one can outline a systematic process for adding new feedback loops in a manner that minimizes sensitivity. For example, for sixth- and eighth-order filters, this method results in the MSF topologies in Fig. 5-41, where the non-FLF loops are shown as dashed lines. FLF is a convenient starting structure in that its dependent variables (**t**) are readily determined from Eq. (5-93a). As noted previously, solutions to Eqs.

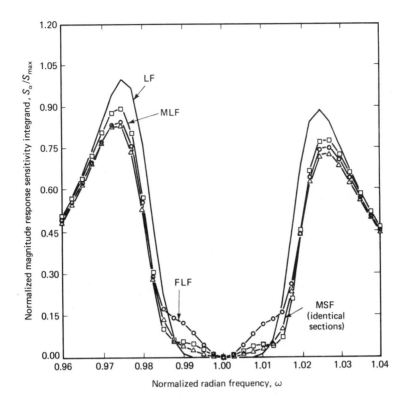

Fig. 5-40 Comparison of the gain response sensitivity S_α versus frequency for LF, MLF, FLF, and MSF (with identical sections) realizations for a sixth-order Butterworth band-pass filter ($I_{\max} = 0.135274$).

(5-93) always exist, and the addition of one or more feedback loops as independent variables does not compromise this important property.

As illustrated in Fig. 5-41, MSF configurations do not possess a repetitive feedback pattern; also, their complexity increases rapidly as order increases. Synthesis equations and sensitivity measures for all-pole sixth- and eighth-order BP designs can be readily derived, but for general high-order filters they become unwieldy. For symmetrical all-pole BP designs, the MSF synthesis equations can be written in a form similar to that in Eq. (5-93):

$$\mathbf{Lt} = \hat{\mathbf{d}} - \mathbf{g} \qquad (5\text{-}160)$$

When $\mathbf{L}$, $\mathbf{t}$, and $\hat{\mathbf{d}}$ are the FLF parameters and $\mathbf{g}$ is an $N \times 1$ vector which takes into account the "pseudo loop gains" of the non-FLF loops, for example, for the sixth-order ($N = 3$ low-pass prototype) shown in Figs. 5-24a and 5-41a,

$$g_1 = 0, \qquad g_2 = t_{23}, \qquad g_3 = \frac{\gamma \hat{Q}}{Q_1} t_{23} = q_1 t_{23} \qquad (5\text{-}161)$$

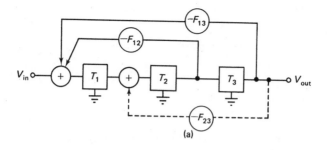

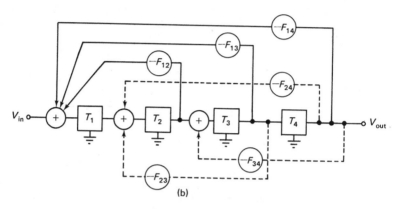

Fig. 5-41 MSF topologies for (a) sixth-order and (b) eighth-order all-pole band-pass realizations.

Corresponding expressions for eighth-order ($N = 4$) are as follows:

$$g_1 = 0 \qquad g_2 = t_{23} + t_{34} \tag{5-162a}$$

$$g_3 = t_{24} + \left(\frac{\gamma\hat{Q}}{Q_1} + \frac{\gamma\hat{Q}}{Q_4}\right)t_{23} + \left(\frac{\gamma\hat{Q}}{Q_1} + \frac{\gamma\hat{Q}}{Q_2}\right)t_{34} \tag{5-162b}$$

$$g_4 = \frac{(\gamma\hat{Q})}{Q_1}t_{24} + \frac{(\gamma\hat{Q})^2}{Q_1 Q_4}t_{23} + \frac{(\gamma\hat{Q})^2}{Q_1 Q_2}t_{34} + t_{12}t_{34} \tag{5-162c}$$

MSF sensitivities are computed using Eqs. (5-82), where for $N = 3$ (Prob. 5.24)

$$\mathcal{L}_1 = D_1 D_2 D_3 + \left(\frac{j\omega}{\gamma\hat{Q}}\right)^2 t_{23}D_1 \tag{5-163a}$$

$$\mathcal{L}_2 = D_1 D_2 D_3 \tag{5-163b}$$

$$\mathcal{L}_3 = D_1 D_2 D_3 + \left(\frac{j\omega}{\gamma\hat{Q}}\right)^2 t_{12}D_3 \tag{5-163c}$$

The determination of $\mathcal{L}_1$, $\mathcal{L}_2$, $\mathcal{L}_3$, and $\mathcal{L}_4$ for $N = 4$ MSF are left as an exercise in Prob. 5.24. Note that t_{ij} are defined as in Eqs. (5-95a) and (5-157):

$$t_{ij} = F_{ij}\prod_{k=i}^{j} H_{Bk}q_k \tag{5-164}$$

High-Order Filter Realization

The independent parameters of MSF synthesis, the Q_i and the non-FLF t_{ij}'s, are selected to minimize sensitivity as discussed in Section 5.5. The sensitivity improvements achievable are also illustrated in Fig. 5-40, which shows the sensitivity spectrum for a PRB-type (i.e., identical T_i blocks) MSF design, together with those equivalent LF, MLF, and FLF circuits. The slightly improved sensitivity behavior in the passband and at the band edge is evident. The design was achieved by optimizing S_α over the band $0.96 \leq \omega \leq 1.04$, which corresponds to twice the filter bandwidth. Full optimization (i.e., nonidentical Q_i's) over the narrower band $0.99 \leq \omega \leq 1.01$ gives different results, which interestingly possess a slightly broader sensitivity minimum than the LF design.

The remaining free parameters of MSF design, H_{Bi}, are selected to equalize the signal level according to Eq. (5-87). No general equations or results are available, because the transfer functions to the internal outputs $H_i(s)$, depend on the "non-FLF" loop gains t_{ij} and thus can be calculated only after the MSF topology has been established by means of sensitivity minimization.

The MSF synthesis procedure can be summarized as follows:

1. Choose the Q_i's and the non-FLF t_{ij}'s to minimize either M_α, M_β, or M_T given by Eqs. (5-82). With each iteration compute the FLF t_{1j}'s according to Eq. (5-160).

2. After the MSF topology has been established in step 1, determine H_{Bi}'s to equalize the signal maxima by means of Eq. (5-87). Compute F_{ij}'s according to Eq. (5-164).

3. Design independently the BP biquads for the parameters Q_i, H_{Bi}, and ω_0.

5.9 DESIGN EXAMPLES AND COMPARATIVE DISCUSSION OF MF DESIGNS

The first step in any all-pole MF design procedure is to find the values of center frequencies, ω_{0i}, pole quality factors Q_i, and gain constants H_{Bi}, so that the biquad sections $T_i(s)$ of the desired type, as discussed in Chapter 4, can be implemented. The feedback factors (or loop gains) are then determined, which result in the prescribed transfer function $H(s)$. In this section a few examples will be presented to provide the reader with some information about the numbers involved and the ease or complexity of the calculations. Before discussing the examples, however, a few general observations regarding MF BP design procedures should be noted:

1. The center frequencies of all sections are the same and equal to the center frequency of the overall filter (i.e., $\omega_{0i} = \omega_0$ for $1 \leq i \leq N$). For the optimization, the actual value of ω_0 is normalized to unity. Frequency and impedance normalizations are discussed in Section 1.8.

2. For symmetrical BP functions, the overall quality factor $\hat{Q}$, which determines the bandwidth of the total filter ($\Delta\omega_{3dB} = \omega_0/\hat{Q}$), is merely a scale factor because only the relative quality factors $q_i = \gamma\hat{Q}/Q_i$ enter the synthesis proce-

dure. $\hat{Q}$ affects the element values in the blocks, $T_i(s)$, and scales the total sensitivity.

3. *Any* cascadable second-order filter can be used to realize the functions $T_i(s)$. The specific topology chosen enters the optimization algorithm only through the values $S_{x_i}^{\omega_{0i}}$. The fabrication technology chosen (discrete, integrated-hybrid, or monolithic) will also influence the type of implementation and enters the analysis and design through the element statistics ρ and σ.

4. The topology of the total filter (FLF, MLF, MSF, etc.) is reflected in the form of the parameters $\mathcal{L}_i$, defined in Eq. (5-84). Thus, the sensitivity measure is separated into section-dependent and topology-dependent parts.

5. The gain constraints H_{Bi} of Eq. (5-81) which satisfy Eq. (5-88) are independent of the section type and of the desired overall gain H_{MN}. They do not affect any prior sensitivity minimization.

6. PRB-type FLF and IFLF do not require sensitivity optimization and can be readily synthesized using the design equations for FLF and IFLF. These topologies will usually yield good, suboptimum, sensitivity behavior. For better performance one can employ varying degrees of sophistication from FLF to MSF to meet tough requirements.

7. A comparison of the dynamic range for different topologies (FLF, IFLF, and MLF) indicates that in addition to the advantage of being easy to design, the FLF filter is the only topology whose output noise can remain constant regardless of the overall filter gain and the type of second-order section used [P28]. Although the passband signals occupy a wider range in FLF than IFLF and MLF (Fig. 5-42), the dynamic range and noise performance of FLF design can be significantly better than cascade, MLF, and IFLF topologies [P30]. It should be noted that the dynamic range of all MF filters can be approximately the same when certain second-order sections are used (e.g., the Tarmy–Ghausi section). In other words, different biquads result in different noise behavior as a function of the overall filter gain, and thus a careful choice of section type is important in all MF filter design except for the FLF type.

8. As in any other active filter, the performance of MF filters can be improved with postfabrication tuning. Since thin-film capacitors can have tolerances on the order of 5%, postfabrication tuning and laser trimming have become somewhat standard practices in the production of precision active filters. Optimum tuning procedures [P24] have been developed for tuning cascaded biquads. With some modification, these procedures are also applicable to MF filters [P29].

9. Typically, the most critical section in a cascade of biquads is the highest-Q section. That is, variations in the response of the highest-Q section will yield a more significant variation in the overall response than equivalent variations in the responses of the lower-Q sections. Unfortunately, the highest-Q section is also the one that is most difficult to control statistically in production. In MF filters, on the other hand, the highest-Q sections are not always the most critical. Most important, the impact of variations in the higher-Q sections on the overall response is much reduced in MF filters. The LF and CB topologies,

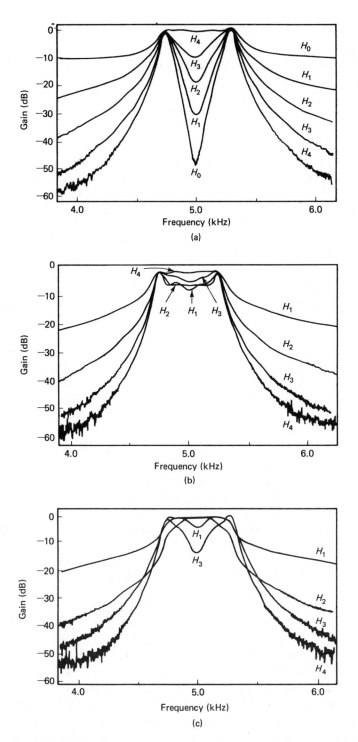

Fig. 5-42 Experimentally determined output voltages V_{oi}, $i \leq 4$, for optimized (a) FLF, (b) MLF, and (c) IFLF realizations of an eighth-order 0.5-dB ripple Chebyshev band-pass filter with $f_0 = 5$ kHz, using Tarmy–Ghausi band-pass biquads to realize all $T_i(s)$.

comprised of $N - 2$ infinite Q_i sections, are excellent examples of this phenomenon.

The examples chosen in the following are Butterworth and 0.5-dB ripple Chebyshev band-pass filters of order 4, 6, and 8. The second-order blocks assumed in the examples are low-sensitivity Tarmy–Ghausi sections (see Fig. 4-34). As emphasized throughout this chapter, other biquad circuits can also be used as sections for MF designs. For the realizations, FLF, LF, and MLF topologies are considered so that the differences and similarities of the approaches and results can be compared. The results for sixth-order MSF are also shown. The design parameters presented in Table 5-7, have been calculated by means of a FORTRAN computer program for realizing minimum [P23] sensitivity, maximum dynamic range FLF, and MLF-type all-pole bandpass filters. SCF, PRB, LF, CB, and IFLF topologies are treated as special cases, as is the sensitivity calculation for cascade structures. The program is based on the FLF synthesis method outlined in this chapter and it is presently limited to handling BP filters up to order 14 for MLF and 16 for FLF topologies. The sensitivity data in Table 5-7 have been normalized to those of the corresponding cascade realizations, denoted by M_α^c, to demonstrate the improvement obtainable with MF synthesis.

The design data and sensitivities in Table 5-7 are based on the element statistics $\sigma = 0.2\%$, and $\rho_{R_iR_j} = \rho_{C_iC_j} = 0.8$ and $\rho_{RC} = 0$. The sensitivity comparisons shown are, however, more general and apply to realizations employing any active biquad for which the assumptions stated in Section 5.5.1 remain valid. In addition, they will be essentially unaltered by changes in σ and ρ_{ij}, as long as the element variations determined by σ do not become excessively large to invalidate the linear approximation.

Listed in Table 5-7 are the values of $Q_i/\hat{Q}$ that minimize the sensitivity M_α of Eq. (5-82a). Each design was carried out twice, once to minimize M_α over the passband only, exclusive of the band edges [i.e., $1 - 1/(4\hat{Q}) \le \omega \le 1 + 1/(4\hat{Q})$], and a second time over the broader band $1 - 1/\hat{Q} \le \omega \le 1 + 1/\hat{Q}$, which includes the transition regions. The passband sensitivity improvements are quite dramatic, particularly for the ladder-derived topologies. Observe that the sensitivity improvement over that of cascade design, when calculated over the broader band, is not as striking; nevertheless, the MF sensitivities are still lower by about a factor 2. This result ought to be expected in view of Fig. 5-25, since now the band edges with their large sensitivity contributions are included in the evaluation. It should be noted that the Q_i and H_{Bi} variations are excluded from the sensitivity measure (per Section 5.5); if these variations are included in the analysis, the ratio M_α/M_α^c for passband evaluations will increase somewhat. This increase will, however, not substantially alter the percent improvement $|M_\alpha^c - M_\alpha|/M_\alpha^c$. It is interesting to compare the sensitivity improvements as a function of filter order. For the examples evaluated, this improvement appears to be a function of response shape, MF topology, and frequency band of interest. For example, the sensitivity improvements with FLF Chebyshev designs can either increase or decrease with filter order, depending on the frequency band of interest. Before drawing general conclusions about the trends

TABLE 5-7 Design examples[a]

	N	Feedback Topology	$\frac{Q_1}{Q}$	$\frac{Q_2}{Q}$	$\frac{Q_3}{Q}$	$\frac{Q_4}{Q}$	M_0	M_1	M_2	M_3	M_4	t_{12}	t_{13}	t_{23}	t_{14}	t_{34}	$1-\frac{1}{4Q}\le\omega$ $\le 1+\frac{1}{4Q}$ M_α/M_α^c	$1-\frac{1}{Q}\le\omega$ $\le 1+\frac{1}{Q}$ M_α/M_α^c
Butterworth	2	All	1.414	1.414	—	—	2.453	1.620	1.000	—	—	0.500	—	—	—	—	0.005	0.567
	3	FLF[b]	1.724	1.786	1.163	—	5.216	2.200	1.218	1.000	—	0.695	0.123	—	—	—	0.050	0.521
	3	LF	1.351	∞	0.794	—	—	0.0794	1.159	1.000	—	0.404	—	0.664	—	—	0.001	0.582
	3	MLF$_1$	1.852	20.000	0.826	—	—	4.295	1.576	1.000	—	0.447	—	0.560	—	—	9×10^{-4}	×
	3	MLF$_2$	2.857	2.857	0.901	—	—	4.179	1.596	1.000	—	0.607	—	0.217	—	—	×	0.512
	3	MSF	1.370	1.961	1.316	—	—	×	×	×	—	0.340	0.207	0.345	—	—	×	0.483
	4	FLF[c]	1.715	1.639	1.639	1.235	7.309	4.023	1.840	1.281	1.000	0.870	0.283	—	0.165	—	0.063	0.530
	4	LF, MLF$_1$	0.665	∞	∞	0.901	—	0.0050	0.0770	1.125	1.000	0.861	—	0.295	—	0.589	3×10^{-4}	0.594
	4	MLF$_2$	0.831	1.250	5.000	2.439	—	4.079	1.623	1.232	1.000	0.239	—	0.269	—	0.640	×	0.500
Chebyshev 0.5 dB	2	All	1.950	1.950	—	—	4.251	2.384	1.000	—	—	1.008	—	—	—	—	0.005	0.509
	3	FLF[d]	3.506	3.244	2.085	—	26.41	8.420	2.694	1.000	—	1.027	0.074	—	—	—	0.190	0.536
	3	LF, MLF	1.816	∞	1.914	—	—	0.1225	1.784	1.000	—	0.570	—	0.573	—	—	0.032	0.429
	3	MSF	∞	0.965	∞	—	—	×	×	×	—	0.724	0.683	0.759	—	—	0.025	×
	4	FLF[e]	4.091	5.202	3.524	2.664	168.3	37.74	8.223	2.853	1.000	1.190	0.067	—	0.193	—	0.097	0.642
	4	LF, MLF	1.861	∞	∞	1.791	—	0.0080	1.319	1.899	1.000	0.488	—	0.354	—	0.516	0.004	0.361

[a] Each example is optimized according to sensitivity and dynamic range per Eqs. (5-82) and (5-87).
[b] For PRB $Q_1/Q = Q_2/Q = Q_3/Q = 1.500$.
[c] For PRB $Q_1/Q = Q_2/Q = Q_3/Q = Q_4/Q = 1.531$.
[d] For PRB $Q_1/Q = Q_2/Q = Q_3/Q = 2.800$.
[e] For PRB $Q_1/Q = Q_2/Q = Q_3/Q = Q_4/Q = 3.650$.

suggested by the sensitivities in Table 5-7, more careful investigations, extended to higher order, need to be made.

Since, with the exception of the FLF topology, the actual section gains H_{Bi} can be determined only after the BP section and summer implementation are determined [compare Eqs. (5-144) and (5-159)], Table 5-7 lists only the values M_i, defined in Eqs. (5-99) and (5-158), which are needed for calculating H_{Bi}. M_0, required for the FLF topology with input summer (Fig. 5-27), is also included.

The equalization of the maximum signal levels is demonstrated by the experimental results shown in Fig. 5-42, obtained for eighth-order 0.5-dB ripple Chebyshev FLF, MLF, and IFLF BP filters with $f_0 = 5\,\mathrm{kHz}$ and a 0.5-dB bandwidth of 500 Hz, designed using the parameters in Table 5-7. Although the output V_4 is of course the same for Fig. 5-42a–c, it is noted that the internal voltages V_i, $i < 4$, occupy widely differing signal ranges for the three topologies.

5.10 MINIMUM OP AMP MF REALIZATIONS

As mentioned previously, both noise and power dissipation can be lowered by reducing the number of op amps used in an active-RC realization. In MF realizations we see that several op amps, which are in addition to those required to implement the biquad sections, are used to sum and invert section output voltages.

A scheme that eliminates the need for these op amps in the formation of the external feedback paths is illustrated in Fig. 5-43 for a two-section filter. In keeping with goal of minimizing the number of op amps, the BP biquad sections are the single-amplifier biquads (SABs) shown in Fig. 4-17b. The scheme to eliminate the inverting and summing amplifiers is to feed the output of the second section back into the noninverting terminal of the initial SAB op amp. Implementing the feedback in this manner allows both the summation and sign inversion to be performed by the SAB op amp. The SAB networks that correspond to $T_1(T_2)$ and T_1' are shown in Figs. 5-44 and 5-45, respectively. Analyzing the circuit in Fig. 5-44 yields the following BP transfer function (see Section 4.2.6):

$$T_1 = T_2 = \frac{-\dfrac{K_1}{R_1 C_2}(1 + R_a/R_b)s}{s^2 + \left(\dfrac{C_1 + C_2}{C_1 C_2 R_2} - \dfrac{1}{R_1 C_2}\dfrac{R_a}{R_b}\right)s + \dfrac{1}{R_1 R_2 C_1 C_2}} \qquad (5\text{-}165a)$$

$$= \frac{-H_B b_1 s}{s^2 + b_1 s + b_0} \qquad (5\text{-}165b)$$

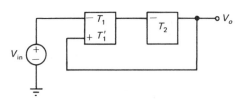

Fig. 5-43 Two-op amp coupled biquad fourth-order band-pass filter. The output of single amplifier biquad T_2 is fed into the positive terminal of the op amp in single-amplifier biquad T_1. Note that T_1' is the transfer function from the input to the positive op amp terminal of T_1 to the output of T_1.

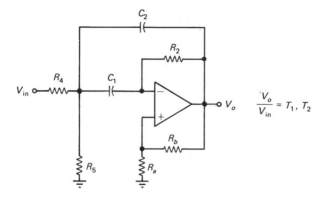

Fig. 5-44   Band-pass single-amplifier biquad (Fig. 4-17) realization for transfer functions T_1 and T_2.

Fig. 5-45   Circuit for determining the transfer function T'_1.

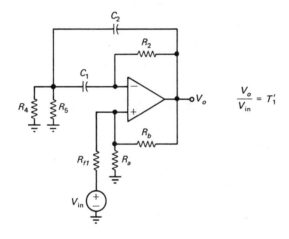

Similarly, analyzing the circuit in Fig. 5-45 yields

$$T'_1 = \frac{\dfrac{R_a}{R_a + R_{F1}}\left[s^2 + \left(\dfrac{R_1 + R_2}{C_2 R_1 R_2} + \dfrac{1}{C_1 R_2}\right)s + \dfrac{1}{R_1 R_2 C_1 C_2}\right]}{s^2 + \left(\dfrac{C_1 + C_2}{C_1 C_2}\dfrac{1}{R_2} - \dfrac{1}{R_1 C_2}\dfrac{R_d}{R_b}\right)s + \dfrac{1}{R_1 R_2 C_1 C_2}} \tag{5-166a}$$

$$= \frac{K_F(s^2 + C_F s + b_0)}{s^2 + b_1 s + b_0} \tag{5-166b}$$

$$= K_F + \frac{K_F(C_F - b_1)s}{s^2 + b_1 s + b_0} \tag{5-166c}$$

where

$$R_1 = R_4 \,\|\, R_5, \qquad K_1 = \frac{R_5}{R_4 + R_5}, \qquad K_f = \frac{R_a}{R_a + R_{f1}}$$

It is noted that the roots of the numerator in Eq. (5-166) are always real and negative; hence, they play only a negligible rôle in determining the response of the overall feedback system in Fig. 5-43.

Let us now write the transfer function for the MF network in Fig. 5-43 in terms of the block transfer functions T_1, T_1' and T_2:

$$H = \frac{V_o}{V_{in}} = \frac{T_1 T_2}{1 - T_1' T_2} \tag{5-167}$$

Substituting Eqs. (5-165b) and (5-166c) into Eq. (5-167) yields

$$H = \frac{H_B^2 b_1^2 s^2 / (s^2 + b_1 s + b_0)^2}{1 + \dfrac{H_B b_1 s}{s^2 + b_1 s + b_0} \left(K_F + \dfrac{K_F(C_F - b_1)s}{s^2 + b_1 s + b_0} \right)} \tag{5-168}$$

With the narrow band approximation $s \simeq j\sqrt{b_0}$ we have

$$H \simeq \frac{H_B^2 b_1^2 s^2}{(s^2 + b_1 s + b_0)^2 + K_F H_B b_1 C_F s^2} = \frac{N}{D} \tag{5-169}$$

Comparing Eq. (5-169) with that of a two-section FLF with $D = (s^2 + b_1 s + b_0)^2 + F_{12} H_B^2 b_1^2 s^2$, we observe that the role of the FLF feedback has been preserved, and

$$K_F = \frac{F_{12} H_B b_1}{C_F} > 0 \tag{5-170}$$

To realize transmission zeros, feedforward summation is conveniently used, as illustrated in Fig. 5-46. This network is straightforwardly analyzed to determine the transfer function

$$H(s) = \frac{k_2 T_1 T_2 + k_1 T_1 + k_0 - k_0 T_1' T_2}{1 - T_1' T_2} \tag{5-171}$$

Fig. 5-46 Three-op amp coupled biquad fourth-order band-pass filter with finite transmission zeros. The poles are realized using the topology in Fig. 5-43 and the finite transmission zeros are realized using feedforward summation.

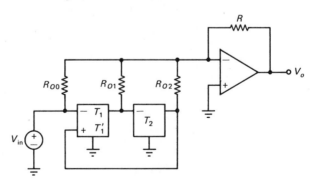

High-Order Filter Realization

when $k_i = R/R_{0i}$ and T_1, T_2, T_1' are given in Eqs. (5-165) and (5-166). Substituting Eqs. (5-165) into Eq. (5-171) yields

$$H(s) = \frac{c_4 s^4 + c_3 s^3 + c_2 s^3 + c_1 s + c_0}{s^4 + d_3 s^3 + d_2 s^2 + d_1 s + d_0} \qquad (5\text{-}172\text{a})$$

where

$$c_4 = k_0/(1 + K_F) \qquad (5\text{-}172\text{b})$$

$$c_3 = (2k_0 b_1 + k_0 H_B K_F b_1^2 - K_1 H_B b_1^2)/(1 + K_F) \qquad (5\text{-}172\text{c})$$

$$c_2 = (K_2 H_B^2 b_1^4 + k_0(b_1^2 + 2b_0) + k_0 H_B K_F b_1^2 C_F - k_1 H_B b_1^3)/(1 + K_F) \qquad (5\text{-}172\text{d})$$

$$c_1 = b_0(2k_0 b_1 + k_0 H_B K_F b_1^2 - k_1 H_B b_1^2)/(1 + K_F) \qquad (5\text{-}172\text{e})$$

$$c_0 = k_0 b_0^2/(1 + K_F) \qquad (5\text{-}172\text{f})$$

Also,

$$d_3 = 2b_1 \qquad (5\text{-}172\text{g})$$

$$d_2 = 2b_0 + b_1^2 + \frac{K_F H_B b_1}{1 + K_F}(C_F - b_1) \qquad (5\text{-}172\text{h})$$

$$d_1 = 2b_0 b_1 \qquad (5\text{-}172\text{i})$$

$$d_0 = b_0^2 \qquad (5\text{-}172\text{j})$$

It is seen that Eq. (5-172a) can realize exactly any fourth-order symmetric BP transfer function.

The design procedure for this MF implementation is as follows:

1. Determine K_F to minimize sensitivity. In this regard K_F can be computed from an optimum FLF F_{12} using Eq. (5-170).
2. The feedforward gain constants $k_0, k_1,$ and k_2 are computed in order to obtain the desired numerator coefficients [e.g., Eqs. (5-172b) to (5-172f)].

The symmetric BP filter, characterized by the LP prototype given previously in Eq. (5-132) and realized by GFLF and FLF networks in Figs. 5-30 and 5-31, was realized using the network in Fig. 5-46. The details of this design are left as an exercise for the reader. The transfer functions T_1 and T_2, with ω_0 normalized to unity, are characterized by the following parameters:

SAB Section 1: $\omega_0 = 1$, $Q_1 = 37.01$, and $H_{B1} = 2.985$

SAB Section 2: $\omega_0 = 1$, $Q_2 = 37.01$, and $H_{B2} = 1.496$

Note that $Q_1 = Q_2 = 37.01$ is somewhat higher than what is commonly considered to be the upper limit ($Q \leq 30$) for precision SAB filters.

Four models of this design were implemented, and the resulting measurements are summarized in Figs. 5-47 and 5-48. The filters were tuned by using the measured capacitor values of each SAB section to calculate the desired resistor values via the BP subset of the synthesis equations Eqs. (4-66). The final

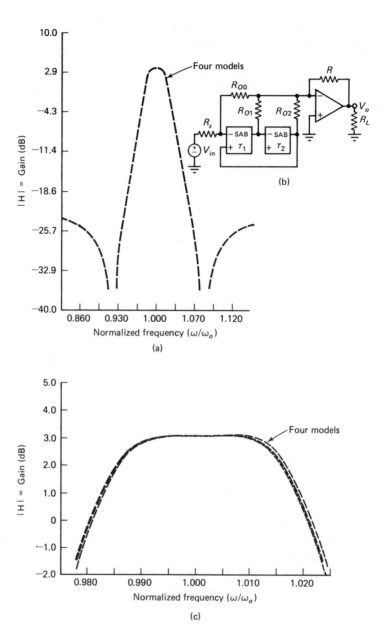

Fig. 5-47 Overall (a) and passband (c) response for four models of the fourth coupled biquad filter (b) realizing the transfer function in Eq. (5-132).

step is to laser-trim the SAB resistors to these values. Laser trimming is not perfect and trim tolerances on the order of 0.2% are typical. Figures 5-47 and 5-48 show the resulting responses for the four pairs of SAB sections which comprise the four MF BP filters shown schematically in Fig. 5-47b. Note the variations in the sections; the spread about the center frequency is about 0.4 dB

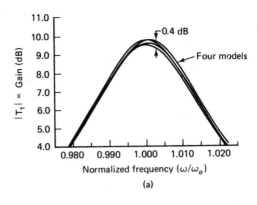

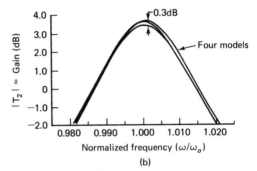

Fig. 5-48 Experimental responses of the four single-amplifier biquad (SAB) sections (Fig. 5-44) used to implement (a) section 1 and (b) section 2 of the four filters in Fig. 5-47.

(± 0.2 dB) for each section. After the sections are measured, they were randomly combined to realize the four MF filters with responses shown in Figs. 5-47a and c. Note especially the experimental curves in Fig. 5-47c where the four MF filter responses are plotted to the same scales (1 dB/div and 0.005 Hz/div) as the SAB section responses in Fig. 5-48. The deviation in the MF responses at and about the center frequency is almost indiscernible. In particular, the response deviations over the band $0.99 < \omega/\omega_0 < 1.01$ are much smaller than the deviation of any one section. If these sections had been cascaded, the individual section variations at each frequency would sum to yield the overall response variations, which would be expected to be on the average significantly larger than those shown in Fig. 5-47c for the MF realization. Note also the four responses plotted over a wider frequency range in Fig. 5-47a. The stop band performance, including the notch areas, are not degraded by the feedforward summation. These experimental results illustrate that the sensitivity trends in Table 5-7 do occur in practice. The extent to which the dramatic improvements cited in Table 5-1 are achieved in practice depends on the design requirements, the quality of the passive and active components, and the degree to which the assumptions stated in Section 5.5 are valid.

The coupling scheme presented in Fig. 5-43, for two sections, can be generalized to N sections using either the LF, MLF, and IFLF topologies, as shown in Figs. 5-49a and b. In each of these MF realizations N op amps are used to realize a $2N$th-order filter. In addition to the fully coupled MF topologies, one

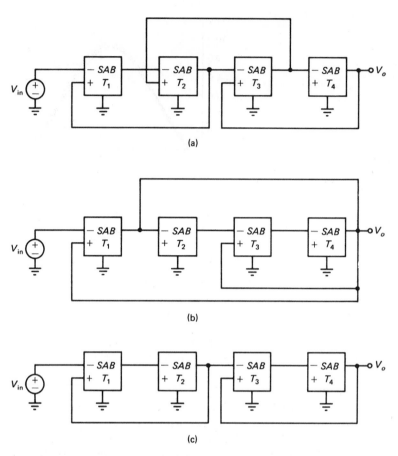

Fig. 5-49 Minimum-op amp MF topologies using band-pass SAB sections (shown for eighth-order band-pass realization). (a) LF, MLF. (b) IFLF. (c) Mixed coupled cascade (MCC).

can use partially coupled topologies, such as that shown in Fig. 5-49c. Partially coupled realizations do not require sophisticated computer aids for their design and achieve improved sensitivity performance over cascade. In particular, the mixed coupled cascade (MCC) MF topology requires only the appropriate pairing of the biquad sections and then repeated use of Eq. (5-170) to design each coupled pair. The derivations of the transfer functions for the MF topologies in Fig. 5-49 are left as exercises for the reader.

5.11 SUMMARY

In this chapter we have presented numerous alternative means for realizing high-order filters. These techniques are:

1. Direct synthesis.
2. Cascade synthesis.

3. GIC immittance simulation.

4. Multiple-loop feedback (and feedforward) synthesis.

The technique that will serve a given application best is the one that meets the frequency-domain and statistical requirements with minimum hardware and engineering time. For all but the most tolerant applications, the direct approach is impractical. To minimize op amps, designs based on the use of single-amplifier biquads (SABs) are encouraged. A simple design method requiring minimum engineering time is the cascade approach. In most applications we find that cascade designs, with a 0.2% resistor tuning capability, provide sufficient precision. To meet more challenging requirements, the SAB MF topologies in Fig. 5-49 are recommended. When the sections are assigned identical Q's, the FLF (PRB) design will provide lower sensitivities than will a corresponding cascade design. In fact, all MF designs can provide lower sensitivities than the cascade within the passband. However, lower in-band sensitivities will be achieved with either an LF or MLF design. Since LF designs can be derived directly from a properly terminated LC ladder network, the very low LF sensitivities can often be obtained with modest engineering effort. FLF filter is not only easy to design, it is the only topology whose output noise can remain independent of the overall filter gain and the type of second-order section used [P27]. Finally, when the LF, MLF, and FLF designs do not quite satisfy requirements, an optimized MSF design should be tried.

REFERENCES

Books

B1. SEDRA, A. S., AND P. O. BRACKETT, *Filter Theory and Design*. Portland, Oreg.: Matrix, 1978.

B2. DARYANANI, G., *Principles of Active Network Synthesis and Design*. New York: Wiley, 1976.

B3. HUELSMAN, L. P., *Active RC Filters: Theory and Application*. Stroudsberg, Pa.: Dowden, Hutchinson & Ross, 1976.

B4. GHAUSI, M. S., *Electronic Circuits*. New York: Van Nostrand Reinhold, 1971.

B5. SHEAHAN, D. F., AND R. A. JOHNSON, eds., *Modern Crystal and Mechanical Filters*. New York: IEEE Press, 1977.

B6. SCHAUMANN, R., M. A. SODERSTRAND, AND K. R. LAKER, eds., *Modern Active Filter Design*, New York: IEEE Press, 1981.

Papers

P1. HALFIN, S., "An Optimization Method for Cascaded Filters," *Bell Syst. Tech. J.*, 49, no. 2 (February 1970), 185–190.

P2. LUEDER, E., "A Decomposition of a Transfer Function Minimizing Distortion and Inband Losses," *Bell Syst. Tech. J.*, 49, no. 3 (March 1970), 455–469.

P3. LUEDER, E., "Optimization of the Dynamic Range and the Noise of *RC*-Active

Filters by Dynamic Programming," *Int. J. Circuit Theory Appl.*, 3 (1975), 365–370.

P4. ANTONIOU, A., "Realization of Gyrators Using Operational Amplifiers and Their Use in *RC*-Active Network Synthesis," *Proc. IEE* (Lond.), November 1969 1838–1850.

P5. SHEAHAN, D. F., "Gyrator-Floatation Circuit," *Electron. Lett.*, 3, (1967), 39–40.

P6. HOLMES, W. H., S. GRUETZMANN, AND W. E. HEINLEIN, "Direct-Coupled Gyrators with Floating Ports," *Electron. Lett.*, 3, (1967), 46–47.

P7. BRUTON, L. T., "Network Transfer Functions Using the Concept of Frequency Dependent Negative Resistance," *IEEE Trans. Circuit Theory*, CT-16 (August 1969), 406–408.

P8. GORSKI-POPIEL, J., "*RL*-Active Synthesis Using Positive-Immittance Converters," *Electron. Lett.*, 3 (August 1967), 381–382.

P9. BRUTON, L. T., "Multiple Amplifier *RC*-Active Filter Design with Emphasis on GIC Realizations," *IEEE Trans. Circuits Syst.*, CAS-25 (October 1978), 830–845.

P10. GIRLING, F. E. J., AND E. F. GOOD, "Active Filters: Part 12. The Leap Frog or Active Ladder Synthesis," *Wireless World*, July 1970, 341–345.

P11. GIRLING, F. E. J., AND E. F. GOOD, "Active Filters: Part 13. Applications of Active Ladder Synthesis," *Wireless World*, September 1970, 445–450.

P12. GIRLING, F. E. J., AND E. F. GOOD, "Active Filters: Part 14. Bandpass Types," *Wireless World*, October 1970, 505–510.

P13. SZENTIRMAI, G., "Synthesis of Multiple-Feedback Active Filters," *Bell Syst. Tech. J.*, 52 (April 1973), 527–555.

P14. LAKER, K. R., M. S. GHAUSI, AND J. J. KELLY, "Minimum Sensitivity Active (Leap Frog) and Passive Ladder Bandpass Filters," *IEEE Trans. Circuits Syst.*, CAS-22 (August 1975), 670–677.

P15. DUBOIS, D., AND J. J. NEIRYNCK, "Synthesis of a Leap Frog Configuration Equivalent to an *LC*-Ladder Filter between Generalized Terminations," *IEEE Trans. Circuits Syst.*, CAS-24, (November 1977), 590–597.

P16. LAKER, K. R., AND M. S. GHAUSI, "Synthesis of a Low Sensitivity Multiloop Feedback Active *RC* Filter," *IEEE Trans. Circuits Syst.*, CAS-21 (March 1974), 252–259.

P17. TOW, J., "Design and Evaluation of Shifted Companion Form Active Filters," *Bell Syst. Tech. J.*, 54, no. 3 (March 1975), 545–568.

P18. SCHAUMANN, R., W. A. KINGHORN, AND K. R. LAKER, "Minimizing Signal Distortion in FLF Active Filters," *Electron. Lett.*, 12, no. 9 (April 1976), 211–213.

P19. LAKER, K. R., AND M. S. GHAUSI, "Computer Aided Analysis and Design of Follow-the-Leader Feedback Active *RC* Filters," *Int. J. Circuit Theory Appl.*, 4 (April 1976), 177–187.

P20. LAKER, K. R., AND M. S. GHAUSI, "A Comparison of Active Multiple-Loop Feedback Techniques for Realizing High Order Bandpass Filters," *IEEE Trans. Circuits Syst.*, CAS-21 (November 1973), 774–763.

P21. ACAR, C., M. S. GHAUSI, AND K. R. LAKER, "Critical Block Analysis for the Transfer, Gain, and Phase Functions in Multiple-Loop Feedback Networks," *Int. J. Circuit Theory Appl.*, 6, no. 1 (January 1978), 89–104.

P22. GADENZ, R. N., "On Low Sensitivity Realizations of Band Elimination Active Filters," *IEEE Trans. Circuits Syst.*, CAS-24 (April 1977), 175–183.

P23. LAKER, K. R., R. SCHAUMANN, AND M. S. GHAUSI, "Multiple-Loop Feedback Topologies for the Design of Low-Sensitivity Active Filters," *IEEE Trans. Circuits Syst.*, CAS-26 (January 1979), 1–22.

P24. LOPRESTI, P. V., "Optimum Design of Linear Tuning Algorithms," *IEEE Trans. Circuits Syst.*, CAS-24 (March 1977), 144–151.

P25. TOW, J., "Comments on 'Lower Bounds on the Summed Absolute and Squared Voltage Transfer Sensitivities in *RLC* Networks'," *IEEE Trans. Circuits Syst.*, CAS-26 (March 1979), 209–211.

P26. LAKER, K. R., AND M. S. GHAUSI, "Design of Minimum Sensitivity Multiple Loop Feedback Band-pass Active Filters," *J. Franklin Inst.*, vol. 310 (1980), 51–64.

P27. PADUKONE, P., J. MULAWKA, AND M. S. GHAUSI, "An Active-*RC* Biquadratic Section with Reduced Sensitivity to Operational Amplifiers," *J. Franklin Inst.*, vol. 310 (1980), 27–40.

P28. CHIOU, C., AND R. SCHAUMANN, "Comparison of Dynamic Range Properties of High-Order Active Band-pass Filters," *Proc. 1980 Int. Symp. Circuits Syst.*, Houston, Tex., 1980, 816–819.

P29. LOPRESTI, P. R., AND K. R. LAKER, "Optimum Tuning of Multiple-Loop Feedback Active *RC* Filters," *Proc. 1980 Int. Symp. Circuits Syst.*, Houston, Tex., 1980, 812–815.

P30. SCHAUMANN, R., K. R. LAKER, W. A. KINGHORN, "The Maximization of Dynamic Range of FLF Bandpass Filters by Use of Stagger Tuning," *Proc. 1977 IEEE Int. Symp. Circuits Syst.*, Pheonix, Ariz., April 1977, 450–453.

PROBLEMS

5.1 Use a single op amp to realize a Butterworth low-pass third-order filter with $\omega_{3dB} = 10^4$ rad/sec. Show the designed circuit.

5.2 For the circuit shown in Fig. 5-1:
(a) Derive Eq. (5-1).
(b) With $Y_3 = Y_5 = G = 1$, design the circuit to realize the following transfer function:

$$\frac{V_o}{V_i} = \frac{s^2 + 1}{(s + 2)(s^2 + s + 1)}$$

5.3 Design the circuit shown in Fig. 4-1 to realize

$$\frac{V_o}{V_i}(s) = \frac{0.5(s^2 + 1.5)}{(s + 1)(s^2 + 0.5s + 1.0)}$$

5.4 A single-op amp circuit that can be used for the realization of a general high-order transfer function is shown in Fig. P5-4.
(a) Show that

$$\frac{V_o}{V_i} = \frac{Y_1(Y_3 + Y_4 + Y_6) - Y_3(Y_1 + Y_2 + Y_5)}{Y_6(Y_1 + Y_2 + Y_5) - Y_5(Y_3 + Y_4 + Y_6)}$$

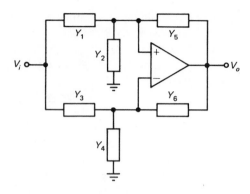

Fig. P5-4

(b) Let $Y_1 + Y_2 + Y_5 = Y_3 + Y_4 + Y_6$, and realize the transfer function given in Prob. 5.1.

5.5 The pole–zero location of a fifth-order elliptic filter, used in telephone D-channel bank systems, is given by Eq. (1-57), repeated below.

$$H(s_n) = \frac{100[s_n^2 + (29.2)^2][s_n^2 + (43.2)^2]}{(s_n + 16.8)(s_n^2 + 19.4s_n + 400.34)(s_n^2 + 4.72s_n + 507.33)}$$

where $s_n = s/10^3$. Show a complete designed circuit using the cascade procedure by using two single-op amp second-order blocks.

5.6 The element sensitivities for the two biquads are listed in Table 5-2.
(a) Verify the results for $S_{R_1}^{T_i}$.
(b) Verify the results for $S_{C_2}^{T_i}$.

5.7 A high-order band-pass filter is required having a maximally flat magnitude response and with the following additional specifications:

$$f_0 = 5 \text{ kHz}, \qquad \Delta f_{3dB} = 500 \text{ Hz}, \qquad \Delta f_{30dB} \leq 1.3 \text{ kHz}$$

The overall gain at the center frequency is arbitrary.
(a) Find the low-pass prototype transfer function that will meet the foregoing requirement.
(b) Find the band-pass transfer function and show a cascade realization by simply denoting ω_{oi} and Q_i of each block.

5.8 Repeat Prob. 5.7 by realizing the design in PRB configuration Fig. 5-24b.

5.9 Repeat Prob. 5.7 by realizing the design in FLF configuration. Show only the values of ω_{oi}, Q_i, H_i, and F_{1i}.

5.10 Show a cascade realization of Eq. (5-41a) using single-op amp biquads.

5.11 Show a cascade realization of Eq. (5-41c) using two-op amp GIC-derived sections shown in Fig. 4-28a–f.

5.12 Show a cascade realization of Eq. (5-40) with optimum pole–zero pairing and using the three-op amp biquads shown in Fig. P4-24 and verify the circuit element values given in Table 5-4.

5.13 The circuit shown in Fig. P5-13 can be used to realize a third-order elliptic func-

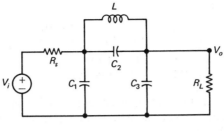

Fig. P5-13

tion of the form

$$\frac{V_o}{V_i}(s) = \frac{H(s^2 + a_1)}{(s + p)(s^2 + b_{11}s + b_{10})}$$

(a) If $p = 0.08878$, $b_{11} = 0.3821$, $H = 0.5058$, $b_{10} = 1.1654$, and $a_1 = 2.0455$, determine the circuit element values and sketch the frequency response.

(b) Show an inductance simulation of Fig. P5-13 indicating all the element values.

(c) For the FDNR realization shown in Fig. 4-2c, can the number of FDNRs be reduced by some circuit manipulation in Fig. 4-2a prior to FDNR substitution? Show the circuit.

5.14 A sixth-order elliptic band-pass filter with center frequency 2804 Hz, 0.1-dB pass band ripple, a passband of 80 Hz, with a minimum attenuation of 30 dB in the stopband (which extends above 2919.8 Hz and below 2694.8 Hz) is specified as shown in Fig. P5-14a. The transfer function which realizes this specification is

Fig. P5-14

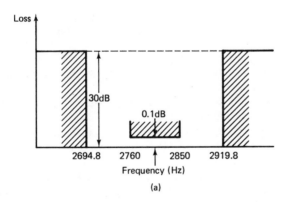

(a)

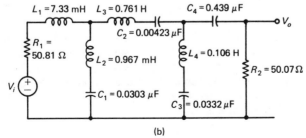

(b)

given by [B2, P17]

$$T(s) = \frac{597s}{s^2 + 597s + 3.106172(10)^8} \times \frac{0.36[s^2 + 2.834248(10)^8]}{s^2 + 233.33s + 2.987363(10)^8}$$
$$\times \frac{0.562[s^2 + 3.404184(10)^8]}{s^2 + 242.61s + 3.229704(10)^8}$$

and a passive-RLC ladder realization of this circuit is shown in Fig. P5-14b. Show an inductance simulation of Fig. P5-14b, using Fig. 2-33 to realize the inductors.

5.15 Show a cascade realization of Prob. 5.14 using the two-op amp GIC-derived biquads.

5.16 Show an FDNR realization of the circuit shown in Fig. P5-14b using the circuit in Fig. 2-34.

5.17 For the circuit of Fig. 5-27, show that the transfer function is given by Eq. (5-90).

5.18 Design an eighth-order FLF band-pass filter with the following specifications: $H_{MN} = 2, f_0 = 10$ kHz, $\hat{Q} = 30$, and the LP prototype is the fourth-order Butterworth response. The parameters are to be chosen for maximized dynamic range and minimized M_α over the frequencies $\omega_0(1 - 1/4\hat{Q}) \le \omega \le \omega_0 (1 + 1/4\hat{Q})$. Show the circuit realization in the block diagram form as in Fig. 5-24, clearly indicating the individual block H_{Bi}, ω_{0i}, and Q_i as well as the feedback coefficients. Note that the design information can be obtained from Table 5-7 and you may use it.

5.19 Repeat Prob. 5.18 for IFLF configuration.

5.20 Design the filter of Prob. 5.18 using the PRB technique.

5.21 It is required to design a fourth-order GFLF band-pass filter having $\hat{Q} = 25$ and $f_0 = 1$ kHz based on the following LP prototype function:

$$H(p) = \frac{0.1(p^2 + 5.1532)}{p^2 + 0.4341p + 1.0106}$$

The biquad pole Q's are to be equal.

(a) Show the circuit block diagram indicating the biquad parameters and the feedback factors.

(b) If the transfer function above is to be realized by the cascade method, what are the required values of pole Q?

(c) Find the ratio $(Q_c)_{max}/Q_{GFLF}$, where $(Q_c)_{max}$ is the maximum pole Q of the biquad blocks in the cascade design and Q_{GFLF} is the pole Q if the biquad blocks in the GFLF design.

5.22 Using Table 5-7, design a sixth-order MLF band-pass filter for the specifications given below: LP prototype: 0.5-dB band-pass ripple Chebyshev response, $f_0 = 5$ kHz, $\hat{Q} = 40$, $H_{MN} = 1$. The design parameters are to be chosen to minimize the gain sensitivity, M_α, over the frequency range $(1 - 1/\hat{Q}) \le \omega \le (1 + 1/\hat{Q})$.

5.23 Design a sixth-order bandpass filter with Butterworth response characteristics in the MSF configuration (Fig. 5-24g) but using identical Q blocks. The desired center frequency and overall $\hat{Q}$ are: $f_0 = 1$ kHz, $\hat{Q} = 25$. It has been found [P26] that the gain sensitivity M_α minimization over the band $0.96 \le \omega/\omega_0 \le 1.04$ is achieved if the individual block Q's are $Q_1 = Q_2 = Q_3 = 37.5$. Determine the feedback parameters of this design.

5.24 Derive the MSF $\mathcal{L}_i$'s for $i = 1, \ldots, N$, where
   (a) $N = 3$ (see Eq's (5-163))
   (b) $N = 4$

5.25 Repeat Prob. 5.23 for the PRB design.

5.26 For the mixed coupled cascade circuit shown in Fig. P5-26, determine the overall transfer function of the circuit if the individual blocks are characterized by

$$T_i(s) = \frac{H_{Bi}s}{s^2 + (\omega_{0i}/Q_i)s + \omega_{0i}^2}$$

Fig. P5-26

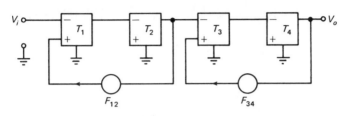

5.27 Design the circuit of Fig. P5-26 for an eighth-order Butterworth band-pass filter response with $\omega_0 = 1$ and the overall $\hat{Q} = 50$. Assume identical blocks (i.e., $T_1 = T_2 = T_3 = T_4 = T$).

chapter six

....

Active Switched Capacitor
Sampled-Data Networks

6.1 INTRODUCTION

Over the past few years the MOS integrated-circuit technology had found wide
usage in industry because of its superior logic density, as compared to that
achievable with bipolar technologies. Today, with large-scale integration (LSI),
tens of thousands of MOS transistors can be placed on a single chip. With
feature sizes in the vicinity of 1 μm invisioned for VLSI, the packing density
will no doubt continue to increase with subsequent decreases in the cost for
complex digital/analog MOS integrated circuits.

   A property unique to MOS integrated circuits is the ability to store charge
on a node for several milliseconds and to sense this stored charge continuously
and nondestructively. This storage in MOS integrated circuits comes about
naturally and cheaply. This property was first utilized in dynamic random-access
memories and dynamic logic (e.g., dynamic shift registers). This charge storage
feature was first used in analog signal processing with the development of the
bucket-brigade shift register. In this device the analog signal is converted into a
series of charge packets, with each packet proportional to sampled values of the
input analog signal. These charge packets are then passed serially through a
chain of MOS capacitors, with charge being passed from one capacitor to
another by a switching operation achieved with externally clocked MOS transistors. The charges on these capacitors can be nondestructively sensed and
summed in a weighted fashion to realize a sampled-data (sometimes referred to
as sampled analog) transversal or finite impulse response (FIR) filter. The subject
of transversal filters and the wave shapes typical of sampled-data filters were
treated briefly in Chapter 1. The charge-coupled device (CCD), developed
subsequent to the bucket-brigade circuits, provided a similar analog function

but offered much improved packing density and performance [P1]. A detailed treatment of CCD filters is beyond the scope of this book. Those readers interested in learning more about this subject are referred to the literature [B1, P1].

More recently, it has been found more attractive to implement MOS analog sampled-data filters as active switched capacitor (SC) filters [P2]. SC filters are recursive or infinite impulse response (IIR) filters; therefore, high Q's and flat passbands can be realized efficiently, much like the active-RC realizations studied in Chapters 4 and 5. Furthermore, sampling rates in excess of 100 kHz can be routinely used. The use of high sampling rates serves to lessen the burden placed on the continuous anti-aliasing filter which must band-limit the input spectrum to one-half the sampling frequency. As noted in Chapter 4, sufficient anti-aliasing can usually be achieved with a relatively imprecise second-order continuous active filter. These are some of the features that render switched capacitor filters more attractive implementations than CCD filters for many applications.

In addition, SC filters take full advantage of the inherent precision achieved by MOS processing. As we will see later, the transfer function coefficients are completely determined by a single, precise crystal-controlled clock frequency and ratioed capacitors. It has been shown that capacitor ratios can be held to about 0.3%, and with appropriate circuit techniques, capacitances as small as 0.5 pF can be used. Furthermore, MOS capacitors are nearly ideal, with very low dissipation factors and good temperature stability. Of economical importance, precision SC filters can be fabricated using memory-like NMOS and CMOS processing. Thus, analog and digital circuitry can be placed on the same chip, providing the capability of integrating complete systems on a single chip.

In this chapter we describe and demonstrate techniques for the analysis and design of active switched capacitor networks. Because of their sampled-data character, switched capacitor networks are most conveniently analyzed and designed, in the z-transform domain, like digital filters. However, SC filters are analog networks; thus, the analog concepts of impedance and loading, which are absent in digital filters, are retained. In this chapter a library of z-domain equivalent circuits will be developed to facilitate the analysis, design, and, more importantly, to aid the reader's understanding of switched capacitor networks. With these equivalent circuits, the familiar circuit theory tools, which we use routinely in the analysis of continuous active networks, are extended for use in switched-capacitor networks. Like active-RC networks, there are many SC topologies that can be used to realize a given z-domain transfer function. We will examine some of the more interesting and useful ones.

6.2 APPLICATIONS OF z-TRANSFORM TECHNIQUES TO THE ANALYSIS OF SWITCHED-CAPACITOR NETWORKS WITH BIPHASE SWITCHES

The purpose of this section is to briefly discuss the basic assumptions regarding the sampled-date nature of SC networks and the fundamentals that are germane to the derivations and procedures given in succeeding sections. This discussion also provides an opportunity to define the symbols and to acquaint the reader with the notation employed.

6.2.1 Sampled-Data Filter Systems

Let us begin with a general discussion of a sampled-data filter system which is suitable for use in a continuous analog (i.e., analog input/analog output) environment. This represents, from a hardware point of view, the most severe environment for an SC filter, or for that matter, any sampled-data filter. The system given in Fig. 6-1 shows the analog signal being passed through a con-

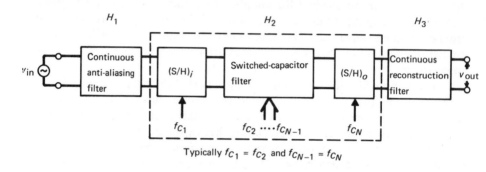

Typically $f_{C_1} = f_{C_2}$ and $f_{C_{N-1}} = f_{C_N}$

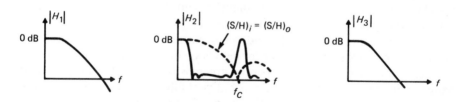

Fig. 6-1 Sampled-data filter system for continuous analog input and smooth analog output.

tinuous anti-aliasing (low-pass) filter, an input sample-and-hold circuit $(S/H)_1$ which samples the band-limited analog input at intervals of $1/f_{C_1}$, the switched capacitor filter, an output sample-and-hold circuit $(S/H)_0$ which resamples the output of the SC filter at intervals $1/f_{CN}$, and a final continuous (low-pass) reconstruction filter which serves to smooth the sharp transitions in the sampled-data waveform. The SC filter is shown to be controlled by clocks of multiple frequencies (f_{C2} through f_{CN-1}). To minimize the silicon area, it is often desirable to clock the low-pass sections at a high rate in order to lessen the burden on the continuous anti-aliasing filter, and to clock the high-pass sections at a lower rate, in order to reduce their total capacitance. It we order the sections so that the low-pass sections, sampled at f_{C2}, precede the high-pass sections, which are sampled at $f_{C3} < f_{C2}$, then the low-pass sections provide the needed anti-aliasing protection up to $f_{C2}/2$. It is noted that decreasing the sampling rate (or decimation), as shown in Fig. 6-1, typically requires no additional circuitry. However, increasing the sampling rate (or interpolation) is a smoothing operation which typically requires additional low-pass filtering. Typically, the clock frequencies are related by binary division [i.e., $f_{Ci} = f_{MACL}/(2)^i$]. This can be achieved by passing the master clock f_{MACL} through a simple binary count-down chain [B2]. It is noted that in many SC networks the sample-and-hold operations shown in Fig. 6-1 are inherently performed by the SC filter. Thus, in these cases, the SC filter includes the three blocks enclosed by the dashed rectangle in Fig. 6-1. Typical frequency responses for the various blocks in Fig. 6-1 are shown below each block for the case $f_{C1} = f_{C2} = f_{CN} = f_C$.

It is noted that when the input or output is interfaced with digital or sampled-data circuitry, such as D/A or A/D converters, some of this hardware is no longer needed. For example, when the output is to be interfaced with a digital environment, the continuous reconstruction filter is no longer needed and the sample-and-hold circuit is typically incorporated with the digital circuitry. Although the need for filtering is reduced, interfacing with digital or sampled-data circuits requires synchronization between the clocks that control the SC filter and those that control the external sampling operations. This is accomplished by passing synchronization pulses between the SC network and external samplers. One reason for the synchronization is to ensure that the SC network output is sampled after all transients have settled and the output is truly held constant.

Now that we have discussed in general terms the hardware used in sampled-data filter systems, let us focus in greater detail on the SC portion of this system.

6.2.2 Operation of Ideal SC Networks

Consider now the operation of an ideal SC network, comprised of ideal capacitors, ideal switches, and ideal voltage-controlled voltage sources (i.e., finite, frequency-independent gain, or infinite-gain op amps) when excited by sampled-data voltage inputs. It is noted that MOS op amps have been designed which settle to within 0.1 % of final value in 2 μ sec and achieve dc gains (a_0) greater than 60 dB [P3].

Therefore, for sampling rates of less than 250 kHz, a good approximation of an actual MOS op amp is a voltage-controlled voltage source with gain of a_0. Typically, the switches are controlled by a two-phase, nonoverlapping clock of frequency $f_C = \frac{1}{2}T$, as shown in Fig. 6-2. Note that ϕ^e is used to denote the even clock phase, which instantaneously closes the e switch on the even $2nT$ times. Similarly, ϕ^o denotes the odd clock phase, which instantaneously closes the o switch on the odd $(2n + 1)T$ times. The switches are assumed to have a 50 % duty cycle with equal (T second) on and off time periods. The upper and lower limits on clock frequency are usually dictated by the settling time of the operational amplifier and various other requirements (e.g. noise, capacitor leakage, Nyquist rate, and anti-aliasing specs). Switched capacitor filters have been described in the literature which are clocked at frequencies that range from

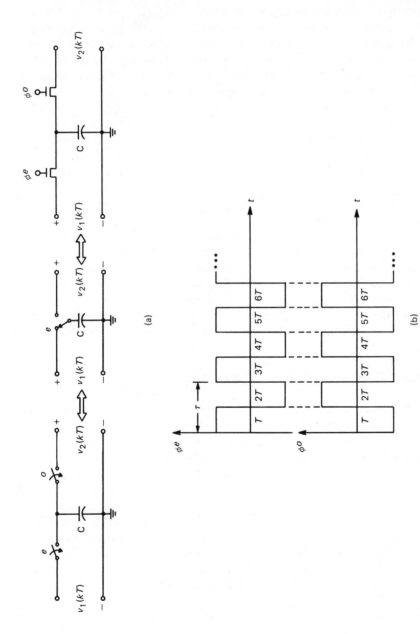

Fig. 6-2 (a) Simple switched capacitor network with (b) biphase nonoverlapping clocks.

256 kHz to as low as 8 kHz. Since MOS op amps can be designed with 0.1 % settling times on the order of 0.5 μsec, clock rates up to 1 MHz seem feasible. As noted briefly in the discussion of Eq. (2-34), as the clock rate increase, the capacitor ratios, hence the silicon area, increases. Therefore, in practice the clock rate is typically chosen no higher than is required to achieve the desired degree of anti-aliasing protection with a second-order continuous filter of sufficiently high cutoff frequency to render its main passband variation acceptably small. The 50% duty cycle assumption is merely for simplification. The behavior of switched capacitor filters is strongly dependent on the clock period and typically is insensitive to the duty cycle (or it can be made to be so). In fact, in practice, to ensure that the e and o switches are never turned on simultaneously, the clocks are made nonoverlapping (i.e., the duty cycle is slightly less than 50%). It is noted that turning both the ϕ^e and ϕ^o switches off simultaneously does not affect the behavior of the circuit; however, turning both switches on will cause improper circuit function. It should be noted that there is no standard symbol for a switch. Consequently, in the literature several different switch symbols have been used in switch capacitor network schematics. Three of the more popular switch symbols are illustrated in Fig. 6-2a. To encourage the reader to become familiar with these different symbols, all three will be used interchangeably throughout this chapter.

Along with the form of the clocks, it will be further assumed that both the input and output of the SC network are sampled-data signals which change in value only at the switching instants kT. Thus, in their most general from, the voltage sources and internal circuit voltages are assumed to be sampled at times kT and held over a one-half clock period interval (T), as shown in Fig. 6-3a. With this assumption, we can apply [P4, P5] z-transform techniques to the general analysis and synthesis of SC networks. The z-transform [B3], $z = e^{s\tau}$, where s is the complex analog frequency variable and $\tau = 2T$ is the clock period, then provides us with a convenient means for performing frequency-domain analysis. The z-transform is discussed in Section 1.9.

As noted in Section 1.9, the use of the z-transform implies a discrete, impulse form of sampling as in a digital filter. Thus, the z-transform will accurately predict the input–output relationship of an SC filter on a sample-by-sample basis. How the input and output waveforms behave between the samples, which is the analog character of the switched capacitor filter, is not yet taken into account. The analog character referred to is the held, staircase-like wave shape shown in Fig. 6-3a. To restore the analog character to our z-transform-computed frequency response, this response must be modified [B4] by a multiplicative $(\sin \omega\tau/2)/(\omega\tau/2)$, where τ is the sampling period. For high sampling rates, where $\omega\tau \ll 1$, the passband of the frequency response is left virtually uneffected. It should be pointed out that in general, when continuous signals are applied to the SC network, a rigorous analysis [P6] is considerably more complex.

As we demonstrate later, the switching action described in Fig. 6-2 provides a time-varying nature to the SC network. That this is the output observed will in general depend upon when, and how often, the output is sampled. Topologically, this time-varying nature results in a network graph that changes

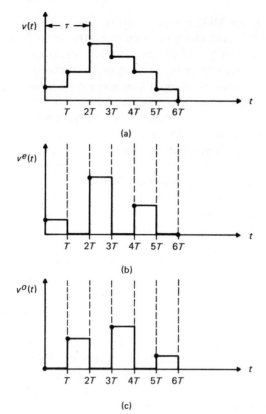

Fig. 6-3 (a) Sampled-data voltage waveforms partitioned into (b) even and (c) odd components.

with time, alternating between two topologies as the switches open and close. One topology corresponds to the even clock phase and a second topology to the odd clock phase. Thus, one can view the time varying SC network, with biphase switches, as two interrelated time-invariant networks [P4–P6]. In view of this fundamental approach, it is mathematically convenient to partition the sampled-data voltage waveform in Fig. 6-3a into its even and odd components as shown, respectively, in Fig. 6-3b and c. Comparing v^e and v^o to the clock waveforms ϕ^e and ϕ^o, we observe that v^e is nonzero only when the e switch is closed, and v^o is nonzero only when the o switch is closed. This fundamental observation [P4] has opened the door to a rigorous understanding of switched-capacitor networks and has resulted in several methods for their analysis.

One way to interpret the relationship between the even and odd topologies is to consider them topologically decoupled, with the states of one determining the initial conditions for the other [P6]. This interpretation results in two distinct circuits coupled together via dependent sources which establish the aforementioned initial conditions. This formulation has been found to be particularly convenient for computer-aided analysis. Another interpretation [P4, P5] is to combine the even and odd networks topologically into a single z-domain equivalent circuit. In general, an n-port bi-phase SC network will

Active Switched Capacitor Sampled-Data Networks

require a $2n$-port equivalent circuit, (i.e., n-ports for the even clock phase and n-ports for the odd clock phase). It is this interpretation that provides the kinds of valuable insight that Laplace transform techniques have provided for linear time-invariant networks. Although in this text we restrict our consideration to bi-phase SC networks, the concepts are extendable to the poly-phase case.

Since SC networks can be most rigorously characterized in terms of charge-transfer operations, discrete-time voltages $v_i(kT)$ and discrete-time charge variations or transfers $\Delta q_i(kT)$ are used as port variables. At the switching times, kT, charges are instantaneously redistributed, with the principle of charge conservation maintained at every node in the network. It is this principle that allows us to write nodal charge equations similar to the way Kirchhoff's current law is used in continuous networks. In general, due to the biphase switching operation, two distinct, but coupled, nodal charge equations are required to characterize the charge conservation condition at a particular node for all time instants, kT; one equation for the even sampling instants and a second equation for the odd sampling instants are required. These equations are written, for some node p, as follows (recall that $\tau = 2T$):

$$\Delta q_p^e(kT) = \sum_{i=1}^{M_{ep}} q_{pi}^e(kT) - \sum_{i=1}^{M_{ep}} q_{pi}^o[(k-1)T] \qquad \text{for } k \text{ an even integer} \qquad (6\text{-}1a)$$

$$\Delta q_p^o(kT) = \sum_{i=1}^{M_{op}} q_{pi}^o(kT) - \sum_{i=1}^{M_{op}} q_{pi}^e[(k-1)T] \qquad \text{for } k \text{ an odd integer} \qquad (6\text{-}1b)$$

or equivalently in the z-domain,

$$\Delta Q_p^e(z) = \sum_{i=1}^{M_{ep}} Q_{pi}^e(z) - z^{-1/2} \sum_{i=1}^{M_{ep}} Q_{pi}^o(z) \qquad (6\text{-}2a)$$

$$\Delta Q_p^o(z) = \sum_{i=1}^{M_{op}} Q_{pi}^o(z) - z^{-1/2} \sum_{i=1}^{M_{op}} Q_{pi}^e(z) \qquad (6\text{-}2b)$$

where q_{pi}^e, q_{pi}^o and Q_{pi}^e, Q_p^o denote, respectively, the instantaneous charges stored on the ith capacitor connected to node p for the even and odd kT time instants and their z-transforms. Also, M_{ep} and M_{op} denote, respectively, the total number of capacitors connected to node p during the even and odd clock phases. When the charge is sensed during the even clock phase, $q_{pi}^e(kT)$ represent the charges that reside on the capacitors C_{pi} at a particular even kT instant of time and $q_{pi}^o[(k-1)T]$ represent the capacitor charges retained from the previous odd $(k-1)T$ instant of time. Hence, for a given even value of k, $q_{pi}^o[(k-1)T]$ serve as the initial condition for circuit action observed at this even time instant. An analogous description can be given for charge sensed during the odd clock phase.

For single-capacitor SC blocks (e.g., Fig. 6-2a), z-transformed nodal charge equations [P5] lead directly to simple z-domain equivalent circuits described in Section 6.3. To characterize a complex SC network we simply substitute, one for one, the appropriate z-domain block equivalent circuit for each SC element in the network schematic. As demonstrated in Section 6.4, transformed nodal charge equations for each node in the network are then written by inspection from the equivalent circuit. The desired voltage transfer function(s) is then obtained by algebraically manipulating these z-domain equations in the usual manner. By this process we are able to analyze SC networks using familiar methods for dealing with linear time-invariant circuits.

6.2.3 Sampled-Data Waveforms

It should be noted that there are several sampled-data waveforms which can be modeled as special cases of the waveform depicted in Fig. 6-3. These waveforms and their respective even and odd components are shown in Fig. 6-4a. One can immediately invoke the z-transform to mathematically describe these waveforms. The return-to-zero waveforms in Fig. 6-4a and b can be expressed mathematically as follows:

$$V_a(z) = V_a^e(z) + V_a^o(z) \qquad (6\text{-}3a)$$

where

$$V_a^o(z) = 0 \qquad (6\text{-}3b)$$

and

$$V_b(z) = V_b^e(z) + V_b^o(z) \qquad (6\text{-}4a)$$

where

$$V_b^e(z) = 0 \qquad (6\text{-}4b)$$

In a similar manner we can characterize the full-clock-period sample-and-hold (S/H) waveforms in Fig. 6-4c and d as follows:

$$V_c(z) = V_c^e(z) + V_c^o(z) \qquad (6\text{-}5a)$$

where

$$V_c^o(z) = z^{-1/2} V_c^e(z) \qquad (6\text{-}5b)$$

Equation (6-5b) simply states that $v_c^o(kT)$ is the delayed replica of $v_c^e(kT)$. Also,

$$V_d(z) = V_d^e(z) + V_d^o(z) \qquad (6\text{-}6a)$$

where

$$V_d^e(z) = z^{-1/2} V_d^o(z) \qquad (6\text{-}6b)$$

If a capacitor of value C is placed across the terminals of the voltage source $v_c(t)$, one observes that the charge on the capacitor changes in value only once per clock cycle, i.e., at the even kT time instants when $v_c(t)$ changes. At the odd kT time instants $v_c(t)$, the capacitor voltage is unchanged; thus, the charge remains constant. This phenomenon is described analytically, for the even and odd clock phases, in the following manner:

$$\Delta Q_c^e(z) = C V_c^e(z) - C z^{-1/2} V_c^o(z) = C(1 - z^{-1}) V_c^e(z) \qquad (6\text{-}7a)$$
$$\Delta Q_c^o(z) = C V_c^o(z) - C z^{-1/2} V_c^e(z) = 0 \qquad (6\text{-}7b)$$

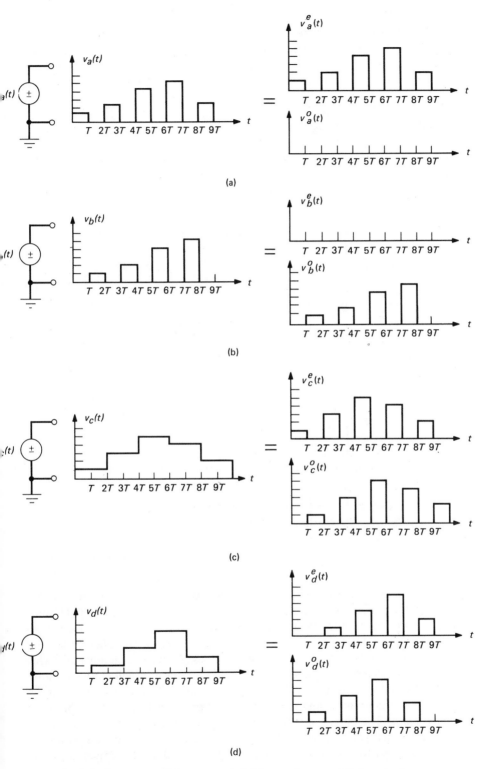

Fig. 6-4 Common special case sampled-data waveforms and their respective even and odd components.

It is noted that the condition $\Delta Q_c^o(z) = 0$ can also be obtained by disconnecting V_c from the capacitor with a switch that is open during the odd clock phase. Thus, in the sense that no charge is transferred, this open-circuit condition also implies a full-cycle sample-and-hold operation. Corresponding relations can also be written for the full-cycle S/H waveform in Fig. 6-4d.

It is useful to note that the return-to-zero waveforms, $v_a(t)$ and $v_b(t)$, can be obtained by processing $v(t)$ in Fig. 6-3 with simple switch networks as shown in Fig. 6-5. When the switches in Fig. 6-5 are ideal, v_a and v_b are ideal, zero-impedance voltage sources with waveforms as depicted in Fig. 6-4a and b, respectively.

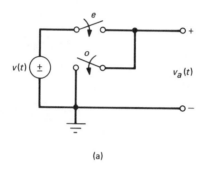

(a)

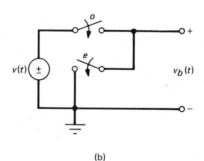

(b)

Fig. 6-5 Switch networks for return-to-zero voltage generation.

The S/H waveforms, with full clock period (Fig. 6-4c and d) and one-half clock period (Fig. 6-3a) hold intervals, can be derived from a continuous waveform using the circuits in Fig. 6-6. In Fig. 6-6a the switch is normally open and only closes briefly at the sampling instants. Ideally, the duty cycle for ϕ should be as short as possible so that $v(t)$ does not change significantly while the switch is closed. When the switch closes, the capacitor will charge instantaneously to the value of $v(t)$ at the sampling instant. When the switch opens, the charge will remain on C and the output will remain unchanged until the switch closes again. The simple circuit in Fig. 6-6a cannot be connected to a low-impedance dissipative load such as a large-valued switched capacitor or a low-valued resistor. In such cases the charge will leak off the capacitor during

Active Switched Capacitor Sampled-Data Networks

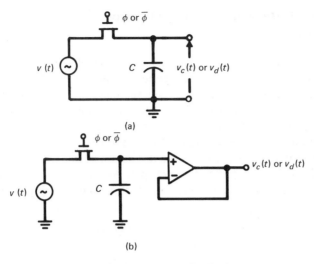

Fig. 6-6 Sample-and-hold circuits.

the periods when the switch is open and the output voltage will not remain constant. By buffering the holding capacitor from the load with a unity-gain amplifier, as shown in Fig. 6-6b, this difficulty is overcome.

The $\sin x/x$ frequency response, mentioned earlier for the S/H operation, is derived directly from the impulse response for the circuit in Fig. 6-6. Assuming that $v(t)$ is constant during the period the switch is closed, we can write the impulse function as follows:

$$h(t) = \frac{1}{\tau} u_1(t) - \frac{1}{\tau} u_1(t - \tau) \qquad (6\text{-}8)$$

where $u_1(\cdot)$ is the unit step function and τ is the time duration between samples (i.e., between switch closings). Taking the Laplace transform of Eq. (6-8) yields the following transfer function:

$$H(j\omega) = \frac{1 - e^{-\tau s}}{s\tau}\bigg|_{s=j\omega} = \frac{\sin \omega\tau/2}{\omega\tau/2} e^{-j\omega\tau/2} \qquad (6\text{-}9)$$

Thus, $|H(j\omega)| = (\sin \omega\tau/2)/(\omega\tau/2)$, where $\tau = T$ for the sampling operations defined in Figs. 6-3a, 6-4a, and b, and $\tau = 2T$ for the sampling defined by Fig. 6-4c and d.

The various sampled-data waveforms considered in this section can be generated externally (i.e., by independent voltage source) or at any internal node by an appropriate combination of switches and capacitors. It is often crucial, particularly at the network output, where one may either resample or couple to another SC network, to identify the waveform type of internal node voltages. As we will see later, using the properties described in this section, this identification process is quite straightforward.

6.2.4 SC Network Transfer Function Relations

In previous chapters we observed the value of the Laplace-transformed voltage transfer function in the characterization and synthesis of continuous active filters. As we see in succeeding sections, the z-transformed voltage transfer function plays an equivalently important role in specifying and designing active SC filters. At this point it should be again emphasized that SC network transfer functions are most conveniently written in the z-domain.

Let us, for simplicity, confine the discussion in this subsection to two-port SC networks with one input and one output. As noted previously, the two-port can be represented by an equivalent four-port, as shown in Fig. 6-7. In general,

Fig. 6-7 Four-port z-domain equivalent circuit.

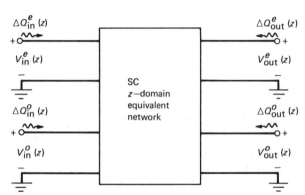

a 2×2 transfer matrix is required to fully characterize the input–output relations for this four-port network:

$$\begin{bmatrix} V_{out}^e(z) \\ V_{out}^o(z) \end{bmatrix} = \begin{bmatrix} H_1(z) & H_2(z) \\ H_3(z) & H_4(z) \end{bmatrix} \begin{bmatrix} V_{in}^e(z) \\ V_{in}^o(z) \end{bmatrix} \tag{6-10}$$

where, by superposition,

$$V_{in}(z) = V_{in}^e(z) + V_{in}^o(z) \tag{6-11a}$$

$$V_{out}(z) = V_{out}^e(z) + V_{out}^o(z) \tag{6-11b}$$

It is noted that at the even kT times, $V_{in}^o(z) = 0$ and $V_{out}^o = 0$, and at the odd kT times, $V_{in}^e(z) = 0$ and $V_{out}^e(z) = 0$. In essence, Eq. (6-10) is a statement of the time-varying character of SC networks; that is, if we sample the output at the even kT times only, we observe

$$V_{out}^e(z) = H_1(z)V_{in}^e(z) + H_2(z)V_{in}^o(z) \tag{6-12a}$$

Active Switched Capacitor Sampled-Data Networks

However, if we sample the output at the odd kT times only, we observe

$$V_{out}^{o}(z) = H_3(z)V_{in}^{e} + H_4(z)V_{in}^{o}(z) \tag{6-12b}$$

Furthermore, if we sample at both the even and odd time instants, we observe

$$V_{out}(z) = V_{out}^{e}(z) + V_{out}^{o}(z) = [H_1(z) + H_3(z)]V_{in}^{e}(z) \\ + [H_2(z) + H_4(z)]V_{in}^{o}(z) \tag{6-12c}$$

In general $V_{out}^{e}(z) \neq z^{-1/2}V_{out}^{o}(z) \neq V_{out}(z)$. We note that even and odd samples are separated by a one-half clock period time delay; therefore, equality among even and odd components implies that $V^{o}(z) = z^{-1/2}V^{e}(z)$. This is, as we noted earlier, a mathematical statement for a full-clock period sample-and-hold.

As our later examples will illustrate, the signal conditioning performed at the input and output imposes constraints on the form of the transfer relations. For example, consider the application of the return-to-zero source in Fig. 6-4a to the SC network in Fig. 6-7. Substituting $V_{in}^{o}(z) = 0$, obtained from Eq. (6-3b), into Eq. (6-10) yields the following transfer relations:

$$V_{out}^{e}(z) = H_1(z)V_{in}^{e}(z) \tag{6-13}$$

$$V_{out}^{o}(z) = H_3(z)V_{in}^{e}(z) \tag{6-14}$$

Thus, depending on whether $v_{out}(kT)$ is sampled at the even kT times or the odd kT times, the voltage transfer function is either $H_1(z)$ or $H_3(z)$, respectively. However, if v_{out} is sampled at all kT times, then

$$V_{out}(z) = [H_1(z) + H_3(z)]V_{in}^{e}(z) \tag{6-15}$$

In general, $H_1(z) \neq H_3(z)$; however, they involve the same components. Therefore, in general, one is not able to independently synthesize $H_1(z)$ and $H_3(z)$.

In practice, by appropriately conditioning the input and output signals, we can realize an SC network that is completely characterized by a single transfer function. Equations (6-13) or (6-14) describe examples of this class of SC network. We can then synthesize SC networks of this type directly in the z-domain, using digital filter [B3] synthesis techniques. Several examples of multi-transfer and single-transfer function SC networks are provided in Section 6.4.

6.2.5 Frequency Transformations Between s and z Domains

The initial step in the synthesis of an SC network is to obtain an appropriate z-domain transfer function. Since filters are typically specified by frequency-domain requirements, it is convenient to have a mathematical expression that allows us to transform rational s-domain transfer functions to rational z-domain transfer functions. To be generally useful, such an expression should, in addition, possess two qualities:

1. Stable s-domain transfer functions map into stable z-domain transfer functions.
2. The imaginary $j\omega$-axis of the s-plane map onto the unit circle of z-plane.

Item 1 ensures that our transformed z-domain transfer functions will be stable. Item 2 ensures that not only will they be stable but that the shape of the gain response can be preserved.

There are four transformations which have been used to synthesize SC networks. They are as follows [B3, P7]:

1. Backward difference (BD)

$$\frac{1}{s} = \frac{\tau}{1 - z^{-1}} \tag{6-16a}$$

or

$$z = \frac{1}{1 - s\tau} \tag{6-16b}$$

2. Forward difference (FD)

$$\frac{1}{s} = \frac{\tau z^{-1}}{1 - z^{-1}} \tag{6-17a}$$

or

$$z = 1 + s\tau \tag{6-17b}$$

3. Bilinear

$$\frac{1}{s} = \frac{\tau}{2} \frac{1 + z^{-1}}{1 - z^{-1}} \tag{6-18a}$$

or

$$z = \frac{(2/\tau) + s}{(2/\tau) - s} \tag{6-18b}$$

4. Lossless discrete integrator (LDI)

$$\frac{1}{s} = \tau \frac{z^{-1/2}}{1 - z^{-1}} \tag{6-18c}$$

or

$$z = \frac{1}{2}[(2 + s^2\tau^2) \pm \sqrt{s^2\tau^2(4 + s^2\tau^2)}] \tag{6-18d}$$

Note that z in Eq. (6-18d) is a bi-valued function of s. Either value can be assigned to z with equal validity.

Let us now examine each of these transformations in more detail. First, setting $s = -\sigma + j\omega$ and evaluating $|z|$, we can determine whether items 1

Active Switched Capacitor Sampled-Data Networks

and 2 are satisfied. For the BD we see that

$$|z| = \sqrt{\frac{1}{(1 + \sigma\tau)^2 + \omega^2\tau^2}} < 1 \qquad (6\text{-}19)$$

From Eq. (6-19) we observe that the BD maps the $j\omega$-axis inside the unit circle. Thus, item 1 is satisfied; however, item 2 is not. This mapping is illustrated in Fig. 6-8a. The net result of Eq. (6-19) is that high-Q s-plane poles and zeros are transformed into the z-plane at lower Q values, depending on the value of $\omega\tau$. Only when $\omega\tau \ll 1$ is the character of the gain response approximately preserved.

Applying the same procedure to Eq. (6-17b) for the FD, we obtain

$$|z| = \sqrt{(1 - \sigma\tau)^2 + \omega^2\tau^2} > 1 \qquad (6\text{-}20)$$

Equation (6-20) implies that the FD maps the $j\omega$-axis into a straight line outside the unit circle, as illustrated in Fig. 6-8b. Thus, the FD satisfied neither item 1

Fig. 6-8 (a) Backward difference (BD) mapping of the $j\omega$-axis. (b) Forward difference (FD) mapping of the $j\omega$-axis.

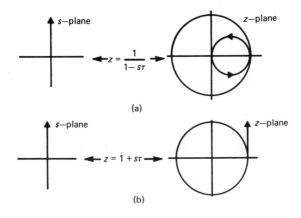

(a)

(b)

nor 2. It can be concluded that FD maps high-Q s-plane poles and zeros into higher-Q z-plane poles and zeros. More to the point, the resulting z-plane poles may be unstable. On the other hand, with the bilinear transform, evaluating $|z|$ with $s = j\omega$ yields

$$|z| = \left| \frac{(2/\tau) + j\omega}{(2/\tau) - j\omega} \right| = 1 \qquad (6\text{-}21)$$

Similarly, we can show for the LDI that

$$|z| = \frac{1}{2}\sqrt{(2 - \omega^2\tau^2)^2 + \omega^2\tau^2(4 - \omega^2\tau^2)} = 1 \qquad (6\text{-}22)$$

for either value of z.

Both the bilinear and LDI transformation satisfy item 2 per Eqs. (6-21) and (6-22), and can be shown to satisfy item 1.

We have thus far determined that the FD is unsuitable and that the BD marginally suitable (i.e., for $\omega\tau \ll 1$) for transforming s-domain transfer functions to z-domain functions. Let us now further investigate the utility of the bilinear and LDI transformations by examining the forms of the resulting z-domain functions. Applying the bilinear transformation to some arbitrary nth-order s-domain transfer function ($m \leq n$),

$$
H(s) = \frac{a_m s^m + a_{m-1} s^{m-1} + \ldots + a_1 s + a_0}{s^n + b_{n-1} s^{n-1} + \ldots + b_1 s + b_0} = \frac{a_0 + \sum\limits_{i=1}^{m} a_i s^i}{b_0 + \sum\limits_{k=1}^{n-1} b_k s^k + s^n}
\tag{6-23}
$$

we obtain

$$
H(z) = \frac{a_0(1 + z^{-1})^n + \sum\limits_{i=1}^{m} a_i \left[\dfrac{2}{\tau}(1 - z^{-1})\right]^i (1 + z^{-1})^{n-i}}{b_0(1 + z^{-1})^n + \sum\limits_{k=1}^{n-1} b_k \left[\dfrac{2}{\tau}(1 - z^{-1})\right]^k (1 + z^{-1})^{n-k} + \left[\dfrac{2}{\tau}(1 - z^{-1})\right]^n}
$$

$$
= \frac{\hat{a}_0 + \hat{a}_1 z^{-1} + \ldots + \hat{a}_{n-1} z^{-n+1} + \hat{a}_n z^{-n}}{\hat{b}_0 + \hat{b}_1 z^{-1} + \ldots + \hat{b}_{n-1} z^{-n+1} + \hat{b}_n z^{-n}}
\tag{6-24}
$$

Note that:

1. nth-order s-domain functions are transformed into nth-order z-domain functions.
2. Both the numerator and denominator of the z-domain function are nth order.
3. Both numerator and denominator are polynomials in integer powers of z^{-1}.
4. We state without proof that Eq. (6-24) is always realizable with SC networks.

Let us now transform Eq. (6-23) using the LDI, which yields

$$
H(z) = \frac{a_0(z^{-1/2})^n + \sum\limits_{i=1}^{m} a_i \left[\dfrac{1}{\tau}(1 - z^{-1})\right]^i (z^{-1/2})^{n-i}}{b_0(z^{-1/2})^n + \sum\limits_{k=1}^{n-1} b_k \left[\dfrac{1}{\tau}(1 - z^{-1})\right]^k (z^{-1/2})^{n-k} + \left[\dfrac{1}{\tau}(1 - z^{-1})\right]^n}
\tag{6-25}
$$

Note that:

1. $H(z)$ is nth order, but the numerator and denominator are polynomials in integer powers of $z^{-1/2}$.
2. To eliminate the powers of $z^{-1/2}$, we can define a new z-domain variable $z^{-1/2} = \hat{z}^{-1}$. Making this substitution in Eq. (6-25) results in a $2n$th-order

$H(\hat{z})$. $H(\hat{z})$, although realizable as an SC network, requires twice as much hardware as that required to realize bilinear transformed transfer functions.

To use the bilinear transform to convert s-domain transfer functions to z-domain functions, there are two further properties that the reader should be aware of. The first property is the well-known frequency warping effect. Frequency warping results from the nonlinear relationship that exists between the analog frequency ω, where $s = j\omega$, and the sampled-data domain frequency $\hat{\omega}$, where $z = e^{j\hat{\omega}T}$. The nature of the nonlinearity is revealed by substituting into Eq. (6-18a) $s = j\omega$ and $z = e^{j\hat{\omega}\tau}$. Solving the resulting equation for ω in terms of $\hat{\omega}$ yields

$$\omega = \frac{2}{\tau}\tan\left(\frac{\hat{\omega}\tau}{2}\right) \tag{6-26}$$

Note when $\hat{\omega}\tau \ll 1$, the warping effect is small. Warping is readily compensated by simply prewarping the cutoff and stop band edge frequencies according to Eq. (6-26). This procedure is demonstrated graphically in Fig. 6-9. Once the prewarped transfer function is obtained, Eq. (6-18a) can be used to derive the z-domain function. To illustrate the procedure, let us consider the following example.

Fig. 6-9   (a) Desired filter requirements. (b) Prewarped equivalent requirements.

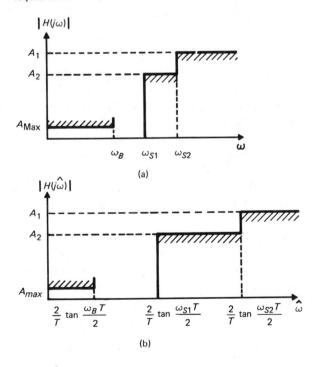

EXAMPLE 6-1

The transfer function to be realized is based on the following s-domain bandpass (BP) transfer function:

$$H(s) = \frac{2027.9s}{s^2 + 641.28s + 1.0528 \times 10^8} \qquad (6\text{-}27)$$

which possess a center frequency at $f_0 = 1633$ Hz, a quality factor $Q = 16$, and a peak gain of 10 dB at f_0. For the SC filter design, let the sampling frequency be 8 kHz (i.e., $\tau = 125$ μsec). The prewarped s-domain function is obtained by prewarping the upper and lower 3-dB frequencies of Eq. (6-27) according to Eq. (6-26), which yields the following prewarped frequencies $\omega_l = 1.145246 \times 10^4$ rps, $\omega_h = 1.245149 \times 10^4$ rps and $\omega_o = (\omega_h + \omega_l)/2 = 1.195198 \times 10^4$ rps. Note that that $\Delta\omega = \omega_h - \omega_l = 999.03$ rps. Once these frequencies are calculated, the prewarped transfer function is determined to be

$$\hat{H}(\hat{s}) = \frac{3159.2\hat{s}}{\hat{s}^2 + 999.03\hat{s} + 1.4285 \times 10^8} \qquad (6\text{-}28)$$

Note that the gain constant in Eq. (6-26) has been modified from its value in Eq. (6-27) in order to preserve the desired 10-dB peak gain. Applying now the bilinear transform Eq. (6-18a) to Eq. (6-28) yields the following z-domain transfer function:

$$H(z) = \frac{0.1219(1 - z^{-1})(1 + z^{-1})}{1 - 0.5455z^{-1} + 0.9229z^{-2}} \qquad (6\text{-}29)$$

●

Other properties of the bilinear transform, thus far not mentioned, are:

1. That all finite analog frequencies ω are mapped into frequencies $\hat{\omega}$ which are bounded by the relation $0 \leq \hat{\omega} < \omega_s/2$, where ω_s (recall that $\omega_s = 2\pi/\tau$) is the sampling frequency. This bound, which is a direct result of Eq. (6-26), implies that the base-band spectrum of $H(z)$ is compressed within $\omega_s/2$. For example, the s-plane zero at infinity in Eqs. (6-27) and (6-28) is mapped into $\omega_s/2$, that is, one-half the sampling frequency (i.e., $z = -1$) in Eq. (6-29). This property enhances the high-frequency attenuation for LP and BP filters. We will say more about this later.

2. Although the gain characteristics of the transfer function can be recovered by prewarping the analog frequencies, the group delay can be shown to be in error by amount $\sec^2(\omega\tau/2)$ (Prob. 6.3). This error can be objectionable in a delay equalizer design. When the group delay is important, the sampling rate can be chosen to make this quantity insignificant.

It should be emphasized that a z-domain transfer function, obtained with the rational s to z transformations, provides a frequency response (with $z = e^{j\omega\tau}$)

which is at best a good approximation to the desired analog response. As the frequency increases toward $\omega_s/2$, the z-domain response becomes less representative. For sampling frequencies that are much higher than the highest pole frequency, as is typically the case in switched capacitor filters, the approximate z-transformed responses are quite adequate.

In Chapter 1 we defined the generic types of filters: the low-pass (LP), band-pass (BP), high-pass (HP), low-pass notch (LPN), high-pass notch (HPN) and all-pass (AP). Let us state, for the second-order case, the z-domain equivalent generic forms. A second-order z-domain transfer function [e.g., Eq. (6-29)] can be expressed, in general, as

$$H(z) = \frac{N(z)}{D(z)} = \frac{\gamma + \epsilon z^{-1} + \delta z^{-2}}{1 + \alpha z^{-1} + \beta z^{-2}} \qquad (6\text{-}30)$$

The numerators $N(z)$ and pole–zero formations, with reference to Eq. (6-30), for the various generic forms are listed in Table 6-1. The LP and BP functions

TABLE 6-1   Generic second-order z-domain transfer functions [P17]

Generic Form	Numerator $N(z)$	z-Plane Poles and Zeros ($\omega\tau \ll 1$)
LP20 (bilinear transform)	$K(1 + z^{-1})^2$	
LP11	$Kz^{-1}(1 + z^{-1})$	
LP10	$K(1 + z^{-1})$	
LP02	Kz^{-2}	
LP01	Kz^{-1}	
LP00	K	
BP10 (bilinear transform)	$K(1 - z^{-1})(1 + z^{-1})$	
BP01	$Kz^{-1}(1 - z^{-1})$	
BP00	$K(1 - z^{-1})$	
HP (bilinear transform)	$K(1 - z^{-1})^2$	
LPN (bilinear transform)	$K(1 + \epsilon z^{-1} + z^{-2})$, $\epsilon > \alpha/\sqrt{\beta}, \beta > 0$	
HPN (bilinear transform)	$K(1 + \epsilon z^{-1} + z^{-2})$, $\epsilon < \alpha/\sqrt{\beta}, \beta > 0$	
AP (bilinear transform)	$K(\beta + \alpha z^{-1} + z^{-2})$	

are particularly interesting in that there are several different forms that can be used. These forms are referred to in Table 6-1 as LPIJ and BPIJ, where I denotes the number of $1 + z^{-1}$ factors and J the number of z^{-1} factors. As already noted, the zeros at $z = -1$ arise only when the bilinear transform is used. These transfer functions, of course, exhibit steeper cutoff in the vicinity of half the sampling rate, but they may not afford the most economical realization. As a rule of thumb, the additional cutoff will become less and less important as the sampling frequency increases with respect to the pole–zero locations (i.e., as $\omega_{p,z}\tau \rightarrow$ small).

To ensure proper pole placement, the denominator, $D(z)$, coefficients in Eq. (6-30) should be obtained via the bilinear transform. For the LP and BP cases, the different $N(z)$'s are obtained by replacing one or both of the zeros at $z = -1$ with either 2 or $2z^{-1}$.[1] The sacrifice made in removing these zeros is to reduce the cutoff at and about the half-sampling frequency $(\omega_s/2)$. With this sacrifice we often achieve a significant savings in the total capacitance required to implement the function as a switched capacitor filter. To illustrate the effect of removing zeros at $z = -1$, let us plot the bilinear BP10 and BP00 z-domain responses derived to approximate the $Q = 16, f_0 = 1633$ Hz band-pass s-domain function given in Eq. (6-27). As in Example 6-1, the sampling frequency has been chosen to be 8 kHz. This sampling frequency represents a minimum value for voice-band ($f < 4$ kHz) applications. The BP10 transfer function was given in Eq. (6-29). The BP00 function, obtained by replacing the zero at $z = -1$ in Eq. (6-29) with unity, is

$$H(z) = \frac{0.1953(1 - z^{-1})}{1 - 0.5455z^{-1} + 0.9229z^{-2}} \qquad (6\text{-}31)$$

The gain responses, corresponding to Eqs. (6-29) and (6-31), are plotted in Fig. 6-10. Note that:

1. The denominators of Eqs. (6-29) and (6-31) are identical; therefore, as expected, the passband gain responses are essentially identical to about 10 dB below the peak.

2. The gain constant was altered in Eq. (6-31) to maintain the 10-dB peak.

3. As observed in Fig. 6-10, the primary difference between the two responses is the high attenuation near 4 kHz for BP10 and the retarded high-frequency roll-off for BP00. These are the effects we referred to as transformation-related approximation errors in Section 6.2.4. Be that as it may, the BP00 achieves nearly 25 dB of attenuation at the half-sampling frequency 4 kHz. The significance of these zeros at $z = -1$ will diminish as the sampling frequency is increased.

As we have seen, z-domain transfer functions can be obtained, via rational s to z transformations, which represent good approximations to desired s-domain response specifications. Once a suitable z-domain transfer function is

[1]Note that the 2 multiplier preserves the dc gain of the bilinear $H(z)$.

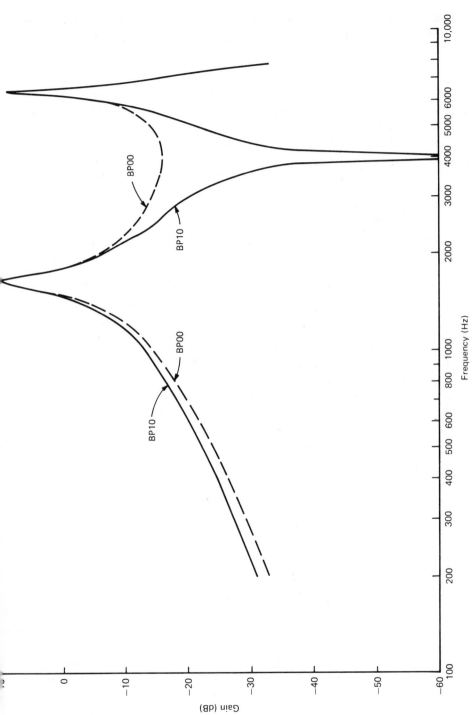

Fig. 6-10 Gain responses for the BP10 and BP00 transfer functions in Eqs. (6-29) and (6-31).

obtained, the SC filter is synthesized by the same coefficient matching techniques used in previous chapters for active-RC design.

6.2.6 Circuit Transformations

In lieu of computing a z-domain transfer function and synthesizing the SC network in the z-domain, analog active-RC circuits can be transformed into z-domain circuits using the frequency transformations given in Section 6.2.5. The object is to transform elemental block transfer functions or admittances into their z-transformed equivalents. Once suitable SC realizations for the elemental blocks are derived, they can be substituted directly into the analog-circuit schematic. This type of synthesis enables us to transform low-sensitivity active-RC filter designs into active-SC realizations of comparable quality.

There are many active-RC topologies and schemes which are based on the use of lossless and lossy integrators. For example, the state-variable biquads in Figs. 4-31a, b and 4-32 employ two such integrators. Low-sensitivity high-order filters can be realized by placing multiple feedback paths around a chain of lossless and lossy integrators, as illustrated in Chapter 5. For these types of active-RC filters. z-domain equivalent networks can be derived by functionally replacing each s-domain integrator with a z-domain integrator. The form of the z-domain integrator is determined by the frequency transformation employed. Lossless and lossy integrators derived from the BD, FD, bilinear, and LDI transformations are tabulated in Table 6-2. The properties of the active-RC

TABLE 6-2 z-domain integrators

Transformation	Lossless Integrator $\left(\dfrac{1}{s}\right)$	Lossy Integrator $\left(\dfrac{p}{s+p}\right)$
$s = \dfrac{1 - z^{-1}}{\tau}$	$\dfrac{\tau}{1 - z^{-1}}$	$\dfrac{\tau p}{(1 + \tau p) - z^{-1}}$
$s = \dfrac{1 - z^{-1}}{\tau z^{-1}}$	$\dfrac{\tau z^{-1}}{1 - z^{-1}}$	$\dfrac{\tau p z^{-1}}{1 - (1 - \tau p)z^{-1}}$
$s = \dfrac{1 - z^{-1}}{\tau z^{-1/2}}$	$\dfrac{\tau z^{-1/2}}{1 - z^{-1}}$	$\dfrac{\tau p z^{-1/2}}{1 + \tau p z^{-1/2} - z^{-1}}$
$s = \dfrac{2}{\tau}\dfrac{1 - z^{-1}}{1 + z^{-1}}$	$\dfrac{\tau}{2}\dfrac{1 + z^{-1}}{1 - z^{-1}}$	$\dfrac{\dfrac{\tau p}{2}(1 + z)^{-1}}{\left(1 + \dfrac{\tau p}{2}\right) - \left(1 - \dfrac{\tau p}{2}\right)z^{-1}}$

prototype are closely approximated by the z-domain equivalent network when the two transformation characteristics cited in Section 6.2.5 are satisfied. Therefore, best agreement is obtained when the bilinear and LDI integrators are used. Note, however, that cascading pairwise BD and FD lossless integrators are equivalent to cascading pairs of LDIs, i.e.:

$$\frac{\tau}{1 - z^{-1}}\frac{\tau z^{-1}}{1 - z^{-1}} = \frac{\tau z^{-1/2}}{1 - z^{-1}}\frac{\tau z^{-1/2}}{1 - z^{-1}} \tag{6-32}$$

Equation (6-32) informs us that BD and FD integrators can be used as circuit functions, despite the deleterious properties cited in Section 6.2.5, when they are appropriately paired.

Once the integrator forms are chosen, the SC network is derived by replacing each z-domain integrator with its SC network realization. In Section 6.4, simple SC realizations for many of the integrator functions listed in Table 6-2 are developed and analyzed. The only function that is unrealizable is the damped LDI. This nonrealizability, which results from the $z^{-1/2}$ term in the denominator, will be demonstrated by example in Section 6.4. This fact may at first seem to doom the use of the LDI. However, this is not the case, since it is often sufficient to approximate (when $\tau p \ll 1$) the lossy LDI with

$$\frac{\tau p z^{-1/2}}{1 + \tau p z^{-1/2} - z^{-1}} \simeq \frac{\tau p z^{-1/2}}{1 + (\tau p - z^{-1})} \qquad (6\text{-}33)$$

Another method for transforming an active-RC network into an equivalent SC network is to scale the network admittances using one of the aforementioned s to z frequency transformations. z-Transformed equivalent admittances for the capacitor, conductance, inductor, and frequency-dependent negative resistor (FDNR), derived using each of the four transformations, are listed in Table 6-3.

TABLE 6-3 z-domain admittances

Transformation	sC	G	$1/sL$	$s^2 D$
		Admittances		
$s = \dfrac{1 - z^{-1}}{\tau}$	$C(1 - z^{-1})$	$G\tau$	$\dfrac{\tau^2}{L} \dfrac{1}{1 - z^{-1}}$	$\dfrac{D}{\tau}(1 - z^{-1})^2$
$s = \dfrac{1 - z^{-1}}{\tau z^{-1}}$	$C(1 - z^{-1})$	$G\tau z^{-1}$	$\dfrac{\tau^2}{L} \dfrac{z^{-2}}{1 - z^{-1}}$	$\dfrac{D}{\tau} \dfrac{(1 - z^{-1})^2}{z^{-1}}$
$s = \dfrac{1 - z^{-1}}{\tau z^{-1/2}}$	$C(1 - z^{-1})$	$G\tau z^{-1/2}$	$\dfrac{\tau^2}{L} \dfrac{z^{-1}}{1 - z^{-1}}$	$\dfrac{D}{\tau} \dfrac{(1 - z^{-1})^2}{z^{-1/2}}$
$s = \dfrac{2}{\tau} \dfrac{1 - z^{-1}}{1 + z^{-1}}$	$C(1 - z^{-1})$	$\dfrac{G\tau}{2}(1 + z^{-1})$	$\dfrac{\tau^2}{4L} \dfrac{(1 + z^{-1})^2}{1 - z^{-1}}$	$\dfrac{2D}{\tau} \dfrac{(1 - z^{-1})^2}{1 + z^{-1}}$

It is important to note that the admittances in Table 6-3 are not admittances in the usual I–V sense. They are, in fact, admittances of the form $y = \Delta Q/V$, which come about naturally in the writing of nodal charge equations [i.e., Eq. (6-2)]. Furthermore, all but the BD transformed admittances have been scaled by common factors. The scaling factors, which are different for each transformation, have been chosen such that the capacitive admittance is always of the form $C(1 - z^{-1})$. As we see in succeeding sections, admittance scaling in this manner yields admittances that are realizable with simple SC networks. Since voltage transfer functions are dimensionless, they are unaffected by the change in admittance definition and the scaling.

To see how the z-transformed admittances are obtained, let us derive the FD equivalent admittances for the resistor, capacitor, and inductor. The varification of the remaining admittance entries in Table 6-3 are left as exercises for the reader.

Resistor   The instantaneous voltage and current for a resistor $R = 1/G$ are related by

$$i = Gv \qquad \text{(6-34a)}$$

If the current and voltage are sampled at discrete instants $k\tau$, we may express Eq. (6-34a) in a discrete-time sense as

$$i(k\tau) = Gv(k\tau) \qquad \text{(6-34b)}$$

However, since $i = dq/dt$, or in the discrete time $i(k\tau) = \Delta q(k\tau)/\tau$,

$$\frac{\Delta q(k\tau)}{\tau} = Gv(k\tau)$$

or

$$\Delta q(k\tau) = \tau Gv(k\tau) \qquad \text{(6-34c)}$$

Taking the z-transform of both sides of Eq. (6-34c) yields

$$\Delta Q(z) = \tau GV(z)$$

or

$$y_R = \frac{\Delta Q(z)}{V(z)} = \tau G \qquad \text{(6-34d)}$$

Capacitor   The instantaneous voltage and current for a capacitor C are related by

$$i = C\frac{dv}{dt}$$

or

$$\frac{dq}{dt} = C\frac{dv}{dt} \qquad \text{(6-35a)}$$

For i and v, sampled at discrete times $k\tau$, we express

$$\frac{dq}{dt} = \frac{\Delta q(k\tau)}{\tau}$$

and, using forward differences (FD), we approximate

$$\frac{dv}{dt} = \frac{v[(k+1)\tau] - v(k\tau)}{\tau}$$

Thus, for discrete time we write for Eq. (6-35a)

$$\frac{\Delta q(k\tau)}{\tau} = C\left[\frac{v[(k+1)\tau] - v(k\tau)}{\tau}\right]$$

or

$$\Delta q(k\tau) = C[v[(k+1)\tau] - v(k\tau)] \tag{6-35b}$$

Taking the z-transform of Eq. (6-35b) yields

$$\Delta Q(z) = C(z-1)V(z)$$

Thus, we may write

$$y_C = \frac{\Delta Q(z)}{V(z)} = \frac{C(1-z^{-1})}{z^{-1}} \tag{6-35c}$$

Inductor   The instantaneous voltage and current for an inductor L are related by

$$v = L\frac{di}{dt}$$

or

$$v = L\frac{d(dq/dt)}{dt} \tag{6-36a}$$

Using forward differences, we can approximate $d(dq/dt)/dt$ as follows:

$$\frac{d(dq/dt)}{dt} = \frac{\Delta q[(k+1)\tau] - \Delta q(k\tau)}{\tau^2}$$

Thus, for i and v sampled at discrete times $k\tau$, Eq. (6-36a) can be written as

$$v(k\tau) = L\left[\frac{\Delta q[(k+1)\tau] - \Delta q(k\tau)}{\tau^2}\right] \tag{6-36b}$$

Taking the z-transform of Eq. (6-36b) yields

$$V(z) = L\left(\frac{z-1}{\tau^2}\right)\Delta Q(z) = L\left(\frac{1-z^{-1}}{\tau^2 z^{-1}}\right)\Delta Q(z)$$

or

$$y_L = \frac{\Delta Q(z)}{V(z)} = \frac{\tau^2}{L}\frac{z^{-1}}{1-z^{-1}} \tag{6-36c}$$

The final step in the derivation is to multiply y_R, y_C, and y_L by z^{-1}. With this scaling we obtain the FD[$s = (1 - z^{-1})/\tau z^{-1}$] admittances listed in Table 6-3.

It is noted that the widely acclaimed resistor equivalence for a switched capacitor stems from the resistor equivalents in Table 6-3. We will examine the validity of this equivalence in Section 6.3. As we show later, the bilinear admittances are readily realizable with SC networks. The other equivalences have not yet been fully explored and appropriate realizations, particularly for the inductor and FDNR, do not exist in all cases. Where possible, circuits will be given in Sections 6.3 and 6.4.

6.3 z-DOMAIN EQUIVALENT CIRCUIT MODELS FOR SC BUILDING BLOCKS

In this section multiport z-domain equivalent circuits are derived for several SC building blocks. Here we see that the capacitor and some conductance equivalents in Table 6-3 come about naturally. The objective here is to provide simple elemental equivalent circuits. When they are interconnected, according to the schematic of an SC network, it is intended that the resulting z-domain equivalent circuit be canonic and analyzable with pencil and paper. We will learn to treat SC elemental circuits, comprised of one capacitor and from one to four switches, as basic circuit elements much like passive R's, L's, and C's in analog circuits. As noted in Section 6.2, the z-domain transfer relations can be derived from the equivalent circuit using familiar network analysis techniques [B5].

Figures 6-11, 6-12, and 6-13 contain listings of the commonly occurring SC elements and their respective z-domain equivalent circuits or building blocks. In addition to the SC building blocks, z-domain models are also given for each of the sampled-data sources discussed in Section 6.2. This library is sufficiently general to accommodate all of the published [P2, P8–P11] SC networks that use nonoverlapping biphase switches. The equivalent circuits in Fig. 6-11 are derived in their most general $2n$-port form, assuming that all voltages update at one-half clock cycle intervals, as per $v(t)$ in Fig. 6-3. The e,o notation refers to the switch phasings, as noted in Section 6.2. Similarly, superscripts e,o and o,e are used to denote the even or odd port-variable (V_i, ΔQ_i) components and the complementary odd or even port-variable components, respectively. This e,o notation conveniently provides the connectivity information for interconnecting the building blocks.

In practice, there are many SC networks in which the charges and voltages update, because of the internal switching action of the SC network, only on full clock cycle intervals. This behavior, which is readily identified on a block-by-block basis, results in $2n$-port equivalent circuits with n open ports. Many of the SC blocks in Fig. 6-11 fall into this category. When properly interconnected, these $2n$-port equivalent circuits can be reduced to the n-port equivalent circuits in Fig. 6-12. SC networks comprised of these type elements sample their terminal voltages once per clock period (i.e., during the even or odd time slots but not both).

There is yet another class of elemental SC networks in which the even and odd signal paths are identical. These elements are seen to achieve perfect even–odd symmetry. When SC networks of this class sample their terminal voltages once per one-half clock period (i.e., during both the even and odd time

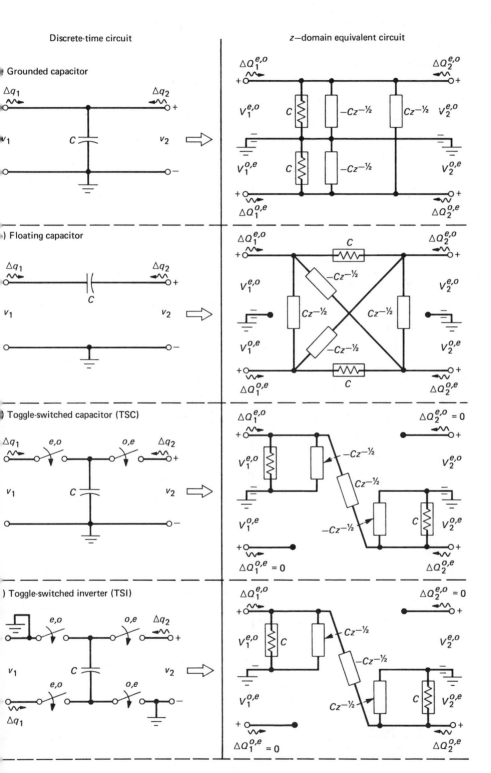

Fig. 6-11 General library of 2*n*-port *z*-domain equivalent circuits for switched-capacitor building blocks [P5].

Discrete-time circuit | z—domain equivalent circuit

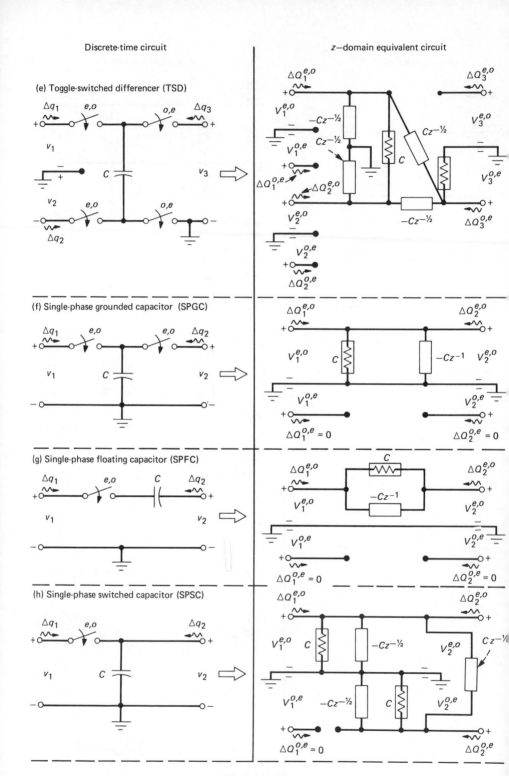

(e) Toggle-switched differencer (TSD)

(f) Single-phase grounded capacitor (SPGC)

(g) Single-phase floating capacitor (SPFC)

(h) Single-phase switched capacitor (SPSC)

Fig. 6-11   (Cont.)

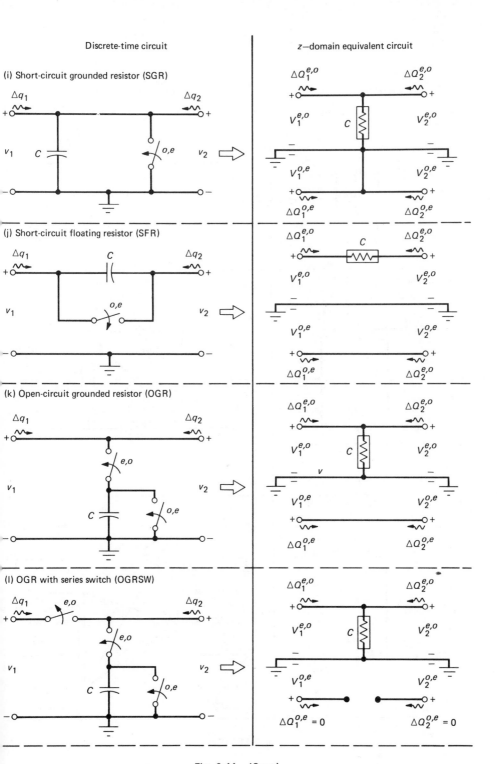

Fig. 6-11 (Cont.)

Discrete-time circuit

z—domain equivalent circuit

(m) Open-circuit floating resistor (OFR)

(n) Bilinear floating resistor (BFR)

(o) Toggle-switched floating four-port (TSFFP)

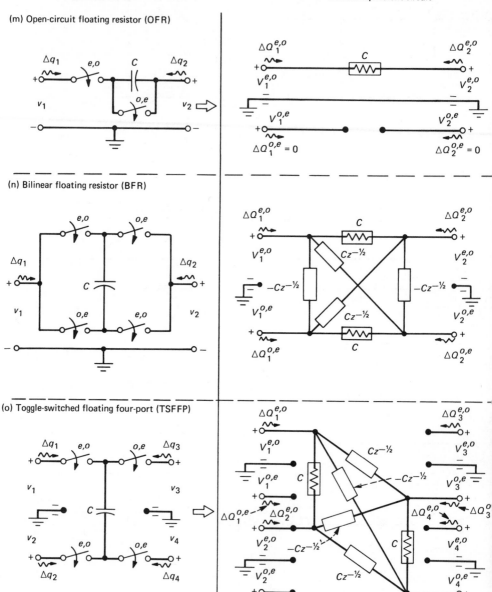

Fig. 6-11   (Cont.)

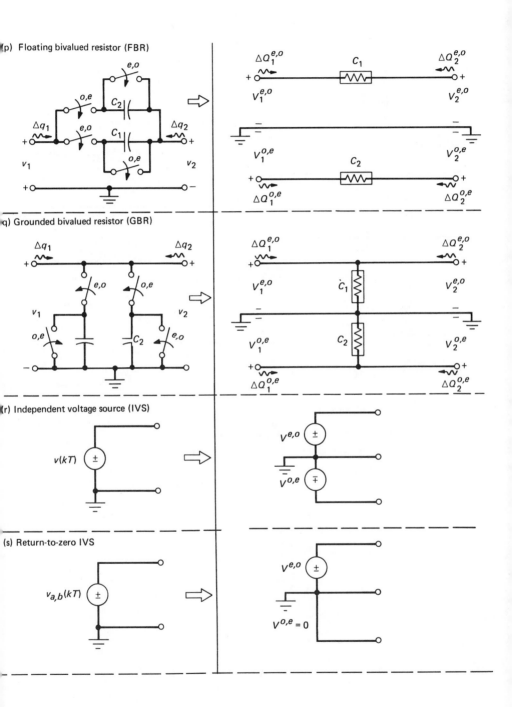

(p) Floating bivalued resistor (FBR)

(q) Grounded bivalued resistor (GBR)

(r) Independent voltage source (IVS)

(s) Return-to-zero IVS

Fig. 6-11   (Cont.)

407

Discrete-time circuit | z—domain equivalent circuit

(t) Full-cycle S/H IVS

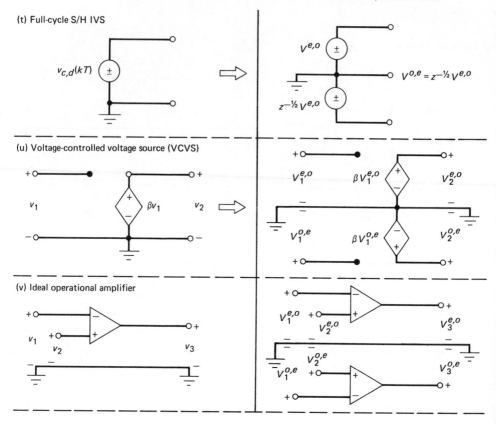

(u) Voltage-controlled voltage source (VCVS)

(v) Ideal operational amplifier

Fig. 6-11 (Cont.)

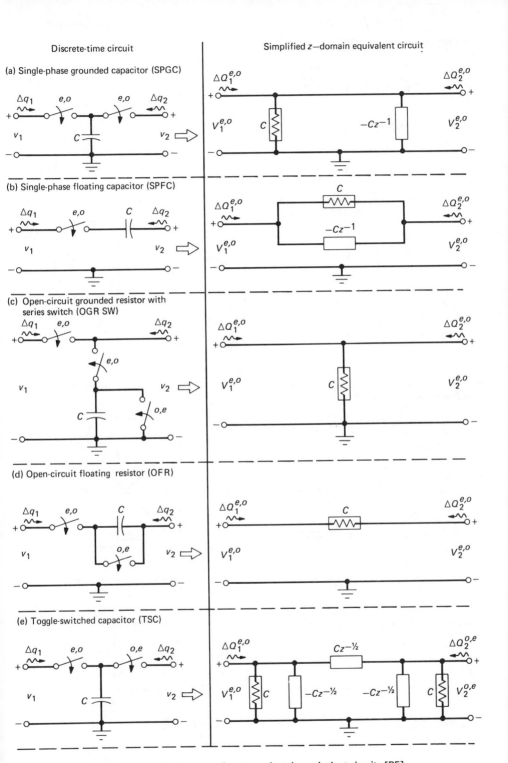

Fig. 6-12 Simplified library of n-port z-domain equivalent circuits [P5].

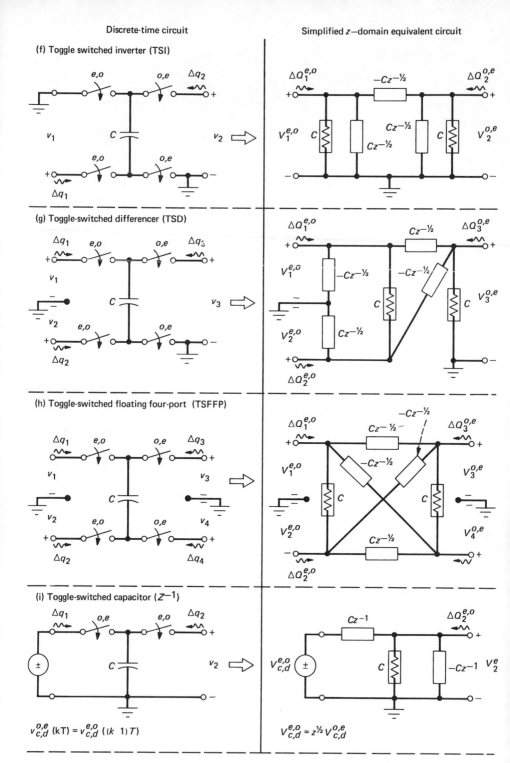

Fig. 6-12   (Cont.)

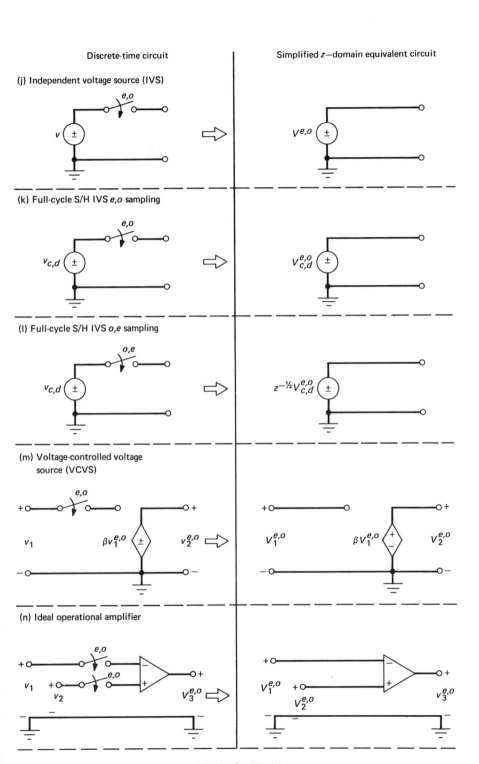

Discrete-time circuit · Simplified z-domain equivalent circuit

(j) Independent voltage source (IVS)

(k) Full-cycle S/H IVS e,o sampling

(l) Full-cycle S/H IVS o,e sampling

(m) Voltage-controlled voltage source (VCVS)

(n) Ideal operational amplifier

Fig. 6-12 (Cont.)

411

(a) TSFFP

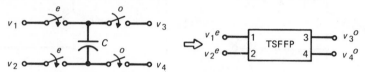

(b) Bilinear resistor (floating)

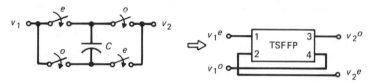

(c) Floating capacitor

(d) TSC

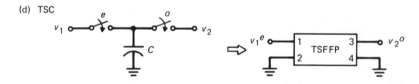

(e) TSI

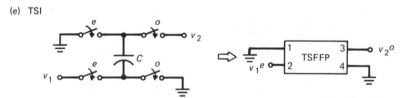

(f) OFR

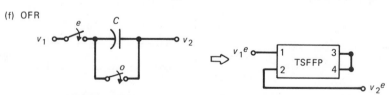

Fig. 6-13 Modeling of switched capacitor building block circuits with terminated TSFFP equivalent circuits.

412

slots), the four-port z-domain equivalent circuits reduce to simple two-port equivalent circuits. Equivalent circuits of this type are given in Fig. 6-16.

6.3.1 2n-Port SC Building Block Equivalent Circuits

In this section derivations are given for several of the SC equivalent circuits in Fig. 6-11. These derivations will be based on z-transformed nodal charge equations which can be derived by inspection from the SC circuit. As noted in Section 6.2, one can write a distinct nodal charge equation for each switch phase. Therefore, an n-port SC block is characterized by $2n$ nodal charge relations. The desired $2n$-port z-domain equivalent circuit evolves directly from these relations. The equivalent circuit for a complex SC network is derived by properly interconnecting the appropriate block equivalent circuits. To avoid boring the reader with excessive repetition, derivations will only be provided for blocks b through f and l of Fig. 6-11. Once these derivations are understood, the validity of the remaining equivalent circuits can be established by inspection. The independent and dependent voltage source equivalent circuits are obtained directly from the relations in Section 6.2.

6.3.1a Floating Capacitor Equivalent Circuit

One can derive the desired equivalent circuit in a straightforward manner directly from the nodal charge equations. In these equations the even and odd voltage components (V^e and V^o) serve as independent variables and the even and odd charge-variation components (ΔQ^e and ΔQ^o) serve as the dependent variables. Since the floating capacitor block in Fig. 6-11b contains no switches, the discrete-time nodal charge equations [see Eq. (6-1)], where v_1^e, v_1^o, v_2^e, and v_2^o are independent discrete-time voltage excitations, are instantly written as follows:

For the even kT times:

$$\Delta q_1^e(kT) = Cv_1^e(kT) - Cv_1^o[(k-1)T] - Cv_2^e(kT) + Cv_2^o[(k-1)T]$$
$$\Delta q_2^e(kT) = Cv_2^e(kT) - Cv_2^o[(k-1)T] - Cv_1^e(kT) + Cv_1^o[(k-1)T]$$

For the odd kT times:

$$\Delta q_1^o(kT) = Cv_1^o(kT) - Cv_1^e[(k-1)T] - Cv_2^o(kT) + Cv_2^e[(k-1)T]$$
$$\Delta q_2^o(kT) = Cv_2^o(kT) - Cv_2^e[(k-1)T] - Cv_1^o(kT) + Cv_1^e[(k-1)T]$$

Invoking the z-transform, where $Z[u(kT)] = U(z)$ and $Z[u[(k-1)T]] = z^{-1/2}U(z)$, we readily obtain the following z-transformed nodal charge equations:

$$\Delta Q_1^e(z) = CV_1^e(z) - Cz^{-1/2}V_1^o(z) - CV_2^e(z) \\ + Cz^{-1/2}V_2^o(z) \tag{6-37a}$$

$$\Delta Q_2^e(z) = CV_2^e(z) - Cz^{-1/2}V_2^o(z) - CV_1^e(z) \\ + Cz^{-1/2}V_1^o(z)$$ (6-37b)

$$\Delta Q_1^o(z) = CV_1^o(z) - Cz^{-1/2}V_1^e(z) - CV_2^o(z) \\ + Cz^{-1/2}V_2^e(z)$$ (6-37c)

$$\Delta Q_2^o(z) = CV_2^o(z) - Cz^{-1/2}V_2^e(z) - CV_1^o(z) \\ + Cz^{-1/2}V_1^e(z)$$ (6-37d)

In succeeding equivalent circuit derivations, we skip the writing of the discrete-time nodal charge equations. Those readers with little or no previous exposure to the z-transform will find it useful during home study to retain this step. After some practice, you should be able to write, for the elemental circuits, the z-transformed nodal charge equations directly from the circuit schematic.

There are perhaps several circuit interpretations for this set of equations. One convenient interpretation is the balanced lattice equivalent circuit shown in Fig. 6-11b. Another circuit interpretation [P4] for these equations is a four-port network comprised of an unbalanced floating link two-port (LTP) coupled to the even and odd transmission paths via ideal transformers. A link two-port is a network that couples the even and odd signal paths of the equivalent circuit. By interpreting Eq. (6-37) as a balanced lattice, one can eliminate the transformers. This balanced lattice will be referred to in this book as a balanced floating LTP; in contrast, the Kurth–Moschytz circuit [P4] will be referred to as an unbalanced floating LTP. Both circuits are equivalent and valid under all port termination conditions.

6.3.1b Toggle-Switched Capacitor (TSC) Equivalent Circuit [P8]

Because of the switching action of the toggle switch, the capacitor C receives charge from v_1 only on the even (odd) times and charge from v_2 only on the odd (even) times. When the switches are open, the corresponding ports are open and $\Delta Q = 0$. These observations are consistent with the following z-transformed nodal charge equations:

$$\Delta Q_1^{e,o}(z) = CV_1^{e,o}(z) - Cz^{-1/2}V_2^{o,e}(z)$$ (6-38a)

$$\Delta Q_2^{e,o}(z) = 0$$ (6-38b)

$$\Delta Q_1^{o,e}(z) = 0$$ (6-38c)

$$\Delta Q_2^{o,e}(z) = CV_2^{o,e}(z) - Cz^{-1/2}V_1^{e,o}(z)$$ (6-38d)

Note again that the superscript e,o refers to even or odd (but not both simultaneously) and the inverse superscript o,e refers to odd or even. These equations lead directly to the four-port equivalent circuit in Fig. 6-11c. As described in [P4], an unbalanced LTP is seen to bridge the 1-e,o and 2-o,e ports. Note that ports 1-o,e and 2-e,o are always open; therefore, no transmission occurs at these ports. This is a property common to all biphase toggle switched SC blocks (e.g., equivalent circuits d and e in Fig. 6-11).

6.3.1c Toggle-Switched Inverter (TSI) Equivalent Circuit [P2]

The operation of this circuit, shown in Fig. 6-11d, is similar to the TSC element, with the exception that in the TSI the voltage is inverted as the charge on C is transferred from port 1 to port 2. This process is described by the following z-transformed nodal charge equations:

$$\Delta Q_1^{e,o}(z) = CV_1^{e,o}(z) + Cz^{-1/2}V_2^{o,e}(z) \qquad (6\text{-}39a)$$

$$\Delta Q_1^{o,e}(z) = 0 \qquad (6\text{-}39b)$$

$$\Delta Q_2^{e,o}(z) = 0 \qquad (6\text{-}39c)$$

$$\Delta Q_2^{o,e}(z) = CV_2^{o,e}(z) + Cz^{-1/2}V_1^{e,o}(z) \qquad (6\text{-}39d)$$

These equations are readily interpreted by the four-port equivalent circuit in Fig. 6-11d. Note for this block that the 1-e,o and 2-o,e ports are bridged by an unbalanced LTP network in which the storage elements ($Cz^{-1/2}$) are all premultiplied by (-1). Since this network serves as both a link between even and odd transmission paths and as a signal inverter, it will be referred to as an unbalanced inverting LTP.

6.3.1d Toggle-Switched Differencer (TSD) Equivalent Circuit [P2]

In this element, shown in Fig. 6-11e, the charge on C is determined by the voltage difference $v_1^{e,o}(kT) - v_2^{o,o}(kT)$ during the e,o switch phase. When the o,e switches close, this voltage difference appears directly across port 3. This operation is described by the following z-transformed nodal charge equations:

$$\Delta Q_1^{e,o}(z) = CV_1^{e,o}(z) - CV_2^{e,o}(z) - Cz^{-1/2}V_3^{o,e}(z) \qquad (6\text{-}40a)$$

$$\Delta Q_1^{o,e}(z) = 0 \qquad (6\text{-}40b)$$

$$\Delta Q_2^{e,o}(z) = CV_2^{e,o}(z) - CV_1^{e,o}(z) + Cz^{-1/2}V_3^{o,e}(z) \qquad (6\text{-}40c)$$

$$\Delta Q_2^{o,e}(z) = 0 \qquad (6\text{-}40d)$$

$$\Delta Q_3^{e,o}(z) = 0 \qquad (6\text{-}40e)$$

$$\Delta Q_3^{o,e}(z) = CV_3^{o,e}(z) - Cz^{-1/2}V_1^{e,o}(z) + Cz^{-1/2}V_2^{e,o}(z) \qquad (6\text{-}40f)$$

The six-port equivalent circuit representation for these equations is given in Fig. 6-11e. Note that three of the six ports are open. The TSD element exhibits yet another form of LTP. In this element two e,o transmission paths are linked to a single o,e path through a differencing operation.

6.3.1e Single-Phase Grounded Capacitor (SPGC)

This SC element, shown in Fig. 6-11f, occurs frequently in SC networks, particularly in low-pass SC filters. In a sense it serves as a companion element for the grounded capacitor in Fig. 6-11a. It is also a special case of the grounded capacitor. The nodal charge relations for this block are readily written as follows:

$$\Delta Q^{e,o}(z) = C(1 - z^{-1})V^{e,o}(z) \qquad (6\text{-}41a)$$

$$\Delta Q^{e,o}(z) = \Delta Q_1^{e,o}(z) + \Delta Q_2^{e,o}(z) \qquad (6\text{-}41b)$$

$$\Delta Q_1^{o,e}(z) = 0 \qquad (6\text{-}41c)$$

$$\Delta Q_2^{o,e}(z) = 0 \qquad (6\text{-}41d)$$

where $V^{e,o}(z) = V_1^{e,o}(z) = V_2^{e,o}(z)$. It should be noted that one can derive Eq. (6-41) from the grounded capacitor equivalent circuit in Fig. 6-11a by setting $\Delta Q_1^{o,e} = \Delta Q_2^{o,e} = 0$, which implies that $V^{o,e}(z) = z^{-1/2}V^{e,o}(z)$. For the SPGC block, $V^{o,e}(z)$ represents the voltage stored and held on capacitor C and no longer refers to port voltages $V_1^{o,e}$ and $V_2^{o,e}$. As noted in Section 6.2.3, this condition is equivalent to a full clock period S/H.

Equations (6-41a) through (6-41d) lead directly to the four-port equivalent circuit in Fig. 6-11f. Because of the switches, two of the ports are open, as described by Eqs. (6-41c) and (6-41d). This network is equivalent to the open-circuit LTP described in [P4]. The equivalent circuit for the floating capacitor is seen to similarly reduce to that in Fig. 6-11g when a series switch is added. Since these blocks occur frequently in complex SC networks, their recognition results in substantially simplified equivalent circuits. We also note that the capacitor equivalent circuits in Fig. 6-11f and g provide the equivalent capacitive admittance $y = \Delta Q/V = C(1 - z^{-1})$ given in Table 6-3. It will become apparent later that these circuits represent FD and BD transformed admittances. The admittance representation above is valid only when the admittance is viewed during the time slot when the switch is closed. When the switch is open $\Delta Q = 0$; hence, $y = 0$.

6.3.1f Open-Circuit Grounded Resistor with Series Switch (OGR/SW) Equivalent Circuit

This block performs a function similar to the SGR in Fig. 6-11i, with the exception that capacitor C is discharged while it is totally disconnected from the circuit. Therefore, the shorted capacitor does not load the circuit during the discharging switch phase. The equivalent circuit, in Fig. 6-11l, for this block is obtained from the following z-transformed nodal charge equations:

$$\Delta Q^{e,o}(z) = CV^{e,o}(z) \qquad (6\text{-}42a)$$

$$\Delta Q_1^{o,e}(z) = 0 \qquad (6\text{-}42b)$$

$$\Delta Q_2^{o,e}(z) = 0 \qquad (6\text{-}42c)$$

where

$$\Delta Q^{e,o}(z) = \Delta Q^{e,o}_1(z) + \Delta Q^{e,o}_2(z)$$

$$V^{e,o}(z) = V^{e,o}_1(z) = V^{e,o}_2(z)$$

This block is memoryless and it is an open circuit during the o,e clock phase when capacitor C discharges. Thus, during the e,o clock phase the block serves as a resistor, and during the o,e clock phase it does nothing. In many SC network arrangements, where it is necessary to transmit during only one clock phase, this block serves as an excellent resistor equivalent, with $y = \Delta Q/V = C$. Comparing this admittance with the BD transformed conductance in Table 6-3, we observe that

$$y = C = G\tau = \frac{\tau}{R} \tag{6-43}$$

that is,

$$R = \frac{\tau}{C} = \frac{1}{Cf_s} \tag{6-44}$$

Equation (6-44) is the much-acclaimed resistor equivalence relation often cited in the literature. The resistor equivalence in Eq. (6-42) is a direct manifestation of the BD transformation in Eq. (6-16a). Therefore, Eq. (6-44) is, in fact, an approximation which is accurate to the extent that $(1 - z^{-1})/T$ approximates the z-transform of the first derivative (d/dt) operator. With reference to Fig. 6-8a, this approximation is seen to be accurate at high sampling rates where $\omega\tau \ll 1$ (or in the z-plane, where $z \simeq 1$). The same comments also apply to BD-derived inductor and FDNR admittance equivalences.

Equivalent circuits of Eqs. 6-11p and q represent straightforward, nevertheless useful, generalizations of the circuits of Fig. 6-11ℓ and m. These SC networks are seen to provide, for the even and odd clock phases, different-value resistor-like components. This type of component suggests the possibility of time-sharing capacitors and operational amplifiers to achieve different even and odd circuit behaviors.

This concludes the derivations for equivalent circuits in Fig. 6-11. At this point the interested reader should be able to derive the remaining SC equivalent circuits easily.

6.3.2 Simplified Equivalent Circuits for SC Blocks That Sample Their Terminal Voltages Once per Clock Period

In Section 6.3.1 it was observed that many of the four-port equivalent circuits result in n (of $2n$) open-circuit ports. Obviously, any signals applied to one or more of these open ports will be neither processed nor transmitted. Therefore, SC networks comprised of these blocks will only provide transmission and filtering when the switches are phased such that the blocks interconnect to provide one nonopen signal path from input to output. Assuming this connection rule, the open ports are nonfunctional and can be removed from the equivalent circuits. The immediate identification of these blocks in a complex SC network results in much labor-saving equivalent circuit simplification. More specifically, $2n$-port equivalent circuits reduce directly to n-port equivalent circuits. To emphasize this point, the appropriate four-port equivalent circuits in Fig. 6-11 have been reconfigured as two-port equivalent circuits in Fig. 6-12. Many complex SC networks can be modeled exclusively with these simplified equivalent circuits. For this class of SC networks, circuit analysis is no more complex than that for continuous (linear) time-invariant networks. Since the blocks listed in Fig. 6-12 perform all the necessary network functions, it is expected that one can synthesize general z-domain transfer functions using only these blocks. This restriction, with little sacrifice in generality, should lead to efficient z-domain synthesis procedures for SC networks.

In addition to the reconfigured equivalent circuits from Fig. 6-11, Fig. 6-12 contains two additional building blocks. These blocks are shown in Fig. 6-12h and i. Let us briefly discuss each of these blocks on an individual basis.

6.3.2a Toggle-Switched Floating Four-Port (TSFFP) Equivalent Circuit [P5]

This element is the most general of the toggle-switched (single) capacitor elements. Thus, the equivalent circuits for the TSC, TSI, and the TSD can be derived directly from the equivalent circuit in Fig. 6-12h, by simply shorting to ground the appropriate port or ports. The z-transformed nodal charge equations for the block are expressed as follows:

$$\Delta Q_1^{e,o}(z) = CV_1^{e,o}(z) - CV_2^{e,o}(z) - Cz^{-1/2}V_3^{o,e}(z) + Cz^{-1/2}V_4^{o,e}(z) \tag{6-45a}$$

$$\Delta Q_1^{o,e}(z) = 0 \tag{6-45b}$$

$$\Delta Q_2^{e,o}(z) = CV_2^{e,o}(z) - CV_1^{e,o}(z) - Cz^{-1/2}V_4^{o,e}(z) + Cz^{-1/2}V_3^{o,e}(z) \tag{6-45c}$$

$$\Delta Q_2^{o,e}(z) = 0 \tag{6-45d}$$

$$\Delta Q_3^{e,o}(z) = 0 \tag{6-45e}$$

$$\Delta Q_3^{o,e}(z) = CV_3^{o,e}(z) - CV_4^{o,e}(z) - Cz^{-1/2}V_1^{e,o}(z) + Cz^{-1/2}V_2^{e,o}(z) \tag{6-45f}$$

$$\Delta Q_4^{e,o}(z) = 0 \tag{6-45g}$$

$$\Delta Q_4^{o,e}(z) = CV_4^{o,e}(z) - CV_3^{o,e}(z) - Cz^{-1/2}V_2^{e,o}(z) + Cz^{-1/2}V_1^{e,o}(z) \tag{6-45h}$$

To be completely general, Eq. (6-45) describes an eight-port equivalent circuit with four open ports. Such an eight-port description is shown in Fig. 6-11o. The more useful four-port equivalent circuit in Fig. 6-12h is obtained by deleting the open ports.

Comparing Figs. 6-11b and 6-12h, one observes that the four nonopen ports of the TSFFP are coupled together via a 90° rotated, balanced floating LTP. In fact, if the TSFFP is rotated 90° with ports 1 and 3 serving as the incoming ports and ports 2 and 4 as the outgoing ports, we indeed have the equivalent circuit for the floating capacitor in Fig. 6-11b. If ports 2 and 4 are then shorted to ground, one can then easily derive the equivalent circuit for the grounded capacitor in Fig. 6-11a. All the toggle-switched elements—the TSC, TSI, and TSD elements—can be readily derived from the TSFFP. For example, if ports 2 and 4 are shorted to ground, the TSFFP equivalent circuit reduces to that of the TSC in Fig. 6-12e. Also, the TSI in Fig. 6-12f is obtained when ports 1 and 4 are shorted to ground. In summary, by providing the proper termination conditions, we can derive the equivalent circuits for any of the single-capacitor SC elements in Fig. 6-11 and 6-12 with the TSFFP. This fact is shown schematically in Fig. 6-13.

6.3.2b Toggle-Switched Blocks Driven by Full-Clock-Cycle S/H Voltage Sources

Equivalent full-cycle-time delays can be experienced when toggle-switched SC blocks are driven with full-cycle S/H voltage sources. A situation of this type is illustrated in Fig. 6-12i. The behavior of this circuit can be described in the following manner. When source $v_{c,d}(t)$ changes to value $v_{c,d}^{e,o}$, the o,e switch is open; thus, the charge on capacitor C remains unchanged. One-half clock period later when switch o,e closes, capacitor C acquires the charge $Cv_{c,d}^{e,o}$. Another one-half clock period later the o,e switch opens, the e,o switch closes, and $v_{c,d}^{e,o}$ appears at the output with a net time delay of one full clock period. Obviously, when the source changes value in synchronism with the initial o,e switch, the net time delay is one-half of a clock period. The SC circuit in Fig. 6-12i can be modeled according to the equivalent circuit in Fig. 6-14. Writing a

Fig. 6-14 Toggle-switched capacitor driven by a full-clock-period S/H voltage source.

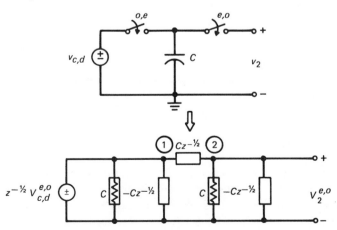

nodal charge equation at node 2 yields

$$CV_2^{e,o}(z) = Cz^{-1/2}[z^{-1/2}V_{c,d}^{e,o}(z)] = Cz^{-1}V_{c,d}^{e,o}(z) \tag{6-46}$$

The equivalent circuit in Fig. 6-12i conveniently characterizes this relationship. Similar equivalent circuits can be derived for the TSI and TSD blocks, as shown in Fig. 6-15a and b, respectively. Full-cycle-time delays can readily occur when appropriately phased toggle-switched blocks are driven by op amp integrator circuits.

Fig. 6-15 (a) Toggle-switch inverter. (b) Toggle-switch differencer, driven by full-clock-period S/H voltage sources.

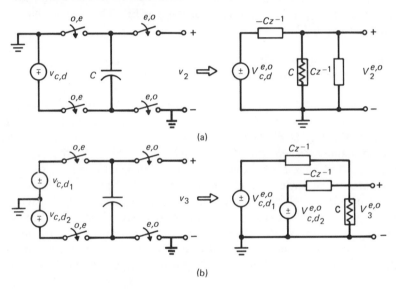

(a)

(b)

6.3.3 Simplified Equivalent Circuits for SC Blocks That Sample Their Terminal Voltages Twice per Clock Period

There are two elemental, single-capacitor elements in Fig. 6-11 which achieve perfect even–odd symmetry; that is, the even and odd signal paths are identical. Circuits of this type are interesting in that, by virtue of their symmetry, they are truly time-invariant. When circuits of this type are interconnected to form an SC filter, the filter is also time-invariant.

6.3.3a Unswitched Capacitor

The first element we discuss that possesses this property is the unswitched capacitor. As should be expected, even–odd symmetry is possessed by both grounded and floating capacitors. To illustrate the property, let us consider the floating capacitor given in Fig. 6-11b and described mathematically by Eqs.

Active Switched Capacitor Sampled-Data Networks

(6-37). For convenient reference, let us rewrite Eqs. (6-37):

$$\Delta Q_1^e(z) = C V_1^e(z) - C z^{-1/2} V_1^o(z) - C V_2^e(z) + C z^{-1/2} V_2^o(z) \qquad (6\text{-}47a)$$

$$\Delta Q_1^o(z) = C V_1^o(z) - C z^{-1/2} V_1^e(z) - C V_2^o(z) + C z^{-1/2} V_2^e(z) \qquad (6\text{-}47b)$$

$$\Delta Q_2^e(z) = C V_2^e(z) - C z^{-1/2} V_2^o(z) - C V_1^e(z) + C z^{-1/2} V_1^o(z) \qquad (6\text{-}47c)$$

$$\Delta Q_2^o(z) = C V_2^o(z) - C z^{-1/2} V_2^e(z) - C V_1^o(z) + C z^{-1/2} V_1^e(z) \qquad (6\text{-}47d)$$

Let us now add Eq. (6-47a) to Eq. (6-47d) and Eq. (6-47c) to Eq. (6-47d). This yields

$$\Delta Q_1^e(z) + \Delta Q_1^o(z) = C(1 - z^{-1/2})[V_1^e(z) + V_1^o(z)]$$
$$- C(1 - z^{-1/2})[V_2^e(z) + V_2^o(z)] \qquad (6\text{-}48a)$$

$$\Delta Q_2^e(z) + \Delta Q_2^o(z) = C(1 - z^{-1/2})[V_2^e(z) + V_2^o(z)]$$
$$- C(1 - z^{-1/2})[V_1^e(z) + V_1^o(z)] \qquad (6\text{-}48b)$$

However, we recall from Section 6.2 that

$$\Delta Q(z) = \Delta Q^e(z) + \Delta Q^o(z) \qquad (6\text{-}49a)$$

$$V(z) = V^e(z) + V^o(z) \qquad (6\text{-}49b)$$

Using the identities in Eq. (6-49), we can rewrite Eq. (6-48) as follows:

$$\Delta Q_1(z) = C(1 - z^{-1/2})V_1(z) - C(1 - z^{-1/2})V_2(z)$$
$$= y_C(z)V_1(z) - y_C(z)V_2(z) \qquad (6\text{-}50a)$$

$$\Delta Q_2(z) = C(1 - z^{-1/2})V_2(z) - C(1 - z^{-1/2})V_1(z)$$
$$= y_C(z)V_2(z) - y_C(z)V_1(z) \qquad (6\text{-}50b)$$

where

$$y_C = C(1 - z^{-1/2}) \qquad (6\text{-}51)$$

The appearance of $z^{-1/2}$ in Eq. (6-51) informs us that the sampling rate has doubled. We can restore y_C in Eq. (6-51) to the form given in Table 6-3 by defining $\hat{z} = e^{s\tau/2}$; then,

$$y_C = C(1 - \hat{z}^{-1}) \qquad (6\text{-}52)$$

The equivalent circuits for the grounded and floating capacitors, when the charge on the capacitor is updated twice per clock period, are given in Fig. 6-16a and b.

It is noted that the equivalent circuits in Fig. 6-16a and b can be used only when combined with other circuits of equivalent symmetry. Another element in Fig. 6-11 that exhibits even–odd symmetry is the bilinear resistor.

6.3.3b Bilinear Resistor [P11]

Let us derive the necessary equations for the floating bilinear resistor; those for the grounded bilinear resistor follow analogously. The z-transformed nodal charge equations are obtained directly from the circuit schematic in Fig. 6-16d;

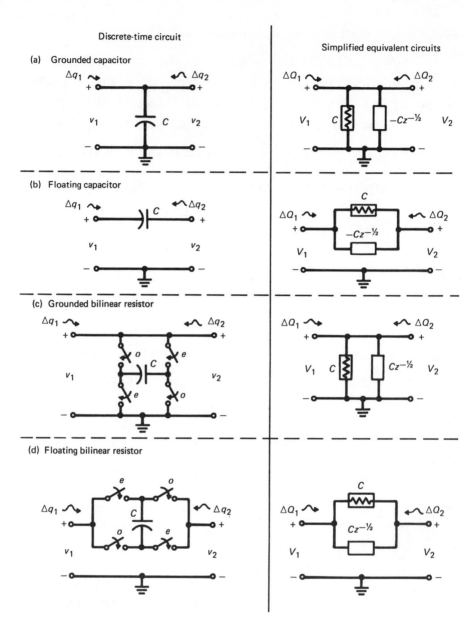

Fig. 6-16 Equivalent circuits for the capacitor and bilinear resistor when capacitors are sampled during both even and odd clock phases.

$$\Delta Q_1^e(z) = C V_1^e(z) + C z^{-1/2} V_1^o(z) - C V_2^e(z) - C z^{-1/2} V_2^o(z) \quad (6\text{-}53a)$$

$$\Delta Q_1^o(z) = C V_1^o(z) + C z^{-1/2} V_1^e(z) - C V_2^o(z) - C z^{-1/2} V_2^e(z) \quad (6\text{-}53b)$$

$$\Delta Q_2^e(z) = C V_2^e(z) + C z^{-1/2} V_2^o(z) - C V_1^e(z) - C z^{-1/2} V_1^o(z) \quad (6\text{-}53c)$$

$$\Delta Q_2^o(z) = C V_2^o(z) + C z^{-1/2} V_2^e(z) - C V_1^o(z) - C z^{-1/2} V_1^e(z) \quad (6\text{-}53d)$$

Active Switched Capacitor Sampled-Data Networks

Adding Eqs. (6-53b) and (6-53d) to Eqs. (6-53a) and (6-53c), respectively, and recombining terms using the identities in Eq. (6-49), we obtain

$$
\begin{align}
\Delta Q_1(z) &= C(1 + z^{-1/2})V_1(z) - C(1 + z^{-1/2})V_2(z) \\
&= y_R(z)V_1(z) - y_R(z)V_2(z)
\end{align}
\tag{6-54a}
$$

$$
\begin{align}
\Delta Q_2(z) &= C(1 + z^{-1/2})V_2(z) - C(1 + z^{-1/2})V_1(z) \\
&= y_R(z)V_2(z) - y_R(z)V_1(z)
\end{align}
\tag{6-54b}
$$

Thus, with $z = e^{s\tau}$,

$$ y_R = C(1 + z^{-1/2}) \tag{6-55a} $$

or with $\hat{z} = e^{s\tau/2}$,

$$ y_R = C(1 + \hat{z}^{-1}) \tag{6-55b} $$

Comparing y_R in Eq. (6-55b) with the resistor equivalents in Table 6-3, we observe that bilinear equivalent resistor is realized exactly by the SC networks in Fig. 6-16c and d. Equating Eq. (6-55b) to the equivalent in Table 6-3 yields the resistor equivalance

$$ R = \frac{\tau}{2C} $$

where τ represents a sampling interval of one clock period. Here, as we recognized earlier, the sample rate has been doubled to once per $\tau/2$ seconds. Doubling the sample rate reduces the equivalent resistance to

$$ R = \frac{\tau}{4C} \tag{6-56} $$

The equivalent circuits for the grounded and floating bilinear resistors are quite simple, as shown in Fig. 6-16c and d.

It is emphasized that the equivalent circuits in Fig. 6-16 can be used only to model a complex SC network comprised of even–odd symetric elements only. For a mixed circuit, the reader is urged to use the general four-port models in Fig. 6-11b and n.

This concludes the discussion of z-domain models for elementary networks. In succeeding sections we provide numerous examples of passive and active SC networks. These examples will serve to demonstrate the use of the equivalent circuits introduced in this section and to illustrate some of the more interesting and useful circuit topologies.

6.4 ANALYSIS AND SYNTHESIS OF FIRST-ORDER SC NETWORKS

In this section the concepts developed in previous sections are applied to the analysis of several first-order passive and active SC networks. Through the understanding of these simple circuits, the reader will increase his/her confidence in applying the equivalent circuits derived in Section 6.3 while building a repertoire of first-order circuit blocks for realizing active-SC networks of arbitrary order. Biquadratic and higher-order networks are reserved for the next section.

6.4.1 Passive-SC Networks

In this section we examine two single-pole passive-SC networks. The equivalent circuits in Fig. 6-11, 6-12, and 6-16 allow us to examine a given SC network under an assortment of input–output conditions, as per Figs. 6-3 and 6-4. As we will see, such an examination can reveal some rather interesting circuit behavior that is not immediately obvious.

6.4.1a First-Order Low-Pass SC Networks

As the initial example, consider the simple first-order low-pass network depicted in Fig. 6-17a. An equivalent circuit for this network can be obtained by simply cascading blocks in Fig. 6-11c–f as shown in Fig. 6-17b. This circuit can obviously be reduced to that in Fig. 6-17c. One could have immediately written the equivalent circuit in Fig. 6-17c by cascading the simplified block equivalent circuits of Fig. 6-12a and e. Writing a single nodal charge equation at node 2,

$$[C_1 - C_1 z^{-1/2} + C_1 z^{-1/2} + C_2 - C_2 z^{-1}] V_{\text{out}}^o(z) = C_1 z^{-1/2} V_{\text{in}}^e(z) \qquad (6\text{-}57)$$

yields the familiar low-pass z-domain transfer function

$$H_3(z) = \frac{V_{\text{out}}^o(z)}{V_{\text{in}}^e(z)} = \frac{C_1 z^{-1/2}}{C_1 + C_2 - C_2 z^{-1}} \qquad (6\text{-}58\text{a})$$

The pole and zero for Eq. (6-58a) are located in the z-plane, as shown in Fig. 6-18a. It is noted that in Fig. 6-18a, and in future pole-zero plots, the one-half clock period delay is denoted as a zero at the origin with a one-half suffix. Comparing Eq. (6-58a) with the approximated damped LDI in Eq. (6-33), we observe that when $\tau p = C_1/C_2$, these functions are identical.

Note that V_{in}^o and V_{out}^e are removed by sampling operations at the input and output, respectively. By removing the output o-phase switch, as shown in Fig. 6-19, we can restore V_{out}^e to V_{out}. Analyzing this circuit, we find (Prob. 6.15) that $V_{\text{out}}^e(z) = z^{-1/2} V_{\text{out}}^o(z)$, or

$$H_1(z) = \frac{V_{\text{out}}^e(z)}{V_{\text{in}}^e(z)} = z^{-1/2} H_3(z) = \frac{C_1 z^{-1}}{C_1 + C_2 - C_2 z^{-1}} \qquad (6\text{-}58\text{b})$$

In the z-plane Eq. (6-58b) is represented as shown in Fig. 6-18b. Comparing with the entries in Table 6-2, we see that Eq. (6-58b) is a FD damped integrator with $\tau p = C_1/C_2$.

It is instructive to return to the time domain to view the significance of Eq. (6-58). Let us assume that v_{in} is a low-frequency sampled sine wave of the form shown in Fig. 6-20a. The sampling rate has been chosen low to make the samples and waveform more visible. The samples at $t = kT$ are displayed as large dots and the sampled-and-held (S/H) equivalent sine wave is the staircase-

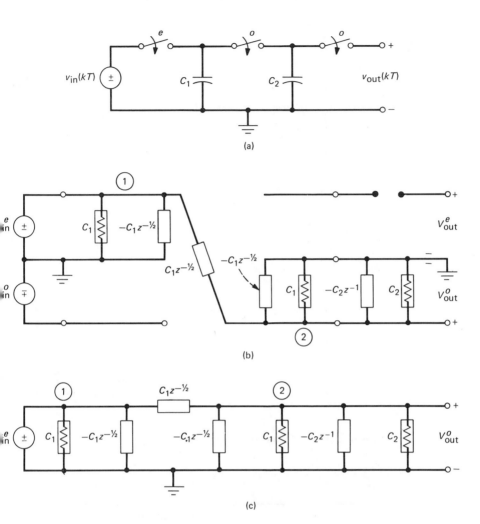

Fig. 6-17 Single-pole "passive" SC low-pass network.

like function superimposed on the sampled function. Note that the S/H sine wave is consistent with our definition of $v(t)$ in Fig. 6-3. Assuming that the frequency for v_{in} is well below the cutoff of H_1 and H_3, we can sketch $v_{out}^e(t)$ and $v_{out}^o(t)$ as shown in Fig. 6-20b and c. To obtain these functions, we use the even sampled values for v_{in}; the dc gain of unity, obtained by setting $z = 1$ in Eq. (6-58); and the definitions for v^e and v^o in Fig. 6-3. Combining the odd and even output components, we obtain the full-clock-period S/H waveform in Fig. 6-20d. If we were to build the circuit in Fig. 6-19 and view v_{out} with an oscilloscope, we would observe a waveform like that in Fig. 6-20d. The voltages v_{out}^e and v_{out}^o are mathematical definitions and do not represent physically measurable waveforms. To view the output of the circuit in Fig. 6-17, we would follow the output o-switch with a small holding capacitor (e.g., Fig. 6-6a). The output of the circuit would again be similar to that in Fig. 6-20d.

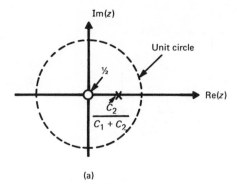

(a)

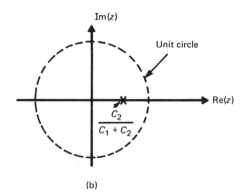

(b)

Fig. 6-18 z-Domain pole and zero locations for (a) H_3 in Eq. (6-58a) and (b) H_1 in Eq. (6-58b).

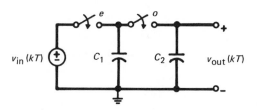

Fig. 6-19 First-order SC low-pass network.

We see from this exercise that we can construct the time-domain output voltages, for simple sinusoidal inputs, from the z-domain transfer functions relating v_{in}^e and v_{in}^o to v_{out}^e and v_{out}^o and the definitions for v^e and v^o in Fig. 6-3. To gain greater appreciation for the physical interpretations for the transfer functions derived from the equivalent circuits, the reader is encouraged to sketch the time-domain output waveforms for the simple circuits in this section as home exercises.

6.4.1b First-Order High-Pass SC Network

The simple first-order high-pass circuit shown in Fig. 6-21a is a rather interesting circuit, as we shall soon see. Its interesting behavior stems from the input-to-output switch free path which permits both V_{in}^e and V_{in}^o to determine the e and o components of V_{out}. To study this circuit, we draw the equivalent circuit in Fig. 6-21b by cascading blocks in Fig. 6-11b and k.

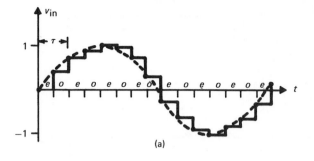

(a)

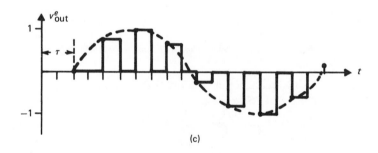

(b)

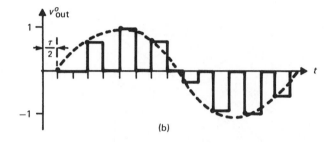

(c)

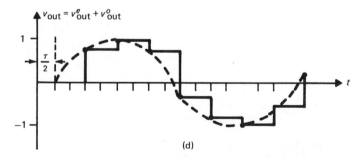

(d)

Fig. 6-20 Time-domain response of the first-order low-pass SC network in Fig. 6-19 to a sinusoidal input. The frequency of the input is assumed to be within the network passband.

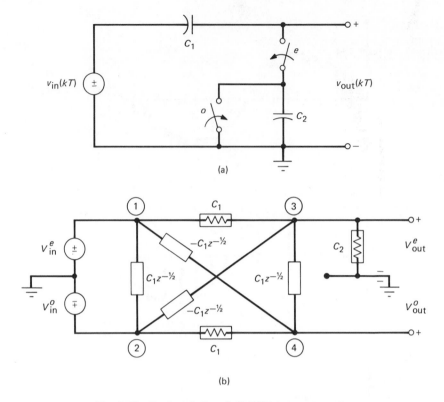

Fig. 6-21 Single-pole "passive" SC high-pass network.

Analysis of the equivalent circuit yields the following relations:

$$V_{\text{out}}^e(z) = \frac{C_1(1 - z^{-1})}{C_1 + C_2 - C_1 z^{-1}} V_{\text{in}}^e(z) + 0 V_{\text{in}}^o(z) \tag{6-59a}$$

$$= H_1(z) V_{\text{in}}^e(z) + H_2(z) V_{\text{in}}^o(z) \tag{6-59b}$$

[note that $H_2(z) = 0$ in Eqs. (6-59)] and

$$V_{\text{out}}^o(z) = -\frac{C_2 z^{-1/2}}{C_1 + C_2 - C_1 z^{-1}} V_{\text{in}}^e(z) + V_{\text{in}}^o(z) \tag{6-60a}$$

$$= H_3(z) V_{\text{in}}^e(z) + H_4(z) V_{\text{in}}^o(z) \tag{6-60b}$$

Note that

$$H_1(z) = \left.\frac{V_{\text{out}}^e(z)}{V_{\text{in}}^e(z)}\right|_{V_{\text{in}}^o = 0} = \frac{C_1(1 - z^{-1})}{C_1 + C_2 - C_1 z^{-1}} \tag{6-61a}$$

is a first-order high-pass function, whereas

$$H_3(z) = \left.\frac{V_{\text{out}}^o(z)}{V_{\text{in}}^e(z)}\right|_{V_{\text{in}}^o = 0} = -\frac{C_2 z^{-1/2}}{C_1 + C_2 - C_1 z^{-1}} \tag{6-61b}$$

is an inverting first-order low-pass function. A most interesting result indeed. From Eq. (6-60) we observe that by forcing $V_{in}^o = 0$, per Fig. 6-22a, this circuit behaves like a first-order high-pass filter when the output is sampled on the even times and like a first-order (inverting) low-pass filter when the output is sampled on the odd times. A circuit that achieves this bifunctional characteristic is shown in Fig. 6-22b. To achieve this behavior, a simple return-to-zero source, of the

Fig. 6-22 Single-pole "passive" SC high-pass/low-pass network.

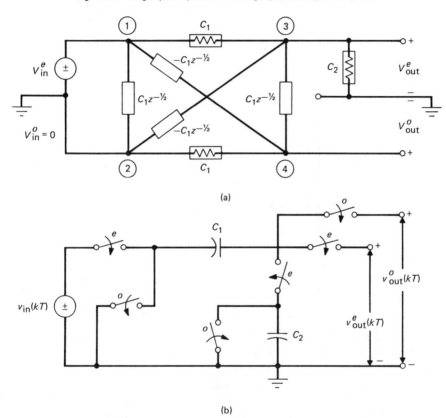

(a)

(b)

form shown in Fig. 6-4a, or equivalently in Fig. 6-5a, is used to drive the high-pass circuit depicted in Fig. 6-21a. We note that in the z-plane the poles and zeros for H_1 and H_3 are located as shown in Fig. 6-23.

6.4.2 First-Order Active-SC Networks

For many of the same reasons that motivate our interest in active-RC networks, we investigate next active-SC networks. It is possible to realize resonant passive-SC networks with multiphase switching [P8, P12]. Those circuits are of questionable value in implementing precision SC filters because of various parasitic capacitances that arise naturally in MOS realizations. A

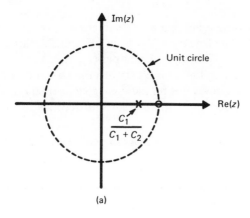

(a)

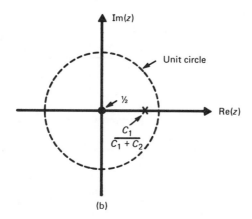

(b)

Fig. 6-23 z-Domain pole and zero locations for (a) H_1 in Eq. (6-61a) and (b) H_3 in Eq. (6-61b).

discussion of parasitics and other practical concerns are reserved for the next section. As we see in the next section, the effects of parasitics can be minimized in active-SC networks. The active-SC networks of concern to us are those realized with capacitors, biphased switches, and operational amplifiers. Many of the active-SC networks in the literature [P2, P9–P11] are comprised of simple SC building blocks of the forms listed in Figs. 6-12 and 6-16, buffered by operational amplifiers. When these operational amplifiers can be assumed to be ideal (infinite gain), the virtual grounds result in further simplifications in the equivalent circuits. Recall that when the noninverting input of an ideal (infinite gain) op amp is connected to ground, a virtual ground ($v^- = 0$) exists at the inverting input. Let us consider next the equivalent circuit representations for the following selection of first-order active-SC networks.

6.4.2a Lossless Integrator

One switched capacitor implementation of a lossless integrator is shown in Fig. 6-24a. The switches turn on and off according to the rise and fall of clocks ϕ^e and ϕ^o in Fig. 6-24b and c, respectively. To verify that this circuit

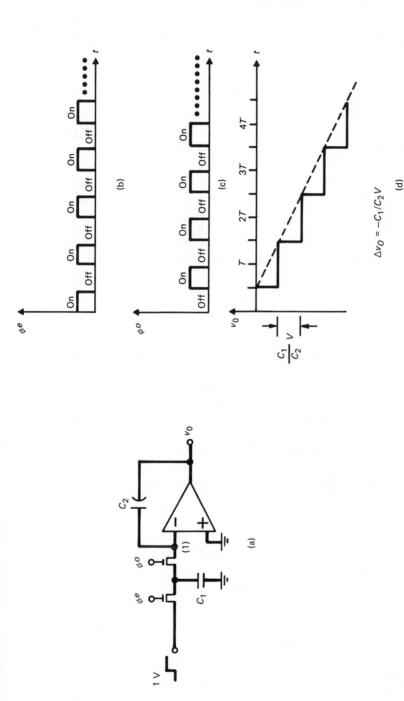

Fig. 6-24 SC lossless integrator time-domain response to a unit-step excitation.

$$\Delta v_O = -C_1/C_2 V$$

431

performs the function of integration, let us derive the response v_0 of the circuit to a unit step input. At $t = 0$, ϕ^e is on and ϕ^o is off; thus, C_1 is now connected across the input and disconnected from the op amp. Capacitor C_1 then charges to $1\ V$ and, in the absence of initial charge on C_2, $v_0 = 0$. At $t = T/2$, ϕ^e turns off and ϕ^o turns on. Since node (v^-) is a virtual ground, the charge on capacitor C_1 transfers immediately to capacitor C_2 and v_0 changes instantaneously to $v_0 = -C_1/C_2(1\ V)$, where $-C_1/C_2$ is the gain of the integrator. After some finite but very short settling time, v_0 reaches steady state and remains constant. At $t = T$, ϕ^e turns on and ϕ^o turns off. Since no additional charge is transferred to C_2 during this phase, v_0 remains constant at $-C_1/C_2(1\ V)$. Capacitor C_1 is now again connected across the input and recharges to $1\ V$. At $t = 3/2T$, ϕ^e turns off and ϕ^o turns on, with the charge on C_1 immediately transferred to capacitor C_2. Voltage v_0, which is sensed nondestructively, experiences another step change of $-C_1/C_2(1\ V)$. Continuing this process (of course, the op amp will saturate when v_0 equals the negative supply voltage), we see in Fig. 6-24d that v_0 is, in the sampled-data sense, a ramp. Thus, integration is achieved. Using the z-domain equivalent circuits derived in Section 6.3, we will now derive a more precise characterization of this integrator circuit.

The equivalent circuit for the lossless integrator in Fig. 6-24a is derived, in full generality, using blocks in Fig. 6-11b, c, and v, as shown in Fig. 6-25b. Of course, one may accommodate the finite dc gain of an actual operational amplifier using the voltage-controlled voltage source, cited in Fig. 6-11 as block u. The rather unwieldy circuit depicted in Fig. 6-25b can be immediately simplified by removing all elements shunting virtual ground points and voltage sources. This network is then redrawn in the form shown in Fig. 6-25c, which can be again reconfigured to yield the circuit in Fig. 6-25d. Finally, the second stage of Fig. 6-25d is noted to be a voltage-controlled voltage source with $\beta = z^{-1/2}$, like the independent source in Fig. 6-14. The final equivalent circuit in Fig. 6-25e implies that the lossless integrator could have been derived directly from the simplified equivalent circuits in Fig. 6-12.

The transfer functions for the lossless integrator are then readily determined to be

$$H_3(z) = \frac{V^o_{\text{out}}(z)}{V^e_{\text{in}}(z)} = -\frac{\dfrac{C_1}{C_2}z^{-1/2}}{1 - z^{-1}} \tag{6-62a}$$

$$H_1(z) = \frac{V^e_{\text{out}}(z)}{V^e_{\text{in}}(z)} = -\frac{\dfrac{C_1}{C_2}z^{-1}}{1 - z^{-1}} \tag{6-62b}$$

The z-plane pole for H_3 and H_1 are identically located as shown in Fig. 6-26a and b, respectively. Observe that:

1. When V_{out} is sampled at the odd times (only), the transfer function is H_3. H_3 in Eq. (6-62a) is seen to realize the LDI given in Table 6-2.
2. When V_{out} is sampled at the even times (only), the transfer function is H_1. H_1 in Eq. (6-62b) is seen to realize the FD integrator in Table 6-2.

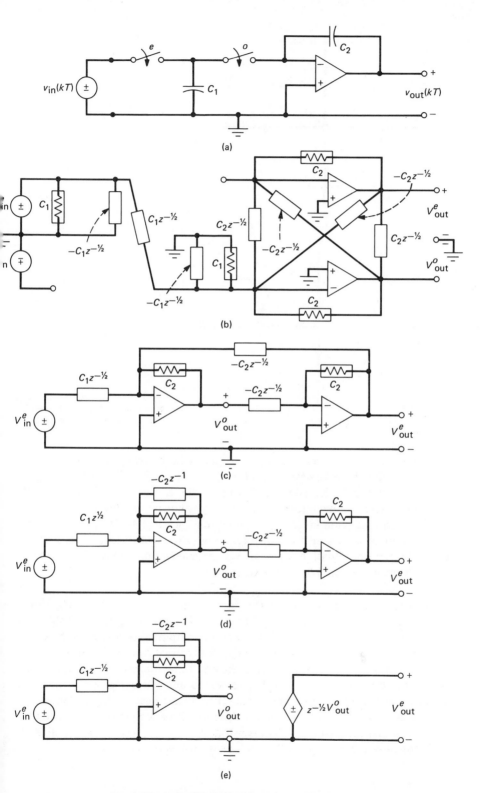

Fig. 6-25 Active-SC lossless integrator (FD/LDI).

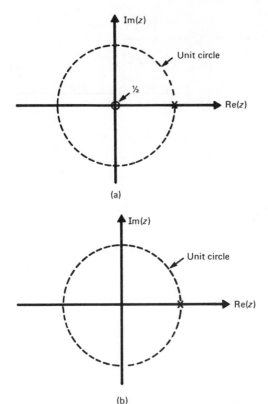

(a)

(b)

Fig. 6-26 z-Domain pole and zero location for (a) H_3 in Eq. (6-62a) and (b) H in Eq. (6-62b).

3. Equations (6-62) describe the operation of an inverting lossless integrator. A noninverting lossless integrator can be realized by simply replacing the TSC in Fig. 6-25a with a TSI (Fig. 6-12f). In active-RC realizations, the noninverting integrator was realized by following the inverting integrator with a unity-gain inverting amplifier (e.g., see Fig. 4-35). In this case, two op amps were used to realize an active-RC noninverting integrator. We see here one situation where an SC implementation requires less hardware than the comparable active-RC realization.

It is further noted, for the TSC element at the input of the integrator in Fig. 6-25a, that it provides $y_R = G\tau z^{-1}$ (the FD resistor given in Table 6-3), when the TSC input is voltage-driven and the output is connected to the virtual ground of an op amp. To obtain z^{-1} in y_R, for the switching shown in Fig. 6-25a, the input must change in value only at the odd time instants (e.g., see Fig. 6-12i). When the output is not connected to an op amp virtual ground, the shunt elements in the TSC equivalent circuit (Fig. 6-12i) load the equivalent circuit for the succeeding SC network stage. One might be tempted to conclude from Fig. 6-25c that the LDI equivalent resistor ($y_R = G\tau z^{-1/2}$) is similarly realizable. This conclusion is only partially valid, as we will soon see.

Active Switched Capacitor Sampled-Data Networks

Let us consider in Fig. 6-27a another integrator structure. Starting with equivalent circuits of Fig. 6-11b, m, and v, we can draw an equivalent circuit for this integrator. Following the simple circuit reduction steps used in Fig. 6-25, the equivalent circuit can be reduced to that in Fig. 6-27b and finally to the circuit in Fig. 6-27c. The circuit in Fig. 6-27c could have been directly obtained using the simplified equivalent circuits in Fig. 6-12b, d, and n and a $z^{-1/2}$ voltage-controlled voltage source.

Fig. 6-27 Active-SC lossless integrator (BD/LDI).

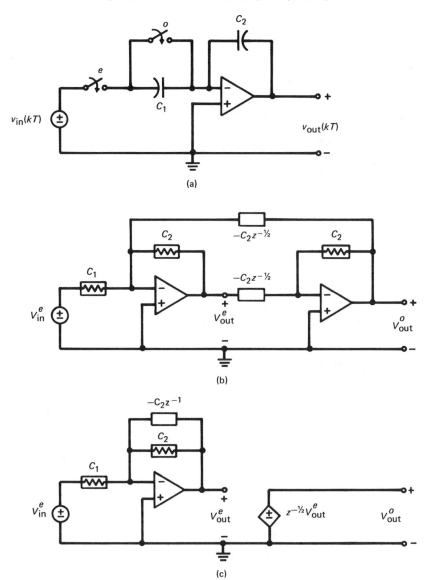

The transfer functions for this circuit are readily derived to be

$$H_1(z) = \frac{V_{\text{out}}^e(z)}{V_{\text{in}}^e(z)} = -\frac{C_1/C_2}{1 - z^{-1}} \qquad (6\text{-}63\text{a})$$

$$H_3(z) = \frac{V_{\text{out}}^o(z)}{V_{\text{in}}^e(z)} = -\frac{(C_1/C_2)z^{-1/2}}{1 - z^{-1}} \qquad (6\text{-}63\text{b})$$

In the z-plane, Eqs. (6-63) are represented as shown in Fig. 6-28. Note that H_1, obtained when V_{out} is sampled periodically during the even time slots, realizes

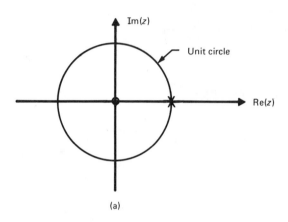

(a)

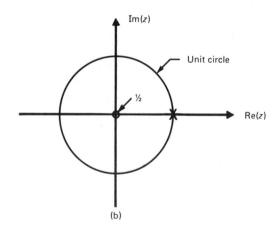

(b)

Fig. 6-28 z-Domain pole and zero locations for (a) H_1 in Eq. (6-63a) and (b) H_3 in Eq. (6-63b).

the BD integrator. But H_3, obtained when V_{out} is sampled during the odd time slots, realizes the LDI integrator. Thus, both integrator structures (Figs. 6-25a and 6-27a) are capable of realizing LDI integrators. In the next section we reveal the practical importance of this seemingly uninteresting duplication of function.

Active Switched Capacitor Sampled-Data Networks

6.4.2b Lossy Integrator with TSC

As should be expected, the equivalent circuit for the lossy integrator, shown in Fig. 6-29a, is similar to that derived for the lossless integrator. To shift the pole to the left of $z = 1 + j0$, a toggle-switched capacitor (TSC) has

Fig. 6-29 Active-SC lossy integrator with TSC damping.

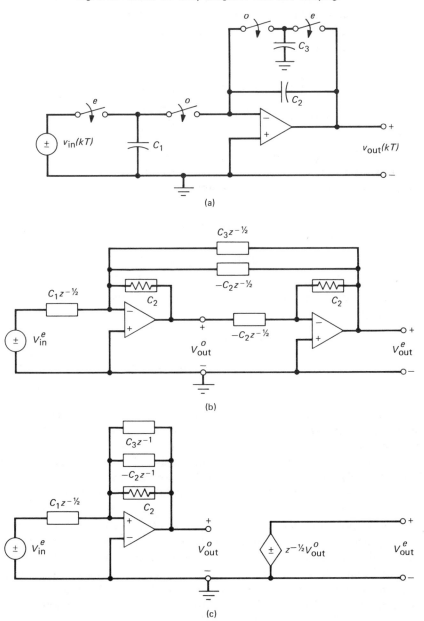

(a)

(b)

(c)

been placed across the feedback capacitor. Obviously, the intent is for the TSC to play a role comparable to a resistor in active-RC lossy integrators. To analyze this network, let us first derive the equivalent circuit. This equivalent circuit can be derived step by step, as was done for the lossless integrator and shown successively in Fig. 6-29b and c. The final equivalent circuit in Fig. 6-29c, like that in Fig. 6-25e, can be readily derived from the simplified equivalent circuits in Fig. 6-12 by direct substitution. With this observation and practice, the equivalent circuits for most active-SC networks can be drawn easily.

The transfer functions for the lossy integrator are then readily obtained from the circuit in Fig. 6-29c:

$$H_3(z) = \frac{V_{out}^o(z)}{V_{in}^e(z)} = -\frac{(C_1/C_2)z^{-1/2}}{1 - (1 - C_3/C_2)z^{-1}} \tag{6-64a}$$

$$H_1(z) = \frac{V_{out}^e(z)}{V_{in}^e(z)} = -\frac{(C_1/C_2)z^{-1}}{1 - (1 - C_3/C_2)z^{-1}} \tag{6-64b}$$

Note that H_1 in Eq. (6-64b) is the FD damped integrator with $\tau p = C_3/C_2$. It is interesting to examine H_1 and H_3 for different values of C_3. Consider the three conditions $C_3 = C_2$, $C_3 = 2C_2$, and $C_3 > 2C_2$. For $C_3 = C_2$:

$$H_1(z) = -\frac{C_1}{C_2} z^{-1} \tag{6-65a}$$

$$H_3(z) = -\frac{C_1}{C_2} z^{-1/2} \tag{6-65b}$$

Equations (6-55a) and (6-55b) represent ideal full- and half-delay elements, respectively. For $C_3 = 2C_2$,

$$H_1(z) = -\frac{(C_1/C_2)z^{-1}}{1 + z^{-1}} \tag{6-66a}$$

$$H_3(z) = -\frac{(C_1/C_2)z^{-1/2}}{1 + z^{-1}} \tag{6-66b}$$

and the circuit oscillates at the half-sampling frequency. Finally, when $C_3 > 2C_2$ the pole of H_1 and H_3 lies outside the unit circle and the circuit is unstable. The root locus of the pole as C_3 is varied, is sketched in Fig. 6-30. Obviously, from this demonstration, the TSC is more than a resistor. In fact, if we examine the feedback equivalent circuitry in Fig. 6-29c, we see that $y_{TSC} = y_R = C_3 z^{-1}$, the FD equivalent resistor. The potential instability that results from the delay in y_R is predictable from the $j\omega$-axis mapping in Fig. 6-8b.

Let us now consider inverting the phases of the switches of the feedback TSC (C_3) per Fig. 6-31a. In view of Eq. (6-64), it might seem that inverting the phasing of these two switches would introduce a $C_3 z^{-1/2}$ term into the denominator of H_1 and H_3. To see that this does not happen, let us analyze this network via the equivalent circuit in Fig. 6-31b. This circuit can be manipulated to yield the reduced equivalent circuit in Fig. 6-31c. The transfer functions for this

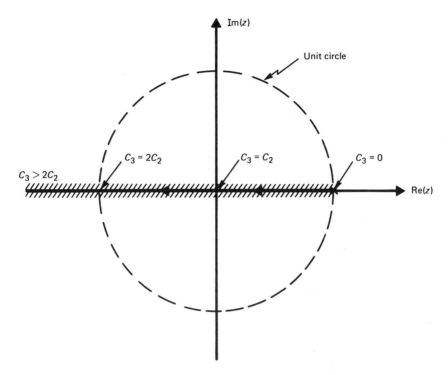

Fig. 6-30 Root locus for pole of H_1 and H_3 in Eq. (6-64) for $C_3 \geq 0$.

circuit are readily determined to be

$$H_3(z) = \frac{V_{out}^o(z)}{V_{in}^e(z)} = -\frac{(C_1/C_2)z^{-1/2}}{1 - (1 - C_3/C_2)z^{-1}} \qquad (6\text{-}67a)$$

$$H_1(z) = \frac{V_{out}^e(z)}{V_{in}^e(z)} = -\frac{\dfrac{C_1}{C_2}\left(\dfrac{C_2 - C_3}{C_2}\right)z^{-1}}{1 - (1 - C_3/C_2)z^{-1}} \qquad (6\text{-}67b)$$

Comparing the pole location for $H_1(z)$ in Eqs. (6-64) and (6-67), one observes that the TSC switch phasing has no effect on this parameter. However, the dc gain for the even component of V_{out} is altered by the factor $(C_2 - C_3)/C_2$. This effect is not surprising, since charge is being injected onto capacitor C_2 at both even and odd times. Note that the denominators of Eq. (6-67) are free of $z^{-1/2}$ terms. Also, as shown in Figs. 6-29 and 6-31, no matter how the switches are phased, the feedback TSC realizes $y_R = G\tau z^{-1}$. This example serves as a demonstration of the nonrealizability of the LDI lossy integrator.

6.4.2c Lossy Integrator with OFR

Another lossy integrator realization is shown in Fig. 6-32a. Of perhaps only theoretical interest is the comparison of the behavior of this circuit with that of its counterpart in Fig. 6-29a. The equivalent circuit, shown in successive

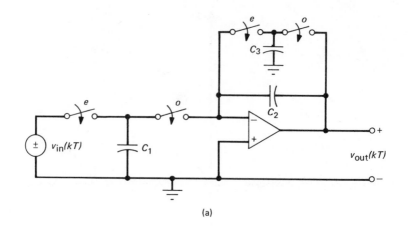

(a)

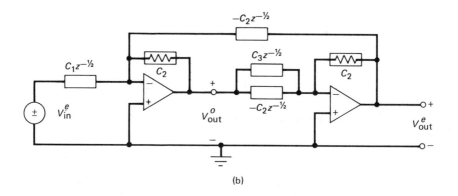

(b)

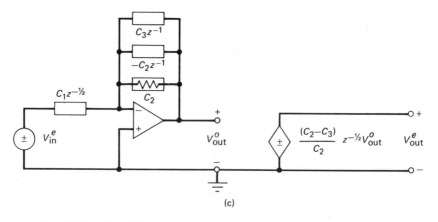

(c)

Fig. 6-31 Active-SC lossy integrator with TSC damping. The switches of the feedback TSC are phased opposite to that shown in Fig. 6-29.

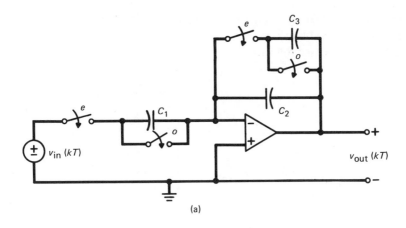

(a)

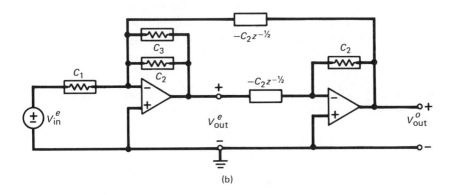

(b)

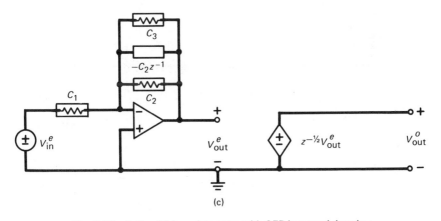

(c)

Fig. 6-32 Active-SC lossy integrator with OFR input and damping.

stages of simplification in Figs. 6-32b and c, respectively, can be derived directly from the simplified block equivalent circuits in Fig. 6-12. The transfer functions for this circuit are as follows:

$$H_3(z) = \frac{V_{out}^o(z)}{V_{in}^e(z)} = -\frac{\dfrac{C_1}{C_2+C_3}z^{-1/2}}{1 - \dfrac{C_2}{C_2+C_3}z^{-1}}$$

(6-68a)

$$H_1(z) = \frac{V_{out}^e(z)}{V_{in}^e(z)} = -\frac{\dfrac{C_1}{C_2+C_3}}{1 - \dfrac{C_2}{C_2+C_3}z^{-1}}$$

(6-68b)

The transfer functions expressed in Eq. (6-68) are seen to be truly representative of lossy integrators. $H_3(z)$ and $H_1(z)$ are absolutely stable for all finite values of C_1, C_2, and C_3. The root locus for the pole of Eq. (6-68) as C_3 increases from 0 to ∞ is shown in Fig. 6-33. Note that H_1 in Eq. (6-68b) is a BD lossy integrator with $\tau p = C_3/C_2$ and that H_3 in Eq. (6-68a) is an approximate LDI lossy integrator [i.e., Eq. (6-33)] with $\tau p = C_3/C_2$.

Fig. 6-33   Root locus for the pole of H_1 and H_3 in Eq. (6-68) as C_3 increases from 0 to ∞.

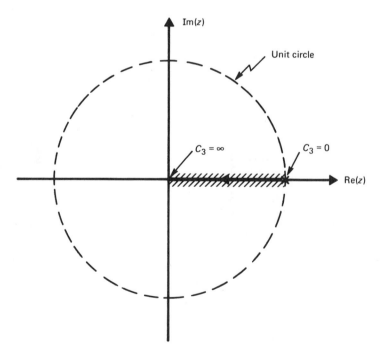

Active Switched Capacitor Sampled-Data Networks

6.4.2d Lossless Bilinear Integrators

Bilinear integration and the advantages of using the bilinear transform were discussed in Section 6.2. There are several ways bilinear integration can be realized [P10–P12] with SC networks, as demonstrated in Figs. 6-34, 6-36, and 6-37. It is interesting to examine the behavior of each of these circuits.

Fig. 6-34 Active-SC bilinear lossless integrator [P12].

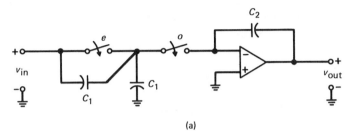

(a)

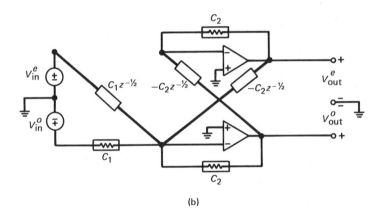

(b)

Let us initially consider the bilinear integrator [P12] shown in Fig. 6-34a. The z-domain equivalent circuit, obtained by interconnecting blocks in Fig. 6-11b, c, m, and v, is shown in Fig. 6-34b. By straightforward nodal charge analysis, the following relations are easily obtained:

$$V_{\text{out}}^o = -\frac{(C_1/C_2)z^{-1/2}}{1-z^{-1}}V_{\text{in}}^e - \frac{C_1/C_2}{1-z^{-1}}V_{\text{in}}^o \qquad (6\text{-}69a)$$

$$V_{\text{out}}^e = z^{-1/2}V_{\text{out}}^o \qquad (6\text{-}69b)$$

From Eq. (6-69a), we observe that the desired bilinear integration is only obtained when

$$V_{\text{in}}^e = z^{-1/2}V_{\text{in}}^o \qquad (6\text{-}70)$$

Substituting Eq. (6-70) into Eq. (6-69a) yields the desired result,

$$H_4(z) = \frac{V^o_{out}}{V^o_{in}} = -\frac{(C_1/C_2)(1 + z^{-1})}{1 - z^{-1}} \tag{6-71a}$$

Also,

$$H_2(z) = \frac{V^e_{out}}{V^o_{in}} = -\frac{(C_1/C_2)z^{-1/2}(1 + z^{-1})}{1 - z^{-1}} \tag{6-71b}$$

The pole and zero for Eq. (6-71a) are located in the z-plane as shown in Fig. 6-35. In summary, this circuit, with the switches phased as shown in Fig. 6-34a, will provide bilinear integration only when the input and output are sampled at the odd $(2k + 1)T$ times and the input is held for the entire clock period.

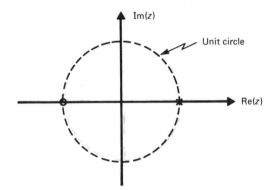

Fig. 6-35 z-domain pole and zero locations for H_4 in Eq. (6-71a).

A second bilinear integrator realization [P10] is shown in Fig. 6-36a. The z-domain equivalent circuit, shown in Fig. 6-36b, is obtained by interconnecting blocks in Fig. 6-11b, e, and v. The transfer relations for this circuit are readily determined to be

$$V^o_{out} = -\frac{(C_1/C_2)(1 + z^{-1})}{1 - z^{-1}} V^o_{in} - \frac{2(C_1/C_2)z^{-1/2}}{1 - z^{-1}} V^e_{in} \tag{6-72a}$$

$$V^e_{out} = -\frac{(C_1/C_2)(1 + z^{-1})}{1 - z^{-1}} V^e_{in} - \frac{2(C_1/C_2)z^{-1/2}}{1 - z^{-1}} V^o_{in} \tag{6-72b}$$

The output, sampled at all (both even and odd) kT times, is obtained by summing Eqs. (6-72a) and (6-72b) according to Eqs. (6-11) and canceling the common factor $(1 + z^{-1/2})$:

$$H(z) = \frac{V_{out}}{V_{in}} = \frac{V^e_{out} + V^o_{out}}{V^e_{in} + V^o_{in}} = -\frac{(C_1/C_2)(1 + z^{-1/2})}{1 - z^{-1/2}} \tag{6-73}$$

Comparing Eqs. (6-73) and (6-71a), we see that the effective sampling rate has been doubled with the circuit in Fig. 6-36a. Note that even–odd symmetry has been achieved by replicating the TSD twice at the input. We can redefine the z-transform according to $\hat{z} = e^{sT/2}$ and rewrite Eq. (6-73) as

Active Switched Capacitor Sampled-Data Networks

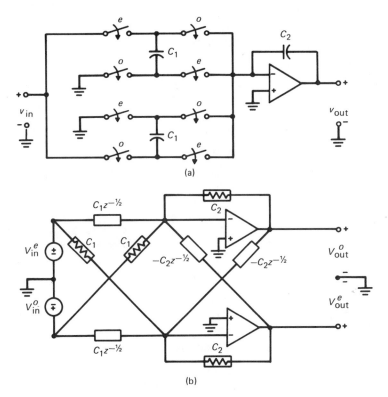

Fig. 6-36 Active-SC lossless bilinear integrator [P10].

$$H(z) = -\frac{(C_1/C_2)(1 + \hat{z}^{-1})}{1 - \hat{z}^{-1}} \qquad (6-74)$$

It was later recognized that Eq. (6-73) could be realized [P11] with a simpler circuit, as shown in Fig. 6-37a. This circuit is transformed to the equivalent circuit in Fig. 6-37b using the elemental equivalent circuits in Fig. 6-16. One can readily show that the transfer function $H(z)$ for this circuit is exactly that given in Eq. (6-73). It is noted that to realize a noninverting bilinear integrator, we must follow an inverting bilinear integrator (Figs. 6-34a, 6-36a, or 6-37a) with a unity-gain inverting amplifier. Thus, two op amps are required to realize a noninverting bilinear integrator with single-input op amps. Recall that LDI and FD noninverting integrators can be realized with a single op amp and a TSI.

6.4.2e Lossy Bilinear Integrator

A lossy bilinear integrator can be realized most efficiently using the circuit in Fig. 6-38a. The z-domain equivalent circuit is given in Fig. 6-38b. The transfer function for this circuit is readily determined to be

$$H(z) = \frac{V_{\text{out}}(z)}{V_{\text{in}}(z)} = -\frac{(C_1/C_2)(1 + z^{-1/2})}{(1 + C_3/C_2) - (1 - C_3/C_2)z^{-1/2}} \qquad (6-75)$$

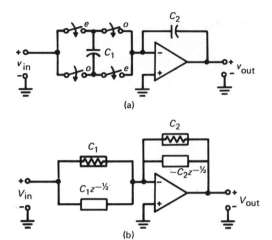

(a)

(b)

Fig. 6-37 Active-SC lossless bilinear integrator [P11].

Fig. 6-38 Active-SC lossy bilinear integrator.

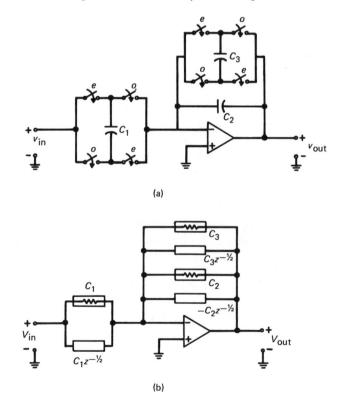

(a)

(b)

446

Comparing Eq. (6-75) with the lossy bilinear integrator in Table 6-2, we observe that they are identical when $\hat{z} = e^{sT/2}$ and $\tau p = C_3/C_2$. The root locus for the pole in Eq. (6-75) in the z-plane as C_3 increases from 0 to ∞ is shown in Fig. 6-39. The circuit is seen to be stable for any value of C_3.

Fig. 6-39 Root locus for the pole of H in Eq. (6-75) as C_3 increases from 0 to ∞.

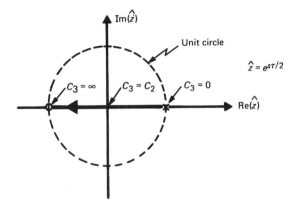

6.4.3 Inductor and FDNR Realizations

In this section we examine some interesting circuits for realizing inductor and FDNR equivalent admittances. The z-domain equivalent admittances for these circuit elements are given in Table 6-3.

6.4.3a Grounded BD Inductor Simulation

One active SC network [P14] which simulates the admittance for a grounded inductor is shown in Fig. 6-40a. The z-domain equivalent circuit for this SC network is given in Fig. 6-40b. Writing nodal charge equations at nodes (1) and virtual ground (2) yield the following circuit equations:

$$\Delta Q_i^e = (C_1 + C_3)V_i^e - C_3 z^{-1/2}V^o \tag{6-76a}$$

$$0 = C_2(1 - z^{-1})V^o + C_1 z^{-1/2}V_i^e \tag{6-76b}$$

Solving Eqs. (6-76) for ΔQ_i^e in terms of V_i^e yields

$$\Delta Q_i^e = \left[(C_1 + C_3) + \frac{C_1 C_3}{C_2}\frac{z^{-1}}{1 - z^{-1}}\right]V_i^e \tag{6-77a}$$

or

$$y = \frac{\Delta Q_i^e}{V_i^e} = C_1 + C_3 + \frac{C_1 C_3}{C_2}\frac{z^{-1}}{1 - z^{-1}} \tag{6-77b}$$

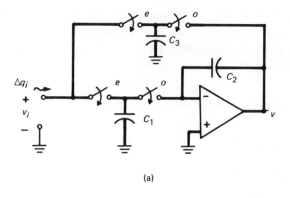

(a)

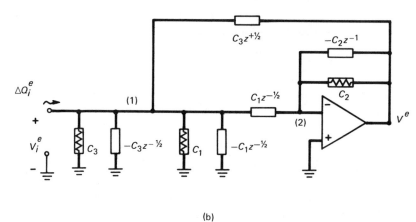

(b)

Fig. 6-40 Active-SC simulation of a grounded BD inductor.

If we add and subtract $C_1 C_3 / C_2$ from the right-hand side of Eq. (6-77b), we obtain

$$y = \frac{\Delta Q_i^e}{V_i^e} = C_1 + C_3 - \frac{C_1 C_3}{C_2} + \frac{C_1 C_3}{C_2}\frac{1}{1 - z^{-1}}$$

$$= y_R + y_L \tag{6-78}$$

Note that y in Eq. (6-78) is the admittance for a lossy inductor. However, by setting

$$C_1 + C_3 - \frac{C_1 C_3}{C_2} = 0$$

or

$$C_2 = \frac{C_1 C_3}{C_1 + C_3} \tag{6-79}$$

the inductor can be made lossless. Comparing the inductive portion of Eq.

(6-78) with the BD inductor admittance in Table 6-3, we obtain

$$\frac{C_1 C_3}{C_2} \frac{1}{1 - z^{-1}} = \frac{\tau^2}{L} \frac{1}{1 - z^{-1}} \tag{6-80}$$

Thus, the inductance simulated is of value

$$L = \frac{\tau^2 C_2}{C_1 C_3} \tag{6-81}$$

To see how the circuit might be used to implement an SC filter, consider the simple *RLC* resonant circuit in Fig. 6-41a. Substituting for each component in Fig. 6-41a its BD equivalent admittance, we obtain the *z*-transformed circuit in Fig. 6-41b. Using the grounded BD inductor simulation in Fig. 6-40, an active SC network that realizes the *z*-transformed circuit in Fig. 6-41b is shown in Fig. 6-41c. Note that the resistive term in *y* of Eq. (6-78) is used to realize R_L. Substituting for switched capacitor C_s and unswitched capacitor C their equivalent circuits from Fig. 6-12, and for the inductor simulation its admittance in Eq. (6-78), we obtain the circuit in Fig. 6-41d. Comparing Fig. 6-41b and d, we see that the desired realization has been achieved.

6.4.3b Grounded Bilinear Inductors and FDNRs

Grounded bilinear inductors and FDNRs can be readily implemented with active-SC networks by simply replacing resistors in active-*RC* GIC implementations with bilinear resistors (Fig. 6-16c and d) [P11]. For example, an SC bilinear inductor simulation can be obtained from the active-*RC* GIC implementation in Fig. 2-34, as shown in Fig. 6-42a. As noted previously, the charge on the bilinear resistor changes twice per clock period τ; thus, the sampling period is $\tau/2$. Using the equivalent circuits in Fig. 6-16 and Eq. (2-63), we readily determine for the grounded bilinear inductor,

$$y_{\text{in}} = \frac{\Delta Q_1}{V_1} = \frac{C_4 C_1 C_3}{C C_2} \frac{(1 + z^{-1/2})^2}{1 - z^{-1/2}} \tag{6-82}$$

Comparing the bilinear equivalent inductive admittance in Table 6-3, with sample interval $\tau \longrightarrow \tau/2$, we equate

$$\frac{\tau^2}{16L} \frac{(1 + z^{-1/2})^2}{1 - z^{-1/2}} = \frac{C_4 C_1 C_3}{C C_2} \frac{(1 + z^{-1/2})^2}{1 - z^{-1/2}} \tag{6-83}$$

Solving for *L* yields

$$L = \frac{\tau^2}{16} \frac{C C_2}{C_4 C_1 C_3} \tag{6-84}$$

The grounded FDNR SC simulation in Fig. 6-42b is similarly obtained

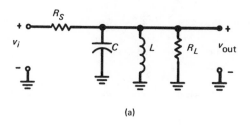

(a)

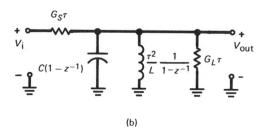

(b)

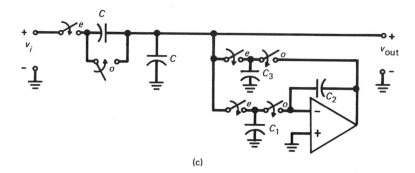

(c)

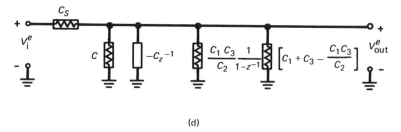

(d)

Fig. 6-41 Active-SC realization of a BD-transformed RLC resonant circuit.

from the active-RC GIC simulation in Fig. 2-35. Using the equivalent circuits in Fig. 6-16 and Eq. (2-75), we readily determine

$$y_{\text{in}} = \frac{\Delta Q_1}{V_1} = \frac{C^2 C_3}{C_2 C_4} \frac{(1 - z^{-1/2})^2}{1 + z^{-1/2}} \tag{6-85}$$

Comparing the bilinear equivalent FDNR admittance in Table 6-3 with sample

Active Switched Capacitor Sampled-Data Networks

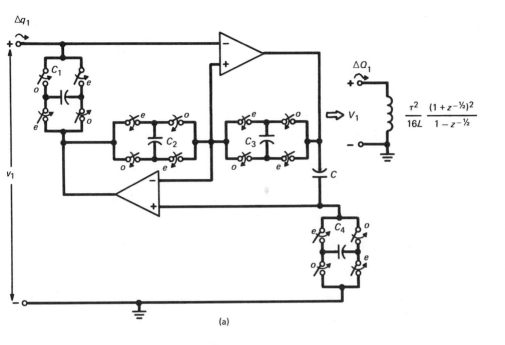

(a)

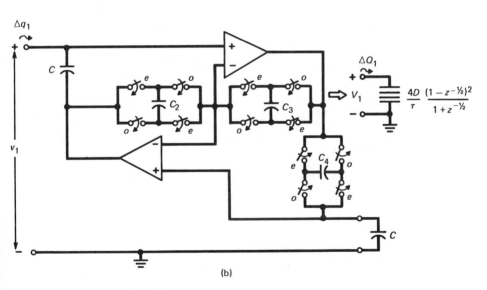

(b)

Fig. 6-42 Active-SC simulations of a grounded (a) bilinear inductor and (b) bilinear FDNR.

interval $\tau \longrightarrow \tau/2$, we determine

$$D = \frac{\tau}{4} \frac{C^2 C_3}{C_2 C_4} \qquad (6\text{-}86)$$

The active-SC circuits in Fig. 6-42, with the bilinear resistor and the unswitched capacitor, are used in SC filter implementations in the same spirit as the active-*RC* counterparts are used to implement active-*RC* filters. Active-*RC* ladder structures, implemented with inductor and FDNR simulation circuits, were considered in Chapter 5. The implementation of active-SC ladders follow analogously.

6.4.3c Floating Inductor and FDNR Simulations

An interesting SC network implementation [P15] for a floating admittance is shown in Fig. 6-43a. The networks SC-η_1 and SC-η_2 are active-SC networks with transfer functions derived according to the admittance to be simulated. The only properties assumed about these networks are that:

1. They accept inputs and present outputs during odd time slots.
2. The networks H_1 and H_2 must be noninteracting; hence, they are best realized as active-SC networks.
3. Their outputs are low-impedance voltage sources (i.e., the outputs of op amps).

Noting the switched capacitors C_1, C_2, and C_3 are TSFFPs, the equivalent circuit is drawn in Fig. 6-43b. Writing nodal charge equations at nodes (1), (2), and (3) of Fig. 6-43b, we obtain, respectively;

$$\Delta Q_a^e = (C_1 + C_2 + C_3)V_a^e - (C_1 + C_2 + C_3)V_b^e \\ - C_1 z^{-1/2}V_0^o - C_2 z^{-1/2}V_1^o - C_3 z^{-1/2}V_2^o \tag{6-87a}$$

$$\Delta Q_b^e = (C_1 + C_2 + C_3)V_b^e - (C_1 + C_2 + C_3)V_a^e \\ + C_1 z^{-1/2}V_0^o + C_2 z^{-1/2}V_1^o + C_3 z^{-1/2}V_2^o \tag{6-87b}$$

$$C_1 V_0^o = C_1 z^{-1/2}V_a^e - C_1 z^{-1/2}V_b^e \tag{6-87c}$$

Also, we know from the input–output relations of SC-η_1 and SC-η_2 that

$$V_1^o = H_1 V_0^o \tag{6-88a}$$

$$V_2^o = H_2 V_1^o \tag{6-88b}$$

From Eqs. (6-88) and (6-87c), we write

$$V_1^o = z^{-1/2}H_1 V_a^e - z^{-1/2}H_1 V_b^e \tag{6-89a}$$

$$V_2^o = z^{-1/2}H_1 H_2 V_a^e - z^{-1/2}H_1 H_2 V_b^e \tag{6-89b}$$

Substituting Eqs. (6-89) into Eqs. (6-87a) and (6-87b) yields

$$\Delta Q_a^e = y_{\text{in}}(z)V_a^e - y_{\text{in}}(z)V_b^e \tag{6-90a}$$

$$\Delta Q_b^e = y_{\text{in}}(z)V_b^e - y_{\text{in}}(z)V_a^e \tag{6-90b}$$

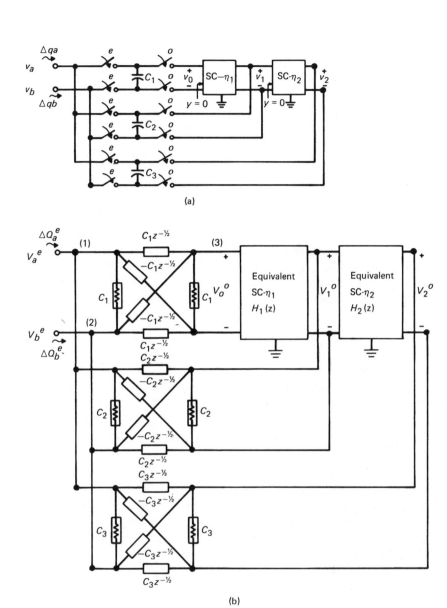

(a)

(b)

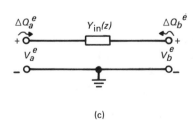

(c)

Fig. 6-43 Active-SC floating admittance simulation.

where

$$y_{in}(z) = (C_1 + C_2 + C_3) - z^{-1}[C_1 + C_2H_1 + C_3H_1H_2] \qquad (6\text{-}90c)$$

Note that if

$$C_1 + C_2 + C_3 = \frac{\tau^2}{4L} \qquad (6\text{-}91a)$$

$$z^{-1}(C_1 + C_2H_1 + C_3H_1H_2) = -\frac{\tau^2}{4L}\frac{3z^{-1} + z^{-2}}{1 - z^{-1}} \qquad (6\text{-}91b)$$

we obtain

$$y_{in}(z) = \frac{\tau^2}{4L}\frac{(1 + z^{-1})^2}{1 - z^{-1}} \qquad (6\text{-}92)$$

the equivalent admittance for a bilinear inductor simulation. To complete the synthesis, we determine C_1, C_2, C_3 and H_1, H_2 according to the constraints of Eqs. (6-91). Let us arbitrarily set $C_1 = C_2 = C_3 = \tau^2/12L$, which satisfies Eq. (6-91a). It can then be shown that choosing

$$H_1 = -\frac{10}{1 - z^{-1}} \quad \text{and} \quad H_2 = \frac{1}{5}z^{-1} \qquad (6\text{-}93)$$

satisfies Eq. (6-91b).

Network realizations, consistent with the assumptions stated earlier, for H_1 and H_2 are given in Fig. 6-44a and b.

6.5 REALIZATION OF PRECISION MOS ACTIVE-SC FILTERS

In previous sections we have considered many different SC networks, emphasizing the interesting theoretical properties of each. Let us now take a practical point of view and discuss those factors that are of concern in the MOS implementation of precision filters. As we will soon see, there are indeed constraints placed on the type of SC network that can be used, which are imposed by the actual properties of MOS circuit elements. Even among useful network structures there are practical considerations that make some preferable to others.

6.5.1 Practical Considerations for MOS Realizations [P2]

There are circuit conditions that must be avoided in the MOS implementation of precision filters. Let us first list those which are particularly treacherous.

1. An unswitched capacitor is the minimum circuitry that can be used to close an op amp feedback path. Since a switched capacitor does not provide a continuous time path, it cannot be used alone to provide the continuous feedback path necessary to stabilize the op amp. In other words, the op amp should not be left open-loop at any time. However, switched capacitors can be used in

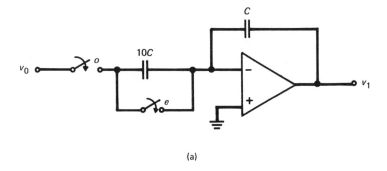

(a)

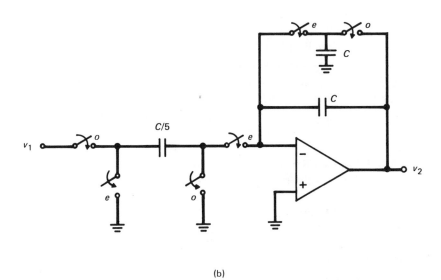

(b)

Fig. 6-44   Active-SC realizations for (a) H_1 and (b) H_2 in Eq. (6-93).

parallel with unswitched capacitors (e.g., the lossy integrators in Figs. 6-29a, 6-30a, 6-32a, and 6-38a).

2. No capacitor-only nodes. All capacitor plates are subject to charge accumulation from a variety of parasitic sources. To ensure circuit integrity, there must be a path either directly, or through a switched capacitor, to a voltage source or ground from every node. In active-C filters, discussed in Chapter 4, large arbitrary resistors were appropriately placed for this purpose.

3. The bottom plate of every capacitor should either be connected to ground, or connected directly or through a switch, to a voltage source. In MOS capacitors there is always a sizable nonlinear parasitic capacitor between the bottom plate and the substrate. This parasitic capacitor can be as large as 15% of the intended capacitor and thus provides an error in the definition of the transfer function coefficients. Furthermore, since it is nonlinear, it is also a source of appreciable nonlinear distortion. Since the substrate is at ac ground,

connecting the bottom plate as prescribed either grounds both plates of this parasitic capacitor or shunts it directly across a voltage source. In either case, it has no effect on the network behavior.

4. The noninverting op amp input ($+$ terminal) should be kept at a constant voltage. If the positive input of the op amp is connected to a signal voltage, the filter response is sensitive to all parasitic capacitances due to switches, bus lines, and substrate that couple to the inverting input. In addition, increased common mode performance would be required of the op amp.

These practical considerations necessitate the avoidance of many topologies as well as some circuit elements for all but very tolerant applications. Examples of topologies and SC elements that are in violation of these considerations are:

1. Single-amplifier structures of order greater than 1. This implies that dual op amp biquads are preferred.

2. Inductor and FDNR networks as shown in Fig. 6-42.

3. The bilinear resistor Fig. 6-16c and d, and the circuits that use, for example, the integrators in Figs. 6-37a and 6-38a.

4. The TSFFP, except applications where the bottom plate is switched to ground as per the floating admittance in Fig. 6-43a.

6.5.2 Effect of Switch and Routing Related Parasitic Capacitances

In addition to the large bottom plate-to-substrate parasitic, there are additional nonlinear parasitic capacitances to the substrate (ac ground) from the source and drain of every switch and from the routing lines which interconnect capacitors, switches and op amps. Although in a good integrated circuit layout these capacitors tend to be smaller than the bottom plate to substrate parasitic, they do set a limit on achievable filter precision. To see the effect of these parasitics on the filter performance, let us analyze the two integrators in Figs. 6-25a and 6-27a in the presence of these parasitic capacitances. Let us first consider the integrator in Fig. 6-25a. This integrator is redrawn in Fig. 6-45a, assuming a voltage source excitation, showing the actual switches and all the switch/routing parasitic capacitances. Since C_{p1} and C_{p4} shunt the voltage source and virtual ground, respectively, they have no effect on the response V_{out}. Lumping C_{p2} and C_{p3} into a single parasitic capacitance C_p, we can reduce Fig. 6-45a to the circuit in Fig. 6-45b. Using the equivalent circuit in Fig. 6-45c, we readily determine the transfer function between V_{out}^o and V^e to be

$$\frac{V_{out}^o}{V^e} = -\frac{\frac{C_1}{C_2}\left(1 + \frac{C_p}{C_1}\right)z^{-1/2}}{1 - z^{-1}} \qquad (6\text{-}94)$$

From Eq. (6-94), the parasitic capacitance is seen to result in a gain error of $1 + C_p/C_1$. For a lossy integrator such as that in Fig. 6-29a, the switch/routing

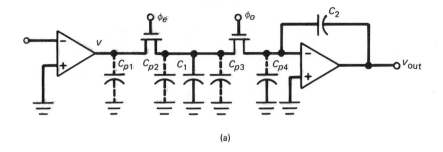

(a)

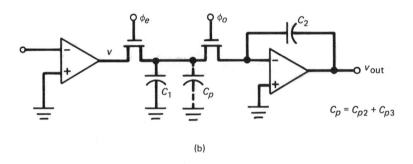

$C_p = C_{p2} + C_{p3}$

(b)

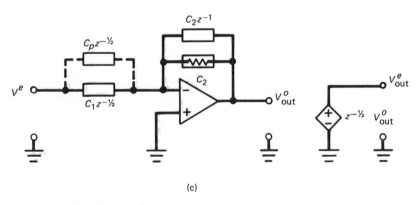

(c)

Fig. 6-45 Lossless inverting integration in Fig. 6-25a with switch parasitic capacitances.

parasitic(s) will result in both a gain and pole position error (see Prob. 6.22).

Let us now similarly analyze the integrator in Fig. 6-27a. This integrator is redrawn in Fig. 6-46a, assuming a voltage source excitation, showing the actual switches and switch/routing parasitics. As before, we can argue that C_{p1} and C_{p4} have no effect on the response. Lumping C_{p2} and C_{p3} into a single parasitic capacitance C_p, we can reduce Fig. 6-46a to the circuit in Fig. 6-46b. Using the equivalent circuit in Fig. 6-46c, we obtain the following transfer

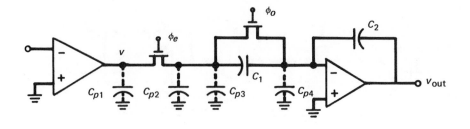

(a)

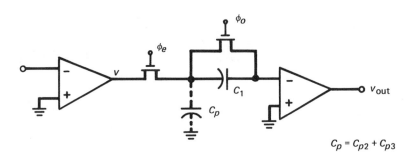

$$C_p = C_{p2} + C_{p3}$$

(b)

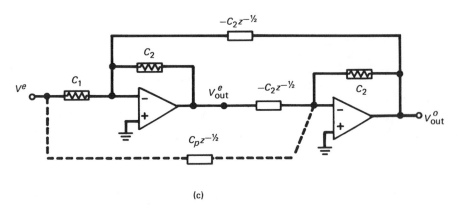

(c)

Fig. 6-46 Lossless inverting integrator in Fig. 6-27a with switch parasitic capacitances.

functions:

$$\frac{V_{out}^e}{V^e} = -\frac{\frac{C_1}{C_2}\left(1 + \frac{C_p}{C_1}z^{-1}\right)}{1 - z^{-1}} \tag{6-95a}$$

$$\frac{V_{out}^o}{V^e} = -\frac{\frac{C_1}{C_2}\left(1 + \frac{C_p}{C_1}\right)z^{-1/2}}{1 - z^{-1}} \tag{6-95b}$$

From Eq. (6-95a), we see that the zero, ideally at the origin for the transfer function V_{out}^e/V^e, has been shifted slightly to the left (i.e., $z = -C_p/C_1$). This term represents a small gain and phase error. Examining Eq. (6-95b), the error introduced in the transfer function V_{out}^o/V^e is a gain error of $1 + C_p/C_1$. This error is identical to that observed in Eq. (6-94).

Let us now look at a third integrator structure, a noninverting integrator using the TSI element. This circuit is shown in Fig. 6-47a with actual switches

Fig. 6-47 Lossless noninverting integrator with switch parasitic capacitances.

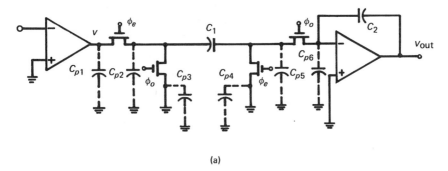

(a)

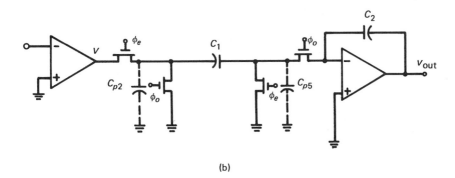

(b)

and all switch/routing parasitic capacitances. Again we assume a voltage source excitation. We can quickly eliminate parasitic capacitances C_{p1}, C_{p6}, C_{p3}, and C_{p4}, since C_{p1} and C_{p6} shunt a voltage source and virtual ground, respectively, and C_{p3} and C_{p4} are shorted to ground. The reduced circuit with only C_{p2} and C_{p5} is redrawn in Fig. 6-47b. To see the effect of these parasitic capacitances, let us walk our way through the circuit as switches ϕ^e and ϕ^o turn on and off during one clock cycle. With ϕ^e on and ϕ^o off, C_{p2} and C_1 are directly across the voltage source and charge to V. Parasitic C_{p5} has been shorted to ground by the closing of switch ϕ^e and does not get charged. When ϕ^o turns on and ϕ^e turns off, the parasitic C_{p2} is discharged to ground, uncharged C_{p5} is connected harmlessly to virtual ground, and C_1 transfers its charge to C_2 without error. In other words, the switch/routing parasitic capacitances have no effect here. Crucial, however, to this demonstration is the buffering of the TSI between a voltage source (or op amp output) and virtual ground. This very desirable parasitic-free property lends further motivation to develop dual op amp biquads and, to be more general, N-op amp Nth-order filters.

We have just demonstrated a parasitic-free noninverting integrator. An inverting integrator with equivalent properties can be realized using a modified OFR (Fig. 6-11m). This circuit will have transfer functions identical to the ideal functions for the integrator in Fig. 6-46. The OFR and the modification [P16, P17] are shown in Fig. 6-48a and b, respectively. One can quickly verify that functionally both circuits are identical. Let us now examine an inverting integrator of the type described above. This circuit, with actual MOS switches and all switch/routing parasitic capacitances, is shown in Fig. 6-49a. As we did in Fig. 6-47, we can quickly eliminate parasitics C_{p1}, C_{p3}, C_{p4}, and C_{p6}, resulting

Fig. 6-48 (a) OFR SC element. (b) OFR SC element modified for parasitic-free operation.

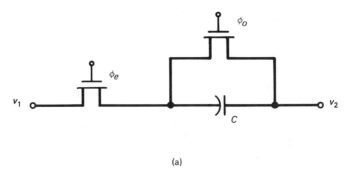

(a)

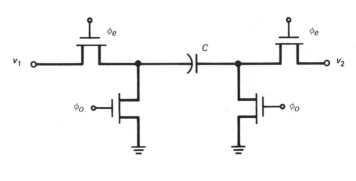

(b)

Active Switched Capacitor Sampled-Data Networks

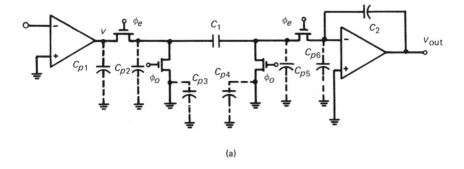

(a)

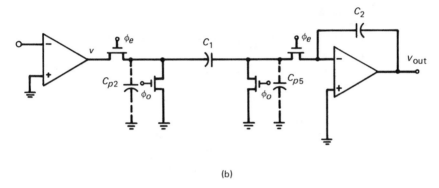

(b)

Fig. 6-49 Modified OFR inverting lossless integrator with switch parasitic capacitances.

in the reduced circuit in Fig. 6-49b. By following the operation of this circuit through one clock cycle, we can similarly verify that the response V_{out} is unaffected by C_{p2} and C_{p5}. With ϕ^e turned on and ϕ^o turned off, parasitic C_{p2} is charged to V and V is also coupled, without error, directly to C_2 through C_1. C_{p5} is connected harmlessly across the virtual ground and is not changed. When ϕ^e turns off and ϕ^o turns on, C_1, C_{p2}, and C_{p5} are all discharged to ground. This circuit, like that in Fig. 6-47, is unaffected by the switch parasitics. In both cases it is preferable to connect the bottom plates of all capacitors directly to, or through a switch, to a voltage source. As in the previous circuit, buffering the switched capacitor between voltage source and virtual ground is crucial to parasitic-free operation.

In addition to lossless integrators, we can also realize parasitic-free lossy inverting and noninverting integrators, as shown in Fig. 6-50a and c, respectively. It is left as an exercise for the reader to show that the responses for these circuits are indeed unaffected by switch/routing parasitics. We can economize on the hardware required to implement these lossy integrators by making the observation that the top plates of C_1 and C_3 synchronously switch between

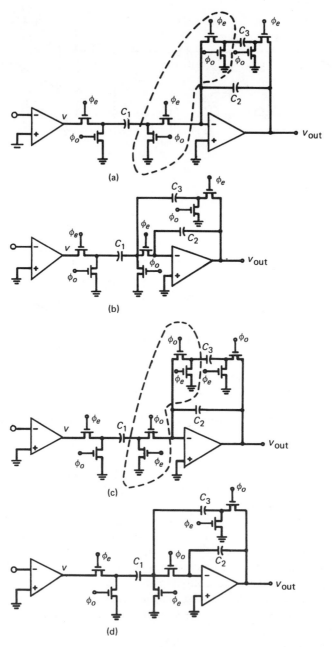

Fig. 6-50 Parasitic-free lossy integrators. (a) Inverting. (b) Switch-reduced inverting. (c) Noninverting. (d) Switch-reduced noninverting.

ground and virtual ground. These switches are enclosed by dashed lines in Fig. 6-50a and c. In this type of switching the synchronous switches can be shared, as shown in Fig. 6-50b and 6-50d, without affecting the circuit behavior. As

Active Switched Capacitor Sampled-Data Networks

shown in Fig. 6-50b and d, this observation results in a savings of two switches. In higher-order SC filters, such savings can become quite significant. The equivalent circuits and transfer functions for parasitic-free lossless and lossy integrators have been listed in Table 6-4 for the reader's convenience. As we shall

TABLE 6-4 Equivalent circuits and transfer functions for parasitic-free lossless and lossy integrators

Circuit Function	Equivalent Circuit	Transfer Functions
Noninverting integrator		$\dfrac{V^o_{out}}{V^e} = \dfrac{(C_1/C_2)z^{-1/2}}{1 - z^{-1}}$ $\dfrac{V^e_{out}}{V^e} = \dfrac{(C_1/C_2)z^{-1}}{1 - z^{-1}}$
Inverting integrator		$\dfrac{V^e_{out}}{V^e} = \dfrac{-C_1/C_2}{1 - z^{-1}}$ $\dfrac{V^o_{out}}{V^e} = \dfrac{-(C_1/C_2)z^{-1/2}}{1 - z^{-1}}$
Lossy noninverting integrator		$\dfrac{V^o_{out}}{V^e} = \dfrac{\dfrac{C_1}{C_2 + C_3}z^{-1/2}}{1 - \dfrac{C_2}{C_2 + C_3}z^{-1}}$ $\dfrac{V^e_{out}}{V^e} = \dfrac{\dfrac{C_1}{C_2 + C_3}z^{-1}}{1 - \dfrac{C_2}{C_2 + C_3}z^{-1}}$
Lossy inverting integrator		$\dfrac{V^e_{out}}{V^e} = \dfrac{-\dfrac{C_1}{C_2 + C_3}}{1 - \dfrac{C_2}{C_2 + C_3}z^{-1}}$ $\dfrac{V^o_{out}}{V^e} = \dfrac{-\dfrac{C_1}{C_2 + C_3}z^{-1/2}}{1 - \dfrac{C_2}{C_2 + C_3}z^{-1}}$

now show, these parasitic-free first-order circuits serve as building blocks for realizing higher-order parasitic-free active-SC filters [P16–P18]. With these circuits the minimum capacitance is no longer limited to be much larger than the parasitic switch capacitances, which can be on the order of 0.1 to 0.2 pF. Thus, we can not only make filters that are more precise but also filters that make more efficient use of the silicon area [P16]. It should be noted that when an insulating substrate is used such as sapphire, all the parasitic capacitances mentioned are negligible. In this case, all the topologies and SC elements, eliminated due to excessive dependence on parasitic capacitances, are usable in precision filters. Although a few companies possess a silicon-on-sapphire (SOS) process, as of this writing there have been no reported SOS switched capacitor filters. This may be due to the higher cost of SOS chips as compared to silicon MOS chips.

6.6 A FAMILY OF PARASITIC-FREE ACTIVE-SC BIQUADS [P18]

In this section we consider a general active-SC biquad topology, recently introduced [P18], which is capable of realizing all stable z-domain biquadratic transfer functions of the form given in Eq. (6-30). For convenience, Eq. (6-30) is rewritten below:

$$H(z) = \frac{N(z)}{D(z)} = \frac{\gamma + \epsilon z^{-1} + \delta z^{-2}}{1 + \alpha z^{-1} + \beta z^{-2}} \tag{6-96}$$

As we will show, all the generic biquad functions given in Table 6-1 are readily realizable.

6.6.1 General Circuit Topology

All the topologies to be considered in this section are special cases of the general active-SC biquad shown in Fig. 6-51a. This circuit bears a close resemblance to the three-amplifier active-RC biquad in Fig. 4-36; however, using the TSI, involving capacitor A, the inverting amplifier in the active-RC biquad is no longer needed. In effect, the circuit consists of two integrators, the first stage is inverting and the second stage noninverting. Damping is provided by the capacitor E and the switched capacitor F. In any particular application, only one of these will be present, leaving a total of nine capacitors, but for now let us keep the two cases together.

The transmission zeros are realized via the multiple feedforward paths consisting of switched capacitors G, H, I, and J. It will be seen later that typically no more than two of these capacitors are needed to realize the useful biquadratic transfer functions. Thus, most often it will be found that only seven capacitors are needed. Although the circuit schematic shown in Fig. 6-51a facilitates an understanding of the circuit, a more efficient implementation can be obtained by allowing similarly switched capacitors to share a common switch, as indicated earlier for the lossy integrators in Fig. 6-50. Rearranging the circuit schematic in this way results in the minimum switch configuration shown in Fig. 6-51b. Implementing the biquad in this way yields a savings of 16 switches. It can be readily verified that the electrical behaviors of both circuits are identical.

As per the inserts in Fig. 6-51a, the biquad is constructed from the parasitic-free switched capacitors described in Section 6.5. This parasitic-free behavior is not altered by the switch sharing in Fig. 6-51b. Note that each pair of ϕ^e, ϕ^o switches in Fig. 6-51 is shown symbolically as a single-pole double-throw switch rather than as two distinct switches. This shortened notation results in considerable labor savings in drawing biquad schematics. Also, the switch positions explicitly shown in the schematic are for the even (e) clock phase. Clearly, the odd (o) phase switch positions are obtained by rotating each symbolic switch to its alternate terminal.

The close analogy of SC filter to the active-RC biquad has already been

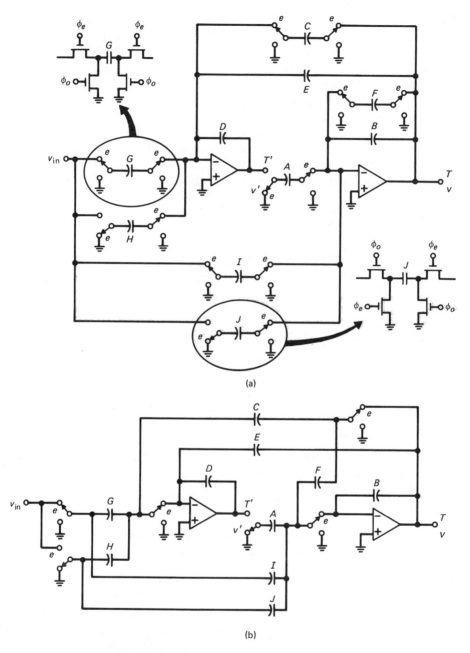

Fig. 6-51   (a) General active-SC biquad with parasitic-free switching. (b) Mini-mum switch configuration.

mentioned. In particular, note that the circuit with $E = 0$ is very similar in spirit to the three-op amp, multiple-input active-RC biquad, except for the absence of the inverter, which is not needed for the SC filter. However, active-SC biquads offer even further versatility, which, to this point, has not been exploited. Because of the inability to trim capacitors and the relative cost factors, practical active-RC biquads are constructed to be canonic in capacitors, namely two. This constraint is unnecessarily placed on active-SC topologies when they are derived from an active-RC topology via a resistor-to-switched capacitor replacement. It will be shown later that we can achieve interesting and beneficial results when this constraint is removed.

Before deriving the transfer functions for the biquad, let us consider the form of the input signal and the timing of the switches. As in all previous circuits in this chapter, ϕ^e and ϕ^o are assumed, for simplicity, to be nonoverlapping clocks with a 50% duty cycle, as in Fig. 6-52. In addition, it is assumed that the

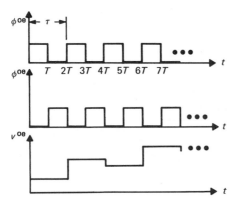

Fig. 6-52 Clock and input waveforms.

input signal is sampled and held over a full clock period, $\tau = 2T$, as shown in Fig. 6-52. In fact, the switch phasings of the SC biquad have been chosen to operate with this kind of input. Under these conditions, we have

$$V_{in}^o(z) = z^{-1/2} V_{in}^e(z) \tag{6-97a}$$

It follows, then, that the output voltages appearing at the op amp outputs are also held for the full clock period and change only at the even sampling instants:

$$V_{out}^o(z) = z^{-1/2} V_{out}^e(z) \tag{6-97b}$$

This fact can be surmised by inspection of Fig. 6-51a. In Section 6.6.4 it is shown that the full-cycle S/H assumption can be relaxed. For the present, since both the input and the outputs are fully held, the even transfer functions provide all the information we need. Therefore, for simplicity, the "even" superscript will be omitted from the equations.

Let us now derive the voltage-transfer functions T' and T (see Fig. 6-51). For analysis, the z-domain equivalent circuit in Fig. 6-53 is readily derived from the biquad schematic in Fig. 6-51a using the equivalent circuits in Fig. 6-12. The

Active Switched Capacitor Sampled-Data Networks

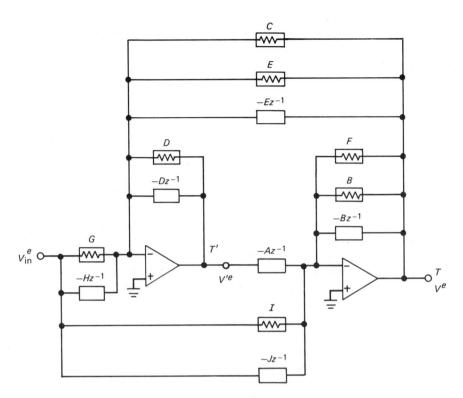

Fig. 6-53 z-domain equivalent circuit for the biquad in Fig. 6-51.

desired transfer functions are then derived using straightforward nodal analysis

$$T = \frac{V}{V_{in}} = -\frac{Az^{-1}(G - Hz^{-1}) + D(1 - z^{-1})(I - Jz^{-1})}{Az^{-1}(C + E - Ez^{-1}) + D(1 - z^{-1})(F + B - Bz^{-1})} \quad (6\text{-}98a)$$

$$= -\frac{DI + (AG - DI - DJ)z^{-1} + (DJ - AH)z^{-2}}{D(F + B) + (AC + AE - DF - 2DB)z^{-1} + (DB - AE)z^{-2}} \quad (6\text{-}98b)$$

$$T' = \frac{V'}{V_{in}} = \frac{(I - Jz^{-1})(C + E - Ez^{-1}) - (G - Hz^{-1})(F + B - Bz^{-1})}{Az^{-1}(C + E - Ez^{-1}) + D(1 - z^{-1})(F + B - Bz^{-1})} \quad (6\text{-}99a)$$

$$= \frac{(IC + IE - GF - GB) + (FH + BH + BG - JC - JE - IE)z^{-1} + (EJ - BH)z^{-2}}{D(F + B) + (AC + AE - DF - 2DB)z^{-1} + (DB - AE)z^{-2}} \quad (6\text{-}99b)$$

The synthesis process can be simplified at the outset by setting $A = B = D = 1$. Later, capacitor A will be used to independently adjust the gain constants associated with T and T' to maximize dynamic range. Gain constant scaling for this purpose was discussed for active-RC biquads in Chapter 4. Capacitors B and D control the admittance levels at the op amp summing junctions. Thus, two groups of capacitors, (C, D, E, G, H) and (A, B, F, I, J), can be arbitrarily and independently scaled without changing the transfer functions T and T'. Setting $A = B = D = 1$ in Eqs. (6-98b) and (6-99b) yields the following reduced transfer functions:

$$T = -\frac{I + (G - I - J)z^{-1} + (J - H)z^{-2}}{(F + 1) + (C + E - F - 2)z^{-1} + (1 - E)z^{-2}} \tag{6-100}$$

$$T' = \frac{(IC + IE - FG - G) + (FH + H + G - JC - JE - IE)z^{-1} + (EJ - H)z^{-2}}{(F + 1) + (C + E - F - 2)z^{-1} + (1 - E)z^{-2}} \tag{6-101}$$

Let us first examine the salient features of the transfer function T. Note that its poles are determined by C, E, and F, its zeros by G, H, I, and J. This unique property does not, to our knowledge, arise in typical active-RC biquads (e.g., those in Chapter 4). It is also clear that the three numerator coefficients are independently adjustable, thus permitting arbitrary zeros to be realized. In [P18] it is proven that all stable pole positions are realizable using the circuit with E damping (i.e., $F = 0$). For the biquad with F damping, all stable pole positions, with the exception of the academic case of real poles on alternate sides of $z = 0$, are realizable. In addition, the only unstable realizable pole positions (i.e., poles outside the unit circle) are real. Thus, changes in capacitor values, whatever their cause, will not force high-Q poles to instability. The derivation of these properties is left as an exercise for the reader.

Regarding the transfer function T', we first observe the obvious fact that its poles are identical to those of T. We note, however, that its zeros are formed in a more complicated fashion and they do not have the aforementioned independence property. Nevertheless, there are cases where T' provides a more economical realization of a given transfer function than does T.

A careful examination of the equivalent circuit in Fig. 6-53 reveals some rather interesting alternative realizations for the zero-forming capacitor pairs (I, J) and (G, H). These alternative realizations are shown in Fig. 6-54 for I and J. Clearly, G and H can be substituted for I and J, respectively, in Fig. 6-54. A crucial assumption in deriving these equivalences is that terminals (1) and (2) be connected to a voltage source and virtual ground, respectively. In addition, it is assumed that the voltage source provides a full-clock-period S/H signal. Later, this assumption will be relaxed. In any event, the biquad in Fig. 6-51 and the equivalences in Fig. 6-54 provide a powerful family of active-SC biquad circuits.

It is interesting to note that when $I = J$, the parallel of two switched capacitors is equivalent to a single unswitched capacitor. In view of the switched capacitor–resistance equivalence so commonly assumed in dealing with SC networks, this equivalence is quite fascinating. We note that I (and G) are BD equivalent resistors and that J (and H) are FD equivalent negative resistors. One should not be surprised that such a seemingly strange result is observed when elements from these radically different transformations are mixed together.

These equivalences are not only academically interesting, but they provide a practical degree of freedom to the synthesis of a biquad. For example, when

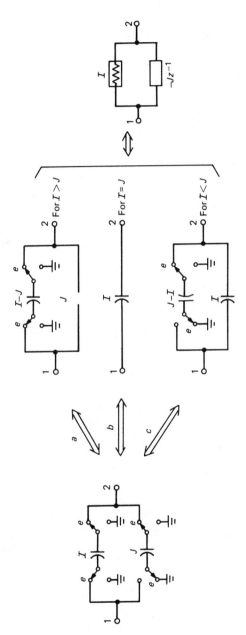

Fig. 6-54 SC element transformations (ports 1 and 2 are assumed to be connected to voltage source and op amp virtual ground, respectively).

$I = J$, not only are one capacitor and eight switches eliminated, but sensitivity is usually improved since I and J now "track" perfectly. Even when $I \neq J$, the transformations may be useful to reduce sensitivity or to lower the total capacitance needed for a given case. For example, if $I = 13$ pF and $J = 12$ pF, transformation (a) can be applied, yielding new element values $I - J = 1$ pF and $J = 12$ pF. The transformation is obviously reversible; thus, the converse transformation can be applied to element pairs C, E and B, F.

6.6.2 E and F Circuits

For synthesis purposes, it is convenient to separate the realizations using E-capacitor damping ($F = 0$) from those using F-capacitor damping ($E = 0$). In [P18] these two cases are referred to as the E-circuit and the F-circuit, respectively. The transfer functions for these two circuits (with $A = B = D = 1$) are readily obtained from Eqs. (6-100) and (6-101). These relations are

$$T_E = -\frac{I + (G - I - J)z^{-1} + (J - H)z^{-2}}{1 + (C + E - 2)z^{-1} + (1 - E)z^{-2}} \tag{6-102a}$$

$$T'_E = \frac{(IC + IE - G) + (H + G - JC - JE - IE)z^{-1} + (EJ - H)z^{-2}}{1 + (C + E - 2)z^{-1} + (1 - E)z^{-2}} \tag{6-102b}$$

$$T_F = -\frac{\hat{I} + (\hat{G} - \hat{I} - \hat{J})z^{-1} + (\hat{J} - \hat{H})z^{-2}}{(\hat{F} + 1) + (\hat{C} - \hat{F} - 2)z^{-1} + z^{-2}} \tag{6-103a}$$

$$T'_F = -\frac{(\hat{G}\hat{F} + \hat{G} - \hat{I}\hat{C}) + (\hat{J}\hat{C} - \hat{F}\hat{H} - \hat{H} - \hat{G})z^{-1} + \hat{H}z^{-2}}{(\hat{F} + 1) + (\hat{C} - \hat{F} - 2)z^{-1} + z^{-2}} \tag{6-103b}$$

The "hats" are placed on the F-circuit elements to distinguish them from the E-circuit elements.

Let us briefly examine these transfer functions. Note that the numerators of T_E and T_F are identical, whereas the numerators of T'_E and T'_F are quite different. Thus, for a given design, in which the desired output is V, the T' as well as the unscaled dynamic range of V' may be quite different for the two networks. The analogous situation is obtained if the desired output is V'. These differences will ultimately affect the final scaled capacitor values and the total capacitance required to realize the circuit. Significant sensitivity differences between the four possible realizations of a given transfer function may also exist.

Let us now consider the synthesis of z-domain biquadratic transfer functions with the E and F biquads. The derivation of z-domain transfer functions from analog filter requirements was considered in Section 6.2.5. We shall refer frequently to this material. As shown in this section, the bilinear transform guaranteed that stable s-domain poles were transformed into stable z-domain poles. In addition, the passband characteristics of the s-domain transfer function were transferred to the z-domain functions. For these reasons, the z-domain pole positions will be obtained via the bilinear transform. Once the poles have been determined, the zeros can be altered as desired.

Let the denominator of the prewarped s-domain transfer function be

$$D(\hat{s}) = \hat{s}^2 + a\hat{s} + b \qquad (6\text{-}104)$$

Then, substituting for $\hat{s}$ the bilinear transform Eq. (6-18a) and equating the coefficients of the resulting z-domain quadratic polynomial with those of the denominator of T_E (or T'_E) in Eq. (6-102), yields

$$E = \frac{a\tau}{1 + a\tau/2 + b(\tau^2/4)} \qquad (6\text{-}105a)$$

$$C = \frac{b\tau^2}{1 + a\tau/2 + b(\tau^2/4)} \qquad (6\text{-}105b)$$

Similar expressions for the F-circuit may be written:

$$\hat{F} = \frac{a\tau}{1 - a\tau/2 + b\tau^2/4} \qquad (6\text{-}106a)$$

$$\hat{C} = \frac{b\tau^2}{1 - a\tau/2 + b\tau^2/4} \qquad (6\text{-}106b)$$

where τ is the full sampling period.

It is noted that when the z-domain transfer function is given [e.g., Eq. (6-96)], relations for C, E and $\hat{C}$, $\hat{F}$ are determined by equating the coefficients of the denominator of Eq. (6-96) with the denominators of T_E and T_F, respectively. These relations are

$$E = 1 - \beta \qquad (6\text{-}107a)$$

$$C = 1 + \beta + \alpha \qquad (6\text{-}107b)$$

$$\hat{F} = \frac{1 - \beta}{\beta} \qquad (6\text{-}108a)$$

$$\hat{C} = \frac{1 + \alpha + \beta}{\beta} \qquad (6\text{-}108b)$$

Once values for C, E or $\hat{C}$, $\hat{F}$ have been determined, values for G, H, I, J or $\hat{G}$, $\hat{H}$, $\hat{I}$, $\hat{J}$ can be determined from knowledge of the desired zero locations or transfer function numerator. At this point T_E and T'_E, and T_F and T'_F, cease to be identical. Complete sets of design equations for determining G, H, I, J and $\hat{G}$, $\hat{H}$, $\hat{I}$, $\hat{J}$ for T_E, T_F, T'_E, and T'_F are given in [P18]. There, design formulas are given for all the generic biquad functions listed in Table 6-1 and for the general case. We will only duplicate here the table for the more useful T_E and T_F designs. These design formulas are given in Table 6-5. For each design considered, a "simple" solution is given. These simple solutions, which are not unique, lead to fewer capacitors by setting as many of the capacitors G, H, I, J and $\hat{G}$, $\hat{H}$, $\hat{I}$, $\hat{J}$ to zero as possible. In addition, where possible, we set $G = H$, $I = J$ or $\hat{G} = \hat{H}$, $\hat{I} = \hat{J}$.

TABLE 6-5 Zero-placement formulas for T_E and T_F[a] [P18]

Filter Type	Design Equations	Simple Solution
LP20	$I = \lvert K \rvert$ $G - I - J = 2\lvert K \rvert$ $J - H = \lvert K \rvert$	$I = J = \lvert K \rvert$ $G = 4\lvert K \rvert,\ H = 0$
LP11	$I = 0$ $G - I - J = \pm\lvert K \rvert$ $J - H = \pm\lvert K \rvert$	$I = 0,\ J = \lvert K \rvert$ $G = 2\lvert K \rvert,\ H = 0$
LP10	$I = \lvert K \rvert$ $G - I - J = \lvert K \rvert$ $J - H = 0$	$I = \lvert K \rvert,\ J = 0$ $G = 2\lvert K \rvert,\ H = 0$
LP02	$I = 0$ $G - I - J = 0$ $J - H = \pm\lvert K \rvert$	$I = J = 0$ $G = 0,\ H = \lvert K \rvert$
LP01	$I = 0$ $G - I - J = \pm\lvert K \rvert$ $J - H = 0$	$I = J = 0$ $G = \lvert K \rvert,\ H = 0$
LP00	$I = \lvert K \rvert$ $G - I - J = 0$ $J - H = 0$	$I = \lvert K \rvert,\ J = 0$ $G = \lvert K \rvert,\ H = 0$
BP10	$I = \lvert K \rvert$ $G - I - J = 0$ $J - H = -\lvert K \rvert$	$I = \lvert K \rvert,\ J = 0$ $G = H = \lvert K \rvert$
BP01	$I = 0$ $G - I - J = \pm\lvert K \rvert$ $J - H = \mp\lvert K \rvert$	$I = 0,\ J = \lvert K \rvert$ $G = H = 0$
BP00	$I = \lvert K \rvert$ $G - I - J = -\lvert K \rvert$ $J - H = 0$	$I = \lvert K \rvert,\ J = 0$ $G = H = 0$
HP	$I = \lvert K \rvert$ $G - I - J = -2\lvert K \rvert$ $J - H = \lvert K \rvert$	$I = J = \lvert K \rvert$ $G = H = 0$
HPN and LPN	$I = \lvert K \rvert$ $G - I - J = \lvert K \rvert \epsilon$ $J - H = \lvert K \rvert$	$I = J = \lvert K \rvert$ $G = \lvert K \rvert \{2 + \epsilon\},$ $H = 0$
AP $(\beta > 0)$	$I = \lvert K \rvert \beta$ $G - I - J = \lvert K \rvert \alpha$ $J - H = \lvert K \rvert$	$I = \lvert K \rvert \beta,\ J = \lvert K \rvert$ $G = \lvert K \rvert (1 + \beta + \alpha)$ $= \lvert K \rvert C$ $H = 0$
General $(\gamma > 0)$	$I = \gamma$ $G - I - J = \epsilon$ $J - H = \delta$	$I = \gamma$ $J = \delta + x$ $G = \gamma + \delta + \epsilon + x$ $H = x \geq 0$

[a] $\hat{G} = G(1 + \hat{F}),\quad \hat{H} = H(1 + \hat{F}),\quad \hat{I} = I(1 + \hat{F}),$ and $\hat{J} = J(1 + \hat{F}).$

Now that the poles and zeros have been realized, what is left is to scale the gain constants of T_E, T'_E (or T_F, T'_F) to maximize dynamic range and to scale the admittance levels at each summing junction. It is convenient to scale the admittances so that the minimum capacitance(s) is unity. The capacitor values determined in this synthesis process represent normalized (unitless) values; the normalization is to whatever the unit capacitance happens to be (e.g., 5 pF).

As described in Chapter 4, the gain-level adjustment can be performed readily with the aid of a circuit analysis program. The object is to adjust the gain levels so that the maximum of V and V' are equalized while realizing the specified gain level. To adjust the voltage level of V' (i.e., the flat gain of T') without affecting T, only the capacitors A and D need to be scaled. More precisely, if it is desired to modify the gain constant associated with V' according to

$$T' \longrightarrow \mu T' \tag{6-109}$$

then it is only necessary to scale A and D as follows:

$$(A, D) \longrightarrow \left(\frac{1}{\mu} A, \frac{1}{\mu} D\right) \tag{6-110}$$

The gain constant associated with T remains invariant under this scaling. The correctness of this procedure follows directly from simple signal-flow-graph concepts. Note that A and D are the only two capacitors that are connected to the first op amp output.

In a similar fashion, it can be shown that if the flat gain associated with V is to be modified,

$$T \longrightarrow v T \tag{6-111}$$

the following capacitors must be scaled:

$$(B, C, E, F) \longrightarrow \left(\frac{1}{v} B, \frac{1}{v} C, \frac{1}{v} E, \frac{1}{v} F\right) \tag{6-112}$$

Once satisfactory gain levels have been obtained at both outputs, it is convenient to scale the admittances associated with each stage so that the minimum capacitance value in the circuit becomes unity. This makes it easier to observe the maximum capacitance ratios required to realize a given circuit and also serves to "standardize" different designs so that the total capacitance required can readily be compared. The two groups of capacitors that may be scaled together are

Group 1: (C, D, E, G, H)

Group 2: (A, B, F, I, J)

Note that capacitors in each group are distinguished by the fact that they are all incident on the same input node of one of the op amps.

This completes the design process for synthesizing practical active-SC biquad networks. In Section 6.6.5 a detailed example is given to demonstrate each step of the design process.

6.6.3 Sensitivity

The sensitivities for the E and F circuits are at least comparable to any active-RC biquad. One can arrive [P18] at this conclusion by examining the Q and ω_0 relations and associated sensitivities for a high-Q resonant response.

For any pair of complex-conjugate poles in the z-domain, one can write the donominator as

$$D(z) = 1 + \alpha z^{-1} + \beta z^{-2} \qquad (6\text{-}113a)$$

$$= (1 - re^{j\theta}z^{-1})(1 - re^{-j\theta}z^{-1})$$

$$= 1 - (2r\cos\theta)z^{-1} + r^2 z^{-2} \qquad (6\text{-}113b)$$

By analogy to the continuous case ($e^{s\tau} = e^{(\sigma + j\omega_p)\tau} = re^{j\theta}$ where $\omega_o \simeq \omega_p$ for high Q), the following equations involving the resonant frequency ω_0 and Q can be written, when $\omega_0\tau \ll 1$ and $Q \gg 1$:

$$\sigma\tau = \ln r \simeq \frac{\theta}{2Q} \qquad \theta \simeq 2\pi\frac{\omega_0}{\omega_s} = \omega_0\tau \qquad (6\text{-}114)$$

i.e.

$$\frac{1}{2Q} \simeq \frac{1-r}{\theta} = \frac{1-r}{\omega_0\tau} \qquad (6\text{-}115)$$

Note that via Taylor series expansion $\ln r \simeq 1 - r$ for $r \simeq 1$. Solving for r in Eq. (6-115) we obtain

$$r \simeq 1 - \frac{\omega_0\tau}{2Q} \qquad (6\text{-}116)$$

Therefore, it follows from Eq. (6-113) through Eq. (6-116) that

$$\alpha \simeq -2\left(1 - \frac{\omega_0\tau}{2Q}\right)\cos\omega_0\tau \qquad (6\text{-}117a)$$

$$\beta \simeq \left(1 - \frac{\omega_0\tau}{2Q}\right)^2 \qquad (6\text{-}117b)$$

The expressions above may be further approximated:

$$\alpha \simeq -2\left(1 - \frac{\omega_0\tau}{2Q}\right)\left(1 - \frac{\omega_0^2\tau^2}{2}\right)$$

$$\simeq -2 + \frac{\omega_0\tau}{Q} + \omega_0^2\tau^2 \qquad (6\text{-}118a)$$

$$\beta \simeq 1 - \frac{\omega_0\tau}{Q} \qquad (6\text{-}118b)$$

Consider first the E-circuit. After suitable manipulations the denominator of Eq. (6-98b) becomes (with $F = 0$)

$$D_E(z) = 1 + \left(-2 + \frac{AC}{DB} + \frac{AE}{DB}\right)z^{-1} + \left(1 - \frac{AE}{DB}\right)z^{-2} \qquad (6\text{-}119)$$

Comparing this to Eq. (6-113a) and Eq. (6-118) immediately yields

$$\frac{\omega_0\tau}{Q} \simeq \frac{AE}{DB} \qquad (6\text{-}120a)$$

$$\omega_0^2\tau^2 \simeq \frac{AC}{DB} \qquad (6\text{-}120b)$$

Therefore,

$$\omega_0\tau \simeq \left(\frac{AC}{DB}\right)^{1/2} \qquad (6\text{-}121a)$$

$$Q \simeq \frac{1}{E}\left(\frac{DBC}{A}\right)^{1/2} \qquad (6\text{-}121b)$$

Similarly, it may be shown that for the F-circuit,

$$\omega_0\tau \simeq \left(\frac{\hat{A}\hat{C}}{\hat{D}\hat{B} + \hat{D}\hat{F}}\right)^{1/2} \qquad (6\text{-}122a)$$

$$Q \simeq \left[\frac{\hat{A}\hat{C}}{\hat{D}\hat{F}}\left(1 + \frac{\hat{B}}{\hat{F}}\right)\right]^{1/2} \qquad (6\text{-}122b)$$

From Eqs. (6-121) and Eqs. (6-122), it is seen ω_0 and Q are controlled by the ratios of four or five capacitors. Furthermore, it is clear that

$$|S_x^{\omega_0}| \leq \tfrac{1}{2} \quad \text{and} \quad |S_x^Q| \leq 1 \qquad (6\text{-}123)$$

where x denotes any capacitor in the E or F circuits. This situation compares favorably to the low-sensitivity active-RC biquads discussed in Chapter 4, where a minimum of four passive elements (i.e., two RC products) determine ω_0. Since in practice, ratios of capacitors can be more tightly controlled than individual resistors and capacitors, the active-SC realization can be expected to be superior to the active-RC case with respect to initial (untuned) response as well as temperature and aging variations. Since τ is normally derived from a very stable crystal controlled clock, it is assumed to be invariant.

Even though in practice the ω_0 variation is usually the most significant contributor to the overall variation in the response (Chapter 3), we note that the Q sensitivities of the active-SC circuit are also as low as those of the low-Q sensitivity active-RC biquads given in Chapter 4.

The overall circuit sensitivity is also affected, of course, by the contribution of the numerator. This aspect of the problem is not amenable to a general analysis and will have to be handled on a case-by-case basis. As a general rule of thumb, though, it may be observed from Eqs. (6-98) and (6-99) that the numerator of T looks simpler than the numerator of T'. Thus, all things being equal, T will tend to provide the lower-sensitivity realization of a given transfer function. But in general no such rule has been rigorously established.

It should be noted that the statistical multiparameter sensitivity measures given in Chapter 3 can be used in the sensitivity analysis of active-SC networks. To apply these measures, we use $H(e^{j\omega\tau})$ with $z^{-k} = e^{-jk\omega\tau}$. The statistically varying parameters here are capacitors or capacitor ratios. Here, of course, individual capacitor variations are highly correlated. For properly designed capacitors [P19], this correlation may be as high as $\rho = 0.9$.

6.6.4 Modifications to the Biquad for S/H Input Signals When the Hold Time Is Less Than One Clock Period

Previously, we have assumed that the input signal is sampled and held for the full clock period. Although this assumption simplifies the analysis, it is by no means necessary. Thus, consider the more general case where the clock period is still τ but the desired input signal is sampled and held only for the interval $\tau_e(\tau_e < \tau)$. The subscript e here is meant to imply the even phase of the clock period. The odd phase of the clock period is referred to as τ_o ($\tau_o = \tau - \tau_e$). The input during this phase is assumed to be "undesirable." These concepts are also shown in Fig. 6-55.

Fig. 6-55 Waveforms when the input is not held for the full clock period, as shown in Fig. 6-52.

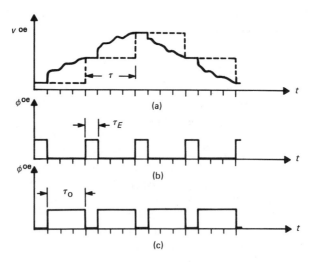

In certain special cases the circuits of Fig. 6-51 will continue to perform correctly even with this less restricted class of inputs. This happens whenever $H = 0$ and $J = 0$. This is readily confirmed by observing that the input voltage during τ_o is coupled into the circuit only via the two capacitors H and J. When these are both absent, the input during the odd clock phase is simply not "seen" by the circuit. Also, when $I = J$ and $H = 0$ or when $G = H$ and $J = 0$, a minor modification to the circuits of Fig. 6-51 can be employed. This modification is shown in Fig. 6-56. By replacing the $I = J$ capacitor or $G = H$ capacitor with a switched capacitor, which opens the otherwise direct feed during the odd

Fig. 6-56 Modified biquad topologies for input signals that are not held for a full clock period. The op amp outputs V and V' are held for the full clock period.

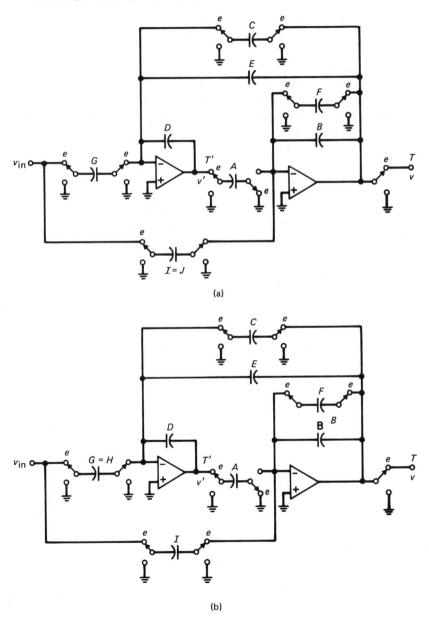

(a)

(b)

phase, we restore the basic properties of the circuits in Fig. 6-51, even though the input signal is not held for the full clock period. This type of capacitor was discussed in Section 6.3. The equivalent circuit for this element is given in Fig. 6-12b. Unfortunately, this solution is not a general one. A general alternative realization will be given shortly.

In general, however, the circuits of Fig. 6-51 can be modified by reversing the switch phasings of the switched capacitors A, H, and J. The resulting active-SC circuits are shown in Fig. 6-57. Note that the topology is so arranged that only the input during the even phase is coupled into the circuit. Thus, the input during the odd phase is again immaterial. One slight constraint on the operation of this circuit is that now the "correct" output is only obtained during the even phase. Thus, if a fully held output signal is desired, the circuits of Fig. 6-57 will have to be followed by a suitable sample-and-hold circuit. Note that the circuit in Fig. 6-57b is the switch-reduced version of the circuit in Fig. 6-57a.

To prove that the circuits in Fig. 6-57 achieve the desired behavior, let us construct the z-domain equivalent circuit using the elemental equivalent circuits in Fig. 6-11. The resulting equivalent circuit is given in Fig. 6-58. Because of the new switch phasings, we see that V_{in}^o does not enter the filter. It is noted that for convenience in analysis the duty cycle is assumed to be 50% (i.e., $\tau_e = \tau_o = \frac{1}{2}\tau$). Since the transfer function we seek has only integer powers of z^{-1} and by definition $\tau_e + \tau_o = \tau$, this does not detract from the generality of the analysis. However, when $\tau_e < \frac{1}{2}\tau$, the desired behavior is obtained by choosing specific non-50% duty cycles for ϕ^e and ϕ^o. Writing nodal charge equations at the four virtual ground nodes of this circuit yields the following system of equations:

$$GV_{in}^e + DV'^e - Dz^{-1/2}V'^o + (C + E)V^e - Ez^{-1/2}V^o = 0 \quad (6\text{-}124)$$

$$-Hz^{-1/2}V_{in}^e + DV'^o - Dz^{-1/2}V'^e + EV^o - Ez^{-1/2}V^e = 0 \quad (6\text{-}125)$$

$$IV_{in}^e + (F + B)V^e - Bz^{-1/2}V^o = 0 \quad (6\text{-}126)$$

$$-Jz^{-1/2}V_{in}^e - Az^{-1/2}V'^e + BV^o - Bz^{-1/2}V^e = 0 \quad (6\text{-}127)$$

Algebraically eliminating V^o and V'^o from these equations results in the following pair of equations:

$$(I - Jz^{-1})V_{in}^e + (F + B - Bz^{-1})V^e - Az^{-1}V'^e = 0 \quad (6\text{-}128)$$

$$(G - Hz^{-1})V_{in}^e + D(1 - z^{-1})V'^e + (C + E - Ez^{-1})V^e = 0 \quad (6\text{-}129)$$

Equations (6-128) and (6-129) can be readily verified to be the nodal equations that characterize the equivalent circuit given in Fig. 6-53. Thus, during the even phase the transfer functions will again be those given in Eqs. (6-98) and (6-99), thus proving our contention. Note carefully, however, that during the odd phase, the transfer functions that relate V_{in}^e to V^o and V'^o are quite different from those that characterize the even phase operation. In fact, we can express the two odd

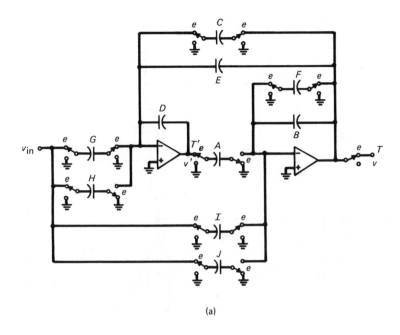

(a)

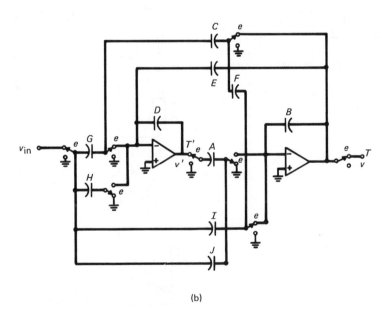

(b)

Fig. 6-57 (a) General active-SC biquad for input signals that are not held constant over the full clock period. (Note new switch phasing for H, J, and A.) (b) Switch-reduced version of the biquad in part (a).

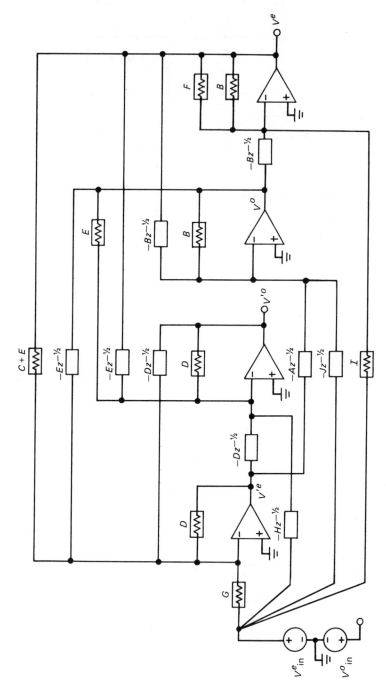

Fig. 6-58   z- domain equivalent circuit for the active-SC biquad in Fig. 6-57.

outputs (V^o and V'^o) as functions of the even outputs (V^e and V'^e) and the even input [V^e_{in}]:

$$V^o = z^{1/2} \left[\left(1 + \frac{F}{B} \right) V^e + \frac{I}{B} V^e_{in} \right] \qquad (6\text{-}130)$$

$$V'^o = z^{1/2} \left[z^{-1} V'^e - \frac{E}{D} \left(1 + \frac{F}{B} - z^{-1} \right) V^e + \left(\frac{EI}{DB} - \frac{H}{D} z^{-1} \right) V^e_{in} \right] \qquad (6\text{-}131)$$

It is noted that for $F = I = 0$, $V^e = z^{-1/2} V^o$; thus, V is held for the full clock period. On the other hand, when $E = H = 0$, V' is held for the full clock period. Thus, in some special cases at least, a fully held output can be obtained. Finally, SC element equivalences, similar to those given in Fig. 6-54, can be used to reduce sensitivity and/or total capacitance. These equivalences and their common z-domain equivalent circuit are given in Fig. 6-59. As in Fig. 6-54, these equivalences are based on the assumption that port 1 is voltage driven and port 2 is connected to an op amp virtual ground. Furthermore, these equivalences maintain the parasitic-free quality of the original biquad in Fig. 6-57.

6.6.5 Examples

In this section some illustrative examples are given. The first example is a low-pass notch network whose design is followed through step by step to illustrate the design procedure. The second example is a band-pass.

6.6.5a Low-Pass Notch Circuit

The transfer function to be realized will be based on the following s-domain transfer function:

$$T(s) = \frac{0.891975 s^2 + 1.140926 \times 10^8}{s^2 + 356.047 s + 1.140926 \times 10^8} = \frac{H_B(s^2 + \omega_z^2)}{s^2 + (\omega_p/Q_p)s + \omega_p^2} \qquad (6\text{-}132)$$

This transfer function provides a notch frequency at $f_z = 1800$ Hz, a peak corresponding to a quality factor of $Q_p = 30$ at $f_p = 1700$ Hz, and 0-dB dc gain. The assigned sampling frequency is 128 kHz (i.e., $\tau = 7.8125$ μsec). The z-domain transfer function is conveniently derived from Eq. (6-132) via the bilinear transform. Because the band edge is much lower than the sampling frequency $\omega_p \tau \ll 1$, it is not necessary to prewarp the $T(s)$ given in Eq. (6-132). Applying the bilinear transform to Eq. (6-132), we obtain the following z-domain transfer function:

$$T(z) = 0.89093 \frac{1 - 1.99220 z^{-1} + z^{-2}}{1 - 1.99029 z^{-1} + 0.99723 z^{-2}} \qquad (6\text{-}133)$$

Note that in obtaining this transfer function, a high degree of numerical precision is required. However, this does not result in high sensitivities, since capacitor ratios are used to realize the departures from -2 and $+1$ (for the z^{-1}

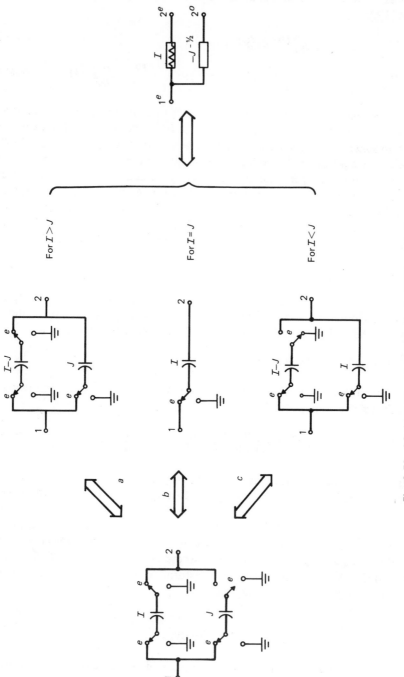

Fig. 6-59   SC element transformations for the SC biquad in Fig. 6-57.

and z^{-2} coefficients, respectively) in both the numerator and denominator of Eq. (6-133).

Only the T_E and T_F realizations of the foregoing circuit will be given here, as these circuits are more economical in the number of capacitors required for realization. The synthesis itself is straightforward. The capacitors C, E and $\hat{C}$, $\hat{F}$ are determined with A, B, D and $\hat{A}$, $\hat{B}$, $\hat{D}$ set to unity, from Eq. (6-133) according to Eqs. (6-107) and (6-108), respectively. The zero-forming capacitors G, H, I, J and $\hat{G}$, $\hat{H}$, $\hat{I}$, $\hat{J}$ are obtained from the "simple solution" LPN entry in Table 6-5. The resulting unscaled capacitor values for the E- and F-circuit designs are listed in Table 6-6. Recall that when $I = J$, the parallel combination of switched capacitors I and J can be replaced by an unswitched capacitor.

TABLE 6-6   Low-pass notch realization

	E-Circuit				F-Circuit		
Capacitor (pF)	Unscaled	Dynamic-Range-Adjusted	Final	Capacitor (pF)	Unscaled	Dynamic-Range-Adjusted	Final
A	1.0000	0.08308	1.0000	$\hat{A}$	1.0000	0.08395	30.1895
B	1.0000	1.0000	12.0365	$\hat{B}$	1.0000	1.0000	359.629
C	0.00694	0.00694	2.5035	$\hat{C}$	0.00696	0.00696	1.0000
D	1.0000	0.08308	29.9613	$\hat{D}$	1.0000	0.08395	12.0591
E	0.00277	0.00277	1.0000	$\hat{E}$	—	—	—
F	—	—	—	$\hat{F}$	0.00278	0.00278	1.0000
G	0.00694	0.00694	2.5035	$\hat{G}$	0.00696	0.00696	1.0000
H	—	—	—	$\hat{H}$	—	—	—
$I = J$	0.89093	0.89093	10.7238	$\hat{I} = \hat{J}$	0.89340	0.89340	321.293
$\sum C$ (pF)	—	—	59.7		—	—	726.1
σ_1 (dB)	—	—	0.068		—	—	0.068
σ_{1700} (dB)	—	—	1.233		—	—	1.271

A computer-aided analysis of the circuit reveals that the maximum gains for T_E and T_F are approximately 10.56 dB; however, the maximum gains for T'_E and T'_F are very low, with $(T'_E)_{max} \simeq -11.05$ dB and $(T'_F)_{max} \simeq -10.96$ dB. Therefore, in accordance with Eq. (6-109), $\mu = 12.0365$ and $\hat{\mu} = 11.9124$. Scaling A, D and $\hat{A}$, $\hat{D}$ in accordance with Eq. (6-110) yields the dynamic range-adjusted capacitor values given in Table 6-6. Finally, the admittance levels at the two summing junctions are scaled so that the minimum capacitance value is 1 pF. These "final" values are also listed in Table 6-6. The final circuit realizations are given in Fig. 6-60. Note that when the input signal is not held for the full clock period, the alternative realization in Fig. 6-56a, with appropriately chosen clocks ϕ^e and ϕ^o (see Fig. 6-55), can be used. Alternatively, the biquad in Fig. 6-57, with equivalence (b) in Fig. 6-59, can be used.

In comparing the "final" realizations, we note that the F-circuit requires roughly 12 times the total capacitance of the E-circuit, in spite of the fact that the initial values were almost identical. It should be noted that other practical examples exist where the F-circuit designs are dramatically more efficient than

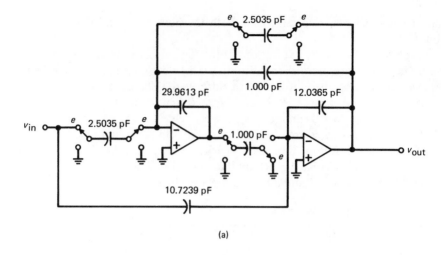

(a)

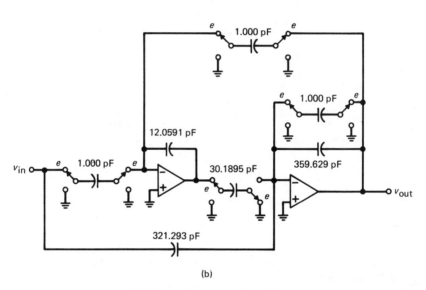

(b)

Fig. 6-60 Realizations for the LPN transfer function in Eq. (6-133). (a) *E*-circuit. (b) *F*-circuit.

the corresponding *E*-circuit designs. Usually, high-*Q* designs are more efficiently realized as *E*-circuits and low-*Q* designs as *F*-circuits. There is, however, no rigorous proof for this trend. Thus, alternative designs must be carried to completion before they can be meaningfully compared.

As a means for comparing the statistical behavior of the two designs, Monte Carlo simulations for the circuits in Fig. 6-60 were performed assuming independent $\pm 1\%$ capacitor variations (Monte Carlo methods were discussed in Section 3.10).

The gain $(\Delta|T|)$ standard deviations at 1 Hz and 1700 Hz, denoted σ_1

and σ_{1700}, are listed in Table 6-6. Comparing these results, we see that the more-silicon-efficient E-circuit design is also slightly less sensitive. The reader should recognize that these statistical performances, although adequate for comparing like designs, are extremely pessimistic for NMOS realizations. We noted earlier that capacitor variations are highly correlated and capacitor ratio variations as small as $\pm 0.1\%$ can be realized with a careful chip layout.

6.6.5b Band-Pass (BP10) Circuit

As a final example, let us consider the biquad realization of the BP10 transfer given in Eq. (6-29), with $\tau = 125\ \mu\text{sec}$. As shown in Fig. 6-10, the bilinear transform places a zero at $z = -1$ to provide infinite attenuation at the half-sampling frequency (4 kHz). In this example, only the final T_E and T_F designs are given. The interested reader can find the T'_E and T'_F designs and the BP00 designs for Eq. (6-30) in [P18]. As shown in [P18], the BP00 designs are about 20% more efficient in silicon usage. As shown in Fig. 6-10, this efficiency is gained at the sacrifice of high-frequency attenuation. As the sampling frequency is raised higher, this sacrifice becomes less significant and the BP00 designs will be preferable.

The final capacitor values for the T_E and T_F designs, the total capacitances, and the gain standard deviation at 1633 Hz σ_{1633} (for independent $\pm 1\%$ capacitor variations) are listed in Table 6-7. The circuit realizations are given in

TABLE 6-7 Final T_E and T_F realizations for the BP10 transfer function, Eq. (6-29)

Capacitor (pF)	E-Circuit (T_E)	F-Circuit (T_F)
A	9.8305	14.9449
B	8.2030	11.9760
C	17.8651	11.2946
D	15.5435	9.4436
E	1.0000	—
F	—	1.0000
$G = H$	1.5812	1.0000
I	1.0000	1.5825
J	—	—
$\sum C\,(\text{pF})$	55.0233	51.2416
σ_{1633} (dB)	0.2738	0.2524

Fig. 6-61. In this example the T_E and T_F circuits are seen to be nearly equivalent. However, experience has shown that for high-Q, high-sample rates, the E-circuit tends to yield the more efficient design. This trend was demonstrated in the previous example via an LPN with $Q = 30$. We note that, when the input signal is not held for the full clock period, the alternative realizations in Fig. 6-56b and Fig. 6-57 can be used. Of course, the number of switches used in the realizations given in Figs. 6-60 and 6-61 can be reduced as per Fig. 6-51b.

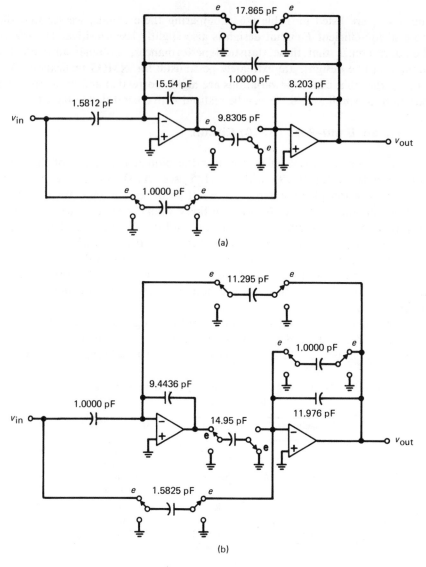

Fig. 6-61   Realizations for the BP10 transfer function in Eq. (6-29). (a) *E*-circuit.
(b) *F*-circuit.

6.7 A FEW COMMENTS ON THE REALIZATION OF HIGH-ORDER ACTIVE-SC FILTERS

All the techniques discussed in Chapter 5 for realizing high-order active-*RC* filters are in principle applicable in high-order active-SC filters. The techniques considered in Chapter 5 are as follows:

1. Cascaded biquads and first-order sections.
2. Admittance-simulated *RLC* ladder networks.

3. Multiple-loop feedback topologies involving biquads or first-order sections.

6.7.1 Cascade Synthesis

Clearly, cascading noninteracting active-SC sections is the most straightforward method to apply. Once again we must first derive an appropriate z-domain transfer function from either the frequency-domain requirements or from a predetermined Nth-order s-domain transfer function. In the latter situation, we may either operate (prewarping and bilinear transformation) on the Nth-order s-domain function or on the individual paired and ordered s-domain second-order sections. Operating on the individual paired and ordered second-order sections is the easier approach and seems to provide perfectly adequate z-domain functions. Another alternative is to synthesize the z-domain transfer function directly from the analog requirements using an approximation program [P20]. For some applications, direct z-domain optimization may yield a transfer function which is more efficient than that obtained via bilinear transformation Once the appropriate z-domain transfer function is obtained, its numerator and denominator must be factored into a product of second-order polynomials. The numerator and denominator factors are then paired and ordered according to procedures which are analogous to those applied in the s-domain case.

As indicated in Fig. 6-1, in cascade synthesis, we can often make most efficient use of the silicon area if we use multiple clock rates. That is, the high-pass sections, which typically have low cutoff and stop-band-edge frequencies, are realized with a lower sampling frequency than the low-pass and band-pass sections, which have higher cutoff and stop-band-edge frequencies. To avoid having to provide anti-aliasing for the low sampling rate, the biquad sections for a multiple-sampling-rate filter are ordered in the cascade as shown in Fig. 6-62a. This method is, of course, most effectively used in an analog input/digital or sampled-data output environment, where reconstruction is not needed. When a smooth analog output is required, the complexity of the continuous reconstruction filter will be dictated by the rejection requirements at frequencies above the lower sampling frequency. In Fig. 6-62a we show three different sampling frequencies: f_{S1} for the low-pass sections, f_{S2} for the band-pass sections, and f_{S3} for the high-pass sections, where $f_{S1} > f_{S2}$ and f_{S3}. To simplify the logic implementation of these clocks f_{S1}, f_{S2}, and f_{S3} are usually binarily related (i.e., $f_{S1} = 2^m f_{S3}$ and $f_{S2} = 2^{m-k} f_{S3}$). In most situations two clock frequencies are sufficient and $f_{S1} = f_{S2}$. To gracefully interface multirate sections, we use the techniques described in Section 6.6.4. Only the nonoverlapping clocks at f_{S1} will usually have near 50% duty cycles. The duty cycles for the nonoverlapping clocks at f_{S2} and f_{S3} are chosen according to the relationship shown in Fig. 6-55. Hypothetical waveforms for the even-phase clocks ϕ_i^e are shown in Fig. 6-62b. The odd-phase clocks ϕ_i^o are the logical inverses of ϕ_i^e. The transfer function, sample to sample, for the cascade is given by

$$H = H_{\mathrm{LP_1}}(z_1) \cdot H_{\mathrm{LP_2}}(z_1) \cdot H_{\mathrm{BP_1}}(z_2) \cdot H_{\mathrm{BP_2}}(z_2) \cdot H_{\mathrm{HP_1}}(z_3) \cdot H_{\mathrm{HP_2}}(z_3) \qquad (6\text{-}134)$$

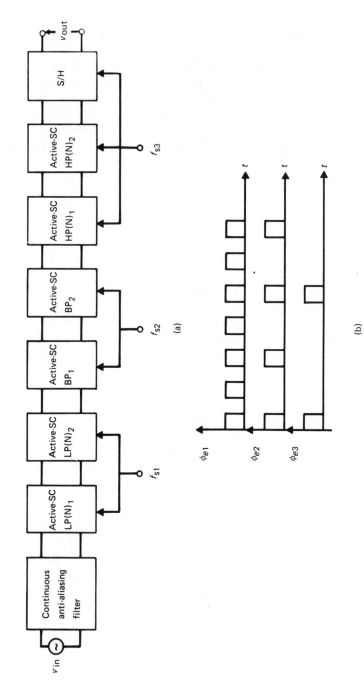

Fig. 6-62 Cascade of six active-SC biquad sections for analog input/digital or sampled-data output environment.

where

$$z_i = e^{s\tau i} \quad \text{and} \quad \tau_i = \frac{1}{f_{si}}$$

As an example of this cascade methodology, consider the tenth-order bandpass filter schematic shown in Fig. 6-63. The waveforms and timing for the clocks are also given. This filter provides the measured voice-frequency bandpass response shown in Fig. 6-64. It is noted that the frequency for nonoverlapping clocks ϕ_1 and $\bar{\phi}_1$ is 100 kHz. The frequency for nonoverlapping clocks $\phi_2, \bar{\phi}_2$ and $\phi_3, \bar{\phi}_3$ is 25 kHz. A basic (50% duty cycle) clock is derived from the 100 kHz clock via a synchronous divide-by-four counter. The duty cycles for $\phi_2, \bar{\phi}_2$ and $\phi_3, \bar{\phi}_3$, shown in Fig. 6-63, are obtained by logical manipulation of the 100 kHz and 25 kHz clocks.

To explain the use of these clock sets, let us examine the SC filter architecture more closely. The first three sections are low-pass sections: a first-order low-pass (LP) section and two low-pass notch (LPN) biquad sections. The switches in these sections are clocked at the full 100 kHz rate. However, the first-order LP section is seen to operate on the continuous anti-aliased input signal during both phases (ϕ_1 and $\bar{\phi}_1$). This operation is equivalent to interpolation and effectively doubles the inherent 100 kHz sampling rate. Doubling the input sampling rate lessens the burden placed on the anti-aliasing section. At the input to biquad LPN1, the sampling rate is immediately decimated to 100 kHz. This decimation is readily verified by examining the switching that occurs on the input side of capacitors G_2 and K_2. With this switching, only one out of every two samples is seen by LPN1, and the output changes in value only once per clock (ϕ_1) period (10 μsec). Since the output of LPN1 is in a full-clock-period sample-and-hold format, the switch that precedes K_2 is not needed for K_3 in LPN2. The output of LPN2 is also held for the full clock period.

The final three sections in the cascade are high-pass sections: two high-pass notch (HPN) biquad sections and a first-order high-pass (HP) section. At the input of HPN1, we decimate to the desired 25 kHz sampling rate. The lower sampling rate in the high-pass sections also results in a 4-to-1 improvement in the capacitor ratios needed for realization. Using the duty cycles shown for $\phi_2, \bar{\phi}_2$, only one of every four samples of the LPN2 output is seen by HPN1. Hence, the outputs of HPN1 and HPN2 change in value once per clock (ϕ_2) period (40 μsec). In addition to the frequency-response specification, it was also desired that the sampled-data output waveform have a return-to-zero format with the output zero when ϕ_3 is low. Clocks $\phi_3, \bar{\phi}_3$ only are needed to fulfill this requirement. With the output of HPN2 held for the full 40 μsec interval, it is convenient to convert to the desired return-to-zero format using the simple high-pass circuit comprised of capacitors X_6, Y_6, and Z_6. This HP section also serves to eliminate the dc offsets assimilated from the preceding sections. Finally, the output is buffered for off-chip routing with a unity-gain amplifier.

The procedure for designing this filter follows closely the cascade design procedure developed in Chapter 5. However, the choice of section ordering was dominated by the desire to lessen the anti-aliasing burden while minimizing the

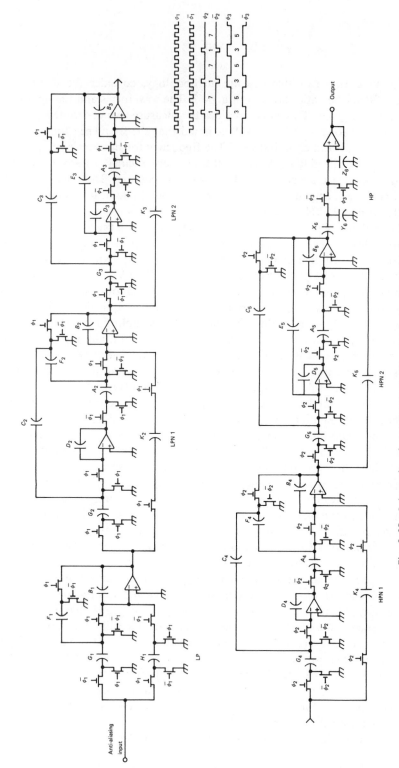

Fig. 6-63 Schematic of a cascade tenth-order voice-frequency band-pass active-SC filter. Also shown are the three pairs of clock signals used in the filter.

Fig. 6-64   Scope trace of the gain response of the tenth-order band-pass filter in Fig. 6-63.

silicon area required to implement the SC filter. This design objective and the output signal format specification determined the clock signal frequencies and waveforms. To conveniently achieve the input interpolation, a first-order low-pass section was extracted from the overall transfer function. A first-order high-pass section was also extracted to achieve the specified return-to-zero output format and to eliminate dc offsets. The remaining four biquad sections were then designed according to the procedures recommended in Section 6.6.

6.7.2 Synthesis Using Multiple-Loop Feedback Topologies

The synthesis of low-sensitivity high-order active-RC filters with multiple-loop feedback topologies involving first- or second-order sections was discussed at length in Chapter 5. In principle, all these topologies and design approaches are directly applicable to active-SC networks. In multiple-loop active-RC synthesis, we observed that symmetrical band-pass filters were conveniently realized with generic band-pass biquad sections. Transmission zeros are conveniently (but not optimally) realized using either multiple-input feedforward or multiple-output feedforward summation. Transmission zeros could also be implemented in cascade fashion (e.g., the generalized FLF). However, putting the zeros into the sections severely complicates the synthesis process.

One method for implementing multiple-loop feedback switched capacitor band-pass filters [P21] is to transform the s-domain flow graph for an active-RC multiple-loop feedback filter into a z-domain flow graph via the bilinear transformation given in Eq. (6-18a). This z-domain flow graph is implemented using switched capacitor biquads (LP20, BP10, HP, LPN, HPN) of the type shown in Fig. 6-51 (including the SC element transformations in Fig. 6-54). Since active-RC multiple-loop feedback implementations of symmetric band-pass filters use BP sections (see Chapter 5), the bilinear transformed equivalent SC implementation uses BP10 sections. For example, consider the s-domain flow graph for a sixth-order band-pass (symmetric) coupled biquad filter, as shown in Fig. 6-65a. Bilinear transforming the blocks of the s-domain flow graph yields the z-domain flow graph in Fig. 6-65b, where (for $i = 1, 2, 3$)

$$\hat{H}_{\beta i} = \frac{H_{\beta i}\tau}{1 + \omega_0\tau/2Q_i + \omega_0^2\tau^2/4} \tag{6-135a}$$

$$\alpha_i = \frac{2 - \omega_0^2\tau^2/2}{1 + \omega_0\tau/2Q_i + \omega_0^2\tau^2/4} \tag{6-135b}$$

$$\beta_i = \frac{1 - \omega_0\tau/Q_i + \omega_0^2\tau^2/4}{1 + \omega_0\tau/Q_i + \omega_0^2\tau^2/4} \tag{6-135c}$$

Note that when $1/Q_i = 0$, as in the second biquad in Fig. 6-65a, $\beta_i = 1$; hence, the poles lie on the unit circle, as we would expect.

To realize the biquad blocks in the z-domain flow graph, both inverting and noninverting BP10 SC biquads are required. Note that coupled biquad implementations typically require high-Q biquad sections. Experience has dictated a preference for E-circuit high-Q realizations. One E-type inverting BP10 realization is shown in Fig. 6-61a. Another convenient inverting BP10 implementation is shown in Fig. 6-66. Note that for this circuit the multiple input feeds are realized with unswitched capacitors, L (for $G = H$) and K (for $I = J$). Substituting $G = H = L$ and $I = J = K$ into the numerator T_E given by Eq. (6-102a), and equating coefficients with the desired numerator function $-\hat{H}_B(1 - z^{-1})(1 + z^{-1})$, yields the design equations

$$K = \hat{H}_B \quad \text{and} \quad L = 2\hat{H}_B \tag{6-136}$$

The complete set of design equations for this biquad then are given by Eqs. (6-136) and (6-107).

Returning to the E-circuit transfer functions in Eqs. (6-102), we note that T_E provides an inverting transfer function, where as T'_E provides a noninverting transfer function. Hence, by permitting the use of both T_E and T'_E realizations, the need for both inverting and noninverting sections is accommodated without any additional amplifiers. A convenient implementation for a noninverting BP10 T'_E, which we shall refer to as an E'-type BP10 biquad, is shown in Fig. 6-67. Note that for T'_E implementations the biquad output is the output of the first op amp. Again the multiple input feeds are realized as unswitched capacitors (L and K). One set of design equations is obtained by equating the coefficients of the numerator of T'_E in Eq. (6-102b), with the substitution $G = H = L$ and $I = J = K$, to the desired numerator function $+\hat{H}_B(1 - z^{-1})(1 + z^{-1})$. Performing the required algebraic operations yields

$$L = \hat{H}_B\left(1 + \frac{2E}{C}\right) \quad \text{and} \quad K = \frac{2\hat{H}_B}{C} \tag{6-137}$$

Active Switched Capacitor Sampled-Data Networks

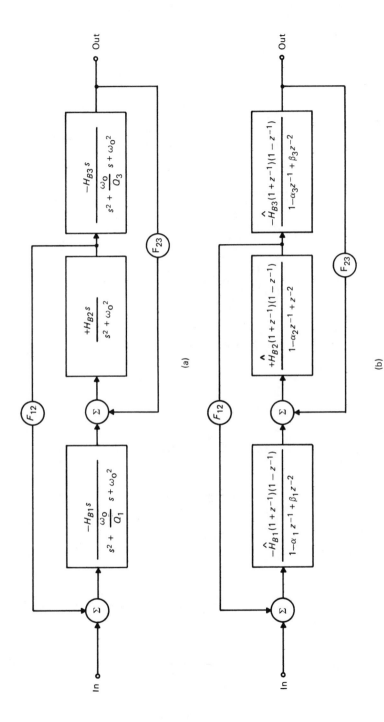

Fig. 6-65 Flow graphs for sixth-order coupled biquad band-pass implementation. (a) s-Domain. (b) z-Domain.

493

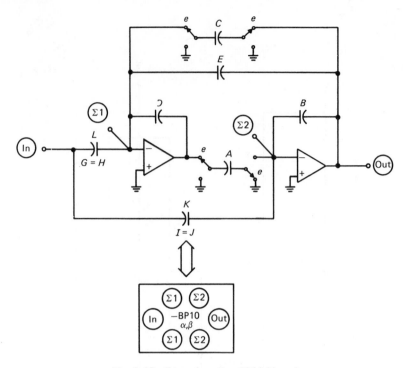

Fig. 6-66 *E*-type inverting BP10 biquad.

Fig. 6-67 *E'*-type noninverting BP10 biquad.

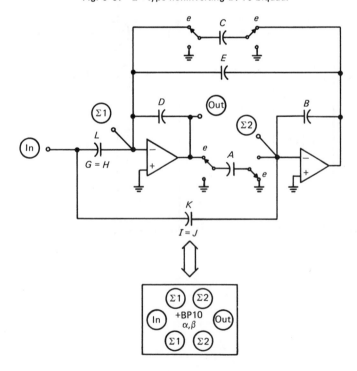

The complete set of design equations are then given by Eqs. (6-107) and (6-137). Note that Eqs. (6-107) are first used to determine the values for capacitors C and E. Also, $\beta_i = 1$ implies that $E = 0$.

To complete the SC coupled biquad implementation, the functional blocks in the z-domain flow graph of Fig. 6-65 are replaced by their SC implementations in Figs. 6-66 and 6-67. The first and third (inverting) BP10 sections are realized according to Fig. 6-66, and the second (noninverting) BP10 section is realized according to Fig. 6-67 with $E_2 = 0$. The external feedback gains F_{12}, F_{23} and associated summing operations are readily incorporated into the internal biquad summers by duplicating the capacitive input feeds to the integrator summing junctions Σ_1 and Σ_2, as shown in Fig. 6-68. Note that, since the input to the

Fig. 6-68   Sixth-order band-pass SC coupled biquad filter.

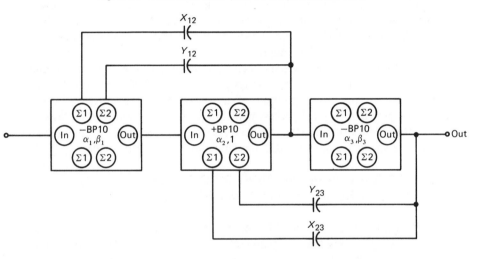

BP10 biquad is capacitively fed to both op amp summing junctions (Σ_1, Σ_2), the external feedback, shown symbolically in Fig. 6-65, must also be capacitively fed to both summing junctions. To obtain the proper amount of feedback, capacitors X_{12}, Y_{12} and X_{23}, Y_{23} are assigned the following values:

$$\frac{X_{12}}{D_1} = 2\hat{H}_{B1}F_{12} \quad \text{and} \quad \frac{Y_{12}}{B_1} = \hat{H}_{B1}F_{12} \qquad (6\text{-}138a)$$

$$\frac{X_{23}}{D_2} = \frac{2\hat{H}_{B2}F_{23}}{C} \quad \text{and} \quad \frac{Y_{23}}{B_2} = \hat{H}_{B2}F_{23}\left(1 + \frac{2E}{C}\right) \qquad (6\text{-}138b)$$

Analogous operations can be performed to implement band-pass FLF, IFLF, MLF, and MSF SC realizations. These multiple-feedback topologies are treated in detail, for active-RC realization, in Chapter 5. Although, we have shown the implementation of multiple-feedback SC band-pass filters with bilinear transformed BP10 biquads, it is possible to achieve more efficient biquad realizations with other band-pass types, such as BP00. By "more efficient" we mean requiring less total capacitance, hence less area on the MOS chip. The best combination of biquad function and multiple-feedback topology to achieve low sensitivity with low total capacitance has not yet been determined.

For low-pass design, the multiple-loop feedback topologies employ first-order lossless and lossy integrators. For this class of filters, multiple-loop feedback topologies such as the leapfrog (LF) topology has been used very productively [P2].

It has been demonstrated that leapfrog active-SC low-pass filters, using the lossless and lossy (modified) LDI integrators in Table 6-2 and Eq. (6-33), respectively, can be implemented efficiently. Furthermore, the active-SC implementations imitate the low-sensitivity properties of their active-RC and passive equivalents. Of practical importance is that parasitic-free leapfrog implementations can be realized using the dual integrator equivalence in Eq. (6-32) and the parasitic-free integrator circuits in Table 6-4. A fourth-order active-SC low pass implementation and its corresponding z-domain equivalent circuit are given in Fig. 6-69. The verification of the equivalent circuit and the analysis of the circuit are left as exercises for the reader.

In Section 6.4 we discussed circuits and techniques for simulating inductor and FDNR admittances of various types. An example was given in Fig. 6-41, demonstrating the transformation of a passive-RLC ladder network to an active-SC network. Typically, these realizations involve unbuffered nodes and/or differential input operational amplifiers. As noted in Section 6.5, the practicality of such networks for silicon MOS implementation is questionable.

6.7.3 Mask-Programmable NMOS Building Block for Realizing High-Order SC Filters

To exploit both the versatility of the general biquad structure in Fig. 6-51 and the ability to pack many biquad sections onto a single MOS chip, an analog building block was developed [P22]. The technology for this development is depletion/enhancement, double-poly NMOS, and it operates from ± 5V supplies. This building block is capable of implementing up to 22 poles (and zeros) of filtering, of which six poles can be allocated to continuous anti-aliasing and smoothing filters. Two polysilicon levels define the top and bottom plates of the capacitors and provide two of three levels of interconnect. The third level of interconnect is aluminum metalization. To implement any particular filter one need only customize the two polysilicon masks. All the other masks are held fixed. With this approach the analog LSI technology is made available to users whose volumes are not sufficiently large to justify a custom development of all the masks and the turnaround time for all users is reduced.

The building block chip, shown in Fig. 6-70, is modest in size (120 $\times$ 240 mils) and is packaged in a hermetic 16-pin DIP. As shown in Fig. 6-70, the specific circuit functions included on the uncommitted building block are 22 operational amplifiers [P3], an oscillator, clock generation logic and drivers, digital input and output buffers, a voltage reference, tapped resistors, 88 analog switches, metal clock and analog signal buses, and second-order continuous

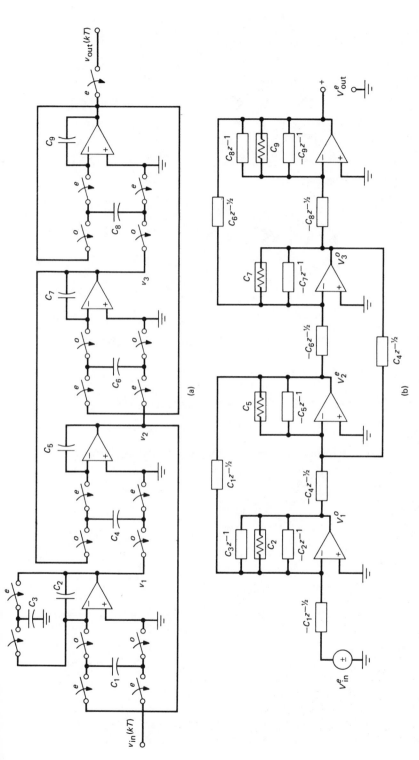

Fig. 6-69 (a) Fourth-order low-pass leapfrog (LF) active-SC filter. (b) z-Domain equivalent circuit.

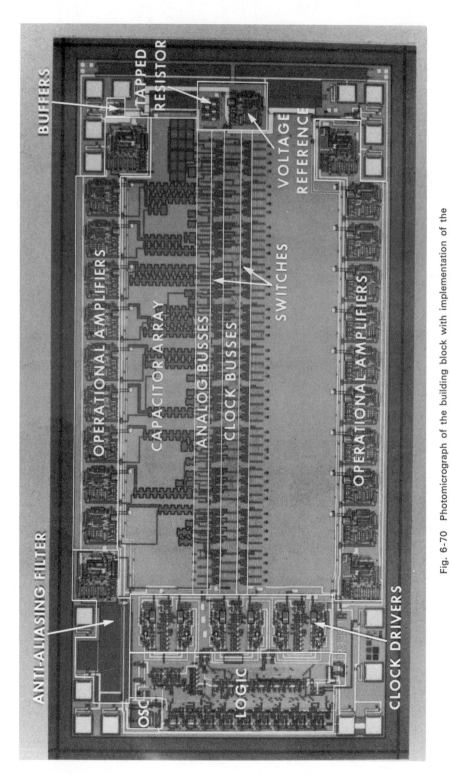

Fig. 6-70 Photomicrograph of the building block with implementation of the tenth-order band-pass filter in Fig. 6-63. It is noted that the building block DIP is shown in Fig. 1-1.

BUFFERS

TAPPED RESISTOR

VOLTAGE REFERENCE

OPERATIONAL AMPLIFIERS

CAPACITOR ARRAY

ANALOG BUSSES

CLOCK BUSSES

SWITCHES

OPERATIONAL AMPLIFIERS

ANTI-ALIASING FILTER

OSC

LOGIC

CLOCK DRIVERS

anti-aliasing/smoothing filters. The customized polysilicon masks determine capacitor arrays and the interconnections to complete circuits for filters and to program the logic. Sufficient metal lines and poly-to-metal windows have been included in the layout to localize the poly routing and to permit wide versatility in interconnections. The vacant areas between the op amps and switches are available for capacitor definition.

The building block can operate from an external precision master clock or derive its own master clock with the on-board oscillator and an external crystal. In addition, it can be interfaced to external digital and sampled-data circuitry by either externally applied or internally generated synchronization signals. Input and output TTL-compatible buffers are included to drive the internal logic circuitry or to route digital signals off-chip. The clock logic is also mask-programmable. It contains nine binary stages of countdown and sufficient logic functions to realize up to three independent sets of nonoverlapping clocks with programmable duty cycles. The duty-cycle programmability is particularly useful in interfacing switching capacitor filter sections which are clocked at different rates, to make efficient use of the capacitor area. The utility of this feature was demonstrated in Section 6.7.2.

Implemented on the top half of the building block in Fig. 6-70 is the tenth-order band-pass filter shown schematically in Fig. 6-63. In this case, as in most applications, only a portion of the building block circuitry is used; hence, to conserve power, all unused circuits are powered down. This power-down feature is also accomplished with the two "custom masks." Although only the implementation of cascade filters has been demonstrated, the building block is sufficiently flexible to permit the realization of multiple-loop feedback filters. In addition, four of the 22 op amps are capable of driving off-chip loads; hence, the filtering capacity can be distributed among as many as four separate filters.

6.8 SUMMARY

In this chapter we have introduced the reader to the analysis and synthesis of switched capacitor filters. For analysis and general understanding, we have relied to great extent on their similarity to active-RC filters. For their synthesis we have taken advantage of their similarity to digital filters. For topological realizations and practical MOS implementation, we have sought to exploit their unique properties. An analysis procedure using z-domain equivalent circuits has been given. The sampled-data behavior of SC networks permits the use of the z-transform, but basically their analog nature allows them to be characterized using basic linear circuit theory.

Many examples have been given which demonstrate the versatility of SC networks to realize filters of all types. Hopefully, this chapter has stimulated the reader's scholarly and practical interests in this new, exciting technology. Testimony to the activity in this area is the rapid emergence of products (see e.g., [P16,] [P22]) using active-SC filters.

REFERENCES

Books

B1. SEQUIN, C. H., AND M. F. TOMPSETT, *Charge Transfer Devices.* New York: Academic Press, 1975.

B2. SCHILLING, D. L., AND H. TAUB, *Digital Integrated Electronics.* New York: McGraw-Hill, 1977.

B3. RABINER, L. R., AND B. GOLD, *Theory and Application of Digital Signal Processing.* Englewood Cliffs, N.J.: Prentice-Hall, 1975.

B4. RAGAZZINI, J. R., AND G. F. FRANKLIN, *Sampled Data Control Systems.* New York: McGraw-Hill, 1958.

B5. DESOER, C. A., AND E. S. KUH, *Basic Circuit Theory.* New York: McGraw-Hill, 1969.

B6. GRAY, P. R., D. A. HODGES, R. W. BRODERSEN, eds., *Analog Mos Integrated Circuits*, IEEE Press, 1980.

B7. SCHAUMANN, R., M. A. SODERSTRAND, K. R. LAKER, eds., *Modern Active Filter Design*, IEEE Press, 1981.

Papers

P1. BUSS, D. D., D. R. COLLINS, W. H. BAILEY, AND C. R. REEVES, "Transversal Filtering Using Charge Transfer Devices," *IEEE J. Solid-State Circuits*, SC-8 (April 1973), 138–146.

P2. BRODERSEN, R. W., P. R. GRAY, AND D. A. HODGES, "MOS Switched Capacitor Filters," *Proc. IEEE*, 67 (January 1979), 61–75.

P3. TSIVIDIS, Y. P., AND D. L. FRASER, JR., "A Process Insensitive NMOS Operational Amplifier," *Dig. 1979 IEEE Int. Solid-State Circuits Sym.*, February 1979, pp. 188–189. TSIVIDIS, Y. P., "Design Considerations in Single Channel MOS Analog Integrated Circuits—A Tutorial," *IEEE J. Solid-State Circuits*, SC-13 (June 1978), 383–391.

P4. KURTH, C. F., AND G. S. MOSCHYTZ, "Nodal Analysis of Switched Capacitor Networks," *IEEE Trans. Circuits Syst.*, CAS-26 (February 1979), 93–104; and "Two-port Analysis of Switched Capacitor Networks Using Four-Port Equivalent Circuits," *IEEE Trans. Circuits Syst.*, CAS-26 (March 1979), 166–180.

P5. LAKER, K. R., "Equivalent Circuits for the Analysis and Synthesis of Switched Capacitor Networks," *Bell Syst. Tech. J.*, 58 (March 1979), 727–767.

P6. LIOU, M. L., AND Y. L. KUO, "Exact Analysis of Switched Capacitor Circuits with Arbitrary Inputs," *IEEE Trans. Circuits Syst.*, CAS-26 (May 1979), 213–223.

P7. BRUTON, L. T., "Low Sensitivity Digital Ladder Filters," *IEEE Trans. Circuits Syst.*, CAS-22 (March 1975), 168–176.

P8. FRIED, D. L., "Analog Sampled Data Filters," *IEEE J. Solid-State Circuits*, SC-7 (August 1972), 302–304.

P9. CAVES, J. T., ET AL., "Sampled Analog Filtering Using Switched Capacitors as Switched Resistors," *IEEE J. Solid-State Circuits*, SC-12 (December 1977), 592–599.

P10. TEMES, G. C., AND I. A. YOUNG, "An Improved Switched Capacitor Integrator," *Electron. Lett.*, 14, no. 9 (April 27 1978), 287–288.

P11. TEMES, G. C., N. J. ORCHARD, AND M. JAHANBEGLOO, "Switched Capacitor Filter Design Using the Bilinear z-Transform," *IEEE Trans. Circuits Syst.*, CAS-25 (December 1978), 1039–1044.

P12. HERBST, D., ET AL., "MOS Switched Capacitor Filters," *Dig. 1979 Int. Solid-State Circuits Conf.*, (February 1979), 74–75.

P13. RAHIM, C. F., ET AL., "A Functional MOS Circuit for Achieving Bilinear Transformation in SC Filters," *IEEE J. Solid-State Circuits*, SC-13 (December 1978), 906-909.

P14. HOSTICKA, B. J., AND G. S. MOSCHYTZ, "Switched Capacitor Simulation of Grounded Inductors and Gyrators," *Electron. Lett.*, 14, no. 24 (November 23 1978), 788–790.

P15. TEMES, G. C., AND M. JAHANBEGLOO, "Switched Capacitor Circuits Bilinearly Equivalent to the Floating Inductor and FDNR," *Electron. Lett.*, 15, no. 3 (February 1979), 87–88.

P16. JACOBS, G. M., ET AL., "Touch Tone Docoder Chip Mates Analog Filters with Digital Logic," *Electronics*, (February 15 1979), 105–112.

P17. MARTIN, K., AND A. S. SEDRA, "Strays Insensitive Switched Capacitor Filters Based on the Bilinear z-Transform," *Electron. Lett.*, 15, no. 13 (July 1979), 365–366.

P18. FLEISCHER, P. E., AND K. R. LAKER, "A Family of Active Switched Capacitor Biquad Building Blocks," *Bell Syst. Tech. J.*, 58 (December 1979), 2235–2269.

P19. McCREARY, J. L., AND P. R. GRAY, "All-MOS Charge-Redistribution Analog to Digital Conversion Techniques—Part I," *IEEE J. Solid-State Circuits*, SC-10 (December 1975), 371–379.

P20. STIEGLITZ, K., "Computer Aided Design of Recursive Digital Filters," *IEEE Trans. Audio Electroacoust.*, Au-18 (June 1970), 123–129.

P21. MARTIN, K., AND A. S. SEDRA, "Exact Design of Switched Capacitor Band-pass Filters Using Coupled Biquad Structures," *IEEE Trans. Circuits Syst.*, CAS-27 (June 1980), 469–475.

P22. FLEISCHER, P. E., K. R. LAKER, D. G. MARSH, J. P. BALLANTYNE, A. A. YIAN-NOULUS, AND D. L. FRASER, "An NMOS Building Block for Telecommunications Applications," *IEEE Trans. Circuits Syst.*, CAS-27 (June 1980), 552–559.

PROBLEMS

6.1 Using Fig. 1-16, determine the minimum order and the cutoff frequency $f_c = \omega_c/2\pi$ for a Butterworth low-pass, anti-aliasing filter with 0 dB gain from dc to 3 kHz and at least 30 dB of attenuation for $f \geq f_s$, where:
(a) $f_S = 8$ kHz
(b) $f_S = 32$ kHz
(c) $f_S = 64$ kHz
(d) $f_S = 128$ kHz

6.2 Consider the second-order low-pass Sallen and Key implementation of the anti-aliasing filter shown in Fig. P6-2, with transfer function given in Eq. (4-5) with $K = 1$. This filter is to be designed for 0 dB gain from dc to 3 kHz and at least 30 dB of attenuation for $f \geq f_s$. If the worst-case variation in the resistors R_1, R_2

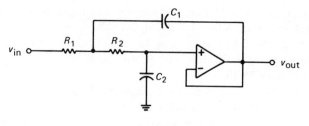

Fig. P6-2

and the capacitors C_1, C_2 are $\Delta R_1/R_1 = \Delta R_2/R_2 = \Delta C_1/C_1 = \Delta C_2/C_2 = \pm 0.5$, determine the minimum value of f_s for which these specifications can be achieved.

6.3 Using Eqs. (6-2), derive the transfer function $V_{\text{out}}^o(z)/V_{\text{in}}^e(z)$ for the switched capacitor network in Fig. P6-3. Note that $\Delta Q_p^e(z) = \Delta Q_p^o(z) = 0$ when there are no independent charge sources and $Q_{pi}^o = C_i V_p^o$ and $Q_{pj}^e = C_i V_p^e$.

Fig. P6-3

$v_{\text{in}} \pm$ e o v_{out}

C_1 C_2

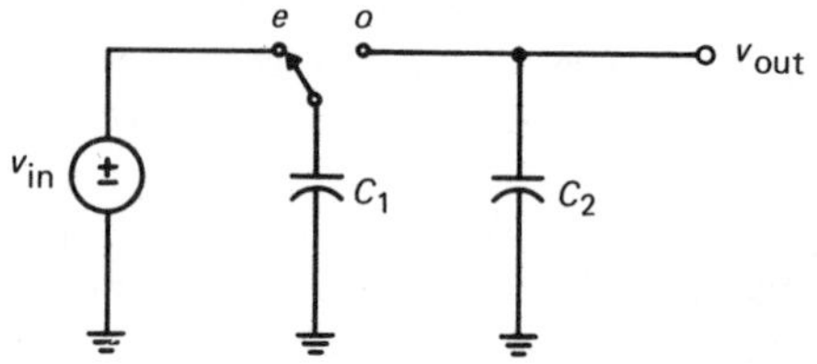

6.4 Determine the transfer functions H_1, H_2, H_3, and H_4 in Eq. (6-10) for the switched-capacitor network in Fig. P6-3.

6.5 Determine the transfer functions H_1, H_2, H_3, and H_4 in Eq. (6-10) for the switched-capacitor network in Fig. P6-5.

Fig. P6-5

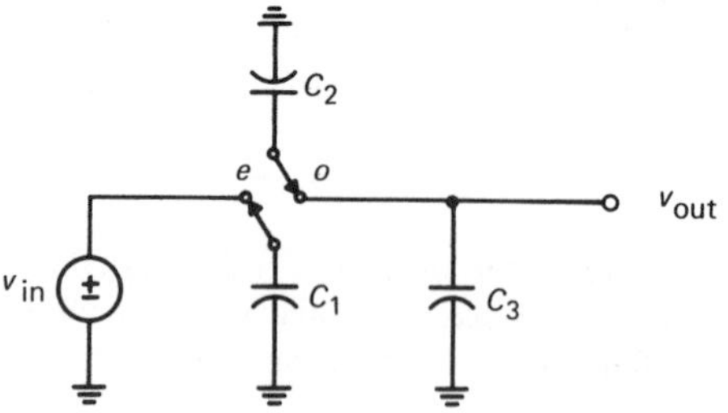

C_2

$v_{\text{in}} \pm$ e o v_{out}

C_1 C_3

6.6 Show that the bilinear transform for integration in Eq. (6-18a) is equivalent to integration by the trapezoidal rule. Note that trapezoidal rule integration of a function $f(x)$ over the limits $0 \le x \le n\tau$ is given by

$$\int_0^{n\tau} f(x)\,dx \simeq \frac{\tau}{2}\Big[f(0) + 2\sum_{k=1}^{(n-1)\tau} f(k\tau) + f(n\tau)\Big]$$

Active Switched Capacitor Sampled-Data Networks

6.7   Derive Eq. (6-26).

6.8   Perform the prewarping for the following transfer function

$$H(s) = \frac{s^2 + 1.4212 \times 10^5}{s^2 + 1004.2s + 6.9833 \times 10^5}$$

and derive $H(z)$ using the bilinear transform.
(a) For the sampling frequency $f_S = 8$ kHz
(b) For the sampling frequency $f_S = 128$ kHz
Discuss the results.

6.9   Derive the z-domain equivalent admittances in Table 6.2.
(a) The BD equivalent admittances
(b) The LDI equivalent admittances
(c) The bilinear transform equivalent admittances

6.10 Scale all the bilinear transform equivalent admittance by the multiplication factor $(1 - z^{-1})$, i.e., $Y_s(z) = (1 - z^{-1})\ Y(z)$. Discuss the form of the resulting
admittance.

6.11 Derive the following equivalent circuits in Figure 6-11.
(a) Equivalent circuit 6-11a
(b) Equivalent circuit 6-11g
(c) Equivalent circuit 6-11h
(d) Equivalent circuit 6-11i
(e) Equivalent circuit 6-11k
(f) Discuss the different circuit implications of circuits 6-11i, 6-11k and 6-11l.

6.12 Draw the schematics for the z-domain equivalent circuits in Fig. 6-13 by manipulating the equivalent circuit for the TSFFP as indicated in the figure. Compare
the results with corresponding equivalent circuits in Figs. 6-11 and 6-12.

6.13 Derive the equivalent circuit for the switched-capacitor circuit in Fig. P6-13.

Fig. P6-13

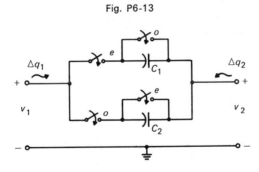

(a) For $C_1 \neq C_2$ and port variables ΔQ_1^e, ΔQ_1^o, ΔQ_2^e, ΔQ_2^o, V_1^e, V_1^o, V_2^e, and V_2^o.
(b) For $C_1 = C_2 = C$ and port variables $\Delta Q_1 = \Delta Q_1^e + \Delta Q_1^o$, $\Delta Q_2 = \Delta Q_2^e$
$+ \Delta Q_2^o$, $V_1 = V_1^e + V_1^o$, and $V_2 = V_2^e + V_2^o$.

6.14 Draw the equivalent circuits for the switched capacitor networks in Figs. P6-14a
and b. Compare these two circuits.

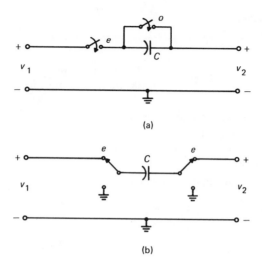

Fig. P6-14

6.15 Derive the transfer functions H_1 and H_3 in Eq. (6-58b) for the switched capacitor network in Fig. 6-19.

6.16 Derive the transfer functions for the switched capacitor integrator in Fig. 6-27, with an op amp of finite dc gain a_0. Compare the results with Eqs. (6-63).

6.17 Repeat Prob. 6.16 for the loss integrator in Fig. 6-32. Let the desired transfer function be

$$H_1(z) = \frac{-\tau p}{(1 + \tau p) - z^{-1}}$$

Defining $C_1 = \alpha C$, $C_2 = C$, and $C_3 = \beta C$, derive the design equations for α and β interims of τp, a_0, and C.

6.18 Derive the expression for the z-transformed even-phase output $V_{out}^e(z)$, for the switched capacitor network in Fig. P6-18. Determine the conditions for bilinear integration.

Fig. P6-18

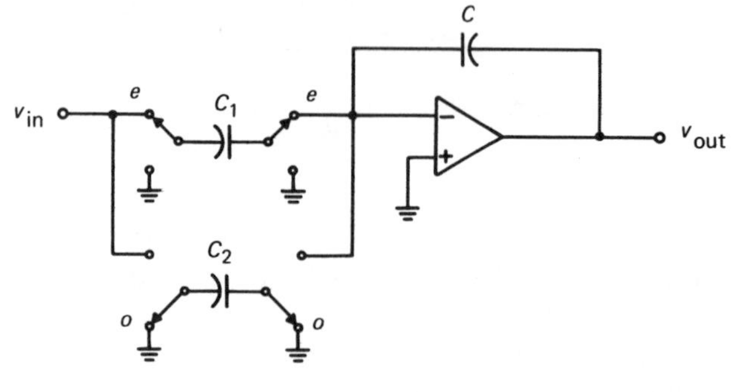

504   Active Switched Capacitor Sampled-Data Networks

6.19 Derive the z-transformed output, $V_{out}^e(z)$, for the switched capacitor network in Fig. P6-19.

Fig. P6-19

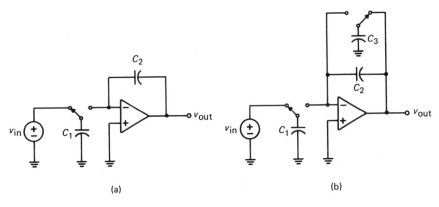

6.20 (a) Derive the analog transfer function for the switched capacitor integrator in Fig. P6-20a when excited with a continuous analog input. Note that the output is sampled and held and that the transfer function for the zero-order hold is given by Eq. (6-9).

(b) Repeat part (a) for the circuit in Fig. P6-20b.

(c) Discuss the results of parts (a) and (b).

Fig. P6-20

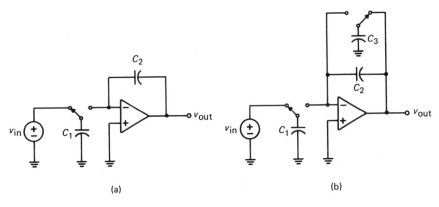

(a)   (b)

6.21 Derive the transfer function for the bilinear integrator in Fig. P6-21 with the top and bottom plate parasitic capacitances C_{pB} and C_{pT}. In typical MOS integrated circuits, $C_{pB} \simeq 0.1C_1$ and $C_{pT} \ll C_{pB}$. C_{pB} is primarily due to the capacitance formed by the bottom plate of C_1 and the substrate; hence, it is proportional to C_1. C_{pT} is a smaller parasitic, because of the capacitances formed by the routing lines (between the switches and the top plate of C_1) and substrate and the sources and drains for the two top plate switches and substrate.

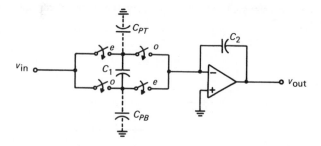

Fig. P6-21

6.22 Derive the z-domain transfer function for the lossy integrator in Fig. 6-29a. Include all relevant parasitic capacitances in your analysis. Determine the gain and pole position errors.

6.23 Construct the z-domain equivalent circuit for the switched capacitor biquad in Fig. P6-23.

Fig. P6-23

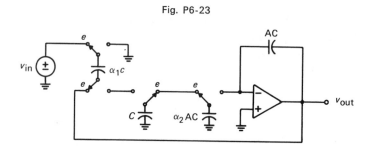

(a) Derive the transfer function $H(z) = V_{out}^e/V_{in}^e$.
(b) Derive the expressions for the selectivity Q and center frequency ω_0.
(c) Insert the top and bottom plate parasitics into the equivalent circuit. Derive the resulting expressions for Q and ω_0 and compare with the ideal Q and ω_0 in part (b).

6.24 Synthesize the transfer function in Prob. 6.8, using the switched capacitor biquad in Fig. 6-51.
(a) Design the E-circuit implementation with $f_S = 8$ kHz.
(b) Design the F-circuit implementation with $f_S = 8$ kHz.
(c) Design the E-circuit implementation with $f_S = 128$ kHz.
(d) Design the F-circuit implementation with $f_S = 128$ kHz.
(e) Compare the total capacitance required for the four realizations.
In the designs, prewarp where necessary, use the transformations in Fig. 6-54 where appropriate, scale the gain to the first op amp to equalize the signal maxima, and scale all capacitances for a minimim of 1 pF.

6.25 Derive the design equations for the E-type biquad in Fig. P6-25. Show that this biquad can realize an all-pass transfer function.

Active Switched Capacitor Sampled-Data Networks

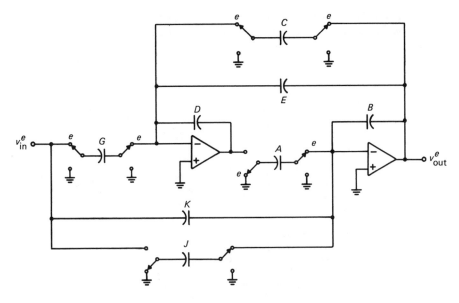

Fig. P6-25

6.26 Synthesize the band-pass biquad specified by the s-domain transfer function given in Eq. (6-27). Let the sampling frequency be $f_S = 128$ kHz. Equalize the signal maxima at both op amp outputs and scale the capcitances for a 1 pF minimum.

(a) Design as a BP10 E-circuit.

(b) Design as a BP10 F-circuit.

(c) Design as a BP01 E-circuit.

(d) Design as a BP01 F-circuit.

(e) Plot and compare the s-domain band-pass, BP10, and BP01 gain responses.

(f) Compare the total capacitances required for designs (a), (b), (c) and (d).

(g) Discuss the results of parts (e) and (f).

6.27 Design a cascade switched capacitor filter to realize the fifth-order low-pass transfer function given by Eqs. (5-35) and (5-36). Choose gain constant K in Eq. (5-35) so that the dc gain is 0 dB. The sampling frequency for this filter is to be $f_S = 128$ kHz. Use the parasitic immune circuits in Figs. 6-50 and 6-51 and the transformations in Fig. 6-54. Given that the input is continuous analog, arrange the circuit so that the filter output is sampled and held for the full period of f_S. In the design, equalize the signal maxima at the outputs of all op amps and scale all capacitors to a minimum value of 1 pF.

(a) Design the biquad sections as E-circuits.

(b) Design the biquad sections as F-circuits.

(c) From parts (a) and (b), determine the minimum capacitance overall design.

6.28 Derive Eq. (6-138).

6.29 Using the techniques described in Section 6.7.2, design a coupled biquad switched capacitor filter to implement the active-RC filter in Fig. 5-31 and Example 5-6. For this design the sampling frequency (normalized) is $f_S = 50$. The s-domain band-pass biquad transfer functions are given in Eq. (5-136).

(a) Show that the feedforward summation function in Fig. 5-31 can be realized using the switched capacitor network in Fig. P6-29.

Fig. P6-29

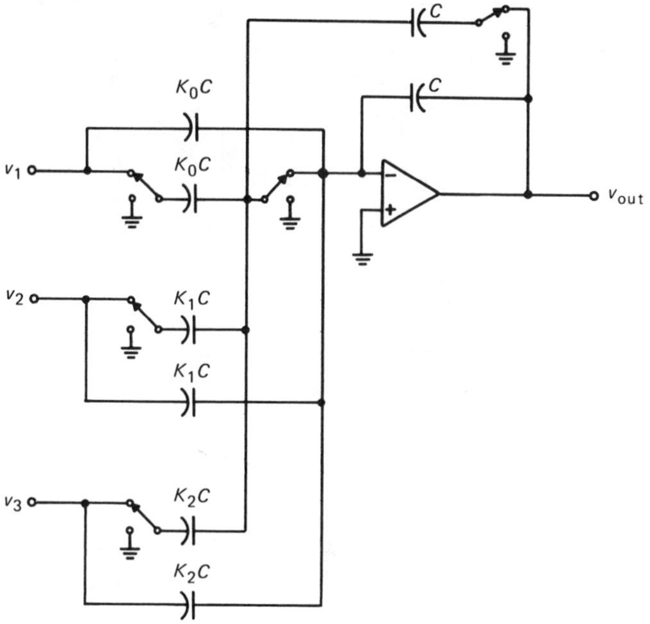

(b) Design the E-circuit BP10 biquads to implement the band-pass T_i's in Eq. (5-136). Equalize the op amp output maxima and scale the capacitances for a minimum value (normalized) of 1.

(c) Show the schematic for the overall filter. In your schematic give all capacitor values (scaled for a minimum value of 1).

6.30 Using the techniques in Section 6.7.2, draw the block diagram for a 16th-order band-pass FLF switched capacitor filter. Show all the external feedback paths, per Fig. 6-68. Give the schematics for the switched capacitor biquads that serve as the blocks for the FLF filter.

6.31 Derive the transfer function for the fourth-order leapfrog low-pass network in Fig. 6-69.

appendix A

....

Selected Topics in
Passive-Network Properties

In the following we discuss briefly some selected topics in passive-network properties. The reader is referred to [B1] and [B2] at the end of this appendix for a detailed treatment of these topics.

In the design of active-RC networks, active devices and passive RC networks are employed. It is helpful to the designer to be familiar with some of the basic properties of RC networks. Before we consider passive-RC networks, it is useful to know some basic properties of passive networks in general.

A.1 BASIS PROPERTIES OF PASSIVE-RLC DRIVING-POINT FUNCTIONS

In Chapter 1 we discussed driving-point functions. A driving-point function is a subclass of a network function where the response and excitation are considered at the same port. The driving-point impedance function $Z(s)$ is the reciprocal of the driving-point admittance function $Y(s)$. It can be shown [B1] that the necessary and sufficient conditions for a driving-point immittance[1] function $H(s)$ to be a passive-RLC network is that $H(s)$ is a positive real (p-r) function. A p-r function is defined as follows:

A function $H(s)$ is p-r if and only if:

1. $H(s)$ is real for real s.
2. Re $H(s) \geq 0$ for Re $s \geq 0$ where Re stands for the real part of.

---

[1]In passive-network-synthesis literature, one often encounters the term "immittance" function, which is a combination of *im*pedance and ad*mittance* functions. In other words, by immittance function we mean admittance and/or impedance function.

A convenient testing criterion for p-r functions is as follows:

1. $H(s)$ is analytic in the right-half s-plane (i.e., no poles in the right half-plane).
2. Poles of $H(s)$ on the $j\omega$-axis are simple, with real positive residues [i.e., no $(s^2 + a^2)^2$ factors in $H(s)$].
3. The real part of $H(j\omega)$ is nonnegative for all ω (i.e., Re $H(j\omega) \geq 0$ for all ω).

The foregoing properties can be heuristically justified as follows. Property (3) follows from the fact that passive networks contain no energy sources and hence dissipate power. A negative value of the real part at any frequency means that power can be delivered. Furthermore, since the network can only dissipate power, it must be stable. Thus, properties (1) and (2) follow from the stability requirements.

●

EXAMPLE A-1

We shall test the following functions to see whether they are p-r.

(a) Consider: $H(s) = \dfrac{s + 1}{s^2 + s + 1}$.

1. $H(s)$ has no poles in the right half-plane.
2. $H(s)$ has no poles on the $j\omega$-axis and the residue test is not needed.
3. Re $H(j\omega) = \dfrac{1}{(\omega^2 - 1)^2 + \omega^2} \geq 0$ for all ω.

Hence, the function is p-r and therefore can be synthesized by a passive-RLC network.

(b) Consider: $H(s) = \dfrac{s^2 + s + 6}{s^2 + s + 1}$.

1. $H(s)$ is analytic in the right half-plane.
2. There are no poles on the $j\omega$-axis (residue test not needed).
3. Re $H(j\omega) = \dfrac{(\omega^2 - 3)^2 - 3}{(\omega^2 - 1)^2 + \omega^2} < 0$ for $\omega^2 - 3 < \sqrt{3}$.

This function is not p-r and not realizable as a driving-point function of a passive-RLC network.

●

For active networks the driving-point function need not be p-r. The only restriction from a practical standpoint is that the network function must be a stable function.

The reader should note that simple methods for testing p-r functions are

Selected Topics in Passive-Network Properties

available (see [B1], [B2]). We shall not go into the synthesis of *RLC* driving-point functions, since (1) we are not interested in networks containing L and (2) these are covered in detail in the references cited. We would like to emphasize, however, that only p-r functions can be synthesized as the driving-point immittance functions of passive networks, since a p-r property is a necessary and sufficient condition for realizing passive-*RLC* networks.

For active networks these restrictions do not hold. In fact, it can be shown that any real rational function of the complex variable s can be realized as a driving-point impedance of a one-port obtained by embedding a single controlled source of unrestricted gain in a transformerless *RC* network [P1]. Of course, there are other ways to synthesize a driving-point immittance function with active *RC* networks [P2].

A.2 PASSIVE-*RC* DRIVING-POINT FUNCTIONS

When the network consists of R's and C's only, we would naturally expect constraints or simplifications in the properties of the driving-point functions. *RC* driving-point impedance functions $Z_{RC}(s)$ satisfy the following properties:

1. The poles and zeros of Z_{RC} lie on the negative real axis, are simple, and alternate.
2. At dc, $Z_{RC}(s)$ is a positive constant or has a pole, whereas at infinity it is a positive constant or has a zero.
3. The residues of $Z_{RC}(s)$ are real and positive.

For the *RC* driving-point admittance function, the following properties hold:

1. The poles and zero of Y_{RC} lying on the negative real axis, are simple, and alternate.
2. At dc, $Y_{RC}(s)$ is a positive constant or has a zero, whereas at infinity it is a positive constant or has a pole.
3. The residues of $(1/s)(Y_{RC})$ are real and positive.

●

EXAMPLE A-2

Consider the synthesis of $Z = \dfrac{(s+2)(s+4)}{(s+1)(s+3)}$.

Note that Z is the driving-point impedance of an *RC* network, as noted above. A partial fraction expansion of $Z(s)$ yields

$$Z(s) = 1 + \frac{\frac{3}{2}}{s+1} + \frac{\frac{1}{2}}{s+3}$$

The circuit is shown in Fig. A-1.

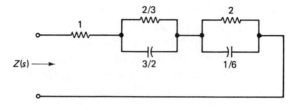

Fig. A-1   Foster first form.

On the admittance basis, we have

$$Y = \frac{1}{Z} = \frac{(s+1)(s+3)}{(s+2)(s+4)}$$

$$\frac{Y(s)}{s} = \frac{\frac{3}{8}}{s} + \frac{\frac{1}{4}}{s+2} + \frac{\frac{3}{8}}{s+4}$$

or

$$Y(s) = \frac{3}{8} + \frac{\frac{1}{4}s}{s+2} + \frac{\frac{3}{8}s}{s+4}$$

The circuit realization is shown in Fig. A-2. Circuit realizations shown in Figs. A-1 and A-2 are referred to as Foster's first and second canonical forms,

Fig. A-2   Foster second form.

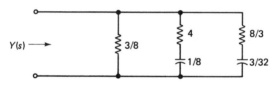

respectively. Other forms, such as ladders, are also possible, as shown in Figs. A-3 and A-4. The ladder forms are referred to as the Cauer canonical forms.

The Cauer forms are obtained by a simple division in the continued fraction form. Consider, for example, the so called Cauer first form

$$Z(s) = \frac{s^2 + 6s + 8}{s^2 + 4s + 3}$$

$$
\begin{array}{r}
1 \leftarrow Z \\
s^2 + 4s + 3 \overline{\smash{\big)}\ s^2 + 6s + 8} \\
\underline{s^2 + 4s + 3} \qquad \frac{1}{2}s \leftarrow Y \\
2s + 5 \overline{\smash{\big)}\ s^2 + 4s + 3} \\
\underline{s^2 + \tfrac{5}{2}s} \qquad \frac{4}{3} \leftarrow Z \\
\tfrac{3}{2}s + 3 \overline{\smash{\big)}\ 2s + 5} \\
\underline{2s + 4} \qquad \frac{3}{2}s \leftarrow Y \\
1 \overline{\smash{\big)}\ \tfrac{3}{2}s + 3} \\
\underline{\tfrac{3}{2}s} \qquad \frac{1}{3} \leftarrow Z \\
3 \overline{\smash{\big)}\ 1}
\end{array}
$$

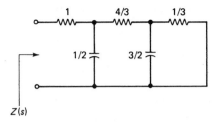

Fig. A-3 Cauer first form.

Fig. A-4 Cauer second form.

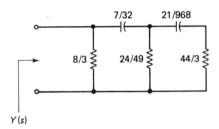

It is called continued fraction because we can write $Z(s)$ as follows:

$$Z(s) = 1 + \cfrac{1}{\frac{1}{2}s + \cfrac{1}{\frac{4}{3} + \cfrac{1}{\frac{3}{2}s + \cfrac{1}{\frac{1}{3}}}}}$$

Similarly, we can rewrite

$$Y(s) = \frac{1}{Z(s)} = \frac{s^2 + 4s + 3}{s^2 + 6s + 8}$$

A simple division, with rearranging, of terms:

$$8 + 6s + s^2 \,\big|\, 3 + 4s + s^2$$

yields the so-called Cauer second form. The Cauer forms of the synthesis are shown in Figs. A-3 and A-4. The reader is referred for details to [B1], [B2].

●

Transmission zeros of an RC two-port transfer function can, of course, be complex. In fact, transmission zeros on the $j\omega$-axis can be obtained by parallel ladders, of which the bridged-T and the twin-T networks are special cases. For a detailed discussion of this topic, see [B3].

A.3 TRANSFER FUNCTIONS OF PASSIVE NETWORKS

In general, there are not many restrictions on the transfer function of a passive-*RLC* transfer function, except that for stability reasons the poles of the transfer function must lie in the left half-plane, the order of the numerator cannot exceed that of the denominator by more than one, and poles on the $j\omega$-axis must be of multiplicity 1 (i.e., simple poles on the $j\omega$-axis). If the network is comprised only of R's and C's, all the poles must lie on the negative real axis. This follows from the fact that the natural frequencies of a network are the poles of the network function, and they are the same whether one considers the driving-point function or the transfer functions. It is for this reason that the polynomials $Q(s)$ in (5-5) are chosen with distinct real roots on the σ-axis.

REFERENCES

Books

B1. VANVALKENBURG, M. E., *"Introduction to Modern Network Synthesis.* New York: McGraw-Hill, 1964.

B2. TEMES, G. C., AND J. W. LAPATRA, *Introduction to Circuit Synthesis and Design.* New York: McGraw-Hill, 1977.

B3. WEINBERG, L., *Network Analysis and Synthesis.* New York: McGraw-Hill, 1962.

Papers

P1. KINARIWALA, B., "Synthesis of Active *RC* Networks", *Bell Syst. Tech. J.*, 38 (September 1959), 1269–1316.

P2. SANDBERG, I. W., "Synthesis of *n*-port Active *RC* Networks," *Bell Syst. Tech. J.*, 40 (January 1961), 329–347.

P3. GOLDMAN, M., AND M. GHAUSI, "Synthesis of Rational Immittances Using Operational Amplifiers and *RC* One-Ports," *IEEE Trans. Circuit Theory*, CT-16 (November 1969), 544–546.

appendix B

···

Tables of Classical
Filter Functions

B.1 ALL-POLE FILTERS

When an ideal low-pass filter characteristic is approximated by an all-pole function of the form

$$H(s) = \frac{H_1}{D_n(s)} = \frac{H_1}{s^n + a_{n-1}s^{n-1} + a_{n-2}s^{n-2} + \cdots + a_1s + a_0}$$

the various demonimator polynomials have the following coefficients.

Butterworth filters Table B-1 gives the polynomials for n up to 10 and a_0 normalized to unity. For $n \leq 4$, see Table 1-1.

Chebyshev filters When the Chebyshev filter requirements call for a $\frac{1}{2}$-dB ripple, the polynomials for n up to 10 and a_n normalized to unity are listed in Table B-2. For $n \leq 4$, see Table 1-3.

The corresponding polynomials for 1-dB ripple are in Table B-3 and for 2-dB ripple, in Table B-4.

Thomson filters The Bessel polynomials for Thomson filters for n up to 10 are given in Table B-5 for $\tau = a_1/a_0 = 1$. For $n \leq 4$, see Table 1-4.

TABLE B-1

n	Butterworth Polynomials
5	$s^5 + 3.2361s^4 + 5.2361s^3 + 5.2361s^2 + 3.2361s + 1$ $(s + 1.0000)[(s + 0.3090)^2 + 0.9511^2][(s + 0.8090)^2 + 0.5878^2]$
6	$s^6 + 3.8637s^5 + 7.4641s^4 + 9.1416s^3 + 7.4641s^2 + 3.8637s + 1$ $[(s + 0.2588)^2 + 0.9659^2][(s + 0.7071)^2 + 0.7071^2][(s + 0.9659)^2 + 0.2588^2]$
7	$s^7 + 4.4940s^6 + 10.0978s^5 + 14.5918s^4 + 14.5918s^3 + 10.0978s^2 + 4.4940s + 1$ $(s + 1.0000)[(s + 0.2225)^2 + 0.9749^2][(s + 0.6235)^2 + 0.7818^2]$ $[(s + 0.9010)^2 + 0.4339^2]$
8	$s^8 + 5.1258s^7 + 13.1317s^6 + 21.8462s^5 + 25.6884s^4 + 21.8462s^3 + 13.1371s^2 +$ $5.1258s + 1$ $[(s + 0.1951)^2 + 0.9808^2][(s + 0.5556)^2 + 0.8315^2][(s + 0.8315)^2 + 0.5556^2]$ $[(s + 0.9808)^2 + 0.1951^2]$
9	$s^9 + 5.7588s^8 + 16.5817s^7 + 31.1634s^6 + 41.9864s^5 + 41.9864s^4 + 31.1634s^3 +$ $16.5817s^2 + 5.7588s + 1$ $(s + 1.0000)[(s + 0.1737)^2 + 0.9848^2][(s + 0.5000)^2 + 0.8660^2][(s + 0.7660)^2$ $+ 0.6428^2][(s + 0.9397)^2 + 0.3420^2]$
10	$s^{10} + 6.3925s^9 + 20.4317s^8 + 42.8021s^7 + 64.8824s^6 + 74.2334s^5 + 64.8824s^4 +$ $42.8021s^3 + 20.4317s^2 + 6.3925s + 1$ $[(s + 0.1564)^2 + 0.9877^2][(s + 0.4540)^2 + 0.8910^2][(s + 0.7071)^2 + 0.7071^2]$ $[(s + 0.8910)^2 + 0.4540^2][(s + 0.9877)^2 + 0.1564^2]$

TABLE B-2

n	0.5-dB Ripple Chebyshev Filter ($\epsilon = 0.3493$)
5	$s^5 + 1.17251s^4 + 1.9374s^3 + 1.3096s^2 + 0.7525s + 0.1789$ $(s + 0.3623)[(s + 0.1120)^2 + 1.0116^2][(s + 0.2931)^2 + 0.6252^2]$
6	$s^6 + 1.1592s^5 + 2.1718s^4 + 1.5898s^3 + 1.1719s^2 + 0.4324s + 0.0948$ $[(s + 0.0777)^2 + 1.0085^2][(s + 0.2121)^2 + 0.7382^2][(s + 0.2898)^2 + 0.2702^2]$
7	$s^7 + 1.1512s^6 + 2.4126s^5 + 1.8694s^4 + 1.6479s^3 + 0.7556s^2 + 0.2821s + 0.0447$ $(s + 0.2562)[(s + 0.0570)^2 + 1.0064^2][(s + 0.1597)^2 + 0.8001^2][(s + 0.2308)^2 +$ $0.4479^2]$
8	$s^8 + 1.1461s^7 + 2.6567s^6 + 2.1492s^5 + 2.1840s^4 + 1.1486s^3 + 0.5736s^2 +$ $0.1525s + 0.0237$ $[(s + 0.0436)^2 + 1.0050^2][(s + 0.1242)^2 + 0.8520^2][(s + 0.1859)^2 + 0.5693^2]$ $[(s + 0.2193)^2 + 0.1999^2]$
9	$s^9 + 1.1426s^8 + 2.9027s^7 + 2.4293s^6 + 2.7815s^5 + 1.6114s^4 + 0.9836s^3 +$ $0.3408s^2 + 0.0941s + 0.0112$ $(s + 0.1984)[(s + 0.0345)^2 + 1.0040^2][(s + 0.0992)^2 + 0.8829^2][(s + 0.1520)^2 +$ $0.6553^2][(s + 0.1864)^2 + 0.3487^2]$
10	$s^{10} + 1.1401s^9 + 3.1499s^8 + 2.7097s^7 + 3.4409s^6 + 2.1442s^5 + 1.5274s^4 +$ $0.6270s^3 + 0.2373s^2 + 0.0493s + 0.0059$ $[(s + 0.0279)^2 + 1.0033^2][(s + 0.0810)^2 + 0.9051^2][(s + 0.1261)^2 + 0.7183^2]$ $[(s + 0.1589)^2 + 0.4612^2][(s + 0.1761)^2 + 0.1589^2]$

TABLE B-3

n	1.0-dB Ripple Chebyshev Filter ($\epsilon = 0.5089$)

5    $s^5 + 0.9368s^4 + 1.6888s^3 + 0.9744s^2 + 0.5805s + 0.1228$
$(s + 0.2895)[(s + 0.0895)^2 + 0.9901^2][(s + 0.2342)^2 + 0.6119^2]$

6    $s^6 + 0.9282s^5 + 1.9308s^4 + 1.2021s^3 + 0.9393s^2 + 0.3071s + 0.0689$
$[(s + 0.0622)^2 + 0.9934^2][(s + 0.1699)^2 + 0.7272^2][(s + 0.2321)^2 + 0.2662^2]$

7    $s^7 + 0.9231s^6 + 2.1761s^5 + 1.4288s^4 + 1.3575s^3 + 0.5486s^2 + 0.2137s + 0.0307$
$(s + 0.2054)[(s + 0.0457)^2 + 0.9953^2][(s + 0.1281)^2 + 0.7982^2][(s + 0.1851)^2 + 0.4429^2]$

8    $s^8 + 0.9198s^7 + 2.4230s^6 + 1.6552s^5 + 1.8369s^4 + 0.8468s^3 + 0.4478s^2 + 0.1073s + 0.0172$
$[(s + 0.0350)^2 + 0.9965^2][(s + 0.0997)^2 + 0.8448^2][(s + 0.1492)^2 + 0.5644^2]$
$[(s + 0.1759)^2 + 0.1982^2]$

9    $s^9 + 0.9175s^8 + 2.6709s^7 + 1.8815s^6 + 2.3781s^5 + 1.2016s^4 + 0.7863s^3 + 0.2442s^2 + 0.0706s + 0.0077$
$(s + 0.1593)[(s + 0.0277)^2 + 0.9972^2][(s + 0.0797)^2 + 0.8769^2][(s + 0.1221)^2 + 0.6509^2][(s + 0.1497)^2 + 0.3463^2]$

10    $s^{10} + 0.9159s^9 + 2.9195s^8 + 2.1079s^7 + 2.9815s^6 + 1.6830s^5 + 1.2445s^4 + 0.4554s^3 + 0.1825s^2 + 0.0345s + 0.0043$
$[(s + 0.0224)^2 + 0.9978^2][(s + 0.1013)^2 + 0.7143^2][(s + 0.0651)^2 + 0.9001^2]$
$[(s + 0.1277)^2 + 0.4586^2][(s + 0.1415)^2 + 0.1580^2]$

TABLE B-4

n	2-dB Ripple Chebyshev Filter ($\epsilon = 0.7648$)

5    $s^5 + 0.7065s^4 + 1.4995s^3 + 0.6935s^2 + 0.4593s + 0.0817$
$(s + 0.2183)[(s + 0.0675)^2 + 0.9735^2][(s + 0.1766)^2 + 0.6016^2]$

6    $s^6 + 0.7012s^5 + 1.7459s^4 + 0.8670s^3 + 0.7715s^2 + 0.2103s + 0.0514$
$[(s + 0.0470)^2 + 0.9817^2][(s + 0.1283)^2 + 0.7187^2][(s + 0.1753)^2 + 0.2630^2]$

7    $s^7 + 0.6979s^6 + 1.9935s^5 + 1.0392s^4 + 1.1444s^3 + 0.3825s^2 + 0.16661s + 0.0204$
$(s + 0.1553)[(s + 0.0346)^2 + 0.9867^2][(s + 0.0968)^2 + 0.7912^2][(s + 0.1399)^2 + 0.4391^2]$

8    $s^8 + 0.6961s^7 + 2.2423s^6 + 1.2117s^5 + 1.5796s^4 + 0.5982s^3 + 0.3587s^2 + 0.0729s + 0.0129$
$[(s + 0.0265)^2 + 0.9898^2][(s + 0.0754)^2 + 0.8391^2][(s + 0.1129)^2 + 0.5607^2]$
$[(s + 0.1332)^2 + 0.1969^2]$

9    $s^9 + 0.6947s^8 + 2.4913s^7 + 1.3837s^6 + 2.0767s^5 + 0.8569s^4 + 0.6445s^3 + 0.16844s^2 + 0.0544s + 0.0051$
$(s + 0.1206)[(s + 0.0209)^2 + 0.9919^2][(s + 0.0603)^2 + 0.8723^2][(s + 0.0924)^2 + 0.6474^2][(s + 0.1134)^2 + 0.3445^2]$

10    $s^{10} + 0.6937s^9 + 2.7406s^8 + 1.5557s^7 + 2.6363s^6 + 1.1585s^5 + 1.0389s^4 + 0.3178s^3 + 0.1440s^2 + 0.0233s + 0.0032$
$[(s + 0.0170)^2 + 0.9935^2][(s + 0.0767)^2 + 0.7113^2][(s + 0.0493)^2 + 0.8962^2]$
$[(s + 0.0967)^2 + 0.4567^2][(s + 0.1072)^2 + 0.1574^2]$

n	Bessel Polynomials
5	$s^5 + 15s^4 + 105s^3 + 420s^2 + 945s + 945$ $(s + 3.6467)[(s + 3.3520)^2 + 1.7427^2][(s + 2.3247)^2 + 3.5710^2]$
6	$s^6 + 21s^5 + 210s^4 + 1260s^3 + 4725s^2 + 10395s + 10395$ $[(s + 4.2484)^2 + 0.8675^2][(s + 3.7356)^2 + 2.6263^2][(s + 2.5159)^2 + 4.4927^2]$
7	$s^7 + 28s^6 + 378s^5 + 3150s^4 + 17325s^3 + 62370s^2 + 135135s + 135135$ $(s + 4.9718)[(s + 4.7583)^2 + 1.7393^2][(s + 4.0701)^2 + 3.5172^2][(s + 2.6857)^2 + 5.4207^2]$
8	$s^8 + 36s^7 + 630s^6 + 6930s^5 + 51975s^4 + 270270s^3 + 945,945s^2 + 2,027,025s + 2,027,025$ $[(s + 5.5879)^2 + 0.8676^2][(s + 2.8390)^2 + 6.3539^2][(s + 4.3683)^2 + 4.1444^2]$ $[(s + 5.2048)^2 + 2.6162^2]$
9	$s^9 + 45s^8 + 990s^7 + 13,860s^6 + 135,135s^5 + 945,945s^4 + 4,729,752s^3 + 16,216,200s^2 + 34,459,425s + 34,459,425$ $(s + 6.2970)[(s + 6.1294)^2 + 1.7378^2][(s + 5.6044)^2 + 3.4982^2][(s + 4.6384)^2 + 5.3173^2][(s + 2.9793)^2 + 7.2915^2]$
10	$s^{10} + 55s^9 + 1,485s^8 + 25,740s^7 + 315,315s^6 + 2,837,835s^5 + 18,918,900s^4 + 91,891,800s^3 + 310,134,825s^2 + 654,729,075s + 645,729,075$ $[(s + 6.9220)^2 + 0.8677^2][(s + 3.1089)^2 + 8.2327^2][(s + 6.6153)^2 + 2.6116^2]$ $[(s + 5.9675)^2 + 4.3850^2][(s + 4.8862)^2 + 6.2250^2]$

B.2 APPROXIMATION WITH POLES AND ZEROS (ELLIPTIC FILTERS)[a]

The coefficients of the elliptic filters for one zero pair and two poles, one zero pair and three poles, and two pairs of zeros and four poles are given in Tables B-6, B-7, and B-8, respectively [P1]. The parameters A_1, A_2, and ω_s are defined in Fig. B-1. The reader is referred to many excellent tabulations for elliptic filter data [B1–B4].

Fig. B-1 Gain response for an elliptic low-pass filter, illustrating parameters A_1, A_2, and ω_s.

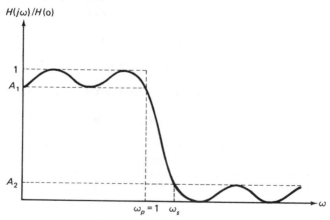

[a]Tables (B-6 to B-8) are reprinted with permission from Int. J. Comput. Elect. Eng. [P1], Copyright 1973, Pergamon Press Ltd.

TABLE B-6 Elliptic function parameters[b]

$$H(s) = \frac{H(s^2 + a_0)}{s^2 + b_1 s + b_0}$$

ω_s \ A_1	0.7	0.75	0.8	0.85	0.9	0.95	0.99
2.0	0.597566	0.672335	0.761953	0.87093	1.09079	1.28475	1.70530
	0.748566	0.807532	0.889100	1.01055	1.21614	1.67671	3.39116
	7.46410	7.46393	7.46393	7.46393	7.46410	7.46394	7.46437
	0.070208	0.081143	0.095295	0.115081	0.146639	0.213409	0.449766
1.8	0.586497	0.658788	0.744765	0.852101	0.996903	1.22172	1.49664
	0.761473	0.821030	0.903240	1.02526	1.23061	1.68388	3.25137
	5.93375	5.93377	5.93375	5.93377	5.93399	5.93377	5.93385
	0.089828	0.103773	0.121775	0.146865	0.186645	0.269588	0.542449
1.6	0.568640	0.636848	0.716947	0.814969	0.942467	1.12264	1.21673
	0.780727	0.840896	0.923621	1.04564	1.24863	1.68414	3.01139
	4.55831	4.55832	4.55842	4.55832	4.55832	4.55832	4.55836
	0.119892	0.138356	0.162086	0.194981	0.246530	0.350990	0.654022
1.4	0.535956	0.596787	0.666375	0.747996	0.845981	0.956021	0.859430
	0.811695	0.872098	0.954382	1.07401	1.26798	1.65969	2.61881
	3.33173	3.33166	3.33173	3.33167	3.33172	3.33167	3.33171
	0.170533	0.196320	0.229155	0.274010	0.342509	0.473248	0.778164
1.3	0.507505	0.562111	0.622959	0.691325	0.766598	0.829058	0.658076
	0.835122	0.894952	0.975687	1.09135	1.27415	1.62319	2.34888
	2.76980	2.76979	2.76981	2.76979	2.76982	2.76980	2.76972
	0.211054	0.242333	0.281803	0.334915	0.414008	0.556729	0.839569
1.2	0.461178	0.506162	0.553873	0.603117	0.647985	0.656811	0.450447
	0.867873	0.925546	1.00200	1.10866	1.26987	1.55118	2.02432
	2.23597	2.23595	2.23597	2.23595	2.23591	2.23595	2.23595
	0.271698	0.310453	0.358503	0.421457	0.511141	0.659054	0.896281
1.1	0.372652	0.401509	0.428498	0.450238	0.457760	0.420582	0.244714
	0.916613	0.967014	1.03128	1.11609	1.23375	1.41020	1.63605
	1.71409	1.71408	1.71409	1.71405	1.71408	1.71409	1.71394
	0.374317	0.423125	0.481308	0.553478	0.647782	0.781571	0.945011
1.05	0.285907	0.302274	0.314810	0.320161	0.310931	0.266561	0.142026
	0.951232	0.991713	1.04130	1.10322	1.18267	1.28875	1.40382
	1.43865	1.43866	1.43866	1.43866	1.43867	1.43867	1.43868
	0.462837	0.516984	0.579033	0.651806	0.739845	0.850992	0.966006

[b]Coefficients in each group, from top to bottom, are b_1, b_0, a_0, and $H(= A_2)$.

TABLE B-7 Third-order elliptic function parameters[a]

$$H(s) = \frac{H(s^2 + a_1)}{(s + p)(s^2 + b_{11}s + b_{10})}$$

A_1 \ ω_s	1.05	1.10	1.15	1.20	1.30	1.40	1.60
0.99	3.00155	2.38167	2.04962	1.84049	1.59035	1.44703	1.29096
	0.085439	0.164793	0.239930	0.308389	0.423881	0.514156	0.641363
	1.17110	1.27550	1.35412	1.41484	1.50033	1.55605	1.62210
	1.20541	1.37031	1.53363	1.69962	2.04551	2.41363	3.22359
	2.91611	2.21688	1.80968	1.53210	1.16647	0.932875	0.649595
	0.835656	0.702859	0.585336	0.488077	0.346305	0.254634	0.150824
0.95	1.39312	1.16920	1.05238	0.978047	0.887854	0.834076	0.773178
	0.128759	0.208146	0.267535	0.313860	0.382084	0.429629	0.491552
	1.09401	1.12638	1.14386	1.15422	1.16538	1.17039	1.17419
	1.20541	1.37031	1.53359	1.69962	2.04548	2.41363	3.22359
	1.26436	0.961054	0.784827	0.664187	0.505756	0.404446	0.281626
	0.550613	0.393746	0.298822	0.235598	0.158103	0.113422	0.066000
0.90	0.989843	0.850207	0.776613	0.729373	0.671004	0.635840	0.595349
	0.131719	0.197936	0.243978	0.278588	0.327809	0.361387	0.404241
	1.04501	1.05130	1.05182	1.05044	1.04612	1.04183	1.03480
	1.20541	1.37031	1.53360	1.69962	2.04550	2.41363	3.22359
	0.858124	0.652271	0.532635	0.450785	0.343166	0.274453	0.191108
	0.408593	0.279160	0.207846	0.162349	0.108001	0.077236	0.044840
0.85	0.796463	0.692067	0.636652	0.600774	0.556216	0.529099	0.497668
	0.125887	0.182370	0.220415	0.248520	0.287999	0.314599	0.348306
	1.01490	1.00922	1.00267	0.996551	0.986391	0.978507	0.967482
	1.20541	1.37031	1.53361	1.69962	2.04549	2.41362	3.22359
	0.670582	0.509696	0.416235	0.352254	0.268216	0.214500	0.149361
	0.330211	0.221527	0.163808	0.127519	0.084606	0.060434	0.035059
0.80	0.672056	0.588202	0.543408	0.514269	0.477902	0.455659	0.429765
	0.117935	0.167009	0.199450	0.223182	0.256259	0.278406	0.306340
	0.993886	0.981243	0.970727	0.962023	0.948671	0.938912	0.925798
	1.20541	1.37031	1.53361	1.69962	2.04549	2.41362	3.22359
	0.554123	0.421193	0.343958	0.291087	0.221642	0.177253	0.123426
	0.277706	0.184503	0.135945	0.105649	0.069994	0.049969	0.028976
0.75	0.580701	0.510790	0.473303	0.448768	0.418123	0.399259	0.377272
	0.109462	0.152592	0.180768	0.201218	0.229618	0.248517	0.272306
	0.978189	0.960951	0.947897	0.937548	0.922206	0.911278	0.896878
	1.20541	1.37031	1.53361	1.69962	2.04549	2.41362	3.22359
	0.471237	0.358197	0.292535	0.247550	0.188505	0.150742	0.104965
	0.238726	0.157652	0.115915	0.089987	0.059570	0.042510	0.024646
0.70	0.508377	0.448812	0.416747	0.395725	0.369369	0.353124	0.334131
	0.100986	0.139154	0.163859	0.181717	0.206412	0.222806	0.243387
	0.965965	0.945461	0.930629	0.919159	0.902442	0.890734	0.875469
	1.20541	1.37033	1.53361	1.69962	2.04549	2.41363	3.22360
	0.407390	0.309657	0.252888	0.214008	0.162958	0.130318	0.090743
	0.207883	0.136715	0.100376	0.077873	0.051520	0.036758	0.021308

[a]Coefficients in each group, from top to bottom, are p, b_{11}, b_{10}, a_1, H, and A_2.

TABLE B-8 Fourth-order elliptic function parameters[a]

$$H(s) = \frac{H(s^2 + a_1)(s^2 + a_2)}{(s^2 + b_{11}s + b_{10})(s^2 + b_{21}s + b_{20})}$$

A_1 \ ω_s	1.05	1.075	1.1	1.15	1.20	1.25	1.3
0.99	1.24184	1.36998	1.43445	1.48461	1.49416	1.48971	1.48067
	1.76639	1.64271	1.52732	1.35343	1.23106	1.14128	1.07349
	1.15362	1.22234	1.29092	1.42978	1.57242	1.71971	1.87203
	0.073511	0.104589	0.132384	0.179552	0.218090	0.250180	0.277321
	1.09961	1.12737	1.14901	1.18196	1.20610	1.22469	1.23958
	3.31266	3.85083	4.34993	5.29789	6.22434	7.15325	8.09589
	0.503140	0.389504	0.309376	0.209067	0.150187	0.112478	0.086921
0.97	1.11694	1.15719	1.17084	1.17132	1.16143	1.14961	1.13807
	1.17034	1.05721	0.974214	0.860463	0.785950	0.733037	0.693369
	1.15363	1.22235	1.29093	1.42979	1.57244	1.71971	1.87203
	0.082382	0.109829	0.132841	0.169869	0.198774	0.222152	0.241517
	1.06041	1.07297	1.08228	1.09550	1.10463	1.11140	1.11664
	3.31252	3.85097	4.34995	5.29782	6.22442	7.15325	8.09588
	0.315149	0.233744	0.182127	0.120698	0.086036	0.064240	0.049553
0.95	1.00616	1.02456	1.02740	1.01897	1.00707	0.994981	0.983063
	0.949099	0.853237	0.785841	0.695368	0.637238	0.595833	0.563827
	1.15362	1.22235	1.29090	1.42980	1.57240	1.71971	1.87203
	0.081657	0.106240	0.126428	0.158254	0.182889	0.202523	0.218409
	1.04043	1.04722	1.05200	1.05833	1.06243	1.06526	1.06699
	3.31238	3.85097	4.34973	5.29772	6.22421	7.15328	8.09613
	0.245492	0.180326	0.139862	0.092248	0.065716	0.049015	0.037702
0.925	0.903747	0.911065	0.909212	0.897844	0.885261	0.873655	0.863732
	0.800608	0.718662	0.662419	0.587723	0.539585	0.505447	0.480059
	1.15363	1.22235	1.29093	1.42979	1.57243	1.71971	1.87204
	0.078638	0.100616	0.118456	0.146316	0.167498	0.184309	0.198090
	1.02465	1.02747	1.02921	1.03102	1.03182	1.03207	1.03216
	3.31249	3.85095	4.34992	5.29794	6.22440	7.15331	8.09589
	0.198567	0.145089	0.112302	0.073994	0.052617	0.039221	0.030239
0.9	0.825168	0.827693	0.822969	0.811096	0.799091	0.788335	0.779239
	0.709059	0.636774	0.587138	0.521773	0.479903	0.450211	0.428090
	1.15363	1.22235	1.29041	1.42981	1.57240	1.71971	1.87203
	0.075144	0.095172	0.111155	0.136283	0.155159	0.170095	0.182294
	1.01374	1.01413	1.01394	1.01299	1.01180	1.01058	1.00947
	3.31240	3.85096	4.34581	5.29794	6.22422	7.15326	8.09589
	0.169268	0.123463	0.095443	0.062763	0.044651	0.033285	0.025661
0.85	0.708408	0.706199	0.700475	0.688058	0.676966	0.667721	0.659911
	0.598882	0.537875	0.496416	0.442241	0.407509	0.383057	0.364750
	1.15363	1.22235	1.29092	1.42978	1.57243	1.71971	1.87203
	0.068333	0.085454	0.099089	0.120059	0.135776	0.148163	0.158248
	0.999163	0.996486	0.993986	0.989660	0.986076	0.983077	0.980558
	3.31250	3.85096	4.34987	5.29786	6.22434	7.15335	8.09589
	0.133093	0.096781	0.074682	0.049106	0.034894	0.026019	0.020059
0.8	0.621079	0.617140	0.611017	0.599431	0.589557	0.581212	0.574306
	0.532447	0.478486	0.442216	0.394408	0.363967	0.342428	0.326332
	1.15363	1.22235	1.29041	1.42982	1.57240	1.71971	1.87204
	0.062131	0.077101	0.088880	0.107103	0.120623	0.131234	0.139842
	0.989514	0.984974	0.981121	0.974761	0.969765	0.965686	0.962302

521

A_1 \\ ω_s	1.05	1.075	1.1	1.15	1.20	1.25	1.3
	3.31250	3.85096	4.34613	5.29772	6.22423	7.15327	8.09589
	0.110293	0.080095	0.061824	0.040580	0.028851	0.021504	0.016576
0.75	0.550687	0.546069	0.540306	0.529453	0.520419	0.513005	0.506861
	0.487251	0.438112	0.405011	0.361820	0.334160	0.314640	0.300027
	1.15363	1.22235	1.29093	1.42979	1.57243	1.71971	1.87204
	0.056514	0.069762	0.080214	0.096136	0.107970	0.117237	0.124738
	0.982530	0.976711	0.971880	0.964177	0.958218	0.953427	0.949476
	3.31251	3.85096	4.34992	5.29791	6.22439	7.15327	8.09589
	0.093955	0.068176	0.052571	0.034540	0.024536	0.018289	0.014096
0.7	0.491207	0.486393	0.480919	0.470946	0.462799	0.456159	0.450655
	0.454281	0.408647	0.377979	0.337967	0.312352	0.294272	0.280747
	1.15363	1.22235	1.29092	1.42979	1.57244	1.71972	1.87203
	0.051378	0.063179	0.072459	0.086558	0.097012	0.105183	0.111795
	0.977203	0.970443	0.964919	0.956218	0.949563	0.944251	0.939898
	3.31247	3.85102	4.34989	5.29795	6.22448	7.15335	8.09607
	0.081317	0.058963	0.045465	0.029861	0.021211	0.015810	0.012188

[a]Coefficients in each group, from top to bottom, are b_{11}, b_{10}, a_1, b_{21}, b_{20}, a_2, and $H(=A_2)$.

●

EXAMPLE B-1

As an illustrative example of the use of Tables B-6 to B-8, consider the following specifications. Let $\omega_s/\omega_p = 1.1$, the minimum stop-band attenuation be ≥ 17 dB, and the ripple in the passband $\delta = 0.45$ dB. Recall from Chapter 1 the following relations:

$$A_1 = \frac{1}{\sqrt{1 + \epsilon^2}}, \qquad \epsilon = \sqrt{\frac{0.1\delta}{10 - 1}}, \qquad \delta = -20 \log_{10} A_1$$

Hence,

$$A_1 = 10^{-0.45/20} \simeq 0.949511 \simeq 0.945$$
$$A_2 = 10^{-17/20} \simeq 0.141254 \simeq 0.141$$

From the values above corresponding to $\omega_s = 1.1$ and $A_1 = 0.95$, we see that Tables B-6 and B-7 will not satisfy the requirements, as they yield $A_2 = 0.781571$ and 0.393746. However, from Table B-8 we find (see dashed box) that $A_2 = 0.139862$, which is close to and lower than 0.141. Hence, the transfer function that meets the requirements is

$$H(s) = \frac{0.13986(s^2 + 1.29090)(s^2 + 4.34973)}{(s^2 + 1.02740s + 0.785841)(s^2 + 0.126428s + 1.05200)}$$

A plot of the magnitude response for this transfer function is shown in Fig. B-2.

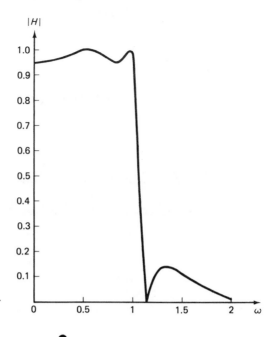

Fig. B-2 Gain response for Example B-1.

●

REFERENCES

Books

B1. SAAL, R., *Handbook of Filter Design*. Berlin, West Germany: AEG-Telefunken, 1979.

B2. HUMPHREYS, D. S., *The Analysis, Design and Synthesis of Electrical Filters*. Englewood Cliffs, N.J.: Prentice-Hall, 1970.

B3. CHRISTIAN, E., AND F. EISENMAN, *Filter Design Tables and Graphs*. New York: Wiley, 1966. 2nd ed., Raleigh, N.C., 1977 (paperback).

B4. ZVEREV, A. I., *Handbook of Filter Synthesis*. New York: Wiley, 1967.

B5. HUELSMAN, L. P., AND P. E. ALLEN, *Introduction to the Theory and Design of Active Filters*. New York: McGraw-Hill, 1980.

Papers

P1. STICHT, D. J., AND L. P. HUELSMAN, "Direct Determination of Elliptic Network Functions," *Int. J. Comput. Elec. Eng.*, 1 (1973), 277–280.

P2. WATANABE, H., "Approximation Theory for Filter Networks," *IRE Trans. Circuit Theory*, CT-8 (September 1961), 341–356.

appendix C

···

Op Amp Terminologies and Selected Data Sheets

In the following, we first define some basic terminologies and specifications used in op amp literature and then provide data sheets for a few typical op amps: the Fairchild μA 741 and National LM 348.

C.1 OP AMP SPECIFICATIONS AND DEFINITIONS

All manufacturers do not, in general, specify the same parameters. However, a number of the parameters defined below are generally specified by most manufacturers. The definitions below will also provide some understanding of the terminologies used in op amp data sheets.

Input offset voltage The voltage V_i that must be applied between the input terminals to make the output voltage zero (see Fig. C-1a). This quantity may be specified over a range of temperatures by the *input offset voltage drift*.

Fig. C-1 Circuits for determining op amp offset voltages, offset currents, and bias currents.

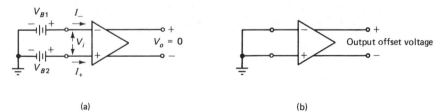

(a)   (b)

Output offset voltage The difference between the dc voltage present at the output terminals (or the output terminal and ground for the single-ended output) when the two input terminals are grounded (see Fig. C-1b).

Input offset current The difference between the separate currents in the input terminals, with the output at zero volts. (In Fig. C-1a, the input offset current is $I_- - I_+$ when $V_0 = 0$.) The *input offset current drift* is the change in $(I_- - I_+)$ due to changes in temperature. This specification is usually provided by the manufacturer.

Input bias current The average separate currents entering the input terminals when $V_0 = 0$ [i.e., $I_a = (I_- + I_+)/2$ in Fig. C-1a].

Common-mode rejection ratio The ratio of the differential gain to the common-mode gain [B1].

Supply voltage sensitivity The change in input offset voltage per unit change in power-supply voltage. Most often, the reciprocal of this quantity, which is referred to as the *supply-voltage rejection ratio*, is given.

Input common-mode range The common mode input signal range for which an op amp remains linear.

Output common-mode range The maximum output signal without significant distortion for a given load resistance.

Input differential range The maximum safe differential signal.

Input resistance Incremental input resistance, usually specified for both differential (between inputs) and common-mode (either input to ground) signals.

Output resistance Incremental output resistance measured without feedback.

Bandwidth specification Most manufacturers will specify the unity-gain frequency. Some may give the 3-dB bandwidth of the open-loop gain (see Chapter 2).

Open-loop gain The voltage gain of the amplifier with no external feedback applied, that is, the ratio of the output signal voltage to the input differential signal voltage, in its linear region, when the input signal varies very slowly (i.e., very low frequencies). Since the open-loop gain a_0 is very high (of the order of 10^5), care should be exercised not to saturate the amplifier.

A technique for measuring a_0 is given in Fig. C-2. To avoid output saturation, the effect of the input offset voltage is canceled and an attenuator is used to ensure linear operation. The input is a sinusoidal signal for which we can measure the open-loop gain and bandwidth. Note that the open-loop bandwidth p_0 is small, usually less than 100 Hz. For a properly designed (i.e., one-pole roll-off) or internally compensated op amp, the open-loop gain multiplied by the 3-dB bandwidth $a_0 p_0$ is equal to the unity-gain frequency. The unity gain–

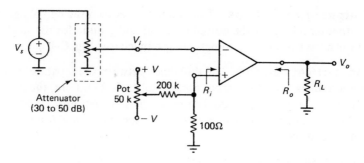

Fig. C-2 Technique for measuring op amp dc open-loop gain (a_0).

bandwidth parameter ($a_0 p_0$) is best measured in a closed-loop configuration such as shown in Fig. C-3. For example, measured data using LM 348 op amps are shown in Fig. 2-18. The $a_0 p_0$ parameter is very important in active-R filters, since this parameter controls the frequency response of the filters.

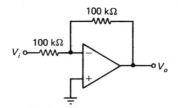

Fig. C-3 Op amp in unity-gain inverting configuration for measuring gain–bandwidth.

Full-power bandwidth The maximum frequency over which the full output voltage swing can be obtained.

Slew rate The maximum time rate of change of output voltage for a step input. Sometimes the specification is given as the maximum frequency at which a given amplitude sinusoidal output remains undistorted. The slew rate is the effect of limiting the dynamic range of the op amp at high frequencies. At frequencies above the full-power bandwidth, the slew rate determines, for a specified operating frequency, the maximum allowable input-signal level before operation becomes nonlinear. The minimum signal level is, of course, determined by the input noise voltage. For low frequencies, well within the full-power bandwidth, the slew rate is not an important parameter. On the other hand, for frequencies above $a_0 p_0 / 5$, linear op amp behavior will be limited by slew rate.

Overload recovery time Time required for the output stage to return to the active region when driven into hard saturation.

C.2 OP AMP DYNAMIC RANGE

The dynamic range of an op amp is determined by the supply voltages or the slew rate specification. Let us consider these effects in greater detail.

C.2.1 Supply Voltages

As the input to a closed loop op amp is increased, the output continues to increase until eventually the op amp saturates and the output becomes clipped. The reason for the clipping is that the output voltage cannot swing beyond the supply voltages. For example, if the supply voltages are ± 12 V, the output voltage swing is constrained to be less than ± 12 V. To prevent clipping, the maximum input amplitude and op amp closed-loop gain must be appropriately limited. Of course, the lowest signal level that can be reliably processed by an op amp is determined by internally generated noise. This internally generated noise is typically low-frequency noise with a $1/f$-type magnitude characteristic. For a detailed discussion of this topic, the reader is referred to [B1, B2].

C.2.2 Slew-Rate Limiting

If the frequency of the input signal is increased, the output signal will eventually distort in the manner shown in Fig. C-4. This type of distortion is referred to as slew-rate limiting and is of particular concern in high-frequency

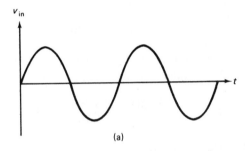

(a)

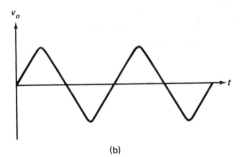

Fig. C-4 (a) Undistorted sine-wave input signal. (b) Distorted output due to slew-rate limiting.

(b)

filters (e.g., active-R and active-C filters). Slew-rate limiting is caused by the inability of capacitances in the op amp to charge or discharge fast enough to follow the signal.

The rate at which voltage changes across a capacitor is given by $dv/dt = i/C$. Since the currents available to the op amp are limited (i.e., to i_{max}), the rate of change in voltage is also limited:

$$\left.\frac{dv}{dt}\right|_{max} = \frac{i_{max}}{C} = \text{slew rate} \qquad (V/\mu\text{sec}) \qquad \text{(C-1)}$$

Slew rates are always given in the op-amp spec sheet and are usually expressed in V/μsec. For example, the typical slew rate for a Fairchild μA741 op amp is 0.5 V/μsec.

To illustrate the limitation in operation imposed by slew-rate limiting, consider the following example. Let the input to a closed-loop inverting op amp be A cos ωt and the output by $-KA$ cos ωt, where K corresponds to the amplifier's closed-loop gain. The maximum rate of change for this output is readily obtained to be

$$\left.\frac{dv_o(t)}{dt}\right|_{max} = KA\omega < \text{slew rate} \qquad (V/\mu\text{sec}) \qquad \text{(C-2)}$$

Thus, if $\omega = 2\pi \times 100$ kHz, $A = 0.1$ V, and the slew rate is 0.5 V/μsec, to prevent slew-rate limiting,

$$K < \frac{\text{slew rate}}{A\omega} = \frac{0.5 \times 10^6}{(0.1)(2\pi \times 10^5)} = 7.96 \qquad \text{(C-3)}$$

On the other hand, if we want to use $K = 10$, with an input amplitude of $A = 1$ V, the maximum frequency is given by

$$\omega_{max} < \frac{\text{slew rate}}{K} = \frac{0.5 \times 10^6}{10} \text{ rad/sec} = 5 \times 10^4 \text{ rad/sec} \qquad \text{(C-4)}$$

C.3 Data Sheets

Spec sheets included are for the Fairchild μA741 and National Semiconductor LM 148 and LM 149 op amps. For other data sheets the reader may consult the manufacturer's data books.

µA741
FREQUENCY-COMPENSATED OPERATIONAL AMPLIFIER
FAIRCHILD LINEAR INTEGRATED CIRCUIT

GENERAL DESCRIPTION — The µA741 is a high performance monolithic Operational Amplifier constructed using the Fairchild Planar* epitaxial process. It is intended for a wide range of analog applications. High common mode voltage range and absence of latch-up tendencies make the µA741 ideal for use as a voltage follower. The high gain and wide range of operating voltage provides superior performance in integrator, summing amplifier, and general feedback applications. Electrical characteristics of the µA741A and E are identical to MIL-M-38510/10101.

- NO FREQUENCY COMPENSATION REQUIRED
- SHORT CIRCUIT PROTECTION
- OFFSET VOLTAGE NULL CAPABILITY
- LARGE COMMON MODE AND DIFFERENTIAL VOLTAGE RANGES
- LOW POWER CONSUMPTION
- NO LATCH-UP

ABSOLUTE MAXIMUM RATINGS

Supply Voltage
µA741A, µA741, µA741E	±22 V
µA741C	±18 V

Internal Power Dissipation (Note 1)
Metal Can	500 mW
Molded and Hermetic DIP	670 mW
Mini DIP	310 mW
Flatpak	570 mW
Differential Input Voltage	±30 V
Input Voltage (Note 2)	±15 V

Storage Temperature Range
Metal Can, Hermetic DIP, and Flatpak	−65°C to +150°C
Mini DIP, Molded DIP	−55°C to +125°C

Operating Temperature Range
Military (µA741A, µA741)	−55°C to +125°C
Commercial (µA741E, µA741C)	0°C to +70°C

Lead Temperature (Soldering)
Metal Can, Hermetic DIPs, and Flatpak (60 s)	300°C
Molded DIPs (10 s)	260°C
Output Short Circuit Duration (Note 3)	Indefinite

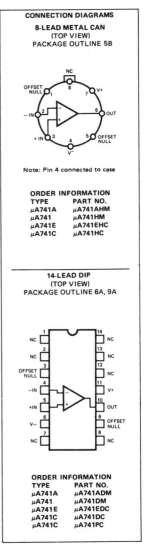

CONNECTION DIAGRAMS

8-LEAD METAL CAN
(TOP VIEW)
PACKAGE OUTLINE 5B

Note: Pin 4 connected to case

ORDER INFORMATION
TYPE	PART NO.
µA741A	µA741AHM
µA741	µA741HM
µA741E	µA741EHC
µA741C	µA741HC

14-LEAD DIP
(TOP VIEW)
PACKAGE OUTLINE 6A, 9A

ORDER INFORMATION
TYPE	PART NO.
µA741A	µA741ADM
µA741	µA741DM
µA741E	µA741EDC
µA741C	µA741DC
µA741C	µA741PC

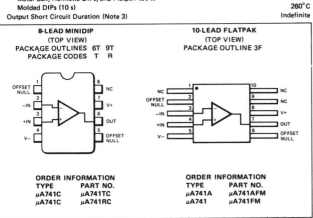

8-LEAD MINIDIP
(TOP VIEW)
PACKAGE OUTLINES 6T 9T
PACKAGE CODES T R

10-LEAD FLATPAK
(TOP VIEW)
PACKAGE OUTLINE 3F

ORDER INFORMATION
TYPE	PART NO.
µA741C	µA741TC
µA741C	µA741RC

ORDER INFORMATION
TYPE	PART NO.
µA741A	µA741AFM
µA741	µA741FM

*Planar is a patented Fairchild process.

[1]Reprinted courtesy of Fairchild Camera and Instrument Corporation.

μA741A

ELECTRICAL CHARACTERISTICS ($V_S = \pm 15V$, $T_A = 25°C$ unless otherwise specified)

PARAMETERS (see definitions)		CONDITIONS	MIN	TYP	MAX	UNITS
Input Offset Voltage		$R_S \leqslant 50\Omega$		0.8	3.0	mV
Average Input Offset Voltage Drift					15	μV/°C
Input Offset Current				3.0	30	nA
Average Input Offset Current Drift					0.5	nA/°C
Input Bias Current				30	80	nA
Power Supply Rejection Ratio		$V_S = +10, -20; V_S = +20, -10V, R_S = 50\Omega$		15	50	μV/V
Output Short Circuit Current			10	25	35	mA
Power Dissipation		$V_S = \pm 20V$		80	150	mW
Input Impedance		$V_S = \pm 20V$	1.0	6.0		MΩ
Large Signal Voltage Gain		$V_S = \pm 20V$, $R_L = 2k\Omega$, $V_{OUT} = \pm 15V$	50			V/mV
Transient Response	Rise Time			0.25	0.8	μs
(Unity Gain)	Overshoot			6.0	20	%
Bandwidth (Note 4)			.437	1.5		MHz
Slew Rate (Unity Gain)		$V_{IN} = \pm 10V$	0.3	0.7		V/μs
The following specifications apply for $-55°C \leqslant T_A \leqslant +125°C$						
Input Offset Voltage					4.0	mV
Input Offset Current					70	nA
Input Bias Current					210	nA
Common Mode Rejection Ratio		$V_S = \pm 20V$, $V_{IN} = \pm 15V$, $R_S = 50\Omega$	80	95		dB
Adjustment For Input Offset Voltage		$V_S = \pm 20V$	10			mV
Output Short Circuit Current			10		40	mA
Power Dissipation		$V_S = \pm 20V$, $-55°C$			165	mW
		$+125°C$			135	mW
Input Impedance		$V_S = \pm 20V$	0.5			MΩ
Output Voltage Swing		$V_S = \pm 20V$, $R_L = 10k\Omega$	±16			V
		$R_L = 2k\Omega$	±15			V
Large Signal Voltage Gain		$V_S = \pm 20V$, $R_L = 2k\Omega$, $V_{OUT} = \pm 15V$	32			V/mV
		$V_S = \pm 5V$, $R_L = 2k\Omega$, $V_{OUT} = \pm 2 V$	10			V/mV

NOTES
1. Rating applies to ambient temperatures up to 70°C. Above 70°C ambient derate linearly at 6.3mW/°C for the metal can, 8.3mW/°C for the DIP and 7.1mW/°C for the Flatpak.
2. For supply voltages less than ±15V, the absolute maximum input voltage is equal to the supply voltage.
3. Short circuit may be to ground or either supply. Rating applies to +125°C case temperature or 75°C ambient temperature.
4. Calculated value from: $BW(MHz) = \dfrac{0.35}{Rise\ Time\ (\mu s)}$

μA741

ELECTRICAL CHARACTERISTICS (V_S = ±15 V, T_A = 25°C unless otherwise specified)

PARAMETERS (see definitions)		CONDITIONS	MIN	TYP	MAX	UNITS
Input Offset Voltage		$R_S \leqslant 10\ k\Omega$		1.0	5.0	mV
Input Offset Current				20	200	nA
Input Bias Current				80	500	nA
Input Resistance			0.3	2.0		MΩ
Input Capacitance				1.4		pF
Offset Voltage Adjustment Range				±15		mV
Large Signal Voltage Gain		$R_L \geqslant 2\ k\Omega$, V_{OUT} = ±10 V	50,000	200,000		
Output Resistance				75		Ω
Output Short Circuit Current				25		mA
Supply Current				1.7	2.8	mA
Power Consumption				50	85	mW
Transient Response (Unity Gain)	Rise time	V_{IN} = 20 mV, R_L = 2 kΩ, $C_L \leqslant$ 100 pF		0.3		μs
	Overshoot			5.0		%
Slew Rate		$R_L \geqslant 2\ k\Omega$		0.5		V/μs

The following specifications apply for −55°C $\leqslant T_A \leqslant$ +125°C:

Input Offset Voltage	$R_S \leqslant 10\ k\Omega$			1.0	6.0	mV
Input Offset Current	T_A = +125°C			7.0	200	nA
	T_A = −55°C			85	500	nA
Input Bias Current	T_A = +125°C			0.03	0.5	μA
	T_A = −55°C			0.3	1.5	μA
Input Voltage Range			±12	±13		V
Common Mode Rejection Ratio	$R_S \leqslant 10\ k\Omega$		70	90		dB
Supply Voltage Rejection Ratio	$R_S \leqslant 10\ k\Omega$			30	150	μV/V
Large Signal Voltage Gain	$R_L \geqslant 2\ k\Omega$, V_{OUT} = ±10 V		25,000			
Output Voltage Swing	$R_L \geqslant 10\ k\Omega$		±12	±14		V
	$R_L \geqslant 2\ k\Omega$		±10	±13		V
Supply Current	T_A = +125°C			1.5	2.5	mA
	T_A = −55°C			2.0	3.3	mA
Power Consumption	T_A = +125°C			45	75	mW
	T_A = −55°C			60	100	mW

TYPICAL PERFORMANCE CURVES FOR μA741A AND μA741

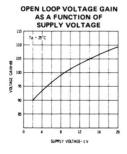

OPEN LOOP VOLTAGE GAIN
AS A FUNCTION OF
SUPPLY VOLTAGE

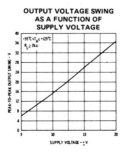

OUTPUT VOLTAGE SWING
AS A FUNCTION OF
SUPPLY VOLTAGE

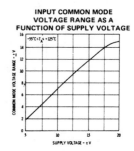

INPUT COMMON MODE
VOLTAGE RANGE AS A
FUNCTION OF SUPPLY VOLTAGE

µA741E

ELECTRICAL CHARACTERISTICS ($V_S = \pm15V$, $T_A = 25°C$ unless otherwise specified)

PARAMETERS (see definitions)	CONDITIONS	MIN	TYP	MAX	UNITS
Input Offset Voltage	$R_S \leqslant 50\Omega$		0.8	3.0	mV
Average Input Offset Voltage Drift				15	µV/°C
Input Offset Current			3.0	30	nA
Average Input Offset Current Drift				0.5	nA/°C
Input Bias Current			30	80	nA
Power Supply Rejection Ratio	$V_S = +10, -20; V_S = +20, -10V, R_S = 50\Omega$		15	50	µV/V
Output Short Circuit Current		10	25	35	mA
Power Dissipation	$V_S = \pm20V$		80	150	mW
Input Impedance	$V_S = \pm20V$	1.0	6.0		MΩ
Large Signal Voltage Gain	$V_S = \pm20V$, $R_L = 2k\Omega$, $V_{OUT} = \pm15V$	50			V/mV
Transient Response	Rise Time		0.25	0.8	µs
(Unity Gain)	Overshoot		6.0	20	%
Bandwidth (Note 4)		.437	1.5		MHz
Slew Rate (Unity Gain)	$V_{IN} = \pm10V$	0.3	0.7		V/µs
The following specifications apply for $0°C \leqslant T_A \leqslant 70°C$					
Input Offset Voltage				4.0	mV
Input Offset Current				70	nA
Input Bias Current				210	nA
Common Mode Rejection Ratio	$V_S = \pm20V$, $V_{IN} = \pm15V$, $R_S = 50\Omega$	80	95		dB
Adjustment For Input Offset Voltage	$V_S = \pm20V$	10			mV
Output Short Circuit Current		10		40	mA
Power Dissipation	$V_S = \pm20V$			150	mW
Input Impedance	$V_S = \pm20V$	0.5			MΩ
Output Voltage Swing	$V_S = \pm20V$, $R_L = 10k\Omega$	±16			V
	$R_L = 2k\Omega$	±15			V
Large Signal Voltage Gain	$V_S = \pm20V$, $R_L = 2k\Omega$, $V_{OUT} = \pm15V$	32			V/mV
	$V_S = \pm5V$, $R_L = 2k\Omega$, $V_{OUT} = \pm2 V$	10			V/mV

EQUIVALENT CIRCUIT

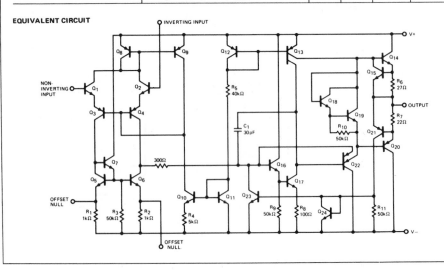

μA741C

ELECTRICAL CHARACTERISTICS (V_S = ±15 V, T_A = 25°C unless otherwise specified)

PARAMETERS (see definitions)	CONDITIONS		MIN	TYP	MAX	UNITS
Input Offset Voltage	$R_S \leqslant 10$ kΩ			2.0	6.0	mV
Input Offset Current				20	200	nA
Input Bias Current				80	500	nA
Input Resistance			0.3	2.0		MΩ
Input Capacitance				1.4		pF
Offset Voltage Adjustment Range				±15		mV
Input Voltage Range			±12	±13		V
Common Mode Rejection Ratio	$R_S \leqslant 10$ kΩ		70	90		dB
Supply Voltage Rejection Ratio	$R_S \leqslant 10$ kΩ			30	150	μV/V
Large Signal Voltage Gain	$R_L \geqslant 2$ kΩ, V_{OUT} = ±10 V		20,000	200,000		
Output Voltage Swing	$R_L \geqslant 10$ kΩ		±12	±14		V
	$R_L \geqslant 2$ kΩ		±10	±13		V
Output Resistance				75		Ω
Output Short Circuit Current				25		mA
Supply Current				1.7	2.8	mA
Power Consumption				50	85	mW
Transient Response (Unity Gain)	Rise time	V_{IN} = 20 mV, R_L = 2 kΩ, $C_L \leqslant 100$ pF		0.3		μs
	Overshoot			5.0		%
Slew Rate	$R_L \geqslant 2$ kΩ			0.5		V/μs

The following specifications apply for $0°C \leqslant T_A \leqslant +70°C$:

Input Offset Voltage					7.5	mV
Input Offset Current					300	nA
Input Bias Current					800	nA
Large Signal Voltage Gain	$R_L \geqslant 2$ kΩ, V_{OUT} = ±10 V		15,000			
Output Voltage Swing	$R_L \geqslant 2$ kΩ		±10	±13		V

TYPICAL PERFORMANCE CURVES FOR μA741E AND μA741C

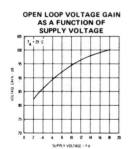

OPEN LOOP VOLTAGE GAIN
AS A FUNCTION OF
SUPPLY VOLTAGE

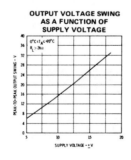

OUTPUT VOLTAGE SWING
AS A FUNCTION OF
SUPPLY VOLTAGE

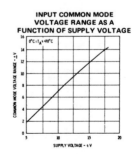

INPUT COMMON MODE
VOLTAGE RANGE AS A
FUNCTION OF SUPPLY VOLTAGE

TYPICAL PERFORMANCE CURVES FOR μA741A, μA741, μA741E AND μA741C

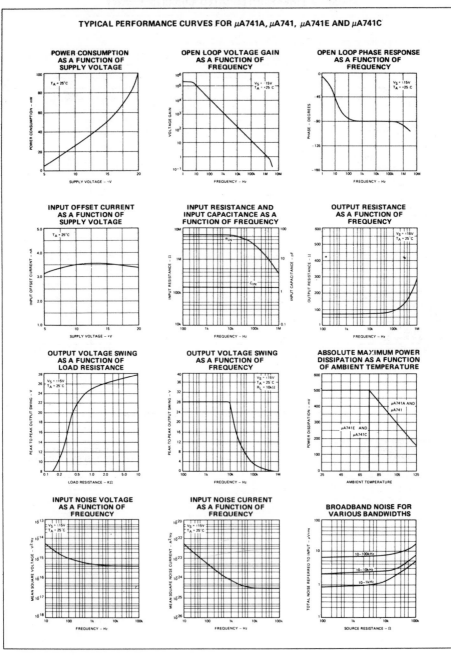

POWER CONSUMPTION AS A FUNCTION OF SUPPLY VOLTAGE

OPEN LOOP VOLTAGE GAIN AS A FUNCTION OF FREQUENCY

OPEN LOOP PHASE RESPONSE AS A FUNCTION OF FREQUENCY

INPUT OFFSET CURRENT AS A FUNCTION OF SUPPLY VOLTAGE

INPUT RESISTANCE AND INPUT CAPACITANCE AS A FUNCTION OF FREQUENCY

OUTPUT RESISTANCE AS A FUNCTION OF FREQUENCY

OUTPUT VOLTAGE SWING AS A FUNCTION OF LOAD RESISTANCE

OUTPUT VOLTAGE SWING AS A FUNCTION OF FREQUENCY

ABSOLUTE MAXIMUM POWER DISSIPATION AS A FUNCTION OF AMBIENT TEMPERATURE

INPUT NOISE VOLTAGE AS A FUNCTION OF FREQUENCY

INPUT NOISE CURRENT AS A FUNCTION OF FREQUENCY

BROADBAND NOISE FOR VARIOUS BANDWIDTHS

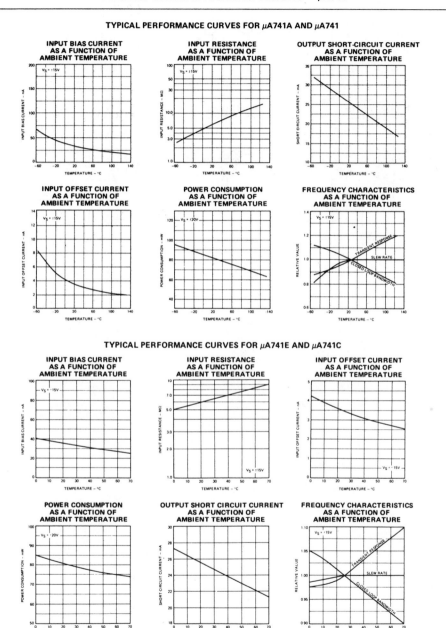

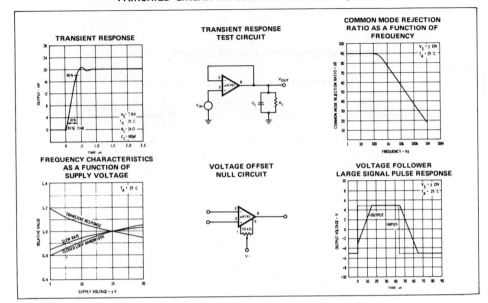

TRANSIENT RESPONSE

TRANSIENT RESPONSE TEST CIRCUIT

COMMON MODE REJECTION RATIO AS A FUNCTION OF FREQUENCY

FREQUENCY CHARACTERISTICS AS A FUNCTION OF SUPPLY VOLTAGE

VOLTAGE OFFSET NULL CIRCUIT

VOLTAGE FOLLOWER LARGE SIGNAL PULSE RESPONSE

Operational Amplifiers/Buffers

LM148, LM149 quad 741 op amps

LM148/LM248/LM348 quad 741 op amps

LM149/LM249/LM349 wide band decompensated $(A_{V(MIN)} = 5)$

general description

The LM148 series is a true quad 741. It consists of four independent, high gain, internally compensated, low power operational amplifiers which have been designed to provide functional characteristics identical to those of the familiar 741 operational amplifier. In addition the total supply current for all four amplifiers is comparable to the supply current of a single 741 type op amp. Other features include input offset currents and input bias current which are much less than those of a standard 741. Also, excellent isolation between amplifiers has been achieved by independently biasing each amplifier and using layout techniques which minimize thermal coupling. The LM149 series has the same features as the LM148 plus a gain bandwidth product of 4 MHz at a gain of 5 or greater.

The LM148 can be used anywhere multiple 741 or 1558 type amplifiers are being used and in applications where amplifier matching or high packing density is required.

features

- 741 op amp operating characteristics

- Low supply current drain   0.6 mA/Amplifier

- Class AB output stage—no crossover distortion

- Pin compatible with the LM124

- Low input offset voltage   1 mV

- Low input offset current   4 nA

- Low input bias current   30 nA

- Gain bandwidth product
  LM148 (unity gain)   1.0 MHz
  LM149 ($A_V \geq 5$)   4 MHz

- High degree of isolation between amplifiers   120 dB

- Overload protection for inputs and outputs

schematic and connection diagrams

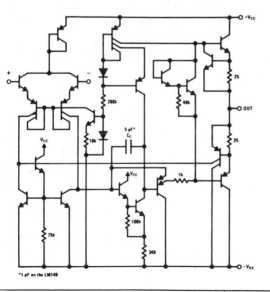

*1 pF on the LM149

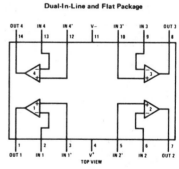

Dual-In-Line and Flat Package

TOP VIEW

Order Number LM148D, LM248D, LM348D,
LM149D, LM249D or LM349D
See Package 1
Order Number LM248J, LM348J, LM249J
or LM349J
See Package 16
Order Number LM148F or LM149F
See Package 4
Order Number LM348N or LM349N
See Package 22

[2]Reprinted courtesy of National Semiconductor Corporation.

absolute maximum ratings

	LM148/LM149	LM248/LM249	LM348/LM349
Supply Voltage	±22V	±18V	±18V
Differential Input Voltage	±44V	±36V	±36V
Input Voltage	±22V	±18V	±18V
Output Short Circuit Duration (Note 1)	Continuous	Continuous	Continuous
Power Dissipation (P_d at 25°C) and Thermal Resistance (θ_{jA}), (Note 2)			
Molded DIP (N) P_d	–	–	500 mW
θ_{jA}	–	–	150°C/W
Cavity DIP (D) (J) P_d	900 mW	900 mW	900 mW
θ_{jA}	100°C/W	100°C/W	100°C/W
Flat Pack (F) P_d	675 mW	–	–
θ_{jA}	185°C/W	–	–
Maximum Junction Temperature (T_{jMAX})	150°C	110°C	100°C
Operating Temperature Range	$-55°C \le T_A \le +125°C$	$-25°C \le T_A \le +85°C$	$0°C \le T_A \le +70°C$
Storage Temperature Range	$-65°C$ to $+150°C$	$-65°C$ to $+150°C$	$-65°C$ to $+150°C$
Lead Temperature (Soldering, 60 seconds)	300°C	300°C	300°C

electrical characteristics (Note 3)

PARAMETER	CONDITIONS	LM148/LM149 MIN	TYP	MAX	LM248/LM249 MIN	TYP	MAX	LM348/LM349 MIN	TYP	MAX	UNITS
Input Offset Voltage	$T_A = 25°C$, $R_S \le 10\,k\Omega$		1.0	5.0		1.0	6.0		1.0	6.0	mV
Input Offset Current	$T_A = 25°C$		4	25		4	50		4	50	nA
Input Bias Current	$T_A = 25°C$		30	100		30	200		30	200	nA
Input Resistance	$T_A = 25°C$	0.8	2.5		0.8	2.5		0.8	2.5		MΩ
Supply Current All Amplifiers	$T_A = 25°C$, $V_S = \pm15V$		2.4	3.6		2.4	4.5		2.4	4.5	mA
Large Signal Voltage Gain	$T_A = 25°C$, $V_S = \pm15V$ $V_{OUT} = \pm10V$, $R_L \ge 2k\Omega$	50	160		25	160		25	160		V/mV
Amplifier to Amplifier Coupling	$T_A = 25°C$, $f = 1\,Hz$ to 20 kHz (Input Referred) See Crosstalk Test Circuit		−120			−120			−120		dB
Small Signal Bandwidth	$T_A = 25°C$ LM148 series		1.0			1.0			1.0		MHz
	LM149 series		4.0			4.0			4.0		MHz
Phase Margin	$T_A = 25°C$ LM148 series ($A_V = 1$)		60			60			60		degrees
	LM149 series ($A_V = 5$)		60			60			60		degrees
Slew Rate	$T_A = 25°C$ LM148 series ($A_V = 1$)		0.5			0.5			0.5		V/μs
	LM149 series ($A_V = 5$)		2.0			2.0			2.0		V/μs
Output Short Circuit Current	$T_A = 25°C$		25			25			25		mA
Input Offset Voltage	$R_S \le 10\,k\Omega$			6.0			7.5			7.5	mV
Input Offset Current				75			125			100	nA
Input Bias Current				325			500			400	nA
Large Signal Voltage Gain	$V_S = \pm15V$, $V_{OUT} = \pm10V$, $R_L > 2k\Omega$	25			15			15			V/mV
Output Voltage Swing	$V_S = \pm15V$, $R_L = 10\,k\Omega$	±12	±13		±12	±13		±12	±13		V
	$R_L = 2\,k\Omega$	±10	±12		±10	±12		±10	±12		V
Input Voltage Range	$V_S = \pm15V$	±12			±12			±12			V
Common-Mode Rejection Ratio	$R_S \le 10\,k\Omega$	70	90		70	90		70	90		dB
Supply Voltage Rejection	$R_S \le 10\,k\Omega$	77	96		77	96		77	96		dB

Note 1: Any of the amplifier outputs can be shorted to ground indefinitely; however, more than one should not be simultaneously shorted as the maximum junction temperature will be exceeded.

Note 2: The maximum power dissipation for these devices must be derated at elevated temperatures and is dictated by T_{jMAX}, θ_{jA}, and the ambient temperature, T_A. The maximum available power dissipation at any temperature is $P_d = (T_{jMAX} - T_A)/\theta_{jA}$ or the 25°C P_{dMAX}, whichever is less.

Note 3: These specifications apply for $V_S = \pm15V$ and over the absolute maximum operating temperature range ($T_L \le T_A \le T_H$) unless otherwise noted.

typical performance characteristics

LM148, LM149

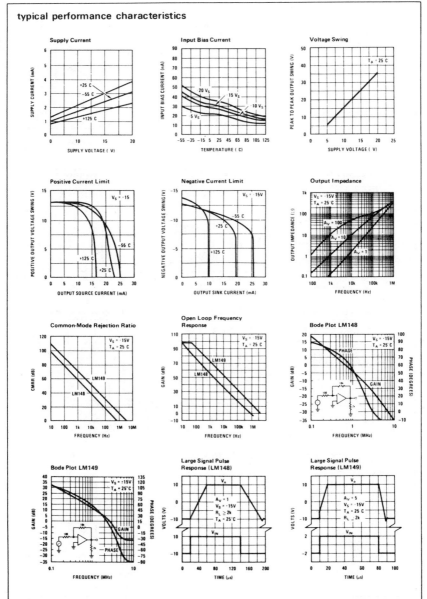

539

.

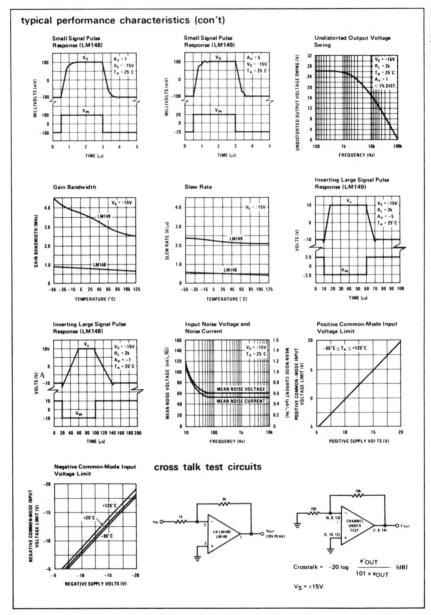

typical performance characteristics (con't)

cross talk test circuits

$$\text{Crosstalk} = -20 \log \frac{e'_{OUT}}{101 \times e_{OUT}} \quad (dB)$$

$V_S = \pm 15V$

540

REFERENCES

B1. ROBERGE, J. K., *Operational Amplifiers: Theory and Practice.* New York: Wiley, 1975.

B2. MILLMAN, J., AND C. C. HALKIAS, *Integrated Electronics.* New York: McGraw-Hill, 1972, Chap. 15.

B3. GREBENE, A. B., *Analog Integrated Circuit Design.* New York: Van Nostrand Rinehold, 1972, Chap. 5.

B4. MEYER, R. G., ed., *Integrated-Circuit Operational Amplifiers.* New York: IEEE Press, 1978.

B5. National Semiconductor Corporation, *Linear Application Handbook*, 1978.

B6. Fairchild Semiconductor, *Linear Integrated Circuits Data Book,* 1976.

Index